I0056495

Agricultural Soil: Classification, Ecology and Technology

Agricultural Soil: Classification, Ecology and Technology

Editor: Molly Ismay

www.callistoreference.com

Callisto Reference,
118-35 Queens Blvd., Suite 400,
Forest Hills, NY 11375, USA

Visit us on the World Wide Web at:
www.callistoreference.com

ISBN: 978-1-63239-791-1 (Hardback)

The publisher's policy is to use permanent paper from mills that operate a sustainable forestry policy. Furthermore, the publisher ensures that the text paper and cover boards used have met acceptable environmental accreditation standards.

Printed in the United States of America.

Cataloging-in-publication Data

Agricultural soil : classification, ecology and technology / edited by Molly Ismay.
p. cm.
Includes bibliographical references and index.
ISBN 978-1-63239-791-1
1. Soil science. 2. Soils--Classification. 3. Soil ecology. 4. Agriculture. I. Ismay, Molly.
S591 .A37 2017
631.4--dc23

Table of Contents

Preface

As a part of soil science, agricultural soil refers to the study of soil as used in the agricultural sector. The areas dealt within are soil erosion, soil classification, plant nutrition, soil fertility, among others. This book explores all the important aspects of agricultural soil in the present day scenario. It strives to provide a fair idea about this discipline and to help develop a better understanding of the latest advances within this field. The objective of this book is to give a general view of the different areas of this discipline and its applications. The various studies that are constantly contributing towards advancing technologies and evolutions of this field are examined in detail. Scientists and students actively engaged in this field will find this book full of crucial and unexplored concepts. It will serve as a beneficial resource guide for readers.

This book has been an outcome of determined endeavour from a group of educationists in the field. The primary objective was to involve a broad spectrum of professionals from diverse cultural background involved in the field for developing new researches. The book not only targets students but also scholars pursuing higher research for further enhancement of the theoretical and practical applications of the subject.

It was an honour to edit such a profound book and also a challenging task to compile and examine all the relevant data for accuracy and originality. I wish to acknowledge the efforts of the contributors for submitting such brilliant and diverse chapters in the field and for endlessly working for the completion of the book. Last, but not the least; I thank my family for being a constant source of support in all my research endeavours.

Editor

1

Prediction of Soil Organic Carbon at the European Scale by Visible and Near InfraRed Reflectance Spectroscopy

Antoine Stevens[1]*, Marco Nocita[1,2], Gergely Tóth[2], Luca Montanarella[2], Bas van Wesemael[1]

1 Georges Lemaître Centre for Earth and Climate Research, Earth and Life Institute, UCLouvain, Louvain-la-Neuve, Belgium, **2** SOIL Action, Land Resource Management Unit, Institute for Environment and Sustainability, Joint Research Centre of the European Commission, Ispra, Italy

Abstract

Soil organic carbon is a key soil property related to soil fertility, aggregate stability and the exchange of CO_2 with the atmosphere. Existing soil maps and inventories can rarely be used to monitor the state and evolution in soil organic carbon content due to their poor spatial resolution, lack of consistency and high updating costs. Visible and Near Infrared diffuse reflectance spectroscopy is an alternative method to provide cheap and high-density soil data. However, there are still some uncertainties on its capacity to produce reliable predictions for areas characterized by large soil diversity. Using a large-scale EU soil survey of about 20,000 samples and covering 23 countries, we assessed the performance of reflectance spectroscopy for the prediction of soil organic carbon content. The best calibrations achieved a root mean square error ranging from 4 to 15 g C kg^{-1} for mineral soils and a root mean square error of 50 g C kg^{-1} for organic soil materials. Model errors are shown to be related to the levels of soil organic carbon and variations in other soil properties such as sand and clay content. Although errors are ~5 times larger than the reproducibility error of the laboratory method, reflectance spectroscopy provides unbiased predictions of the soil organic carbon content. Such estimates could be used for assessing the mean soil organic carbon content of large geographical entities or countries. This study is a first step towards providing uniform continental-scale spectroscopic estimations of soil organic carbon, meeting an increasing demand for information on the state of the soil that can be used in biogeochemical models and the monitoring of soil degradation.

Editor: Han Y.H. Chen, Lakehead University, Canada

Funding: The funders had no role in study design, data collection and analysis, decision to publish, or preparation of the manuscript. The research in this paper is funded by the European Commission (FP7-ENV-2007-1) under the DIGISOIL project (n°211523). A.S. is postdoctoral researcher of the Fonds de la Recherche scientifique-FNRS (F.R.S.-FNRS, Belgium).

Competing Interests: The authors have declared that no competing interests exist.

* E-mail: antoine.stevens@uclouvain.be

Introduction

Human pressure on the soil has now reached the extent to which vital ecosystem services, such as food and fiber production or buffering against increases in greenhouse gas concentrations are at risk [1–3]. Soil Organic Carbon (SOC) is recognized as one of the key soil properties reflecting the state of the soil resource [3]. Existing soil maps and inventories are rarely adequate to assess the trends in SOC over time and determine the main driving forces at the scale of a country [4] let alone a continent [5], as the spatial resolution is generally low and many maps are based on outdated and imprecise methods [3]. Hence, high-throughput and cost-effective methods of SOC analysis should be developed to support the implementation of effective soil inventories and production of digital soil maps at the continental scale from which the state of the SOC can be determined in a consistent manner.

Visible and Near InfraRed (Vis-NIR) diffuse reflectance spectroscopy has been applied in soil analysis over the last 20 years [6] and has been demonstrated to accurately measure several soil attributes at minimal costs [7] and with satisfactory analytical errors [8]. Vis-NIR spectroscopy is currently used in laboratory conditions, but its application in-situ and even on air- or space-borne platforms is growing [9]. Vis-NIR reflectance carries information on the organic and inorganic composition of the soil [10] and, due to its integrative nature, has also been proposed as a screening tool for soil quality and fertility diagnosis [11]. Inference is based on multivariate calibration models developed from digital libraries linking Vis-NIR spectral data with reference laboratory measurements [12]. Obviously, these empirical calibrations are only applicable to samples having similar soil composition and spectral characteristics as those in the library and generally and cannot be extrapolated to other soil types [13].

When applying Vis-NIR to assess soil properties in a region of interest, a spectral library representing the local soil diversity needs to be constructed. As a consequence, many local, purpose-specific libraries are being built independently by different research groups using different protocols for soil and spectral analyses. This can produce good results for individual studies, but extrapolation to other areas is difficult. *Mutatis mutandis* the process will have to be repeated over and over again for each study area. This is not efficient and a waste of resources compared to reference methods of soil analyses. This considerably limits the field of applications of local scale spectral libraries so that national and international databases have been or are being developed [14–16].

Because soils are extremely variable and the relationship between Vis-NIR spectra and soil attributes can be complex and can vary in space, such databases require very large numbers of samples to be collected to adequately cover soil variation at continental scales [14]. To minimize calibration errors, samples should be analyzed by means of high-standard reference soil analyses and using a standardized spectroscopic measurement protocol [17]. Development costs of such databases can be prohibitive, so that there currently exist only few large scale soil spectral libraries and there are even fewer examples of the use of these libraries as an operational tool for routinely measuring soil properties. However, some initiatives have recently been launched [18,19]. The world soil spectral library presented in Brown et al. [14] includes 3,794 samples analyzed for SOC content using the Walkley and Black method [20], most of them originating from North America. The ICRAF-ISRIC spectral library contains 4,436 samples from 785 soil profiles distributed across the five continents (only 3,643 samples were analyzed for both chemical properties and soil texture) [16]. SOC was analyzed with the Walkley and Black method. Shepherd and Walsh [12] collected around 1,000 samples for the spectral library of eastern and southern Africa. The Africa Soil Information Service is currently collecting a large number of samples (more than 17,000 so far) from 60 sentinel sites of 100 square km in sub-Saharan Africa that are measured using both soil reference methods and vis-NIR spectroscopy [21]. The Australian library [22] contains 10,677 samples analyzed for total organic carbon using different methods.

In Europe, a large scale soil spectral library has been developed in the framework of the European Land Use/Cover Area frame Statistical Survey (LUCAS) during which ~20,000 geo-referenced top-soil samples were collected in order to assess the state of the soils across Europe. Thirteen chemical and physical properties, including Vis-NIR reflectance, were analyzed. The database is characterized by a higher sampling density than that of other large scale libraries. All samples were collected following the same sampling protocol and analyzed in a single ISO certified laboratory. SOC content was measured using an automated CN analyzer. To our knowledge, the LUCAS database constitutes to date the most complete and consistent soil spectral library at continental scale. The accuracy of spectroscopic models being limited by *(i)* the number and representativity of the calibration samples, and *(ii)* the quality and consistency of the reference methods, the LUCAS library represents a unique opportunity to evaluate the accuracy of continental-scale soil spectroscopic models. Based on the LUCAS library, we developed the first European-scale calibration models for the prediction of SOC content and analyze model prediction errors.

Materials and Methods

Ethics Statement

The LUCAS survey is part of the Community Statistical Programme 2008–2012, based on the decision No 1578/2007/EC of the European Parliament and the Council of the European Union of 11 December 2007 [23]. Data Confidentiality policy is based on the Regulation (EC) No 223/2009 on European statistics (recital 24 and Article 20(4)) of 11 March 2009 [24]. The policy on soil sampling included the clause that upon denial of access, the given point was skipped and a pre-selected alternative location was sampled instead. Field sampling did not involve endangered or protected species.

The LUCAS SOIL Database

The soil database was compiled as part of the LUCAS survey. Its primary goal was to provide harmonized data on land use/cover in 23 member states of the European Union (EU) by collecting observations in ~250,000 survey points [25]. About 10% of the points were selected based on environmental variables and accessibility criteria [26] and composite samples consisting of five sub-samples of the top soil (0–30 cm) were taken with a spade following a standardized protocol [26]. Vegetation cover and residues, stones and litter in case of forest land cover were removed before taking a sample. The sampling campaign resulted in the collection of about 20,000 topsoil samples for which geographical coordinates, land use/cover, management and other environmental attributes were retrieved. The mean sample density per European level 1 territorial units (NUTS 1) varies between 11 and 77 samples per 10,000 km^2, with a mean of 48 samples per 10,000 km^2 over the 23 EU countries of the survey (Figure 1). While samples are distributed over all land use/cover types, more samples were proportionally taken in cropland soils (Table 1). The surveyed area represents about 68% of the European continent and islands (excluding Russia and Turkey). However, LUCAS samples cover all the major soil types in Europe (Table 2), although Chernozems and Albeluvisols, particularly from Eastern Europe, are underrepresented in the database.

Soil Analyses

All samples were sent to an accredited laboratory (Kecskemét, Hungary) where the following properties were analyzed using ISO standard methods: coarse fragments, particle size distribution, pH in $CaCl_2$, pH in water, cation exchange capacity, organic carbon, carbonate, total nitrogen, P, and extractable K content. SOC content (g C kg^{-1}) was measured by dry combustion (ISO 10694:1995) using a vario MAX CN analyzer (Elementar Analysensysteme GmbH, Germany).

Vis-NIR Measurements and Processing

Absorbance spectra of air-dried and sieved (<2 mm) soil samples were measured with a XDS Rapid Content Analyzer (FOSS NIRSystems Inc., Laurel, MD). The spectrometer is equipped with Si (400–1100 nm) and PbS (1100–2500 nm) detectors, offering 4,200 wavelengths in the Vis-NIR region of the electro-magnetic spectrum. Two scans were acquired and subsequently averaged. For each band, standard deviation between scans was calculated and averaged over the wavelengths. Thirteen spectra with an average standard deviation >0.01 might have been improperly measured and were removed. We corrected the spectra for the shift in absorbance at the splice of the two detectors. The beginning of the Vis (400–500 nm) showed instrumental artifacts and was therefore removed. Several mathematical pre-treatments on the spectra were applied to remove physical variability due to light scattering and enhance features of interest [27]. Firstly, we created two pre-treated spectral matrices by applying Savitzky-Golay (SG) smoothing and first derivative filters [28] with a window size of 101 data points and 3rd order polynomial. Secondly, we applied Standard Normal Variate (SNV) transformation on SG-filtered spectral data [29]. Finally, we kept only one band in twenty (i.e. one every 10 nm), leaving ~200 predictor variables for model calibration.

Spectroscopic Models and SOC Predictions

First, samples were separated into mineral and organic soil samples using the FAO definition of organic soil materials [30]. Mineral soil samples were further divided into samples under *(i)* cropland, *(ii)* grassland and *(iii)* woodland land cover. These subsets are based on the land cover classes defined in the LUCAS survey and are consistent with the IPCC/FAO land cover/land use systems [31]. We carried out identical but separate analyses on each of these five subsets (cropland, grassland, woodland, mineral and organic soil samples). We tested several multivariate regression models and spectral pretreatments for predicting the SOC content. Furthermore, we evaluated the potential of *(i)* implementing variable selection procedures through recursive feature

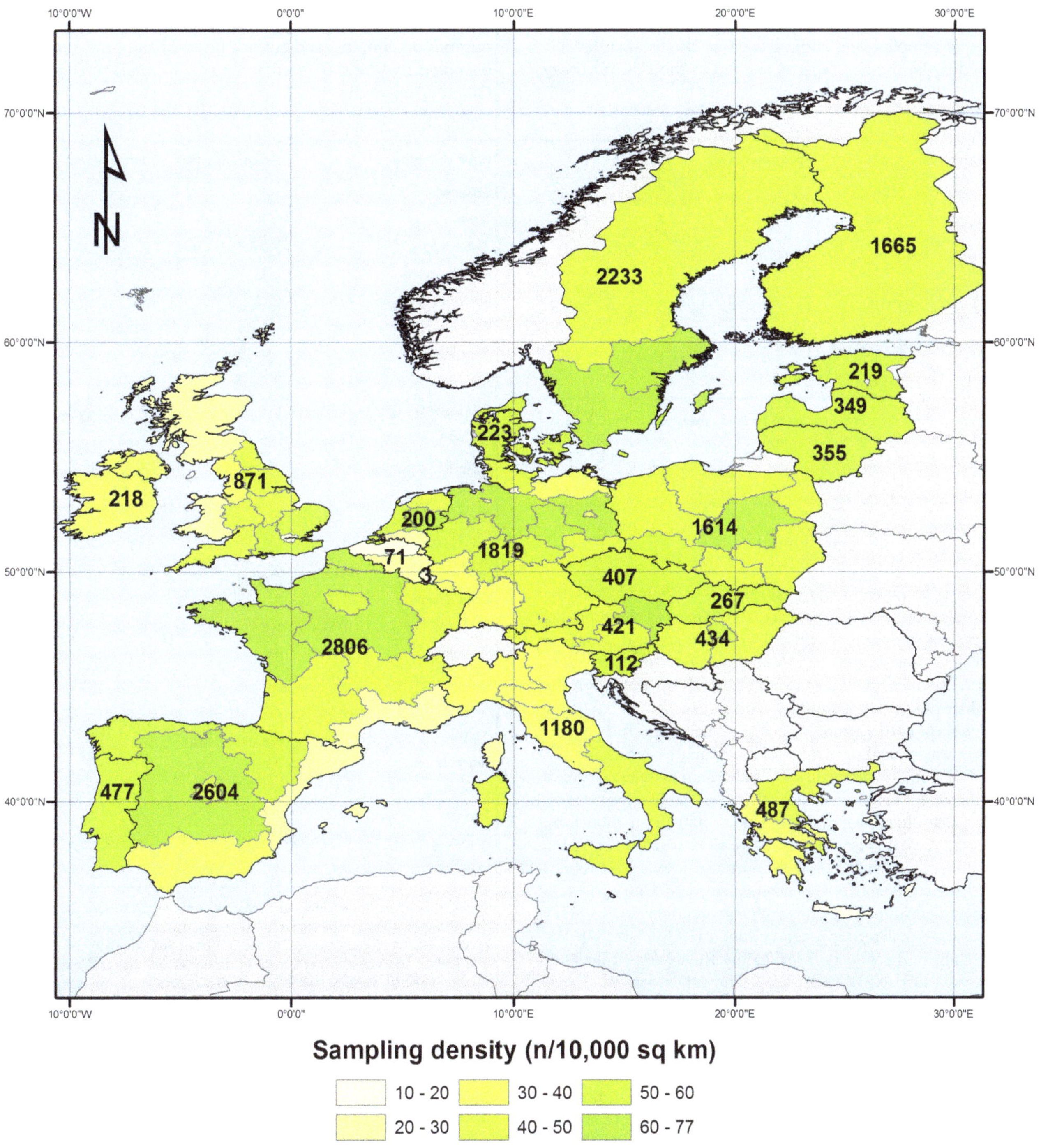

Figure 1. Sampling density of the LUCAS SOIL database per European territorial units, level 1 (NUTS 1). Map labels give the total number of samples per country.

elimination *via* random forest and (*ii*) including sand and clay content as auxiliary predictor into the spectroscopic models.

Sample selection with the Kennard-Stone algorithm. For each subset, two thirds of the samples were selected for training the spectroscopic models using the Kennard-Stone algorithm [32] and the remaining samples were assigned to the test set for assessing the model's performance. Based on a spectral distance measure, the Kennard-Stone algorithm selects a set of samples having a uniform distribution over the predictor space and hence that comprise all sources of variation found in the spectral library. The procedure starts by selecting the pair of points that are the farthest apart. They are put in the training set and removed from the list. Then, the remaining points are iteratively assigned to the training set by computing the distance between each unassigned points i_0 and training points i and finding the point i_0 which is the farthest apart from its closest neighbor i according to:

Table 1. Number of samples (n), frequency (in % of the total number of samples) and surface (in % of the total surface) occupied by land cover type as defined in the LUCAS survey [31].

Land Cover	n	Frequency (%)	Surface (%)[a]
Artificial	39	<1	5
Bare land	346	2	2
Cropland	8426	44	26
Grassland	4205	22	22
Shrubland	444	2	6
Water areas	12	<1	3
Wetland	90	<1	1
Woodland	5473	29	36

[a]Percentage of the total surface occupied by land cover type in the 23 EU countries of the LUCAS survey [25].

$$d = \max_{i_0}\left(\min_i\left(d_{i,i_0}\right)\right) \quad (1)$$

where d is a measure of distance. Here, we defined d as the Euclidean distance in the normalized score space of the principal components explaining more than 99% of the spectral variation. The principal components were computed on the continuum-removed reflectance spectral matrix [33] to select samples on the basis of their absorption features.

Multivariate calibration. Each pre-treated spectral calibration matrix was related to SOC with multivariate regression tools able to deal with high-dimensional and multi-collinear spectral measurements. Using the *caret* package [34] of the R software [35], we ran the following linear and non-linear multivariate models on the training set: partial least square regression, boosted regression tree, random forest, radial-basis support vector machine regression, multivariate adaptive regression splines and Cubist. Details on the latter algorithm, which has shown good prediction accuracy for soil spectral analyses [36], can be found in Quinlan [37]. We used the Cubist GPL C code provided by RuleQuest (RuleQuest Research Pty Ltd, NSW, Australia). The other algorithms are described in Hastie et al. [38] and an overview of their performance for soil spectroscopic inference is given in Viscarra Rossel and Behrens [39]. The models were sequentially developed on a grid of model parameters generated by the *caret* package which provides likely default parameter values. Ten random partitions of the data with a 0.5 selection probability were created and consistently used for leave-one-group-out cross-validation of the models. The best model parameters were determined as the ones producing a model having the smallest value within one standard error of the minimal observed RMSE of cross validation [40].

Recursive feature elimination. We tested the potential of a Recursive Feature Elimination (RFE) algorithm based on random forest to select a small set of optimal (and possibly non-collinear) spectral predictors for model calibration. The RFE procedure, as implemented in the *caret* package, performs a backward selection of the variables by ranking their importance to an initial model run using all the predictors [41]. The algorithm builds several calibration models that use the p_i most important predictors, where p_i is an element of a predetermined sequence $\{p_1, p_2, \ldots, p_n\}$ of possible numbers of predictors. The set of predictors p_i

Table 2. Number of samples (n), frequency (in % of the total number of samples) and surface (in % of the total surface) occupied by World Reference Base (WRB) major soil groups [30].

WRB soil type	n	Frequency (%)	Surface (%)[a]
Unknown	41	<1	-
Town	23	<1	<1
Water body	41	<1	<1
Rock outcrops	3	<1	<1
Albeluvisols	436	2	7
Acrisols	45	<1	<1
Andosols	22	<1	<1
Arenosols	379	2	1
Chernozems	193	1	7
Calcisols	82	<1	<1
Cambisols	6764	36	25
Fluvisols	1178	6	5
Gleysols	502	3	3
Gypsisols	32	<1	<1
Histosols	601	3	4
Kastanozems	0	0	<1
Leptosols	1078	6	8
Luvisols	2949	16	11
Phaeozems	229	1	4
Planosols	75	<1	<1
Podzols	3657	19	16
Regosols	480	3	5
Solonchaks	55	<1	<1
Solonetz	19	<1	<1
Umbrisols	3	<1	<1
Vertisols	148	1	1

[a]Percentage of the total surface occupied WRB major soil groups. Data should be considered approximate: surfaces have been computed using the dominant value of the soil typological units of the European Soil Database [58]. The total land surface considered is the European continent and islands (United Kingdom, Ireland, Iceland, Malta, Sicily, Sardinia, Corsica,...), excluding Russia and Turkey.

producing the best model amongst the candidate models is retained.

Auxiliary predictor. We assessed the possibility to improve the models by adding another predictor to the spectral matrix, which to be useful should be readily available [14]. We tested sand and clay content as auxiliary predictors. Particle size fractions are unlikely to change much over time at the sample location and hence could directly be exploited in models predicting the SOC content of samples collected during a future resampling of the LUCAS database. In order not to overweigh in the multivariate model the spectral data compared to the auxiliary predictor, we computed the principal component scores of the spectral matrix, retained the scores explaining more than 99% of the variation, attached the auxiliary predictor to the scores and scaled the resulting matrix [42]. When RFE is applied, we assumed that the spectral matrix has been reduced to its intrinsic dimensionality so that the predictor matrix was only scaled, without performing the principal component step. This approach was tested only for mineral soils since texture analyses were not realized for organic soil samples.

Assessing model performance. Multivariate models were validated with the test set and their quality assessed by means of the following statistics [43]:

$$bias = \frac{1}{n}\sum_{i=0}^{n} e_i \quad (2)$$

$$SEP_{-b} = \sqrt{\frac{1}{n-1}\sum_{i=0}^{n} (e_i - bias)^2} \quad (3)$$

$$RMSEP = \sqrt{\frac{1}{n}\sum_{i=0}^{n} e_i^2} \quad (4)$$

$$RPD = SD/RMSEP \quad (5)$$

where e_i is the residuals (i.e. prediction error) of sample i in the test set, n is the number of observations, SD is the standard deviation of the observations. The SEP_{-b} (Eq. 3) is the standard error of prediction corrected for bias (i.e. the difference between predicted and observed means, Eq. 2) and is equivalent to the standard deviation of the predicted residuals. The SEP_{-b} and bias represent two independent components of the Root Mean Square of Prediction (*RMSEP*, Eq. 4). The Ratio of Performance to Deviation (*RPD*, Eq. 5) is a way of normalizing *RMSEP*'s to compare calibration models where the measured variables have different ranges or variances. We used the *RMSEP* (Eq. 4) to rank all pre-treatments and multivariate calibration models and choose the best modeling approach.

Reproducibility of the reference and spectral methods. All soil analyses were replicated once for 25 randomly-selected samples which allowed estimating the reproducibility (or intermediate precision) of the reference and spectral analyses methods (i.e. repeatability+between-runs error) using [44]:

$$SEL = \sqrt{\frac{\sum_{i=0}^{n} \varepsilon_i^2}{2n}} \quad (6)$$

where *SEL* is the Standard Error of Laboratory and ε_i is the difference between duplicate measurements/predictions of sample i.

Results and Discussion

Exploratory Analysis of the LUCAS Database

Mineral samples have a mean SOC content of 29 g C kg^{-1}, a median at 19.6 g C kg^{-1} and a highly skewed SOC distribution with 75% of the samples below 35 g C kg^{-1} (Table 3). The SOC values of mineral samples of the LUCAS library are relatively higher than in other large scale spectral library, mainly because many samples were collected in organic-rich soils of northern Europe (Figure 1). The African [12], Australian [22] and world [14] spectral libraries have a median SOC content of respectively 12, 6 and 4.7 g C kg^{-1}. SOC content of organic samples in the database ranges from 156 to 587 g C kg^{-1} with a mean of 387 g C kg^{-1}.

To analyze the spectral variation included in the database, we performed a principal component analysis on the continuum-removed reflectance of mineral and organic samples [33]. The eigenvectors of the three first principal components (PC) show diagnostic variations across the Vis-NIR spectrum that can be linked to soil properties (Figure 2). For mineral soils, the first PC, explaining 56% of the spectral variation, shows important peaks that are associated to overtones of O-H and H-O-H stretch vibrations of free water (1455 and 1915 nm) and overtones and combinations of O-H stretch and metal-OH bends in the clay lattice (1415 and 2207 nm), which express spectral differences between illite and smectite clay minerals [39]. Hence, the scores of PC1 are strongly correlated with soil properties related to clay mineralogy such as clay content ($\rho = -0.45$), cation exchange capacity and pH in water ($\rho = -0.5$; Table 3). The eigenvector of PC2 (explaining 26% of the spectral variation) is dominated by one prominent feature centered at 620 nm that can be attributed to various organic compounds [10], explaining the negative correlation of PC2 scores with OC ($\rho = -0.55$) and N ($\rho = -0.44$; Table 3). While the scores of PC3 show only weak correlations with the measured soil properties (Table 3), the third PC can be related to the absence/presence of iron oxides as indicated by well-defined peaks in the eigenvector of PC3 around 540, 640 and 900 nm (Figure 2) caused by variations in the shape of the absorptions due to electronic transitions of goethite iron oxide at 620 and 920 nm [39].

For organic soils, the first PC scores are strongly correlated with OC ($\rho = -0.46$) while PC2-3 scores show weaker correlations with OC ($\rho = 0.26$–0.35; Table 3). The eigenvectors of PC1-3 display oscillations between 500 and 900 nm that are related to variations in the size and width of the organic matter absorption feature in the visible region (Figure 2). Other important features in the eigenvectors can be distinguished at 1450 nm, 1760 nm and between 1900 and 2500 nm corresponding to vibrations of C-O, O-H, C = O,C-H and N-H bonds that are present in organic compounds such as proteins, starch, cellulose, humic acids and lignin [45]. These absorptions are however difficult to attribute to a single component since they are greatly overlapping in the NIR. Since these organic soils have generally very low clay content, there is no visible feature at 2207 nm due to O-H and metal-O-H bonds in clay mineral lattices. The eigenvector of PC2 shows three local minima around 1450, 2100 and 2300 nm that can be assigned to lignin and cellulose [46] and may therefore account for spectral variation related to the decomposition stage of organic matter (arising e.g. from difference between forest and wetland samples). This preliminary analysis of European-scale spectral variation demonstrates that soil spectra are tightly linked with key soil properties, which supports thereby the development of spectral prediction models.

Multivariate Models for SOC Prediction

We compare here the prediction ability of the different models and pre-treatments tested for the subsets. For cropland, mineral and organic soils, the lowest prediction errors were achieved by models using the first derivative of the spectral matrix, while for grassland and woodland soil samples, the best models used the primary absorbance spectra (Figure S1). Overall, SNV transformation did not noticeably improve the accuracy of the models (Figure S1). Model performance varied greatly with the predictors included (Figure S2). Using sand content in addition to the spectral matrix improved grassland and woodland models compared to models using spectral data only, with a median decrease in *RMSEP* of ~1 g C kg^{-1} for grassland soils and ~4 g C kg^{-1} for woodland soil, while no clear improvement could be observed for cropland

Table 3. Summary statistics of soil properties available in the LUCAS database, for mineral and organic soil materials.

Property	Unit	Mean	SD[a]	Min	Q25[b]	Q50[c]	Q75[d]	Max	Skew	ρ_{PC1}[e]	ρ_{PC2}[e]	ρ_{PC3}[e]	n[f]
Mineral soils													
SOC	g kg^{-1}	29.4	28.9	0.0	12.3	19.6	34.7	199.2	2.67	0.08	−0.55	0.16	17937
N	g kg^{-1}	2.2	1.6	0.0	1.2	1.7	2.6	16.2	2.44	−0.01	−0.44	0.17	17937
clay	g kg^{-1}	18.9	13.0	0.0	8.0	17.0	27.0	79.0	0.91	−0.45	0.23	−0.03	17937
silt	g kg^{-1}	38.2	18.3	0.0	25.0	37.0	51.0	92.0	0.21	0.07	0.27	−0.04	17937
sand	g kg^{-1}	42.9	26.1	1.0	19.0	42.0	64.0	99.0	0.19	0.17	−0.30	0.04	17937
$CaCO_3$	g kg^{-1}	54.6	128.4	0.0	0.0	1.0	16.0	944.0	2.87	−0.32	0.12	0.05	17937
pHw	–	6.3	1.3	3.4	5.2	6.3	7.5	10.1	−0.13	−0.50	0.27	−0.07	17937
CEC	cmol^{+}kg^{-1}	14.1	10.5	0.0	6.8	11.7	18.7	137.0	1.94	−0.50	−0.08	0.13	17937
Organic soils													
SOC	g kg^{-1}	387.1	101.2	156.4	297.3	401.1	475.0	586.8	−0.25	−0.46	0.35	0.26	1099
N	g kg^{-1}	15.5	5.7	3.1	11.2	14.5	19.0	38.6	0.72	−0.53	−0.10	−0.05	1099
$CaCO_3$	g kg^{-1}	2.9	19.9	0.0	0.0	0.0	1.0	418.0	14.63	0.03	0.02	−0.11	1099
pHw	–	4.5	0.7	3.2	4.0	4.3	4.7	7.5	1.35	−0.23	−0.09	−0.54	1099
CEC	cmol^{+}kg−^{1}	42.0	33.0	0.0	23.8	31.8	42.5	234.0	2.54	−0.39	0.03	−0.48	1099

[a]Standard Deviation;
[b]lower quartile;
[c]median;
[d]upper quartile,
[e]correlation of PC1-3 scores with the soil properties;
[f]number of samples.

and mineral soil models (Figure S2). Using clay content and the spectral matrix in the models allowed to decrease *RMSEP* of grassland soil predictions with a median of ~1 g C kg^{-1} but no improvement was observed for other subsets (Figure S2).

Recursive feature elimination provided no overall increase in prediction accuracy for models using spectral data only. This is to be expected since most of the multivariate models that we tested (boosted regression tree, Cubist, multivariate adaptive regression splines) include an internal feature selection method. However, models using RFE in combination with sand and clay content showed clear improvements in accuracy compared to models using the spectral matrix only and the spectral matrix in combination with particle size fractions (Figure S2). This is probably related to the fact that RFE, by reducing the dimensionality of the spectral matrix and by keeping the relevant information for SOC prediction, allowed increasing the relative weight of the auxiliary predictor in the models compared to the spectral matrix. Cubist, closely followed by support vector machine regressions, produced the most accurate predictions (i.e. have lower *RMSEP*) for grassland and woodlands soils, while Cubist regressions performed slightly less for cropland and mineral subsets (Figure S3). This confirms the good performance of support vector machine regression and Cubist in predicting soil properties compared to other multivariate calibration models [36,39]. For organic soils, Cubist and partial least square regression showed the best prediction abilities (Figure S3).

Performance of the Best Spectroscopic Models

Prediction performance statistics of the best models (i.e. having the lowest *RMSEP*'s) with and without auxiliary predictors are given in Table 4. The lowest *RMSEP*'s (Eq. 4) were obtained for cropland soils (4–4.9 g C kg^{-1}), followed by grassland (6.4–9.3 g C kg^{-1}), mineral (7.3–8.9 g C kg^{-1}), woodland (10.3–15 g C kg^{-1}) and organic soils (50.6 g C kg^{-1}; Table 4). The difference in *RMSEP* between the subsets reflected the dependence of the model errors on (*i*) calibration size and (*ii*) the variance of observed SOC values. Hence, cropland, grassland and mineral soils, characterized by a large number of samples and small variance were better predicted than woodland and organic soils (Table 4). The accuracy of spectroscopic models increased with the number of calibration sample [12,47] because a large sample size allows to better describe the soil complexity of a given area. The tendency of *RMSEP* to increase with SOC variance as observed in the LUCAS database is also well documented [8]. Datasets characterized by larger SOC variances usually cover larger areas or areas with an important variation in soil properties, which may be detrimental to SOC prediction models. However, since all subsets cover the same geographical extent, it is more probable that SOC variation itself and SOC concentration rather than soil diversity explain the increase in *RMSEP* from cropland to organic soils [8,48,49].

The bias (Eq. 2) of the spectroscopic models was very low in absolute value (<1.1 g C kg^{-1} for mineral soils, Table 4) compared to the standard error of prediction corrected for bias (SEP_{-b}; Eq. 3), indicating that a large portion of the error was due to the residual variance. Residuals tend to increase with increasing SOC content, except for organic soils (Figure 3). While such increase in model residuals could be attributed to an increase in analytical error with SOC content [12], we rather suggest that this was caused by the skewed distribution of SOC content of mineral soils (Table 3) because predictions at high SOC content were affected by a strong bias (Figure 3). This often occurs when predicted samples are under-represented in the training set [50]. Organic soil samples, having only a small negative skewness (Table 3), did not show an increase in prediction residuals with SOC content.

The accuracy of the models developed from the LUCAS library compared very well to other published results with Ratio of Performance to Deviation (*RPD*; Eq. 5) ranging from 1.74 to 2.88

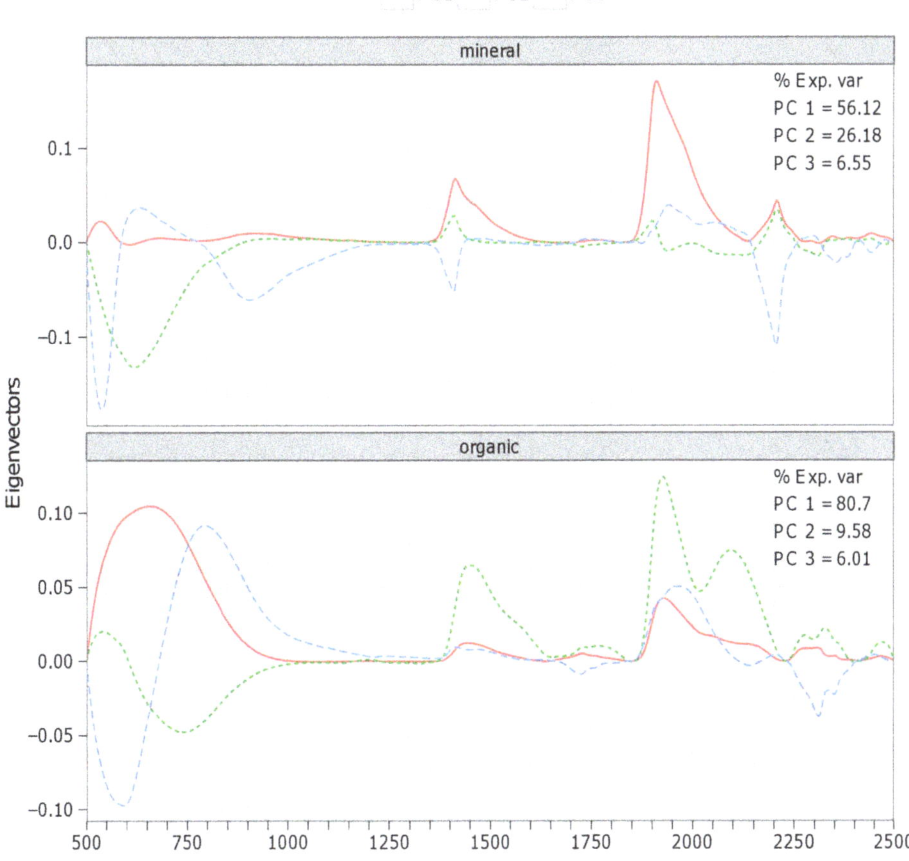

Figure 2. Eigenvectors and eigenvalues of the first three principal components of continuum-removed spectra. The principal component analysis has been realized separately for mineral (top panel) and organic (bottom panel) soil materials.

Table 4. Performance of the best spectroscopic models as measured against the test set.

Subset	Treatment[a]	MVC[b]	Predictor[c]	SD[d]	RMSEP[e]	Bias[f]	SEP_{-b}[g]	RPD[h]	R^2	N[i]
Cropland	SG1	svm	spc	8.6	4.9	0.2	4.9	1.74	0.67	2828
Cropland	SG1+SNV	svm	rfe+clay	8.6	4.0	0.1	4.0	2.17	0.79	2828
Grassland	SG1	svm	spc	17.4	9.3	−0.9	9.3	1.86	0.71	1383
Grassland	SG0	cubist	rfe+sand	17.4	6.4	0.1	6.4	2.70	0.87	1383
Woodland	SG1	svm	spc	29.8	15.0	0.8	15.0	1.99	0.75	1564
Woodland	SG0	cubist	rfe+sand	29.8	10.3	1.1	10.3	2.88	0.89	1564
Mineral	SG1	svm	spc	19.1	8.9	0.2	8.9	2.13	0.78	6053
Mineral	SG1	svm	rfe+sand	19.1	7.3	0.1	7.3	2.62	0.86	6053
Organic	SG1+SNV	cubist	spc	100.8	50.6	−10.9	49.5	1.99	0.76	368

[a]Spectral transformation (SG0 = Savitzky-Golay smoothing; SG1 = Savitzky-Golay first derivative; SNV = standard normal variate);
[b]Multivariate Calibration Model (svm = support vector machine regression; cubist = Cubist);
[c]Predictor used in the models (spc = spectral matrix; rfe = spectral matrix with bands selected by recursive feature elimination);
[d]Standard Deviation of the observations (g C kg^{-1});
[e]Root Mean Square Error of Prediction (g C kg^{-1}; Eq. 4);
[f]Bias (g C kg^{-1}; Eq. 2);
[g]Standard Error of Prediction (g C kg^{-1}; Eq. 3);
[h]Ratio of Performance to Deviation (Eq. 5);
[i]Number of validation samples.

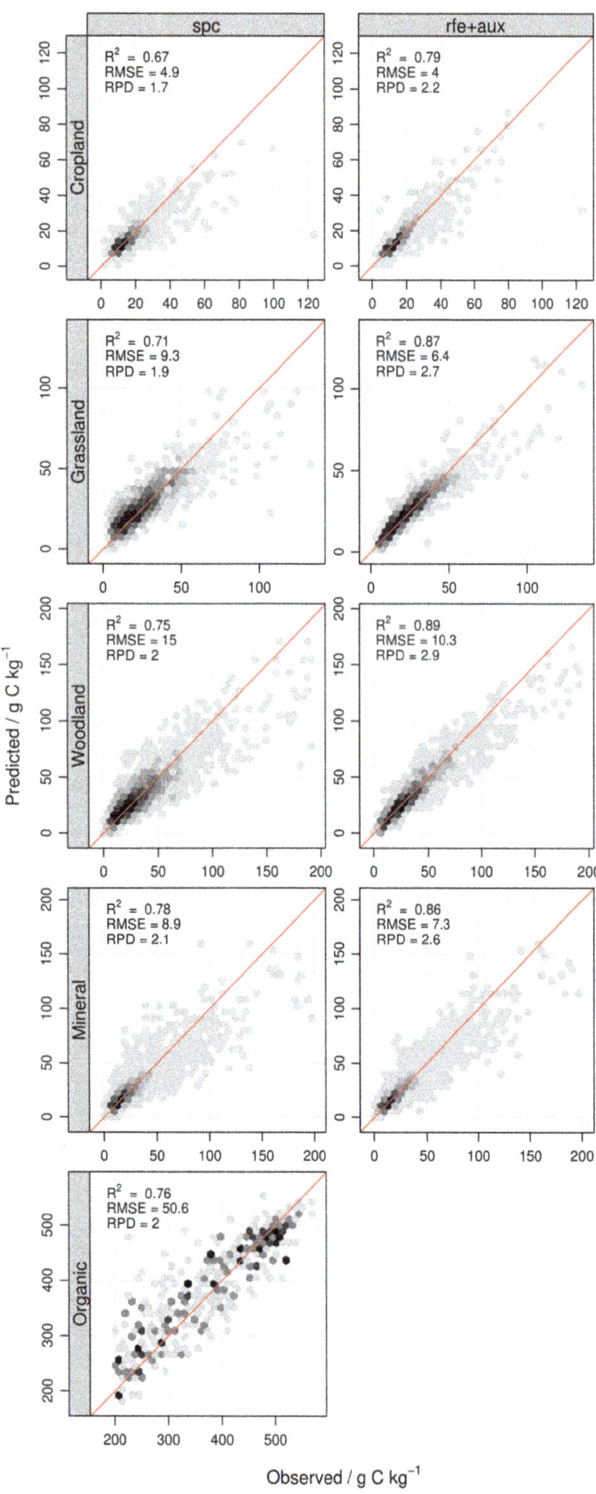

Figure 3. Predicted SOC content as a function of observed SOC content in test sets. Model predictions are shown for models with (rfe+aux, right panels) and without auxiliary predictors (spc, left panels).

and R^2 from 0.67 to 0.89 (Table 3). SOC spectroscopic models reported in the literature achieve an R^2 between predicted and observed values ranging from 0.66 to 0.96 (average of 0.79) and *RPD* values ranging from 1.44 to 4.2 [51,52]. Brown et al. [14] obtained a *RMSEP* of 7.9–9 g C kg^{-1} for global SOC spectroscopic models of mainly mineral soils. The accuracy of spectroscopic models is usually negatively related to the heterogeneity of the samples in a given spectral library [8,12]. This relation explains the relatively low accuracy achieved by large-scale calibrations compared to the one that can be potentially obtained by local-scale (i.e. field, landscape-scale) spectroscopic models. The main reason for the poor performance on heterogeneous soils is that absorption features associated to organic matter can be altered and/or masked by other components of the soil (e.g. iron oxides, clay mineralogy) or can change with the chemical composition or quality of the organic matter [50]. Hence, for heterogeneous soils there is no univocal relationship between SOC content and soil spectra.

To better understand how the spectral response of SOC can be affected by variations in other soil properties, we computed the mean reflectance and continuum-removed reflectance of mineral samples grouped by classes of SOC, sand and clay content (Figure 4–5). Mean reflectance values tend to decrease with both sand and SOC content, so that variation in the spectra that are due to sand content can be confounded with spectral variations due to an increase in SOC content (Figure 4). Sand is a featureless property. However, an increase in sand content typically increases light scattering, which in turn diminishes spectral baseline height and enhances weak absorptions through an increased path length [53]. This effect can be clearly observed in continuum-removed reflectance values (Figure 4): for the same amount of SOC, the absorption feature between 500 and 800 nm that is linked to SOC content is enhanced as sand content increases. Similarly, variations in clay content induce large differences in spectral shape for the same class of SOC content (Figure 5). In each SOC class, one can indeed observe an increase in the depth of absorptions related to O–H and metal–OH in the mineral crystal lattice and O-H in water (1415, 1455, 1915 and 2207 nm) with the increase in clay content. Conversely, the SOC absorption between 500 and 800 nm is progressively masked as the clay content increases. Generally, the albedo of the mean spectra tends to increase with clay content until 60% of clay and decreases thereafter. Differences in albedo are more pronounced for samples with SOC contents below 50 g C kg^{-1} (Figure 5).

The differences in spectral response observed in Figure 4–5 had logically a strong impact on model errors. To illustrate this, we computed the relative *RMSEP* for mineral soil models for intervals of SOC and sand content. The relative *RMSEP* is the *RMSEP* divided by the mean of the observed SOC content in a given class. For models using the spectra only for prediction, the relative *RMSEP* of the models was stable across the SOC content classes but it increased with the sand content (Figure 6). This confirms the results of other studies [48,54] that found larger SOC prediction errors for soils with the highest sand contents. The effect of sand content on SOC prediction accuracy was more pronounced at low SOC content due to the relatively low absorption rates of organic matter and the masking from other soil components ([55]; Figure 4–5). It is therefore expected that spectral libraries of soils characterized by a low SOC content will perform poorly when samples have large variations in particle size distribution. It can be also observed that the use of sand content as auxiliary predictor drastically improved model predictions for sandy soils (Figure 6), explaining the increase in model accuracies compared to models based on the spectral matrix only (Table 4).

Reproducibility Error

We assessed the reproducibility error of the models with a set of duplicate samples. The error of reproducibility (*SEL*, Eq. 6) of the reference method was estimated at 1.5 g C kg^{-1} for cropland,

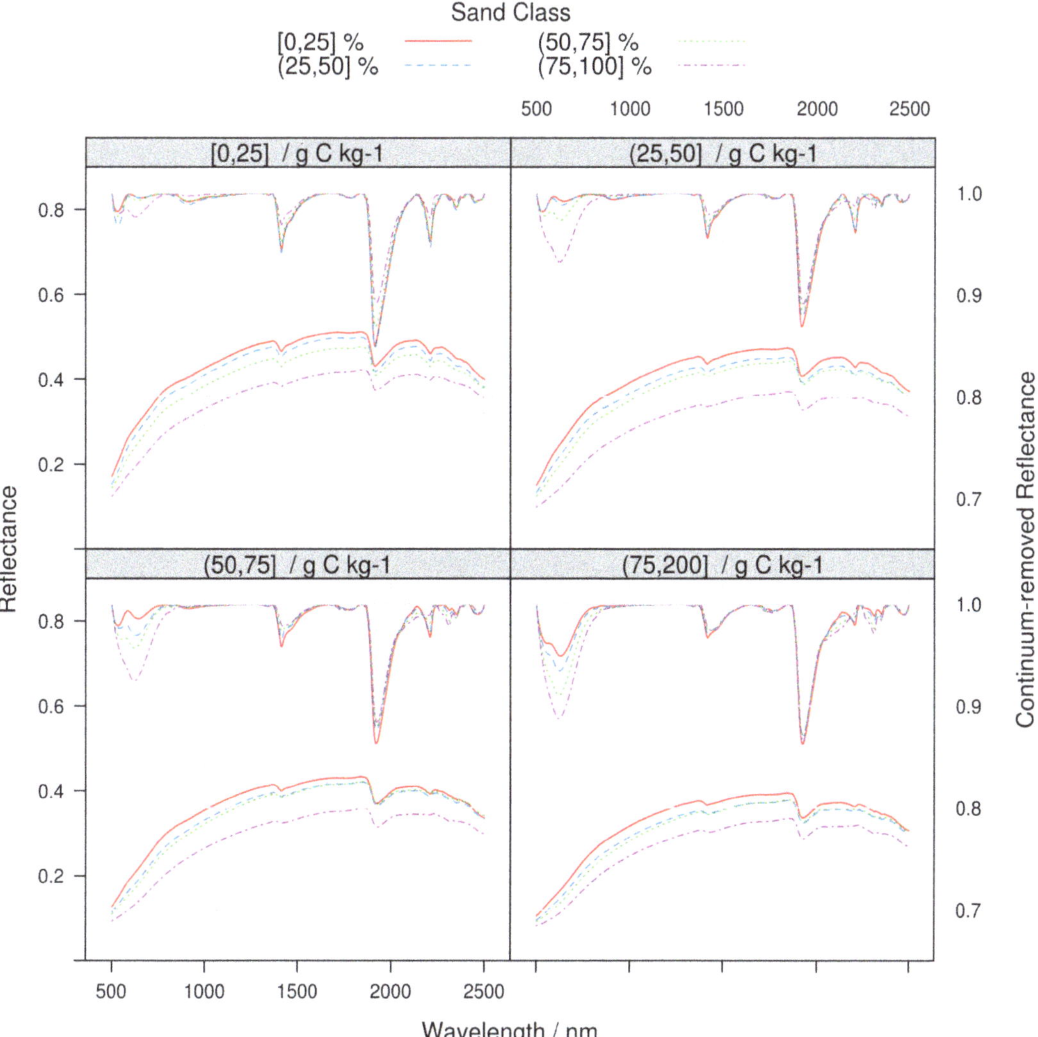

Figure 4. Mean reflectance (left scale) and continuum-removed reflectance (right scale) spectra of LUCAS mineral soil samples, computed for arbitrary sand and SOC classes. The sand classes are 0–25%, 25–50%, 50–75%, 75–100% and the SOC classes are 0–25 g C kg^{-1}, 25–50 g C kg^{-1}, 50–75 g C kg^{-1}, 75–200 g C kg^{-1}. Each panel regroups samples of a given SOC interval.

0.8 g C kg^{-1} for grassland, 2.9 g C kg^{-1} for woodland, 2 g C kg^{-1} for mineral and 11.6 g C kg^{-1} for organic soils (Table 5). The reproducibility of SOC predictions by the spectroscopic models was roughly similar, with values ranging from 1.2 to 9.1 g C kg^{-1} (Table 5). These relatively low values suggested that spectroscopic models are robust through time and only a small component of the error budget can be attributed to measuring conditions (e.g. temperature and humidity in the laboratory). Both analytical techniques showed an increase in reproducibility error with the SOC content.

Implication for SOC Spectroscopic Assessment at Large Scale

The prediction error of the spectroscopic models was ~5 times larger than the reproducibility error of the reference method (Table 3–4) while ratio of *RMSEP* to *SEL* of 1 to 3.5 have been reported in local studies [51]. This clearly indicates that it is currently difficult to produce spectroscopic models of SOC content that are valid for large areas while sufficiently accurate to be useful for most applications at fine scales. However, the models proved to have low biases (Table 4) and hence could be used to estimate the mean SOC content of large areas since the variance of the model residuals is reduced by averaging [6].

Increasing the number of training samples improves the prediction accuracy [12] because a higher number of samples will better describe the soil variability in a given area. Despite the relatively high sampling density of the LUCAS database (Figure 1; Table 1–2), large prediction errors are still observed (Table 4) and it is unlikely that including more samples in the database will significantly improve the prediction for the geographical entities covered by the current LUCAS survey. The two main reasons for this are that: *(i)* soils are complex materials with a strong spatial structure and, as stated above, the relationship linking soil properties with soil spectra is not stationary, so that large-scale spectroscopic models cannot achieve the same level of accuracy as for more homogeneous materials like plants or other agricultural commodities [13] and *(ii)* the natural positive skew of SOC values induces a large model bias at high SOC content. Rather than increasing the number of samples in soil spectral libraries, further efforts should be deployed towards the development of calibration

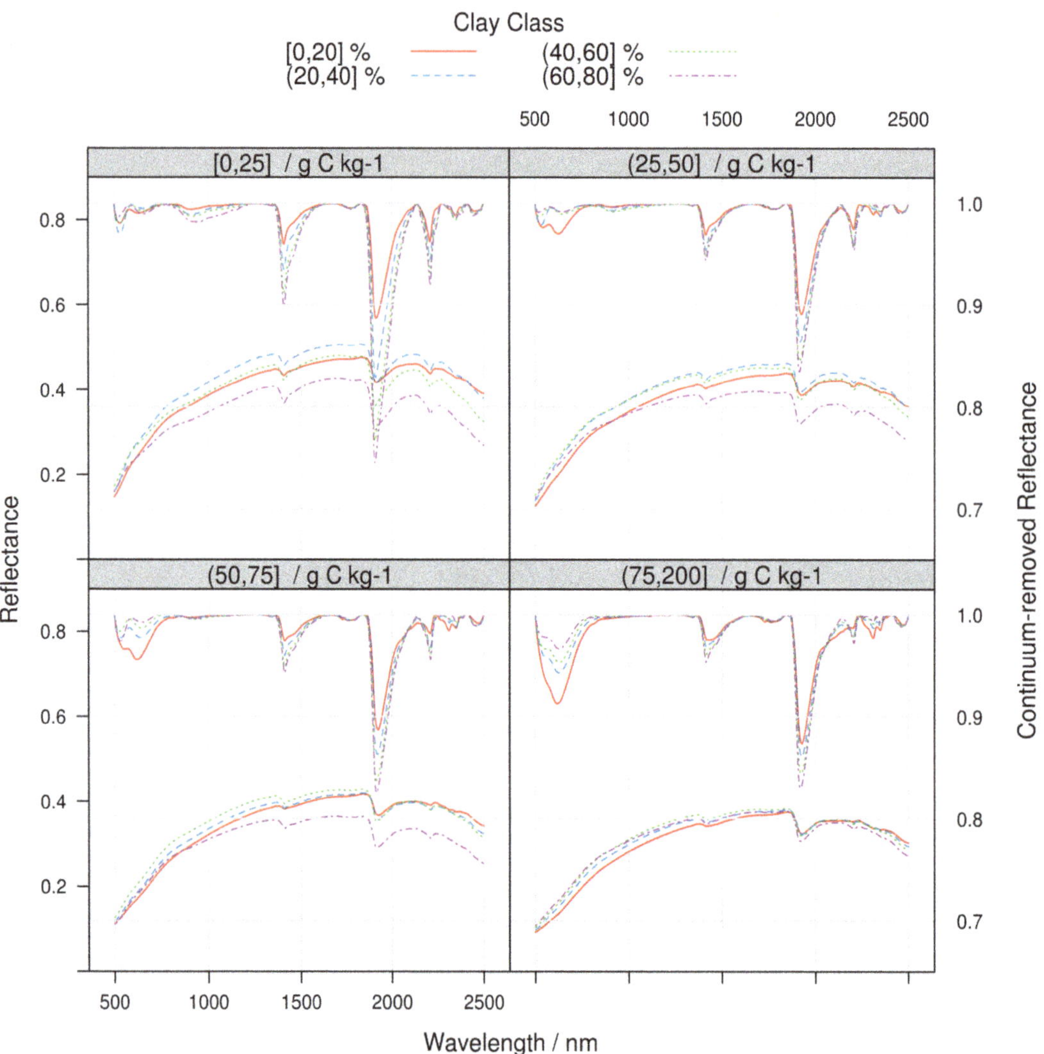

Figure 5. Mean reflectance (left scale) and continuum-removed reflectance (right scale) spectra of LUCAS mineral soil samples, computed for arbitrary clay and SOC classes. The clay classes are 0–20%, 20–40%, 40–60%, 60–80% and the SOC classes are 0–25 g C kg^{-1}, 25–50 g C kg^{-1}, 50–75 g C kg^{-1}, 75–200 g C kg^{-1}. Each panel regroups samples of a given SOC interval.

models that are capable of identifying local patterns of spectral variation in large scale libraries because developing a universal calibration model for SOC prediction is unlikely. We showed that a promising approach consists in including readily available soil covariates in the spectroscopic modeling, although other strategies should be explored, such as simple subsetting of samples by soil type or SOC content [49,56]. Ideally, covariates should be linked with important spectrally-active soil components (e.g. mineralogy, texture, iron content) or with pedogenetic factors such as climate and land cover.

Conclusion

The LUCAS soil database is the result of a harmonized soil survey conducted on a relatively dense sampling grid over 23 member states of the European Union. This database represents currently the most comprehensive soil spectral library at continental scale using a uniform protocol for both chemical and spectral analyses. Still, SOC spectroscopic models showed relatively large errors (>4 g C kg^{-1}) compared to established methods of SOC analysis, suggesting that accurate SOC predictions based on large scale spectral libraries will be hard to achieve. Prediction errors were found to be related to SOC variation, SOC distribution (skewness) and variation in other soil properties such as sand and clay content. These findings strongly suggest that vis-NIR spectral data alone do not contain enough information to get accurate predictions of soil properties at large scales. Hence, to develop spectroscopy as a valuable tool for soil analyses, further research should be directed towards the development of strategies that can address this issue, such as the use of additional predictors in the modeling.

Despite these difficulties, large spectral libraries can be very valuable to (*i*) build local and more accurate spectroscopic models that are specific to a given geographical entity or soil type and (*ii*) develop spectroscopic models able to quickly produce SOC predictions for estimate accurately SOC means across regions or countries, due to the unbiasedness of the method. The LUCAS spectral library will be made publicly available for non-commercial purpose through the European Soil Data Centre (http://eusoils.jrc.ec.europa.eu/projects/lucas/data.html).

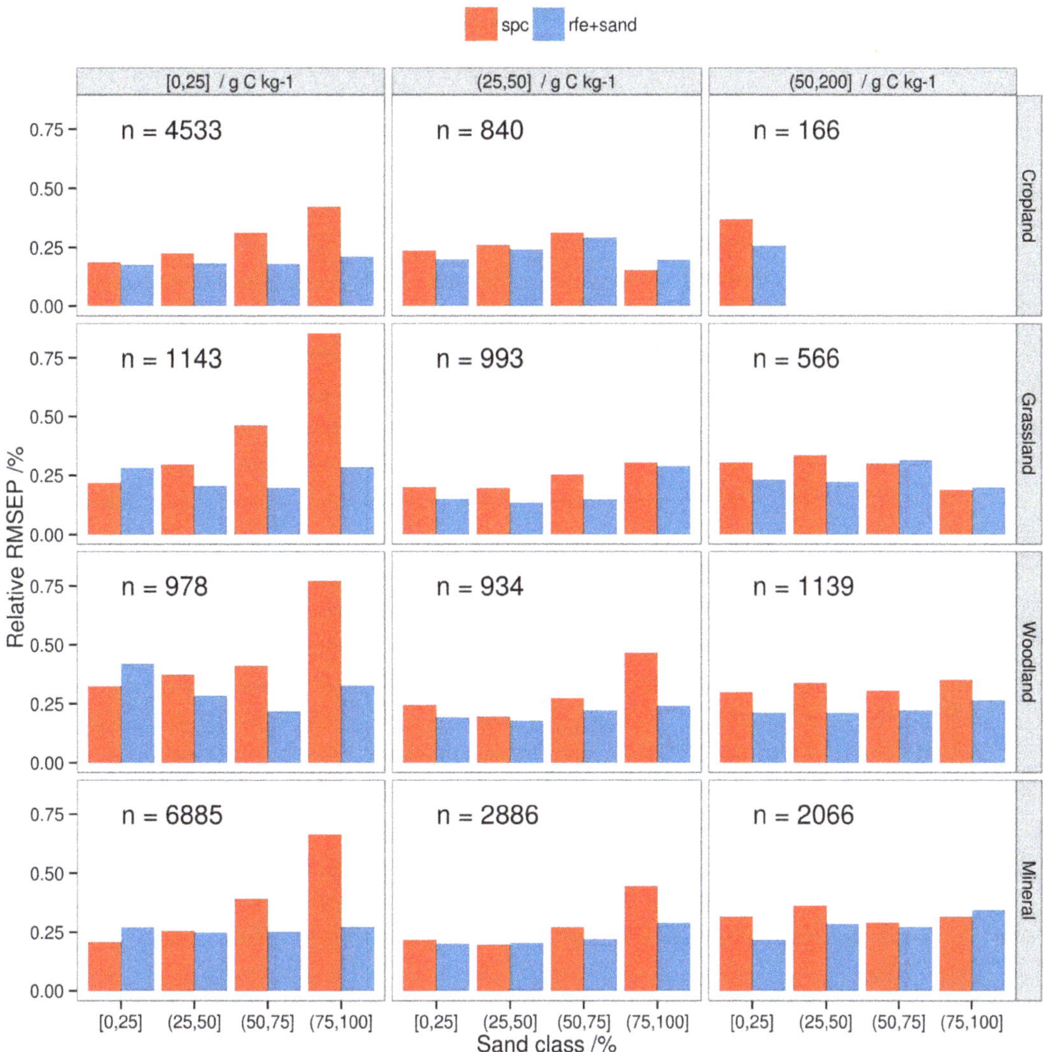

Figure 6. Relative Root Mean Square Error of Prediction (*RMSEP*) per land cover, for arbitrary classes of SOC and sand content. The sand classes are 0–25%, 25–50%, 50–75%, 75–100% and the SOC classes are 0–25 g C kg^{-1}, 25–50 g C kg^{-1}, 50–200 g C kg^{-1}. The relative *RMSEP* is the *RMSEP* divided by the mean of observed SOC values of models developed with (red bars) and without auxiliary predictors (blue bars). Each panel regroups mineral samples of a given SOC interval and land cover type. The number of training samples (n) for each class of SOC content is given in each panel.

Table 5. Reproducibility of SOC estimates (g C kg^{-1}; Eq. 6) of the reference method and the spectroscopic models with (rfe+aux) and without (spc) the use of auxiliary predictors.

Subset	Reference	spc	rfe+aux	n[a]
Cropland	1.5	1.8	0.9	13
Grassland	0.8	1.4	1.7	5
Woodland	2.9	2.1	2.4	4
Mineral	2	1.8	2.4	22
Organic	11.5	9.1	–	3

[a]Number of duplicate samples.

Although not addressed in this paper, there are indications that other key soil properties identified by the Global Soil Map community such as clay content, pH and cation-exchange capacity [3] can simultaneously be analyzed using Vis-NIR spectroscopy [8]. As similar large scale spectral libraries are being developed in USA [57], Africa [16], and Australia [22], spectral analyses will provide consistent soil measurements for a large portion of soils across the globe. However, in order to develop soil spectroscopy into an operational tool producing harmonized data across laboratories and environments, we urgently need consultation and cooperation to define internationally agreed standards for soil spectral analyses, including norms for instrumentation, sampling preparation, reference materials, measuring set-up, quality checks and calibration transfer methods.

Supporting Information

Figure S1 Box-and-whisker plots of the Root Mean Square Error of Prediction (*RMSEP*) as a function of

the pretreatments. Each panel presents separately the results obtained for cropland, grassland, woodland, mineral and organic models. Pretreatments: SG0 = Savitzky-Golay smoothed absorbance; SG1 = Savitzky-Golay first derivative; A = absorbance; SNV = Standard Normal Variate.

Figure S2 Box-and-whisker plots of the Root Mean Square Error of Prediction (*RMSEP*) as a function of the predictors. Each panel presents separately the results obtained for cropland, grassland, woodland, mineral and organic models. Predictors: spc = spectral matrix; rfe = spectral matrix with bands selected by recursive feature elimination.

Figure S3 Box-and-whisker plots of the Root Mean Square Error of Prediction (*RMSEP*) as a function of the multivariate calibration approach. Each panel presents separately the results obtained for cropland, grassland, woodland, mineral and organic models. Multivariate models: pls = partial least square regression; cubist = Cubist; mars = multivariate adaptive regression splines; brt = boosted regression tree; rf = random forest; svm = support vector machine.

Acknowledgments

We thank the SOIL Action (IES, Joint Research Centre, European Commission) for help with the preparation and management of the LUCAS database.

Author Contributions

Conceived and designed the experiments: AS MN GT LM. Performed the experiments: AS MN. Analyzed the data: AS MN BvW. Wrote the paper: AS BvW.

References

1. Lal R (2004) Soil Carbon Sequestration Impacts on Global Climate Change and Food Security. Science 304: 1623–1627.
2. Van Oost K, Quine TA, Govers G, De Gryze S, Six J, et al. (2007) The Impact of Agricultural Soil Erosion on the Global Carbon Cycle. Science 318: 626–629.
3. Sanchez PA, Ahamed S, Carré F, Hartemink AE, Hempel J, et al. (2009) Digital Soil Map of the World. Science 325: 680–681.
4. van Wesemael B, Paustian K, Meersmans J, Goidts E, Barancikova G, et al. (2010) Agricultural management explains historic changes in regional soil carbon stocks. Proceedings of the National Academy of Sciences 107: 14926–14930.
5. Baritz R, Zirlewagen D, Jones R, Arrouays D, Hiederer R, et al. (2011) Carbon in European Soils. Soil Carbon in Sensitive European Ecosystems: From Science to Land Management. Chichester, United Kingdom: Jandl, R., Rodeghiero, M., Olsson, M. 49–84.
6. Bellon-Maurel V, McBratney A (2011) Near-infrared (NIR) and mid-infrared (MIR) spectroscopic techniques for assessing the amount of carbon stock in soils – Critical review and research perspectives. Soil Biology and Biochemistry 43: 1398–1410.
7. O' Rourke SM, Holden NM (2011) Optical sensing and chemometric analysis of soil organic carbon – a cost effective alternative to conventional laboratory methods? Soil Use and Management 27: 143–155.
8. Stenberg B, Viscarra Rossel RA, Mouazen AM, Wetterlind J (2010) Visible and near infrared spectroscopy in soil science. Advances in Agronomy 107: 163–215.
9. Ben-Dor E, Chabrillat S, Demattê JAM, Taylor GR, Hill J, et al. (2009) Using Imaging Spectroscopy to study soil properties. Remote Sensing of Environment 113: S38–S55.
10. Ben-Dor E, Irons J, Epema G (1999) Soil Reflectance. Remote Sensing for the Earth Sciences: Manual of Remote Sensing. Rencz, A.N., Vol. 3. 111–188.
11. Shepherd KD, Walsh MG (2007) Infrared spectroscopy–enabling an evidence-based diagnostic surveillance approach to agricultural and environmental management in developing countries. Journal of Near Infrared Spectroscopy 15: 1–19.
12. Shepherd KD, Walsh MG (2002) Development of reflectance spectral libraries for characterization of soil properties. Soil Science Society of America Journal 66: 988–998.
13. Malley DF, Martin PD, Ben-Dor E (2004) Application in analysis of soils. Agronomy: Near-Infrared Spectroscopy in Agriculture. Agronomy. Madison, WI, USA: Roberts, C.A, Workman, J.W, Reeves, J.B., Vol. 44. 729–784.
14. Brown DJ, Shepherd KD, Walsh MG, Mays MD, Reinsch TG (2006) Global soil characterization with VNIR diffuse reflectance spectroscopy. Geoderma 132: 273–290.
15. Viscarra Rossel R (2009) The Soil Spectroscopy Group and the development of a global soil spectral library. NIR news 20: 14.
16. ICRAF-ISRIC (2010) A Globally Distributed Soil Spectral Library: Visible Near Infrared Diffuse Reflectance Spectra. World Agroforestry Centre (ICRAF) and ISRIC - World Soil Information. Available: http://www.africasoils.net/afsis_files/ICRAF-ISRICSoilVNIRSpectralLibrary.pdf. Accessed 26 March 2012.
17. Ge Y, Morgan CLS, Grunwald S, Brown DJ, Sarkhot DV (2011) Comparison of soil reflectance spectra and calibration models obtained using multiple spectrometers. Geoderma 161: 202–211.
18. Genot V, Colinet G, Bock L, Vanvyve D, Reusen Y, et al. (2011) Near infrared reflectance spectroscopy for estimating soil characteristics valuable in the diagnosis of soil fertility. Journal of near Infrared Spectroscopy 19: 117–138.
19. ICRAF Soil-Plant Spectral Diagnostics Laboratory. Available: http://worldagroforestry.org/research/land-health-management/spectral-diagnostics-laboratory. Accessed 13 May 2013.
20. Walkley A, Black IA (1934) An examination of the Degtjareff method for determining soil organic matter, and a proposed modification of the chromic acid titration method. Soil science 37: 29–38.
21. Africa Soil Information Service (2012) Analyzing Africa's Soils with Infrared Spectroscopy. AfSIS Newsletter: 5.
22. Viscarra Rossel RA, Webster R (2012) Predicting soil properties from the Australian soil visible–near infrared spectroscopic database. European Journal of Soil Science 63: 848–860.
23. Decision No 1578/2007/EC of the European Parliament and of the Council (2007). Available: http://eur-lex.europa.eu/LexUriServ/LexUriServ.do?uri=OJ:L:2007:344:0015:0043:EN:PDF.Accessed 15 January 2013.
24. Regulation (EC) No 223/2009 of the European Parliament and of the Council (2009). Available: http://eur-lex.europa.eu/LexUriServ/LexUriServ.do?uri=OJ:L:2009:087:0164:0173:En:PDF.Accessed 15 January 2013.
25. Eurostat (2009) LUCAS 2009 survey. Available: http://eusoils.jrc.ec.europa.eu/projects/Lucas/. Accessed 8 May 2012.
26. Montanarella L, Tóth G, Jones A (2011) Soil Component in the 2009 LUCAS Survey. Land quality and land use information in the European Union. Luxembourg: Tóth, G and Németh, T. 209–219.
27. Rinnan Å, Berg F van den, Engelsen SB (2009) Review of the most common pre-processing techniques for near-infrared spectra. TrAC Trends in Analytical Chemistry 28: 1201–1222.
28. Savitzky A, Golay MJE (1964) Smoothing and differentiation of data by simplified least squares procedures. Anal Chem 36: 1627–1639.
29. Barnes RJ, Dhanoa MS, Lister SJ (1989) Standard normal variate transformation and de-trending of near-infrared diffuse reflectance spectra. Applied spectroscopy 43: 772–777.
30. FAO IUSS Working Group (1998) World reference base for soil resources. Roma, Italy.
31. Eurostat (2009) LUCAS 2009 (Land Use/Cover Area Frame Survey) - Technical reference document C-3: Land Use and Land Cover Nomenclature. Available: http://epp.eurostat.ec.europa.eu/portal/page/portal/lucas/documents/Nomenclature_LUCAS2009_C_3.pdf. Accessed 9 May 2012.
32. Kennard RW, Stone LA (1969) Computer aided design of experiments. Technometrics 11: 137–148.
33. Clark RN, Roush TL (1984) Reflectance Spectroscopy: Quantitative Analysis Techniques for Remote Sensing Applications. J Geophys Res 89: PP. 6329–6340.
34. Kuhn M, Wing J, Weston S, Williams A, Keefer C, et al. (2012) Caret: classification and regression training. R package, v515. Available: http://cran.r-project.org/web/packages/caret/. Accessed 13 May 2013.
35. R Development Core Team (2012) R: A language and environment for statistical computing. Available: http://www.R-project.org. Accessed 13 May 2013.
36. Minasny B, McBratney AB (2008) Regression rules as a tool for predicting soil properties from infrared reflectance spectroscopy. Chemometrics and Intelligent Laboratory Systems 94: 72–79.
37. Quinlan JR (1993) Combining instance-based and model-based learning. Proceedings of the Tenth International Conference on Machine Learning. 236–243.
38. Hastie T, Tibshirani R, Friedman J (2009) The Elements of Statistical Learning. New York, NY: Springer New York.
39. Viscarra Rossel RA, Behrens T (2010) Using data mining to model and interpret soil diffuse reflectance spectra. Geoderma 158: 46–54.
40. Breiman L, Friedman JH, Olshen RA, Stone CJ (1984) Classification and regression trees. Belmont, USA: Wadsworth International Group.
41. Kuhn M (2012) Variable selection using the caret package. Available: http://cran.r-project.org/web/packages/caret/vignettes/caretSelection.pdf. Accessed 7 May 2012.
42. Fearn T (2010) Combining other predictors with NIR spectra. NIR news 21: 13.

43. Bellon-Maurel V, Fernandez-Ahumada E, Palagos B, Roger J-M, McBratney A (2010) Critical review of chemometric indicators commonly used for assessing the quality of the prediction of soil attributes by NIR spectroscopy. TrAC Trends in Analytical Chemistry 29: 1073–1081.
44. Fearn T (2008) Calculating standard deviations. NIR news 16: 7–8.
45. Ben Dor E, Inbar Y, Chen Y (1997) The reflectance spectra of organic matter in the visible near-infrared and short wave infrared region (400–2500 nm) during a controlled decomposition process. Remote Sensing of Environment 61: 1–15.
46. Elvidge CD (1990) Visible and near infrared reflectance characteristics of dry plant materials. International Journal of Remote Sensing 11: 1775–1795.
47. Kuang B, Mouazen AM (2012) Influence of the number of samples on prediction error of visible and near infrared spectroscopy of selected soil properties at the farm scale. European Journal of Soil Science 63: 421–429.
48. Sørensen LK, Dalsgaard S (2005) Determination of Clay and Other Soil Properties by Near Infrared Spectroscopy. Soil Science Society of America Journal 69: 159.
49. McDowell ML, Bruland GL, Deenik JL, Grunwald S (2012) Effects of Subsetting by Carbon Content, Soil Order, and Spectral Classification on Prediction of Soil Total Carbon with Diffuse Reflectance Spectroscopy. Applied and Environmental Soil Science 2012, ID 294121. Available: http://www.hindawi.com/journals/aess/2012/294121/abs/. Accessed 26 April 2013.
50. Brown DJ, Bricklemyer RS, Miller PR (2005) Validation requirements for diffuse reflectance soil characterization models with a case study of VNIR soil C prediction in Montana. Geoderma 129: 251–267.
51. Nduwamungu C, Ziadi N, Parent L-E, Tremblay GF, Thuriès L (2009) Opportunities for, and limitations of, near infrared reflectance spectroscopy applications in soil analysis: A review. Canadian Journal of Soil Science 89: 531–541.
52. Ladoni M, Bahrami HA, Alavipanah SK, Norouzi AA (2009) Estimating soil organic carbon from soil reflectance: a review. Precision Agriculture 11: 82–99.
53. Clark RN (1999) Spectroscopy of rocks and minerals, and principles of spectroscopy. Manual of Remote Sensing. New York, USA: Rencz, A.N., Vol. 3. 3–58. Available: http://speclab.cr.usgs.gov/PAPERS.refl-mrs/refl4.html. Accessed 27 June 2012.
54. Stenberg B (2010) Effects of soil sample pretreatments and standardised rewetting as interacted with sand classes on Vis-NIR predictions of clay and soil organic carbon. Geoderma 158: 15–22.
55. Ben Dor E, Banin A (1995) Near-Infrared analysis as a rapid method to simultaneously evaluate soil properties. Soil Science Society of America Journal 59: 364–372.
56. Vasques GM, Grunwald S, Harris WG (2010) Spectroscopic models of soil organic carbon in Florida, USA. Journal of Environmental Quality 39: 923–934.
57. West LT, Waltman SW, Wills S, Reinsch TG, Benham EC, et al. (2010) Soil Carbon Stocks in the U.S.: Current Data and Future Inventories. Proceedings of International Workshop on Evaluation and Sustainable Management of Soil Carbon Sequestration in Asian Countries. Bogor, Indonesia. p. 14. Available: http://balittanah.litbang.deptan.go.id/dokumentasi/prosiding/fftc/01-Larry%20Wes_Carbon%20strock.pdf. Accessed 29 August 2012.
58. European Commission and the Europan Soil Bureau Network (2004) The European Soil Database distribution version 2.0. Available: http://eusoils.jrc.ec.europa.eu/ESDB_Archive/ESDB_Data_Distribution/ESDB_data.html. Accessed 13 May 2013.

2

Optimizing Hill Seeding Density for High-Yielding Hybrid Rice in a Single Rice Cropping System in South China

Danying Wang[1], Song Chen[1], Zaiman Wang[2], Chenglin Ji[1], Chunmei Xu[1], Xiufu Zhang[1]*, Bhagirath Singh Chauhan[3]

1 State Key Laboratory of Rice Biology, China National Rice Research Institute, Hangzhou, Zhejiang, China, **2** Key Laboratory of Key Technology on Agricultural Machine and Equipment, South China Agricultural University, Guangzhou, Guangdong, China, **3** Queensland Alliance for Agriculture and Food Innovation, University of Queensland, Queensland, Australia

Abstract

Mechanical hill direct seeding of hybrid rice could be the way to solve the problems of high seeding rates and uneven plant establishment now faced in direct seeded rice; however, it is not clear what the optimum hill seeding density should be for high-yielding hybrid rice in the single-season rice production system. Experiments were conducted in 2010 and 2011 to determine the effects of hill seeding density (25 cm×15 cm, 25 cm×17 cm, 25 cm×19 cm, 25 cm×21 cm, and 25 cm×23 cm; three to five seeds per hill) on plant growth and grain yield of a hybrid variety, Nei2you6, in two fields with different fertility (soil fertility 1 and 2). In addition, in 2012 and 2013, comparisons among mechanical hill seeding, broadcasting, and transplanting were conducted with three hybrid varieties to evaluate the optimum seeding density. With increases in seeding spacing from 25 cm×15 cm to 25 cm×23 cm, productive tillers per hill increased by 34.2% and 50.0% in soil fertility 1 and 2. Panicles per m^2 declined with increases in seeding spacing in soil fertility 1. In soil fertility 2, no difference in panicles per m^2 was found at spacing ranging from 25 cm×17 cm to 25 cm×23 cm, while decreases in the area of the top three leaves and aboveground dry weight per shoot at flowering were observed. Grain yield was the maximum at 25 cm×17 cm spacing in both soil fertility fields. Our results suggest that a seeding density of 25 cm×17 cm was suitable for high-yielding hybrid rice. These results were verified through on-farm demonstration experiments, in which mechanical hill-seeded rice at this density had equal or higher grain yield than transplanted rice.

Editor: Jose Luis Gonzalez-Andujar, Instituto de Agricultura Sostenible (CSIC), Spain

Funding: The authors are grateful for grants from the National Rice Industry Technology System (CARS-01-04A), the 3 Nong 6 Fang Program, and the Five-twelfth National Science and Technology Support Program (2011BAD16B14). The authors state that the funders had no role in study design, data collection and analysis, decision to publish, or preparation of the manuscript.

Competing Interests: The authors have declared that no competing interests exist.

* Email: chais.zju@gmail.com

Introduction

Although transplanting has been a major traditional method of rice establishment in most Asian countries, the rising labor cost and developments in rice production technology have improved the desirability of direct-seeded rice [1]. Direct seeding of rice is a low-cost establishment technology. It avoids nursery raising, seedling uprooting, and transplanting, and has the benefits of saving labor, facilitating timely establishment of rice, and earlier crop maturity (by 7–10 days) [2–4]. Direct seeding helps in solving the labor scarcity problem, which is now very critical in China's agricultural development. Simultaneously, the chemical weed control method has made such a switch technically viable. In 2008, for example, 8.3% of the rice-growing area in China was direct-seeded; in some provinces of south China, it was more than 30% [5].

In the south China irrigated rice ecosystem, broadcasting in wet conditions (wet seeding) is the principal method of rice establishment. In this method, pregerminated seeds are sown onto wet (saturated) puddled soils. But, some difficulties are faced in wet seeding. First, no specific varieties have been developed in China for wet seeding. The existing varieties used for transplanting do not appear to be well-adapted for seedling growth in an initially oxygen-depleted micro environment. Second, the extent of laser leveling in China is currently extremely smaller compared with that in other countries (e.g., 50–80% of the rice land in Australia is laser-leveled) [4]. Due to lack of uniform water distribution associated with unevenness of the land, the problem of excess- or no-water-caused yield variability within a field is common, which leads to poor establishment of direct-seeded rice [3,6]. Third, surface-sown seeds are damaged by birds and rodents. As a result, farmers often resort to the costly practice of increasing the seeding rate for direct-seeded rice, even by two to three times [7]. High seeding rates can result in large yield losses due to excessive number of tillers, increased proportion of ineffective tillers, higher spikelet sterility, and fewer grains per panicle [8]. Moreover, a dense canopy and less ventilation around the plants at high seeding rates can create favorable conditions for diseases (e.g., sheath blight) [9,10] and insects (e.g., brown planthoppers) and can make plants more prone to lodging [11]. High seeding rates using broadcasting restrict individual plant growth, which is not suitable for the production of high-yielding hybrid rice.

High-yielding hybrid rice, which is crucial to meet expanding food demand, had large panicles and high aboveground biomass [12,13]. Hybrid rice varieties grow well with proper planting density, but their lodging resistance decreases dramatically under dense seeding [14]. The use of high seeding rate in a broadcast culture may increase the gap between farmers' potential and actual yield.

Mechanical hill seeding of rice is one of the options used to solve the problem of high seeding rate and uneven plant establishment. Here, rice seeds are sown uniformly in the field at designated spacing and seed number per hill. Combined with a harrow or furrower opener, many hill seeders have land-preparation functions to improve soil evenness. Some hill seeders enable seeding and puddling to be performed simultaneously [15]; others can make furrows and plant the crop at the same time [16]; still others can effectively incorporate seeds into puddled soils at depths of 5–20 mm, which helps decrease losses caused by birds [15]. Even soil conditions and uniform seeding result in an even plant establishment. A well-ventilated micro-environment not only reduces pest and disease incidence but also enhances photosynthesis [17]. In previous studies, lodging resistance of hill-seeded rice was found to be higher than mechanically transplanted rice, row-seeded rice, and broadcast-seeded rice [15,18].

However, the practical use of mechanical hill seeders had not been seen in China until the 2000s. The adoption of this implement was hampered by the belief that broadcasting is not a suitable method for hybrid rice varieties and both government and farmers still have misgivings about the use of mechanical hill seeding for this purpose. Research is thus imperative to shed light on this technology. There is limited information on hill seeding density for high-yielding hybrid rice varieties, and results for optimum hill seeding density in a single-season rice production system varied from 20 cm×14 cm to 30 cm×20 cm spacing [17,19–21]. It is difficult to draw a concrete conclusion about what hill seeding density is optimum for high-yielding hybrid rice varieties.

The aim of this study was to evaluate the optimum hill seeding density to support the development of mechanical hill seeders for hybrid rice. In this study, 2-year field experiments and 2-year on-farm demonstration experiments were conducted in Zhejiang, China. In the field experiments, a high-yielding hybrid rice variety (Nei2you6) was hill-seeded manually at five seeding densities in two fields with different fertility profiles. The objectives were to evaluate the effects of hill seeding density on plant growth and grain yield, the interrelation between seeding density and soil fertility, and the optimum hill seeding density for high-yielding hybrid rice. In the on-farm demonstrations, three high-yielding hybrid rice varieties were planted using three methods: mechanical hill-seeded with a hill seeder at optimum seeding density, broadcasting, and transplanting following standard farmers' practice. The objective was to verify the optimum seeding density in the field by comparing the yields of hill-seeded, broadcast, and transplanted rice.

Materials and Methods

Field experiments and on-farm demonstrations were conducted; field experiments were done in 2010 and 2011 and the on-farm demonstration experiments were carried out in 2012 and 2013. Field experiments were conducted at the experimental farm of the China Rice Research Institute and on-farm demonstration experiments were conducted at the agricultural demonstration base of the China Rice Research Institute in Fuyang and Yuhang. No permission was needed to use fields at these facilities. Authors also confirmed that the study did not involve endangered or protected species.

Field experiments

The 2010 and 2011 experiments were conducted in two paddy fields with different fertility levels (soil fertility 1 and 2) at the China National Rice Research Institute (CNRRI) (120.2′ E, 30.3′ N, 11 m above sea level), Zhejiang Province, China. The area has a subtropical monsoon climate with an annual mean temperature of 13–20°C (range from 2°C in January to 39°C in July). Mean annual precipitation is 1200–1600 mm, with about 80% rainfall occurring between April and September. Soil at the experimental site is classified as ferric-accumulic stagnic anthrosol [22]. The cropping system of soil fertility 1 was monoculture rice-fallow without rice straw incorporation, and the cropping system of soil fertility 2 was rice-potato (*Solanum tuberosum*) with rice and potato straw incorporation. After 10 years of continuous farming, the soil properties in the two fields differed and this was noted at the start of the experiment in 2010. The soil in field 1 had 2.46 g total N kg^{-1}, 0.53 g total P kg^{-1}, 0.25 g available K kg^{-1}, 10.73 C mol kg^{-1} cation exchange capacity (CEC), and pH 5.78 (soil fertility 1). The corresponding soil properties in field 2 were 2.96 g total N kg^{-1}, 0.72 g total P kg^{-1}, 0.31 g available K kg^{-1}, 10.80 C mol kg^{-1} CEC, and pH 5.15 (soil fertility 2). The soil test was based on samples taken from the upper 10 cm of the soil. Soil chemical properties at the 10–20 cm depth were also different [23].

Pregerminated seeds of a high-yielding hybrid rice variety, Nei2you6, were hill-seeded manually in the two fields with different fertility levels using five spacing schemes (between and within rows: 25 cm×15 cm, 25 cm×17 cm, 25 cm×19 cm, 25 cm×21 cm, and 25 cm×23 cm). Each hill was planted with three to five seeds. The experiment was replicated four times and subplot size was 30 m^2. In both years, the fields were puddled in ponded conditions and drained 2 d before seeding. Pre-germinated seeds were sown on the surface of puddled soil on June 2, 2010 and May 30, 2011. Before sowing, seeds were soaked for 24 h and incubated for 24 h for proper germination, and seeds dressed with chemical pesticides to avoid the attack of soil insects, birds, and rodents.

To the crop, 225 kg N ha^{-1} (as urea) was applied in three splits in a ratio of 5:3:2 at basal, tillering, and earing stage, respectively. Potassium at 180 kg K_2O ha^{-1} was applied in two splits: 50% as basal dressing and 50% as topdressing at earing in the form of KCl. Phosphorus at 75 kg P_2O_5 ha^{-1} in the form of calcium superphosphate was applied as basal. Water management followed standard farmers' practice, in which the field was drained 2 d before seeding. There was no standing water on the soil surface from sowing to the three-leaf stage of rice, and then the field was reflooded and a water depth of 1 to 3 cm was maintained until the end of the tillering period. After this, water was drained for 7–10 days to control unproductive tillers and later, a water depth of 3–5 cm was maintained until the grain-filling stage. Weed and insects were intensively controlled by chemicals to avoid biomass and yield loss.

In 2010, the number of tillers from 30 hills of rice was counted at 1-week intervals, starting from 20 d after sowing (DAS) to panicle initiation stage. Tillers with at least one visible leaf were included. In 2011, 30 hills of rice plants were sampled at heading and maturity to calculate average panicle number per hill. In both years, six representative hills of rice plants from each plot were separately sampled at heading and divided into leaf blades, stem plus sheath, and panicles. Green leaves were grouped into the top

Table 1. Grain yield of a high-yielding hybrid rice variety Nei2you6 grown at five seeding spacings in soil fertility 1 and soil fertility 2 at the China National Rice Research Institute, Zhejiang Province, China, in 2010 and 2011.

Year	Field	Seeding spacing	Grain yield (kg ha^{-1})	Grain yield (g hill^{-1})	Grain yield (g shoot^{-1})
2010	Soil fertility 1	25 cm×15 cm	7804 c	29.37 d	3.09 d
		25 cm×17 cm	8348 a	36.01 c	3.49 b
		25 cm×19 cm	8152 ab	38.05 bc	3.41 bc
		25 cm×21 cm	7892 bc	41.65 b	3.36 c
		25 cm×23 cm	7938 b	46.05 a	3.68 a
	Soil fertility 2	25 cm×15 cm	9133 b	34.69 d	3.64 b
		25 cm×17 cm	9433 a	40.63 c	3.84 a
		25 cm×19 cm	8801 c	42.08 bc	3.68 b
		25 cm×21 cm	8437 d	45.09 b	3.50 c
		25 cm×23 cm	8311 d	48.35 a	3.38 d
2011	Soil fertility 1	25 cm×15 cm	7842 c	29.79 e	3.14 d
		25 cm×17 cm	8573 a	36.80 d	3.56 b
		25 cm×19 cm	8218 b	39.61 c	3.55 b
		25 cm×21 cm	8121 b	43.24 b	3.48 c
		25 cm×23 cm	8140 b	47.19 a	3.70 a
	Soil fertility 2	25 cm×15 cm	9380 b	35.39 d	3.71 bc
		25 cm×17 cm	9611 a	41.13 c	3.88 a
		25 cm×19 cm	8942 c	42.91 bc	3.76 b
		25 cm×21 cm	8831 c	47.14 ab	3.66 c
		25 cm×23 cm	8582 d	49.94 a	3.49 d
Analysis of variance					
Seeding density (A)			*	*	*
Soil fertility (B)			*	*	*
Year (C)			NS	NS	NS
A×B			*	*	*
A×C			NS	NS	NS
B×C			NS	NS	NS
A×B×C			*	*	*

Within a column for each year, means followed by the same letters are not significantly different according to the LSD test (0.05).
*Significance at the 0.05 level based on analysis of variance.
NS denotes nonsignificance based on analysis of variance.

three green leaves and others, and their leaf area was measured with a leaf area meter (LI-3000A, LICOR, Lincoln, NE, USA). LAI was calculated as the total green leaf area of the plant, divided by the corresponding ground sampling area. All samples were oven-dried at 80°C until constant weight for determination of aboveground dry weight at flowering. At maturity, 30 hills were sampled diagonally from each subplot, panicle numbers were counted for each hill to determine the number of panicles per m^2. Six representative hills of rice plants were selected to determine aboveground total dry weight and harvest index. Grain yield per hill was measured from the sampled 30 hills and grain yield per shoot was calculated as the ratio of yield per hill to panicle numbers per hill. In each subplot, grain yield was determined from a harvesting area of 30 m^2 and converted to kg ha^{-1} at 14% moisture content. Panicles were hand-threshed and filled spikelets were separated from unfilled spikelets by submerging them in tap water. Dry weights of straw, rachis, and filled and unfilled spikelets were determined after oven drying at 80°C to constant weight. Three subsamples of 20 g of filled spikelets were taken and counted to determine grain weight (20 divided by the counted filled spikelets number); and the number of filled spikelets was calculated as total dry weight of filled spikelets divided by grain weight. All unfilled spikelets were counted and the number of total spikelets (filled and unfilled) was calculated. Aboveground total dry weight at maturity included the total dry matter of straw, rachis, and filled and unfilled spikelets. Dry weight accumulation during the grain-filling phase was calculated as the difference in total aboveground dry weight between flowering and maturity stages. Spikelets per panicle (ratio of total spikelets number to panicles number), grain-filling percentage (ratio of filled spikelets number to total spikelets number), and harvest index (ratio of dry grain yield to total aboveground biomass dry weight) were calculated.

On-farm demonstration experiments

On-farm demonstrations were conducted in 2012 and 2013 in Fuyang (120.0′ E, 30.1′ N; 14 m above sea level) and Yuhang (120.3′ E, 30.4′ N; 40 m above sea level), Zhejiang Province. Soil at both sites is classified as ferric-accumulic stagnic anthrosol [22]. The soil properties in Fuyang were 2.48 g total N kg-1, 0.56 g total P kg-1, 0.26 g available K kg-1, and pH 5.63. The soil

Table 2. Yield components of a high-yielding hybrid rice variety Nei2you6 grown at five seeding spacings in soil fertility 1 and soil fertility 2, China National Rice Research Institute, Zhejiang Province, China, 2010 and 2011.

Year	Field	Seeding spacing	Spikelets panicle^{-1} (no.)	Grain filling (%)	Grain weight (mg)	Panicles m^{-2} (no.)	Panicles hill^{-1} (no.)
2010	Soil fertility 1	25 cm×15 cm	115.52 b	76.82 e	30.30 a	240.97 a	9.03 c
		25 cm×17 cm	125.92 a	77.74 d	30.24 a	228.08 b	9.71 bc
		25 cm×19 cm	126.70 a	78.85 c	30.46 a	223.75 b	10.60 b
		25 cm×21 cm	125.42 a	79.99 b	30.72 a	226.41 b	11.91 a
		25 cm×23 cm	127.55 a	81.51 a	30.52 a	210.76 c	12.11 a
	Soil fertility 2	25 cm×15 cm	135.01 a	81.63 b	30.73 a	247.01 a	9.25 cd
		25 cm×17 cm	135.78 a	83.15 a	30.44 a	238.86 b	10.17 c
		25 cm×19 cm	130.83 a	81.44 b	30.22 a	229.02 b	10.86 c
		25 cm×21 cm	135.80 a	78.06 c	30.56 a	234.84 b	12.36 b
		25 cm×23 cm	133.09 a	74.11 d	30.33 a	236.59 b	13.59 a
2011	Soil fertility 1	25 cm×15 cm	127.68 b	84.90 c	33.48 a	266.33 a	9.98 c
		25 cm×17 cm	142.00 a	86.66 b	34.10 a	257.20 ab	10.95 bc
		25 cm×19 cm	140.04 a	87.15 b	33.66 a	247.31 b	11.72 b
		25 cm×21 cm	135.88 a	86.65 b	33.28 a	245.27 b	12.91 a
		25 cm×23 cm	140.97 a	90.09 a	33.74 a	232.94 c	13.39 a
	Soil fertility 2	25 cm×15 cm	143.37 a	86.67 b	32.63 a	262.29 a	9.83 e
		25 cm×17 cm	147.10 a	90.07 a	32.98 a	258.76 a	11.01 d
		25 cm×19 cm	144.61 a	90.02 a	33.40 a	253.12 a	12.00 c
		25 cm×21 cm	147.12 a	84.56 c	33.10 a	254.42 a	13.40 b
		25 cm×23 cm	147.09 a	81.91 d	33.53 a	261.49 a	15.03 a

Within a column for each site, means followed by the same letters are not significantly different according to the LSD test (0.05).

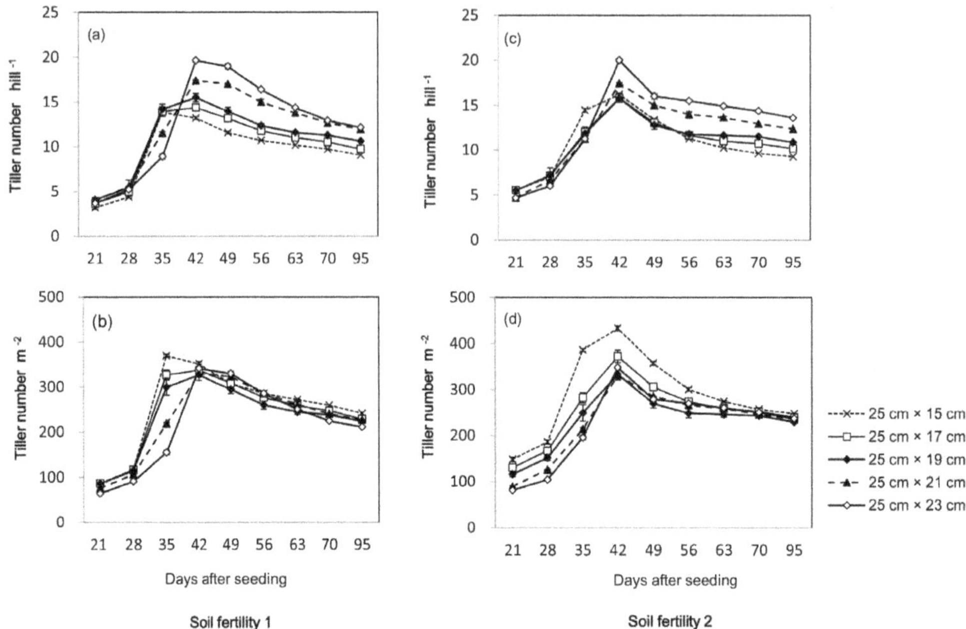

Figure 1. Tillers per hill (a, c) and tillers per m^2 (b, d) of a high-yielding hybrid rice variety Nei2you6 grown at five plant spacings in soil fertility 1 (a, b) and soil fertility 2 (c, d). China National Rice Research Institute, Zhejiang Province, China, 2010.

properties in Yuhang were 2.66 g total N kg-1, 0.62 g total P kg-1, 0.32 g available K kg-1, and pH 5.75. The hybrid rice variety used in Fuyang was Nei2you6. In Yuhang, high-yielding hybrid rice varieties Yongyou9 and Yongyou12 were grown. These high-yielding hybrid rice varieties are used widely in south China under a single rice cropping system. Three treatments were included at each site: mechanical hill seeding, broadcasting, and transplanting. Hill seeding was done by using a newly developed hill-seeder ZB-10 (Wang et al 2010) at a spacing of 25 cm×17 cm, three to five seeds per hill, and the seed rate was 21.0 to 25.5 kg ha^{-1}. This spacing was chosen because the highest grain yield was noted in this treatment in the 2010 and 2011 experiments. Broadcasting and transplanting were standard farmers' practices. In the broadcasting treatment, sprouted seeds were broadcast at a seeding rate of 33.8 kg ha^{-1}. In the transplanting treatment, seedlings with four to five leaves were transplanted at a spacing of 30 cm×20 cm; two to three seedlings per hill. In the mechanical hill seeding and broadcasting treatments, seed treatment was the same as used in field experiments in 2010 and 2011; all varieties were seeded on the same day (June 2) in both years. Because direct seeding could facilitate timely establishment of rice and earlier crop maturity by 7–10 d, the nursery for the transplanting treatment was raised on May 25 and transplanted on June 18 in both years. The planting area of each treatment was 3000 m^2, a total of 9000 m^2 for each variety. Fertilizer and water management in the three treatments were the same as that described for the field experiments in 2010 and 2011. Management operations such as weeding, irrigation, and plant protection measures were done as needed.

At maturity stage, a 5-m^2 area was harvested diagonally from six places in each treatment for yield determination, and grain yield was calculated at 14% moisture content. Ten hills were sampled diagonally from each 5-m^2 harvest area to determine yield components in the mechanical hill seeding and transplanting treatments, and a 0.5-m^2 area from each 5-m^2 harvest area in the broadcasting treatment. Panicles were hand-threshed and filled and unfilled grains were separated. Dry weights of filled and unfilled spikelets were determined after oven drying samples at 80°C to constant weight. Three subsamples of 20 g of filled spikelets were taken and counted to determine weight and total number of filled spikelets. Unfilled spikelets were counted and grain-filling percentage was calculated.

Statistical analyses

In the field experiments, normal distribution of data was tested for grain yield, yield components, productive tiller percentage, aboveground dry weight, and harvest index by Q-Q plots using SPSS 11.5 software, and no data transformation was needed to meet the assumptions of normality and constant variance. Data were analyzed following analysis of variance (ANOVA), and differences among treatments were compared based on the least significant difference (LSD) test at the 5% level of probability. The statistical model used in the analysis of grain yield per hectare, per hill, and per shoot included the sources of variation due to replication, seeding density, soil fertility, year, and the interactions of seeding density×soil fertility, seeding density×year, soil fertility×year, and seeding density×soil fertility×year. The bivariate correlations between grain yield and yield components, including grain weight, panicles per hectare and per hill, spikelets per panicle, and grain-filling percentage, were conducted using the Pearson correlation coefficient (two-tailed).

In the on-farm demonstration experiments, ANOVA and mean comparisons were based on the LSD test at the 0.05 probability level for each variety and year (SPSS 11.5).

Results

Field experiments

In both years, a significant grain yield response to seeding density was observed (Table 1). Grain yield per hill increased consistently with decreases in seeding density. The plants grown at 25 cm× 23 cm spacing had the highest grain yield per hill and

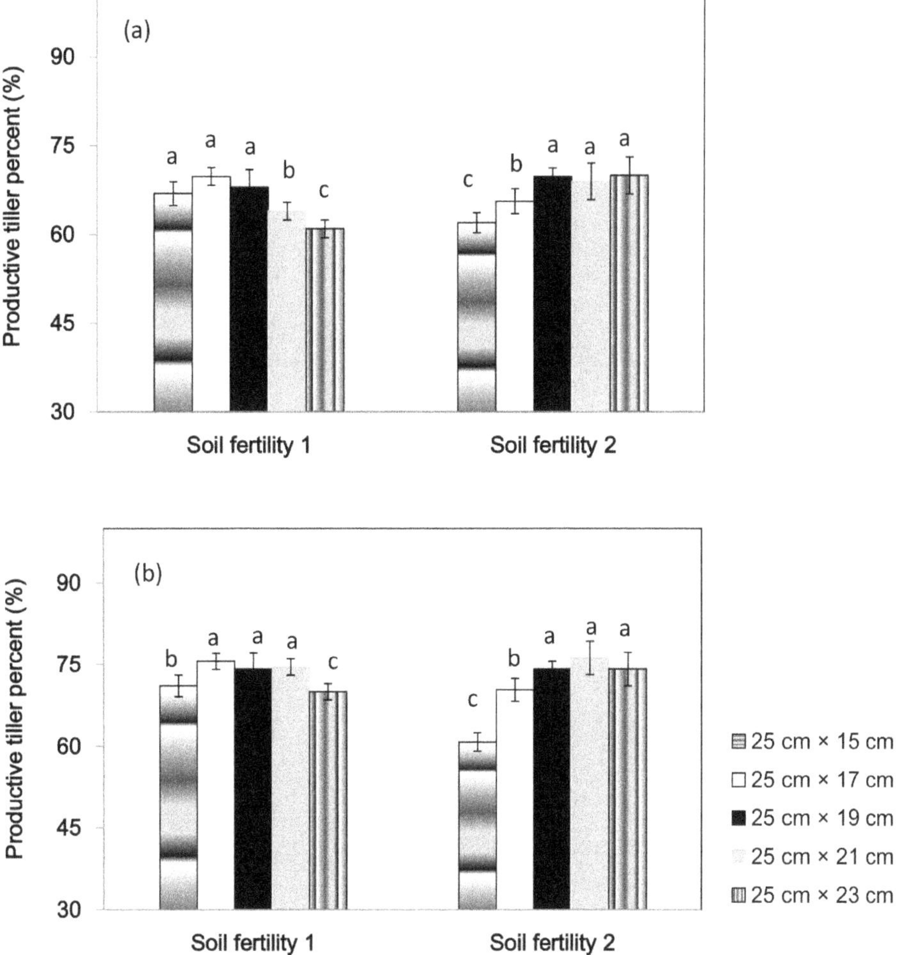

Figure 2. Productive tiller percentage of a high-yielding hybrid rice variety Nei2you6 grown at five seeding spacings in soil fertility 1 and soil fertility 2. China National Rice Research Institute, Zhejiang Province, China, 2010 (a) and 2011 (b).

plants grown at 25 cm×15 cm spacing had the lowest. The rate of change in grain yield per hectare with seeding spacing, however, was different between soil fertility 1 and 2. In soil fertility 1, 25 cm×17 cm spacing had the highest grain yield per ha, while 25 cm×15 cm spacing had the lowest value. In soil fertility 2, highest grain yield per ha was obtained at 25 cm×17 cm spacing, but, beyond this, grain yield decreased with an increase in seeding spacing and rice plants grown at 25 cm× 23 cm produced the lowest grain yield per ha. There were large differences in grain yield per shoot among the five seeding spacings in both soil fertility fields. In soil fertility 1, the spacing of 25 cm× 23 cm produced the maximum grain yield per shoot and the lowest value was observed at 25 cm×15 cm spacing. In soil fertility 2, the response of grain yield per shoot to seeding density was the same as that for grain yield per ha; 25 cm×17 cm had the maximum and 25 cm× 23 cm had the minimum grain yield per shoot. The difference in grain yield between the two years was nonsignificant (Table 1).

In both years, there was no significant difference in grain weight among different spacing treatments in both fields (Table 2). The number of panicles per hill consistently increased with increases in seeding spacing in both years and both fields. Across years, with increases in seeding spacing from 25 cm×15 cm to 25 cm×23 cm, panicles per hill increased by 34.2% in soil fetility 1 and by 50.0% in soil fertility 2. In both years, the number of panicles per m^2 was not significantly different at spacing ranging from 25 cm×17 cm to 25 cm× 21 cm in soil fertility 1; the maximum and minimum number of panicles per m^2 was observed at 25 cm×15 cm and 25 cm× 23 cm spacing, respectively. In soil fertility 2, 25 cm×15 cm spacing had higher number of panicles per m^2 in 2010, while no difference was observed among other treatments in this year and among all spacings in 2011. In both years, 25 cm×15 cm spacing had the minimum number of spikelets per panicle in soil fertility 1, while no difference was found among treatments in soil fertility 2. Grain-filling percentage in 2011 was higher than that in 2010 in both soil fertility fields. The plants grown at 25 cm× 23 cm spacing had the highest and 25 cm×15 cm spacing had the lowest grain-filling percentage in soil fertility 1. Meanwhile, in soil fertility 2, 25 cm× 17 cm spacing had the highest grain filling percentage and 25 cm× 23 cm spacing had the lowest value in both years.

In 2010, maximum tiller number per hill was observed at 25 cm×15 cm spacing at 35 DAS in soil fertility 1, while at 42 DAS at other seeding spacings in soil fertility 1 and all treatments in soil fertility 2 (Figure 1). A decreasing trend of tiller number per hill was observed after the reproductive stage, irrespective of treatment. Maximum tiller numbers per hill and per m^2 were affected by seeding density. In both soil fertility fields, maximum tiller number per hill increased with increases in seeding spacing;

Table 3. Leaf area index (LAI), leaf area per hill, and leaf area per shoot at flowering of a high-yielding hybrid rice variety Nei2you6 grown at five seeding spacings in soil fertility 1 and soil fertility 2, China National Rice Research Institute, Zhejiang Province, China, 2010 and 2011.

Year	Field	Seeding spacing	LAI	Leaf area (m^2 per hill)			Leaf area ($\times 10^{-3}$ m^2 per shoot)		
				Total	Top 3 leaves	Others	Total	Top 3 leaves	Others
2010	Soil fertility 1	25 cm×15 cm	5.72 a	0.21 e	0.12 d	0.11 c	26.87 c	14.16 c	12.71 b
		25 cm×17 cm	5.39 c	0.23 d	0.13 d	0.11 c	27.76 c	15.18 c	12.58 b
		25 cm×19 cm	5.42 bc	0.26 c	0.14 c	0.12 b	28.78 b	16.15 b	12.63 b
		25 cm×21 cm	5.49 b	0.29 b	0.16 b	0.13 a	29.08 b	16.46 b	12.62 b
		25 cm×23 cm	5.53 b	0.32 a	0.18 a	0.13 a	31.18 a	17.80 a	13.35 a
	Soil fertility 2	25 cm×15 cm	7.03 a	0.27 e	0.16 c	0.10 d	35.66 a	21.02 a	13.64 c
		25 cm×17 cm	7.07 a	0.30 d	0.17 bc	0.13 c	35.55 a	20.97 a	14.57 b
		25 cm×19 cm	6.68 b	0.32 c	0.18 b	0.13 c	34.67 b	19.59 b	15.07 b
		25 cm×21 cm	6.66 b	0.35 b	0.18 b	0.16 b	34.06 c	17.89 c	16.17 a
		25 cm×23 cm	6.46 c	0.37 a	0.19 a	0.17 a	32.43 d	17.05 c	15.37 b
2011	Soil fertility 1	25 cm×15 cm	6.20 a	0.23 d	0.12 d	0.11 d	29.11 c	15.34 c	13.77 ab
		25 cm×17 cm	5.61 d	0.23 d	0.13 d	0.11 d	28.90 c	15.80 c	13.10 b
		25 cm×19 cm	5.88 c	0.28 c	0.16 c	0.12 c	31.18 b	17.49 b	13.69 ab
		25 cm×21 cm	5.83 c	0.31 b	0.18 b	0.13 b	31.88 b	17.48 b	13.40 b
		25 cm×23 cm	5.99 b	0.34 a	0.20 a	0.15 a	33.78 a	19.28 a	14.47 a
	Soil fertility 2	25 cm×15 cm	7.77 a	0.29 e	0.18 c	0.12 e	38.30 a	23.24 a	15.08 c
		25 cm×17 cm	7.51 b	0.32 d	0.19 b	0.13 d	37.75 a	22.27 ab	15.47 c
		25 cm×19 cm	7.24 c	0.34 c	0.20 ab	0.15 c	37.55 a	21.23 b	16.33 b
		25 cm×21 cm	7.01 d	0.37 b	0.19 b	0.17 b	35.81 b	18.81 c	17.00 a
		25 cm×23 cm	6.86 d	0.39 a	0.21 a	0.19 a	34.43 b	18.11 c	16.33 b

Green leaves per hill were subdivided into top three green leaves and others. Leaf area per shoot was calculated as the ratio of leaf area per hill to productive tiller number per hill.
Within a column for each site, means followed by the same letters are not significantly different according to the LSD test (0.05).

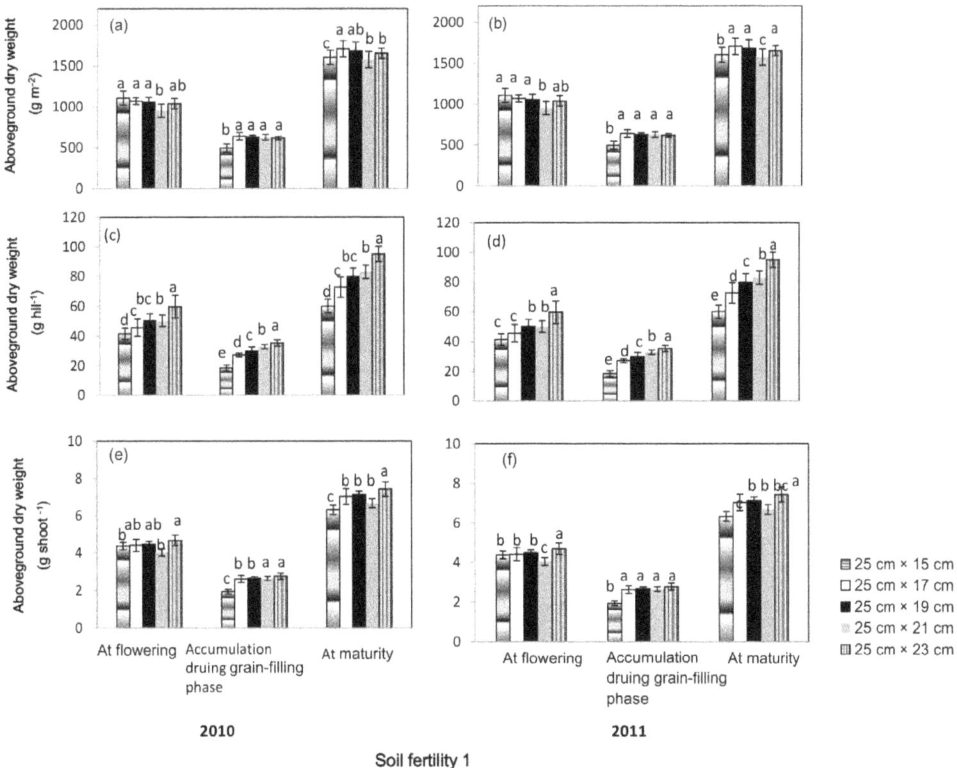

Figure 3. Aboveground dry weight at flowering and maturity, and dry weight accumulation during grain-filling phase of a high-yielding hybrid rice variety Nei2you6 grown at five seeding spacings in soil fertility 1 at the China National Rice Research Institute, Zhejiang Province, China, in 2010 and 2011. Aboveground dry weight was calculated as dry weight per m^2 (a, b), dry weight per hill (c, d), and dry weight per shoot (e, f). Dry weight per shoot was the ratio of dry weight per hill to productive tiller number per hill.

25 cm× 23 cm spacing had the highest tiller number per hill and 25 cm×15 cm spacing had the lowest. However, maximum tiller number per m^2 in both fields decreased with increases in seeding spacing; it was highest at 25 cm×15 cm spacing. From the maximum tillering to flowering stage, the difference in tiller number per m^2 among treatments gradually decreased with the reduction in unproductive tillers.

The effects of seeding density on the percentage of productive tillers were also different between the two soil fertility fields (Figure 2). In soil fertility 1, 25 cm× 23 cm spacing had the lowest productive tiller percentage in both years, and no significant difference was observed among spacings ranging from 25 cm×15 cm to 25 cm×19 cm in 2010 and from 25 cm×17 cm to 25 cm×21 cm in 2011. In both years, in soil fertility 2, 25 cm×15 cm spacing had the lowest productive tiller percent, followed by 25 cm×17 cm spacing, and both these spacings had significantly lower productive tillers than the other ones.

Although leaf area per hill at flowering consistently increased in both years with increases in seeding spacing in both soil fertility fields, the change in leaf area per shoot was different between the two fields (Table 3). In both years, total leaf area per shoot and top three leaf blades per shoot increased with increases in seeding spacing in soil fertility 1, while they decreased in soil fertility 2. The highest leaf area index (LAI) in soil fertility 1 was obtained at 25 cm×15 cm spacing, followed by that at 25 cm×23 cm spacing. The highest LAI in soil fertility 2 was also observed at 25 cm×15 cm spacing, while it gradually decreased with increases in seeding spacing; 25 cm×23 cm spacing had the lowest LAI in both years.

Aboveground dry weight per hill at flowering and maturity, and dry weight accumulation per hill during grain-filling phase increased with increases in seeding spacing; 25 cm× 23 cm spacing had the highest values and 25 cm×15 cm spacing had the lowest in both soil fertility fields in both years (Figures 3 and 4). In terms of aboveground dry weight per shoot, a difference existed between the two soil fertility fields. In both years, the highest aboveground dry weight per shoot at flowering and maturity was observed at 25 cm× 23 cm spacing in soil fertility 1; rice plants grown at 25 cm×15 cm spacing had the lowest dry weight accumulation per shoot during the grain-filling phase and aboveground dry weight per shoot at maturity. In soil fertility 2, no significant difference in dry weight accumulation during the grain-filling phase was observed in spacing ranging from 25 cm×17 cm to 25 cm× 23 cm in both years; and aboveground dry weight per shoot at flowering and maturity decreased with increases in spacing from 25 cm×17 cm to 25 cm× 23 cm. Significant differences also existed in aboveground dry weight per m^2 between the two soil fertility fields. In soil fertility 1, no significant difference was found for aboveground dry weight per m^2 at flowering at spacings ranging from 25 cm×15 cm to 25 cm×19 cm. Rice plants grown at 25 cm×15 cm spacing had the lowest dry weight accumulation during grain filling, and no significant difference was observed at spacings ranging from 25 cm×17 cm to 25 cm×23 cm. Aboveground dry weight at maturity was highest at 25 cm×17 cm spacing in both years and lowest at 25 cm×15 cm spacing in 2010 and at 25 cm×15 cm spacing in 2011. In soil fertility 2, no significant difference in dry weight accumulation during grain-filling phase was observed among the treatments; aboveground dry weights per m^2 at

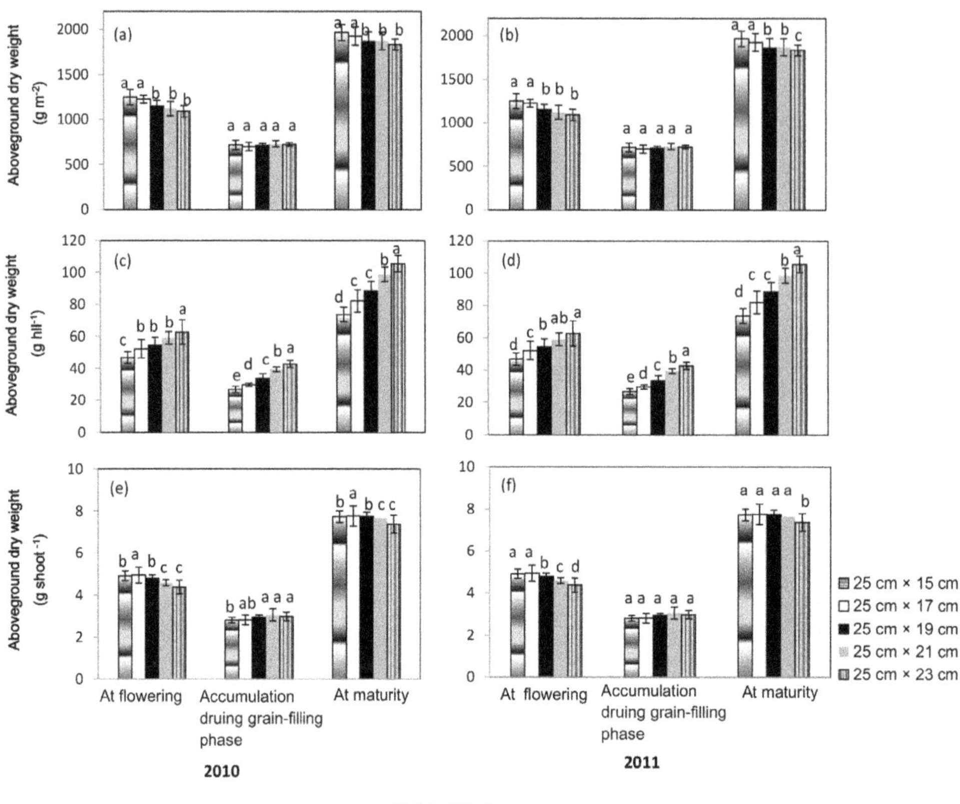

Figure 4. Aboveground dry weight at flowering and maturity, and dry weight accumulation during grain-filling phase of a high-yielding hybrid rice variety Nei2you6 grown at five seeding spacings in soil fertility 2 at the China National Rice Research Institute, Zhejiang Province, China, in 2010 and 2011. Aboveground dry weight was calculated as dry weight per m^2 (a, b), dry weight per hill (c, d), and dry weight per shoot (e, f). Dry weight per shoot was the ratio of dry weight per hill to productive tiller number per hill.

flowering and maturity were significantly higher at 25 cm×15 cm and 25 cm×17 cm spacing than at spacing wider than 25 cm×19 cm.

The effect of seeding density on harvest index (HI) was also different between the two soil fertility fields (Figure 5). In soil fertility 1, 25 cm×23 cm spacing and 25 cm×21 cm spacing had higher HI than that at the other three narrower spacings in both years. In soil fertility 2, plants grown at spacings ranging from 25 cm×17 cm to 25 cm×21 cm had significantly higher HI than plants grown at 25 cm×15 cm and 25 cm×15 cm spacings.

On-farm demonstration experiments

In the on-farm demonstration experiments in 2012 and 2013, mechanical hill seeding, using seeder ZB-10 at a spacing of 25 cm×17 cm, outperformed broadcasting in terms of grain yield for all three high-yielding hybrid rice varieties, increasing grain yield by 10.4% in Yongyou 9 and 12.3% in Yonyou 12 across the 2 years. In 2012, variety Nei2you6 produced 26.6% higher grain yield in mechanical hill seeding compared with broadcasting because more than 30% of the plants lodged in the broadcast culture due to heavy rain during the later grain-filling phase (Figure 6). In both years, no significant difference in grain yield between mechanical hill-seeded rice and transplanted rice was observed for Nei2you 6 and Yonyou 9. Yongyou 12 produced more than 1000 kg ha^{-1} grain yield in the mechanical hill seeding treatment across 2 years, increasing grain yield by 9.1% compared with that of transplanted rice.

Mechanical hill-seeded rice had more spikelets per panicle and fewer panicles per m^2 compared with broadcasting, but fewer spikelets per panicle and more panicles per unit area than transplanted rice (Table 4). Across the years, mechanical hill-seeded hybrid rice varieties Nei2you 6, Yongyou 9, and Yonyou 12 had 36.0%, 28.3%, and 33.1% higher number of spikelets per panicle, and 17.4%, 16.7%, and 14.3% lower number of panicles per m^2, respectively, compared with the broadcast culture. Compared with the transplanted treatment, 7.3%, 15.5%, and 13.4% decline in spikelets per panicle, and 11.2%, 20.7%, and 28.5% increase in panicle number per m^2 were observed in mechanical hill-seeded hybrid rice varieties Nei2you 6, Yongyou 9, and Yonyou 12, respectively.

Discussion

Our study showed that high-yielding hybrid rice variety Nei2you6 produced maximum grain yield at a seeding spacing of 25 cm×17 cm in two soil fertility fields in both 2011 and 2012. Optimal hill seeding density for hybrid rice has also been reported in previous studies: Peilianyou 986 had the highest grain yield at a spacing of 25 cm×18 cm [20]; Jinyou 207 had the highest grain yield when hill-seeded at 30 cm×20 cm spacing [17]; and hill-seeded Peizataifen and Yuyouxiangzhang [19, 20] gave the highest grain yield at 25 cm×14 cm. However, these studies were conducted at one location in a single season and used only one hybrid rice variety with one to three seeding rates. It is thus difficult to conclude which seeding density can fullfill the yield

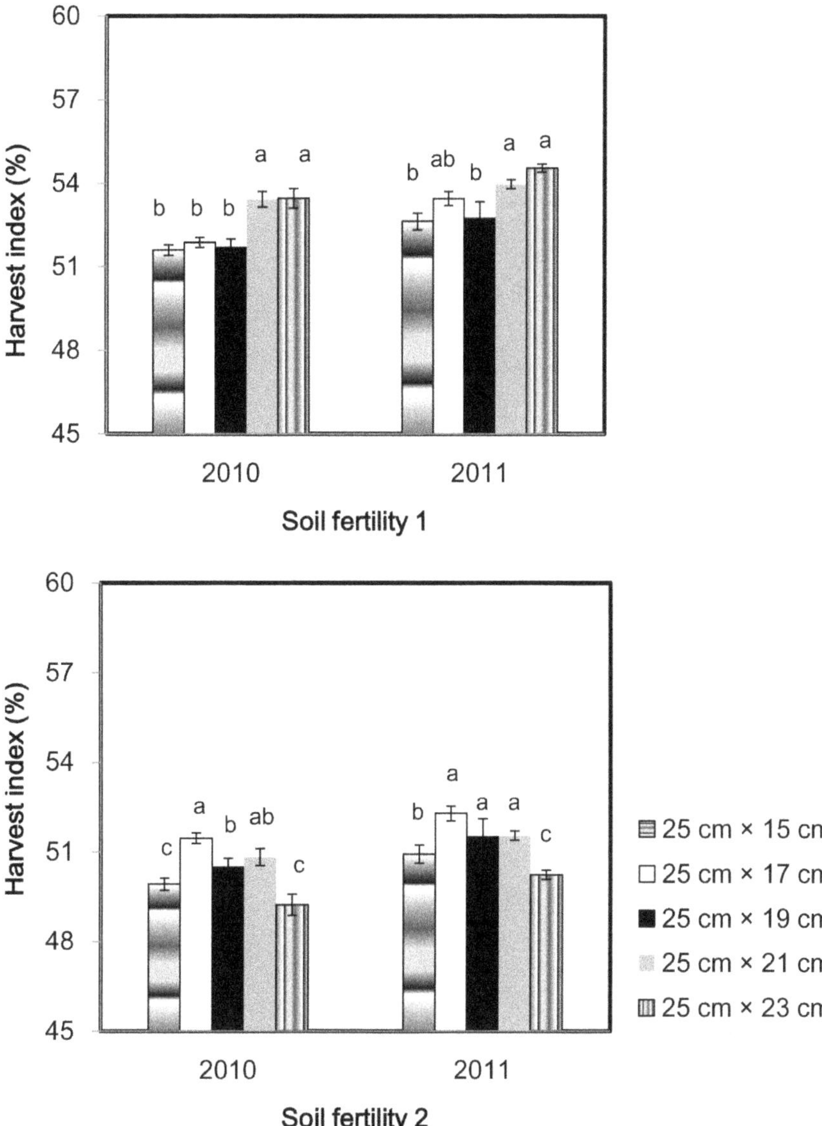

Figure 5. Harvest index of a high-yielding hybrid rice variety Nei2you6 grown at five seeding spacings in soil fertility 1 and soil fertility 2 at the China National Rice Research Institute, Zhejiang Province, China, in 2010 and 2011. Dry weight per shoot was calculated as the ratio of dry weight per hill to productive tiller number per hill.

potential of most high-yielding hybrid rice varieties. In our study, we compared plant yield and dry weight at five hill seeding spacings in two fields of different fertilities during a 2-year period. Besides that, three high-yielding hybrid rice varieties were mechanically seeded with a hill seeder at 25 cm×17 cm spacing in the on-farm demonstration experiments and compared with broadcasting and transplanting. Mechanical hill seeding outperformed broadcasting in terms of grain yield for all three high-yielding hybrid rice varieties; they also had grain yield equal to or higher than that of transplanted rice. We conclude that, for the south China region, a hill seeding spacing of 25 cm×17 cm, with three to five seeds per hill, is the appropriate density to be used for single cropping of high-yielding hybrid rice.

There have been extensive studies on the relationships between yield and plant density under nonstressed conditions [24–29]. For decades, rice reseachers in China have emphasized the importance of regulating the compensatory and competitive relationships among plants to increase grain yield by modifying planting density. In the 1960s, dense planting to increase panicle number per unit area was recommended in inbred rice production [30,31]. In the 1980s, with the development of inbred and hybrid rice varieties with large panicles, the advice on rice planting density changed. It was reported that planting at wide spacing decreased plant number per unit area, which improved the morphological traits of the rice plants such as plant height, tiller angle, and leaf angle [32]. It has been widely accepted at present in high-yielding hybrid rice production systems that increasing panicle size, rather than panicle number per unit area, increases yield [32–34]. In considering best management practices for rice, the basic concept applied is the use of young seedlings at optimal spacing to stimulate individual plant growth [35]. Recently, on-farm surveys conducted in Zhejiang, China, showed that some farmers transplanted hybrid rice using extremely wide spacing because of rising labor cost. They opined that the reduced plant number

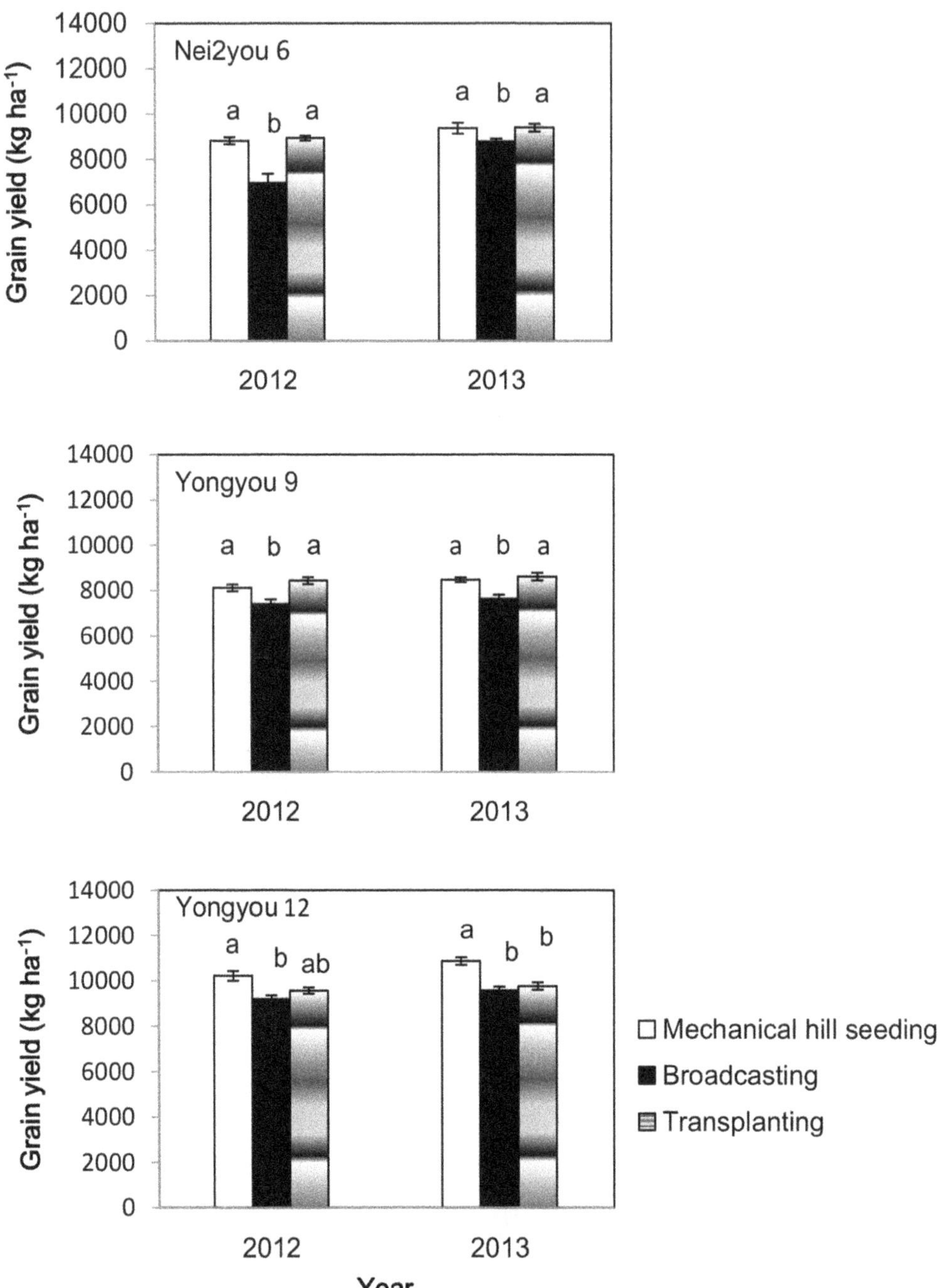

Figure 6. Grain yield of three high-yielding hybrid rice varieties under three plant establishment methods in Fuyang and Yuhang, Zhejiang Province, in 2012 and 2013. Plant spacing for mechanical hill seeding was 25 cm×17 cm; broadcasting and transplanting done according to standard farmer's practice.

caused by low planting density could be compensated for by the improvement in growth of the individual plants [36]. In our study, although grain yield per hill across two years consistently increased with increases in seeding spacing, grain yield per hill of hybrid rice variety Nei2you6 at 25 cm×23 cm spacing was 57.6% and 40.3% higher than at 25 cm×15 cm spacing in soil fertility 1 and 2, respectively. Grain yield per area across two years was highest at 25 cm×17 cm spacing, which was 5.3% and 12.7% higher than at 25 cm×23 cm spacing in soil fertility 1 and 2, respectively. These results suggest that yield per unit area cannot be fully compensated for by the yield increase of individual plants. Therefore, enough seedlings are necessary to achieve high yield. The on-farm surveys in Zhejiang Province also showed that the lowest grain yield was obtained with the widest planting spacing because of the lack of adequate plant population [36].

Rice grain yield per area or per hill is ultimately determined by the number of panicles per area or per hill and the grain yield per panicle. It was reported that yield was positively related to panicle number per unit area [36–37]. In our study, there was a significantly positive correlation between yield per hill and panicles per hill ($r = 0.98^{**}$), and a significantly negative correlation between grain yield per ha and panicles per hill ($r = -0.89^{*}$)

Table 4. Grain yield components of three high-yielding hybrid varieties under three plant establishment methods in Fuyang and Yuhang, Zhejiang Province, China, in 2012 and 2013.

Variety	Year	Treatment	Spikelets panicle^{-1} (no.)	Grain filling (%)	Grain weight (mg)	Panicles m^{-2} (no.)
Nei2you 6	2012	Mechanical hill seeding	176.03 b	83.25 a	26.76 b	235.56 b
		Broadcasting	130.82 c	80.32 b	26.72 b	283.65 a
		Transplanting	193.86 a	84.41 a	27.18 a	213.09 c
	2013	Mechanical hill seeding	174.03 b	85.75 a	27.42 a	250.60 b
		Broadcasting	126.57 c	82.40 b	26.84 b	305.00 a
		Transplanting	183.58 a	86.52 a	27.41 a	224.30 b
Yongyou 9	2012	Mechanical hill seeding	169.87 b	85.16 b	21.88 a	251.52 b
		Broadcasting	133.13 c	83.37 c	21.77 a	303.61 a
		Transplanting	203.08 a	87.63 a	22.17 a	210.60 c
	2013	Mechanical hill seeding	166.79 b	87.55 ab	22.79 a	262.00 b
		Broadcasting	129.37 c	85.49 b	22.45 a	313.00 a
		Transplanting	195.33 a	89.61 a	22.62 a	214.90 c
Yongyou 12	2012	Mechanical hill seeding	388.02 b	82.16 a	19.46 b	171.93 b
		Broadcasting	294.69 c	80.24 b	19.79 ab	202.85 a
		Transplanting	457.15 a	84.45 a	20.21 a	136.71 c
	2013	Mechanical hill seeding	389.09 b	84.46 ab	20.71 a	182.90 b
		Broadcasting	289.35 c	82.40 b	20.62 a	211.30 a
		Transplanting	439.71 a	86.52 a	20.62 a	139.50 c

Plant spacing for mechanical hill seeding was 25 cm×17 cm; broadcasting and transplanting done according to standard farmer's practice.

(Table 5). Panicle number is largely determined by the number of tillers that develop during the vegetative stage [38] and there is a large variation in tillering capacity among rice varieties [39,40]. Environmental factors and agronomic practices also alter tiller production and survival [41–43]. Plant density and fertilization are important factors influencing tiller production in rice [39,44–45]. LAI probably affected tillering by attenuation of light intensity and/or by influencing light quality at the base of the canopy where tiller buds and young tillers are located [46]. Furthermore, there are evidences showing that plant N status and LAI depend on each other in determining tillering [42]. A higher plant N concentration was needed to prevent tillers from dying when LAI was high [42]. In our study, maximum tiller number per hill increased with increases in seeding spacing, suggesting that plant tillering capacity increased with a decrease in seeding density. Compared with soil fertility 1, 17.2% and 10.1% increase in maximun tiller number per hill was obtained at 25 cm×15 cm and 25 cm×17 cm spacing, respectively, in soil fertility 2. However, no difference in maximum tiller number per hill was found between the two soil fertility fields at spacings ranging from 25 cm×19 cm to 25 cm×23 cm, suggesting that the negative effect of dense seeding on plant tillering capacity could be partly compensated for by an increase in nutrition supply in soil fertility 2. However, the rate of change in productive tiller percentage against seeding spacing was different between the two soil fertility fields. The lower productive tiller percentage at 25 cm×15 cm in soil fertility 2 offset its advantage of maximun tiller number per hill, and no significant difference in productive tillers per hill was observed at this spacing between the two fields. A 7% higher productive tillers at 25 cm×23 cm spacing in soil fertility 2 resulted in significantly higher panicle number in this field compared with soil fertility 1. The number of surviving tillers depends mainly on the extent of competition among tillers for carbohydrates produced and N absorbed from the maximum tillering stage to heading [40]. The differences in productive tiller percentage among seeding densities and between two soil fertility fields indicate that nutrition supply could increase panicle number at wide seeding spacing, but it cannot compensate for the negative effect of high seeding density on panicle number per hill because of the reduced productive tiller percentage.

There are compensatory and competitive relationships among tillers and their yield components over a wide range of plant densities [39]. In soil fertility 2, when seeding spacing was increased from 25 cm×17 cm to 25 cm×23 cm, the number of panicles per hill increased by 33.6% in 2010, and by 22.3% in 2011, while grain yield per shoot declined by 12.0% and by 10.0%, respectively. Grain filling percentage declined with increases in seeding spacing in both years in soil fertility 2, and no significant differences in spikelets per panicle and grain weight were found among treatments, suggesting that grain filling percentage decreased with the increase in productive tillers per hill, which resulted in the decline in yield per shoot. However, there was an interaction between seeding density and soil fertility. Less change in grain yield per shoot was in soil fertility 1 than soil fertility 2 at seeding spacings ranging from 25 cm×17 cm to 25 cm×21 cm. Rice plants grown at 25 cm×23 cm spacing had the highest grain yield per shoot in soil fertility 1, but had the the lowest grain yield per shoot in soil fertility 2. It was reported that high-yielding varieties usually have high biomass accumulation during the reproductive phase [47–51]. In our study, although aboveground dry weight per hill at flowering and maturity increased with increases in seeding spacing, aboveground dry weight per shoot was different between the two fertility fields. Rice plants seeded at 25 cm×23 cm spacing had the lowest aboveground dry weight per shoot both at flowering and maturity in soil fertility 2, while they had the highest aboveground dry weight per

Table 5. Correlation coefficients between grain yield and yield components of a high-yielding hybrid rice variety Nei2you6 grown at five seeding spacings at the China National Rice Research Institute, Zhejiang Province, China, in 2010 and 2011.

Yield and yield components	Spikelets per panicle	Grain filling percentage	Grain weight	Panicles per unit area	Panicles per hill	Grain yield per ha	Grain yield per hill
Grain filling percentage	−0.218						
Grain weight	−0.001	−0.802					
Panicles per unit area	0.207	−0.105	−0.449				
Panicles per hill	0.229	−0.842	0.973**	−0.378			
Grain yield per ha	0.159	0.828	−0.947*	0.409	−0.891*		
Grain yield per hill	0.291	−0.729	0.942*	−0.502	0.980**	−0.824	
Grain yield per shoot	0.041	0.965**	−0.791	−0.098	−0.774	0.868	−0.635

*Significant at $p<0.05$.
**Significant at $p<0.01$.

shoot in soil fertility 1. A different response of leaf area per shoot to seeding spacing was one of the factors that influence dry matter accumulation. The top three-leaf area per shoot increased with increases in seeding spacing in soil fertility 1, while it decreased in soil fertility 2. Considering that there are more panicles per hill in soil fertility 2 than in soil fertility 1 at wider seeding spacing and no extra fertilizer application, the lower area of the top three leaves and aboveground dry weight per shoot at 25 cm×23 cm spacing in soil fertility 2 may be attributed to limited N supply and/or N uptake. A recent study reported low N concentrations in shoots due to dilution as a result of more biomass production [43], which suggests that rice plants seeded at wider spacing might need more fertilizer to obtain high yield than those subjected to dense spacing. More fertilizer is needed not only to improve productive tiller percentage at wide spacing but also to increase N concentrations in plants and to improve dry weight accumulation and grain yield per shoot. Nevertheless, we think that the application of extra N to compensate for the reduced grain yield due to low planting density is not a good option because of potential environmental problems and increased cost for rice farmers.

In the on-farm demonstration experiments, mechanical hill seeding of rice at 25 cm×17 cm spacing had equal or higher grain yield than transplanted rice, indicating that this spacing was the optimum seeding density for high-yielding hybrid rice varieties. Hill-seeded rice had more panicles per unit area and less spikelets per panicle compared with transplanted rice.

Acknowledgments

The authors are grateful to Ms. Priscilla Grace Cañas of the International Rice Research Institute, for providing valuable comments on the manuscript.

Author Contributions

Conceived and designed the experiments: DYW. Performed the experiments: DYW SC ZMW CLJ XFZ. Analyzed the data: DYW CMX. Contributed reagents/materials/analysis tools: XFZ. Contributed to the writing of the manuscript: DYW BSC.

References

1. Pandey S, Velasco L (2002) Economics of direct seeding in Asia: patterns of adoption and research priorities. In: Pandey S, Mortimer M, Wade L, Tuong TP, Lopez K, editors. Seeding: research strategies and opportunities. Los Baños: International Rice Research Institute. pp. 3–14.
2. Farooq M, Siddique KHM, Rehman H, Aziz T, Lee DJ, et al. (2011) Rice direct seeding: experiences, challenges and opportunities. Soil Till Res 111: 87–98.
3. Chauhan BS, Mahajan G, Sardana V, Timsina J, Jat ML (2012) Productivity and sustainability of the rice-wheat cropping system in the Indo-Gangetic Plains of the Indian subcontinent: problems, opportunities, and strategies. Adv Agron 117: 315–369.
4. Joshi E, Kumar D, Lal B, Nepalia V, Gautam P, et al. (2013) Management of direct seeded rice for enhanced resource use efficiency. Plant Knowledge J 2: 119–134.
5. Chen F, Chen P (2011) Status and economic evaluation of direct-seeded rice production in south China-based on farmer's investigation. China Rice 17: 1–5 (in Chinese).
6. Chauhan BS (2013) Strategies to manage weedy rice in Asia. Crop Prot. 48: 51–56.
7. Wang D, Zhang X, Lu Y, Xu C, Jin B, et al. (2010) Grain yield difference investigation and development trends discussion of direct seeding rice production in Zhejiang province. China Rice 16: 22–26 (in Chinese).
8. Kabir MH, Saha A, Mollah IU, Kabir MS, Rahman F (2008) Effect of crop establishment methods and weed management practices on the productivity of boro rice in lowland ecosystem. Int J Biol Res 5: 42–51.
9. Mithrasena YJPK, Adikari WP (1986) Effect of density on sheath blight (ShB) incidence. Int Rice Res Notes 11: 20–21.
10. Sittisuang P (1995) Extent of problems and policies in wet-seeded rice production in Thailand. In: Moody K, editor. Constraints, opportunities, and innovations for wet-seeded rice. Los Baños: International Rice Research Institute, pp. 1–5.

11. Islam MF, Sarkar MAR, Islam MS, Parveen S, Hossain MS (2008) Effects of crop establishment methods on root and shoot growth, lodging behavior of Aus rice. Int J Biol Res 5: 60–64.
12. Ying J, Peng S, He Q, Yang H, Yang C, et al. (1998) Comparison of high-yield rice in tropical and subtropical environments. I. Determinants of grain and dry matter yields. Field Crops Res 57: 71–84.
13. Wu W, Zhang H, Qian Y, Cheng Y, Wu G, et al. (2008) Analysis on dry matter production characteristics of super hybrid rice. Rice Sci 15: 110–118.
14. Yang S, Xie L, Zheng S, Li J, Yuan J (2009) Effects of nitrogen rate and transplanting density on physical and chemical characteristics and lodging resistance of culms in hybrid rice. Acta Agron Sin 35: 93–103 (in Chinese, with English abstract).
15. Yoshinaga S (2005) Improved lodging resistance in rice (*Oryza sativa* L.) cultivated by submerged direct seeding using a newly developed hill seeder. Jpn Agric Res Q 39: 147–152.
16. Wang Z, Luo X, Tang X, Ma G, Zhang G, et al. (2010) Precision rice hill-direct-seeding technology and machine based on the combination of agricultural machinery and agronomic technology. J South China Agric Univ 31: 91–95 (in Chinese, with English abstract).
17. Zhu D, Wang J, Xu W, Wang Y, Xu W, et al. (2008) Rice mechanical precision hill seeded experiments. Anhui Agri Sci Bull 14: 133–135 (in Chinese).
18. Lei X, Liu L, Gou W, Ma R, Ren W (2013) Effects of planting methods on culm lodging resistance of indica hybrid rice (*Oryza sativa* L.). Acta Agron Sin 39: 1814–1825 (in Chinese with English abstract).
19. Shu S, Zheng T, Jia X, Luo X, Li G, et al. (2009) Effects of nitrogen fertilizer and density on precision hill-direct-seeding rice I – Yield formation characteristics. Chin Agric Sci Bull 25: 142–146 (in Chinese with English abstract).
20. Cheng J, Luo X, Fan Q, Zhang J, Wu J, et al. (2010) Influence of different planting methods on growth and development characteristics and yield of rice. J Huazhong Agric Univ 29: 1–5 (in Chinese with English abstract).
21. Zeng S, Huang Z, Wang Z, Luo X, Tang X (2014) Effects of different planting density on grain yield of precision hill-drop drilling rice. J Huazhong Agric Univ 33: 12–18 (in Chinese with English abstract).
22. Gong Z, Zhang G, Luo G (1999) Diversify of Anthrosols in China. Pedosphere 9: 193–204.
23. Cheng S, Zheng X, Wang D, Chen L, Xu C, et al. (2012) Effect of long-term paddy-upland yearly rotations on rice (*Oryza sativa*) yield, soil properties, and bacteria community diversity. Sci World J 279641. doi:10.1100/2012/279641.
24. Counce PA, Moldenhauer KAK, Marx DB (1989) Rice yield and plant yield variability responses to equidistant spacing. Crop Sci 29: 175–179.
25. Miller BC, Hill JE, Roberts SR (1991) Plant population effects on growth and yield in water-seeded rice. Agron J 83: 291–297.
26. Gravios KA, Helms RS (1982) Path analysis of rice yield and yield components as affected by seeding rate. Agron J 84: 1–4.
27. Baloch AW, Soomro AM, Javad MA, Ahmed M, Bughio HR, et al. (2002) Optimum plant density for high yield in rice (*Oryza sativa* L.). Asian J Plant Sci 1: 25–27.
28. Oziegbe M, Faluyi JQ (2007) Effects of seeding rate on the yield components of an enhanced rice cultivar (DTPMFe+) *Oryza sativa* Linn. Int J Bot 3: 317–320.
29. Lin X, Zhu D, Chen H, Zhang Y (2009) Effects of plant density and nitrogen application rate on grain yield and nitrogen uptake of super hybrid rice. Rice Sci 16: 138–142.
30. Ding Y (1961) China rice cultivation. Beijing: China Agriculture Press. pp. 53–56 (in Chinese).
31. Chen Y (1959) Chen Yongkang rice cultivation experience. Shanghai: Shanghai Scientific and Technical Publishers. pp. 1–54 (in Chinese).
32. Jiang P, Feng L (1987) The effect of cultivation techniques on plant type in view of the high yield result induced by TFS cultivation method in paddy rice. Chin J Rice Sci 1: 111–117.
33. Lin Q, Zhang H, Cheng G, Zhu J, Lu W (1982) Technical approaches to achieve large-area high-yield cultivation of IR 24 and discuss on the small groups, strong individual cultivation patterns. Jiangsu Agric Sci 9: 1–10 (in Chinese).
34. Ling QH (2000) Crop population quality. Shanghai. Shanghai Scientific and Technical Publishers. pp. 42–107 (in Chinese).
35. Uphoff N (2003) Higher yields with fewer external inputs? The system of rice intensification and potential contributions to agricultural sustainability. Int J Agric Sustain 1: 38–50.
36. Wang D, Zhang X, Zhou C, Zheng G, Zhang G, et al. (2010) Grain yield difference investigation and reasonable planting density analysis of rice production in Zhejiang Province. Acta Agric Zhejiang 22: 330–336 (in Chinese with English abstract).
37. Huang M, Zou Y, Jiang P, Xia B, Ibrahim M, et al. (2011) Relationship between grain yield and yield components in super hybrid rice. Agric Sci China 10: 1537–1544.
38. De Datta SK (1981) Principles and practices of rice production. New York: John Wiley & Sons. pp. 220–358.
39. Wu G, Wilson LT, McClung AM (1998) Contribution of rice tillers to dry matter accumulation and yield. Agron J 90: 317–323.
40. Nuruzzaman M, Yamamoto Y, Nitta Y, Yoshida T, Miyazaki A (2000) Varietal differences in tillering ability of fourteen japonica and indica rice varieties. Soil Sci Plant Nutr 46: 381–391.
41. Hanada K (1993) Tillers. In: Matsuo T, Hoshikawa K, editors. Science of the rice plant. Tokyo: Morphology, Food and Agricultural Policy Center. pp. 222–258.
42. Zhong X, Peng S, Sanico AL, Liu H (2003) Quantifying the interactive effect of leaf nitrogen and leaf area on tillering of rice. J Plant Nutr 26: 1203–1222.
43. Huang M, Zou Y, Jiang P, Xia B, Feng Y, et al. (2011) Yield component differences between direct-seeded and transplanted super hybrid rice. Plant Prod Sci 14: 331–338.
44. Yoshida S, Hayakawa Y (1970) Effects of mineral nutrition on tillering of rice. Soil Sci Plant Nutr 16: 186–191.
45. Fagada SO, De Datta SK (1971) Leaf area index, tillering capacity, and grain yield of tropical rice as affected by plant density and nitrogen level. Agron J 63: 503–506.
46. Zhong X, Peng S, Sheehy JE, Visperas RM, Liu H (2002) Relationship between tillering and leaf area index: quantifying critical leaf area index for tillering in rice. J Agric Sci 138: 269–279.
47. Katsura K, Maeda S, Horie T, Shiraiwa T (2007) Analysis of yield attributes and crop physiological traits of Liangyoupeijiu, a hybrid rice recently bred in China. Field Crops Res 103: 170–177.
48. Horie T, Lubis I, Takai T, Ohsumi A, Kuwasaki K, et al. (2003) Physiological traits associated with high yield potential in rice. In: Mew TW, Brar DS, Peng S, Dawe D, Hardy B, editors. Rice science: innovations and impact for livelihood. Proceedings of the International Rice Research Conference. Beijing: International Rice Research Institute. pp. 117–145.
49. Takai T, Matsuura S, Nishio T, Ohsumi A, Shiraiwa T, et al. (2006) Rice yield potential is closely related to crop growth rate during later reproductive period. Field Crops Res 96: 328–335.
50. Wu W, Zhang H, Qian Y, Cheng Y, Wu G, et al. (2008) Analysis on dry matter production characteristics of super hybrid rice. Rice Sci 15: 110–118.
51. Zhang Y, Tang Q, Zou Y, Li D, Qin J, et al. (2009) Yield potential and radiation use efficiency of "super" hybrid rice grown under subtropical conditions. Field Crops Res 114: 91–98.

3

Spatial Heterogeneity in Human Activities Favors the Persistence of Wolves in Agroecosystems

Mohsen Ahmadi[1], José Vicente López-Bao[2,3], Mohammad Kaboli[1]*

1 Department of Environmental Sciences, Faculty of Natural Resources, University of Tehran, Karaj, Iran, **2** Research Unit of Biodiversity (UO/CSIC/PA), Oviedo University, Mieres, Spain, **3** Grimsö Wildlife Research Station, Dep. of Ecology, Swedish University of Agricultural Sciences (SLU), Riddarhyttan, Sweden

Abstract

As human populations expand, there is increasing demand and pressure for land. Under this scenario, behavioural flexibility and adaptation become important processes leading to the persistence of large carnivores in human-dominated landscapes such as agroecosystems. A growing interest has recently emerged on the outcome of the coexistence between wolves and humans in these systems. It has been suggested that spatial heterogeneity in human activities would be a major environmental factor modulating vulnerability and persistence of this contentious species in agroecosystems. Here, we combined information from 35 den sites detected between 2011 and 2012 in agroecosystems of western Iran (Hamedan province), a set of environmental variables measured at landscape and fine spatial scales, and generalized linear models to identify patterns of den site selection by wolves in a highly-modified agroecosystem. On a landscape level, wolves selected a mixture of rangelands with scattered dry-farms on hillsides (showing a low human use) to locate their dens, avoiding areas with high densities of settlements and primary roads. On a fine spatial scale, wolves primarily excavated dens into the sides of elevated steep-slope hills with availability of water bodies in the vicinity of den sites, and wolves were relegated to dig in places with coarse-soil particles. Our results suggest that vulnerability of wolves in human-dominated landscapes could be compensated by the existence of spatial heterogeneity in human activities. Such heterogeneity would favor wolf persistence in agroecosystems favoring a land sharing model of coexistence between wolves and people.

Editor: Clinton N. Jenkins, Instituto de Pesquisas Ecológicas, Brazil

Funding: The study was supported by Department of Environmental Sciences, Faculty of Natural Resources, University of Tehran, and provincial bureau of Department of Environment, Hamedan province. JVLB was supported by a "Juan de la Cierva" research contract from the Spanish Ministry of Economy and Competitiveness. The funders had no role in study design, data collection and analysis, decision to publish, or preparation of the manuscript.

Competing Interests: The authors have declared that no competing interests exist.

* Email: mkaboli@ut.ac.ir

Introduction

As human populations expand, there is increasing demand and pressure for land (characterized by an increment and expansion in settlements, habitat transformation and extension of agricultural lands, and industrial development) and, consequently, different impacts on wildlife are expected. Under this scenario, behavioural flexibility and adaptation are important processes leading to the persistence of viable animal populations in human-dominated landscapes, including urban environments (e.g. mammalian carnivores [1,2]). For species like large carnivores, with remarkable large spatial requirements, low reproductive rates or low densities [3], as well as a high potential for conflict (e.g. livestock attacks [4,5]), such behavioural processes are key elements determining their persistence in human-dominated landscapes. In fact, the capability of these species to persist in this scenario, and its behavioural, demographic and ecological consequences, have attracted a great attention in recent times [2,6,7,8].

Existing evidence shows how wolves (*Canis lupus*) are able to persist in contrasting human-dominated landscapes [7,9,10,11,12] as soon as legislation is favourable and human pressure is low [13], and minimum food and refuge requirements are fulfilled [3]. Several mechanisms are behind this ability such as the spatio-temporal segregation between wolves and human activities [9,14], their capacity to use different human-related sources of food [15,16] or other behavioural adaptations such as den shifting [17]. All this information suggest that wolves are highly capable to persist in humanized landscapes by perceiving mortality risk associated with humans, adjusting, for instance, the use of the space at different scales over time accordingly [7,17,18] (see [2] for an example with the red wolf). Thus, the spatial and temporal heterogeneity in human activities would emerge as a major environmental factor modulating vulnerability and persistence of wolves in human-dominated landscapes, resulting in wolf persistence even in areas completely transformed by humans [2,7,18,19].

In agroecosystems, ecological systems modified by human beings to produce food, fibre or other agricultural products [20], such heterogeneity in human activities may provide wolves with places of low human use where they can go unnoticed and, more importantly, can reproduce. Although the impact of humans on wolf persistence has been inferred using different surrogates such as human population density, infrastructures, level of transformation of the landscape or the spatial distribution of activities [7,21,22], how these human-related factors interact with the persistence of wolves in agroecosystems remains poorly understood. However, this knowledge becoming particularly important owing to the recent expansion of wolf populations and human activities, particularly agriculture [8,23], being crucial to adopt a balanced landscape planning ensuring both, human needs and wolf persistence [22]. Moreover, understanding the abilities of wolves to persist in each particular local context is a pressing need to reach a context-dependent conservation and management approach in agroecosystems, since heterogeneity is the norm across human-dominated landscapes [24].

Reproductive success is a cornerstone for the persistence of any species. For large carnivores, reproductive success is highly influenced by humans [3]. Because the highest mortality rate of wolves occur in the first months of their life [25,26], selection of the place where to locate the den site is crucial for wolves, being particularly important in human-dominated landscapes [17,27]. Available information suggests that, in agroecosystems, exposure risk to humans will exert the strongest effect on den site selection, with wolves aiming to minimize such risks. As a result, even in completely transformed landscapes wolves may place their den sites in areas where human activities are low [2,18,19]. In addition, the strength of human activities driving the selection of den sites by wolves in these systems may force other natural components of this selection process to the background. For example, in many areas wolves select for sites where they can dig easily [9,28], but in agroecosystems, where intensive cultivation practices are preferable on good soil conditions, wolves may be forced to dig in low-quality sites in terms of soil conditions.

In this study, we aimed to identify patterns of den site selection by wolves in agroecosystems of western Iran (Hamedan province), and provide insights into the behavioural response of wolves to the spatial heterogeneity in human activities. Since large-scale approaches may disregard fine-scale patterns affecting different components of the selection processes we were interested, we evaluated the requirements of denning wolves at large (den area) and fine (den site) spatial scales. In particular, we hypothesized that wolves are able to assess the type and intensity of human activities over a wide geographic range selecting den areas with low human use, minimizing the risk of mortality. Thus, on a landscape level, we first expect that wolves will avoid areas with high densities of infrastructures and humans and, second, we also predict that, in absence of natural dense vegetated areas in this agroecosystem acting as refuge and where to locate the den sites, wolves will select farmlands with the lowest intensity of human activity. On a fine scale, we expect that although wolves will select for den sites fulfilling previous known environmental requirements for the species (e.g. water availability, refuge, human inaccessibility, [9,28,29,30,31]), the strength of humans activities influencing den site selection in agroecosystems may push some components of the selection process into the background as a response to minimize the risk of exposure to humans.

Materials and Methods

Study area

Despite extensive studies on wolf distribution, biology, ecology and behaviour (see review in [7,11,32,33,34]) and conflict with humans (e.g. [4,5]) in Europe, North America or India, wolves are less studied in the Middle East. However, conflicts between wolves and humans are considerable in anthropogenic landscapes of Iran, affecting the attitudes of rural communities and the conservation status of the species [35,36,37].

This study was carried out in Hamedan province, a human-dominated landscape located in western Iran (88 inhabitants/km^2; Fig. 1) [38] and covering an area of 19,546 km^2 (47°34′ – 49°36′ E and 35°25′ – 35°15′ N; Fig. 1). The region has a cold semi-arid climate with an average annual precipitation of 325 mm and a mean annual temperature of 11°C. The landscape in Hamedan province is severely transformed because traditionally rural community has been mostly engaged in agriculture and livestock rearing and husbandry. Consequently, agricultural lands dominate this semi-arid landscape ([39], Fig. 2; Figure S1). The very few (2% of the whole province), and small in size, patches of natural vegetation - composed by shrub species such as *Astragalus* spp. and *Bromus* spp. and with scattered trees such as Persian oak (*Quercus brantii*), Dogwood (*Cornus australis*) or Cherry plum (*Prunus divaricata*) [40] - are distributed within a heterogeneous agricultural matrix composed by intensive irrigated potato and corn farms, dry-farms (cereals) and rangelands – which are used for extensive grazing - with scattered dry-farms (Fig. 2, Figure S2). Landscape transformation has been dramatic in this area in recent times resulting in an increase of agriculture lands from 20,468 ha to 550,264 ha during the past 30 years [39]. Consequently, rangelands covered by perennial bushes and grasses decreased from 539,697 ha to 164,679 ha [39]. The expansion of agriculture lands have significantly reduced the amount of natural refuge for wolves in this open landscape and, at the same time, have also reduced wild prey populations [39], thus increasing human-wolf encounters and associated conflicts [37].

Small variations in topographic attributes - altitude and slope - in this plateau (most of the area ranges between 1,500 and 2,000 m.a.s.l and slope changes between 0 to 41 degrees) strongly determine the use of the landscape by local people. Thus, while flat areas (slope <10 degrees) are the most preferred landscape for settlements, development and human activities (84.5% of the study area), rugged landscapes (slope> 10 degrees) only encompass 15.5% of the whole landscape and is mainly used as rangelands and, sometimes, dry-farms. As a result, human activities are heterogeneously distributed across different types of farmlands. Based on cultivation and livestock practices and land use, intensity of human activities differ across farmlands as follow: irrigated farms> dry farms> rangeland with scattered farms> rangelands. For example, in irrigated farmlands (e.g. potatoes, corn), the use of heavy equipment and mechanized cultivation is quite common and these type of crops requires a continuous human presence during many months of the year, including the peak of reproductive activity of denning wolves. On the other hand, cultivation strategies of other types of farms such as dry-farms require human presence only in two specific periods, plant and harvest, resulting in low human presence especially during denning activities and rearing of immature pups.

Data collection

We used information from 35 den sites detected between 2011 and 2012 (5 den sites in 2011 and 30 in 2012; all den sites were different). Wolf dens were located using information from local sources in the rural areas, especially observations from sheep-herders and game guards of the Department of Environment of Hamedan province, as well as field patrols conducted by motorcycle in those areas where we expected to find wolf dens according to previous local knowledge in the area. Since all issues subject to wildlife care and animal welfare regulations is handled by Department of Environment (DOE) In Iran, as well as the study was in collaborated with Hamedan Provincial Bureau of Department of Environment (43106/140), all our fieldwork procedures was adhered to the animal welfare regulations. Our data sampling was carried on after confirming that wolf packs left their dens. Our field survey did not involve chasing the wolves to locate their dens. We also did not destroy or damage wolf dens.

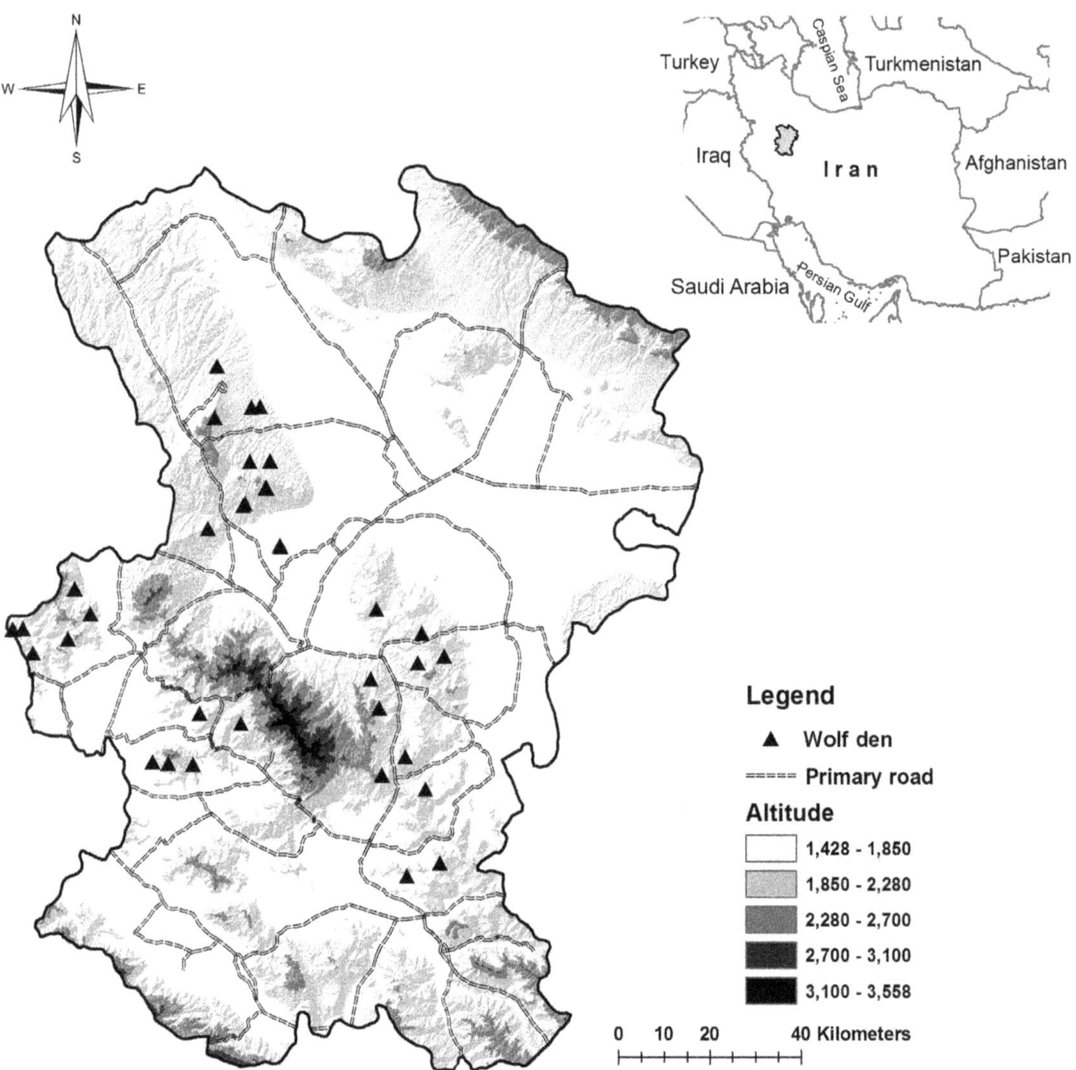

Figure 1. Distribution of gray wolf dens detected between 2011 and 2012 in Hamedan province, Iran. Wolf dens were overviewed in a context of topography and main roads in Hamedan province, Iran.

Since the breeding season is the most sensitive period for wolves [32], once a potential den site was found, we approached to the site when the pups were out of the den (between May and June) to confirm wolf reproduction. After dens were located and absence of wolves and pups was ensured, we took the location of the den sites with a GPS unit and measured the fine-scale variables we were interested (see below).

Data sampling and measurement of environmental variables were performed in two different spatial scales and using different protocols: i) den area (12.5 km^2; landscape scale), where environmental variables were measured by using GIS; and ii) den site (0.01 km^2; fine scale), where variables were measured *in situ*. On a landscape level, we estimated the spatial heterogeneity in human activities around den areas using a 2 km circular buffer centered on the den sites. The lack of information on wolf territory size in the study area confined us to consider a 2 km buffer size based on literature review [28,41], which well-describes landscape characterization of den areas [42]. For non-den areas we randomly selected 100 non-overlapping circular plots with the same radius excluding the largest cities and areas with an altitude of higher than 3,000 m.a.s.l. Because of the extensive movements of wolves, the distance between random and observed (den sites) points was controlled not to fall below 15 km. This conservative distance was selected based on published empirical values of the nearest neighbor distance for active breeding dens of wolves [33,42].

The spatial heterogeneity in human activities was inferred using three different surrogates (Table 1). First, we calculated the proportion of each land use type on a landscape-level (2 km circular buffer) using the Iranian Forests, Range and Watershed Management Organization National land use/land cover map [43]. We focused on four categories of land use representing the above-mentioned gradient in the intensity of human activities (irrigated farms> dry-farms> rangeland with scattered farms> rangelands). We excluded bare lands and rocks areas due to its anecdotic representation in the area (Fig. 2). Second, we used density of settlements and length of roads as a surrogate of human intrusion and risk of mortality in the landscape. These factors are well-known affecting wolf habitat selection in general [7,11,34], and den site selection in particular [9,31]. Density of settlements and length of roads were calculated from topographic military maps of Iran with a 1:25,000 scale. Because of the different response of wolves to road networks with varied level of human activity [11,33], we classified road networks into two categories: primary roads, including national primary roads and highways with bound> 45 m, and secondary roads, including regional and district roads with bound <30 m.

Third, using the Shuttle Radar Topography Mission elevation model with 100 m resolution, we compiled mean altitude and roughness as the main factors describing the topographic context of each area which is expected to be correlated with human activities as mentioned above (human activities decrease with the increment in altitude and roughness; [7,44]). For each den area, we then calculated the mean altitude (m) by averaging altitudes of all raster cells included in this area, and roughness (m) was estimated as the standard deviation of the altitudes of all the 100 m raster cells included in each den area. Both measures reflect different types of human use; i.e. flat areas are preferred for intensive agriculture whereas rough surfaces are more inappropriate to use farm machinery being used for extensive livestock practices and dry-farms. Vegetation types providing structural protection to wolves, such as scrubs or forests, are often selected as refuge [7,34]. But semi-arid agroecosystems of Iran, as well as other open semi-arid landscapes within the wolf's range [18], lacks such suitable cover types to provide concealment for wolves. Hence roughness of terrain that is taken into account in this study could be a representative of concealment for wolf movements [7,35].

On a fine-scale (100 m radius), we measured thirteen variables related to the vulnerability of wolves (vegetation types and slope as surrogates of refuge, human activity – existence of farmlands -), ease to dig (soil/petrology; soil type and rock density can affect den site selection by wolves [17,47]), water availability, which may be a determinant factor to locate the den [28,30], particularly in arid environments, along with solar insulation. These variables were chosen based on their suggested importance for wolf den site selection in other temperate study areas [9,28,29,30,31]. Excepting for solar insulation, all fine-scale variables were measured in five 20 × 20 m plots, one centred at the den opening and the other four plots 50 m far from the den opening in the cardinal directions [29]. We averaged all variables measured in the five plots, excluding water availability and existence of farmlands that were categorized as a binary factor, to get a general overview of the surroundings of the den and to provide a realistic distribution of the selected variables in den sites. We used hillshade as a surrogate of solar insulation [45]. Hillshade was calculated by combining slope and aspect in the den site and using ArcGIS 9.3 [46]. Hillshade values represent the average amount of shade per year received at any point. Thus, warmer slopes (facing southwest) will receive the greatest hillshade values, whereas cooler northeastern slopes will correspond to the lowest hillshade values. Due to the lack of information on accurate home range size of wolves in the study area, we conservatively selected absence plots to measure the same variables for the fine-scale analysis 1 km away from the den in a random direction (i.e. random points; equal number of points per known den sites), where we were ensured of the absence of wolf dens [29,30]. Out of the 35 den sites detected, fine-scale data sampling was carried out in 32 dens (3 den sites were destroyed before we could measure fine scale variables).

Statistical Analyses

In a first step, we carried out univariate analyses (Mann–Whitney U-tests) testing for significant differences between wolf den areas/sites and non-wolf den areas/sites for all the explanatory variables, excepting for the proportion of den sites with water bodies and farmlands within 100 m radius, where Z-proportions tests were used (Table 1, Table 2). At fine scale, we also used principal component analysis (PCA) to extract orthogonal multivariate axes on fine-scale soil-petrologic variables (Table 2). PCs obtained were used to identify the combination

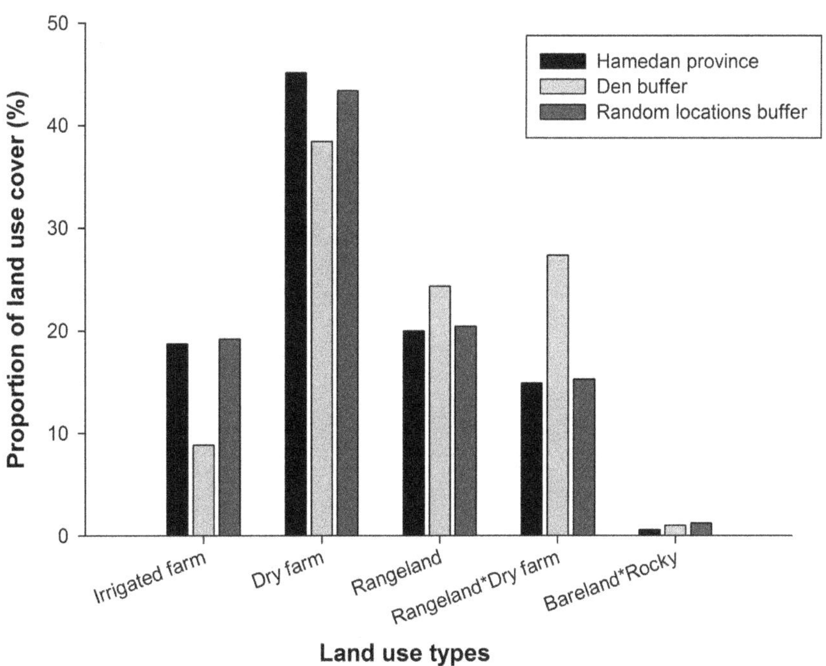

Figure 2. Proportion of land use/land cover categories used in this study. Proportion of each cover type was calculated within 2 km circular buffers around wolf den sites (den areas), random locations (random areas) and proportion of the whole study area (Hamedan province).

Table 1. Mean (SE) values of variables measured at the level of the den area, in 2 km circular buffers with and without wolf dens in Hamedan province, Iran.

Variables (unit)	Abbreviation	Den areas	Random areas	P-value
Dry farms (%)	Dry	39.8 (3.5)	40.3 (3.3)	0.770
Irrigated farms (%)	Irgt	8.5 (2.4)	22.3 (3.2)	0.050
Rangeland (%)	Rng	22.1 (4.4)	19.5 (2.9)	0.259
Rangeland with scattered farms (%)	Rng_Dry	28.3 (4.5)	13.6 (2.3)	0.001
Bareland and Rocks (%)	Bare	0.01 (0.01)	1.3 (0.8)	0.912
Altitude (m)	Alt	2116.0 (20.4)	1999.4 (25.8)	0.000
Roughness (m)	Rough	55.9 (5.1)	49.7 (5.4)	0.022
Length of primary roads (km)	Prim	0.4 (0.2)	1.6 (0.3)	0.010
Length of secondary roads (km)	Scond	2.5 (0.4)	2.2 (0.3)	0.190
Density of settlements (%)	Setl	0.0027 (0.0005)	0.0130 (0.0026)	0.034

Comparisons between den areas and random areas were done by Mann–Whitney U-tests.

of inter-correlated petrologic measurements into organized components that best separate used/unused wolf den sites. We extracted the first two components (PC1soil and PC2soil, Table 2) which explained 73% of soil characteristic variance in measured plots and used them as den site descriptive variables for soil conditions. PC1soil was related to coarse particles of soil and rocks and PC2soil indicated fine soil particles (i.e. optimum areas for cultivation; Table 2).

For both spatial scales, we built separate Generalized Linear Models (GLMs) with binomial error distribution and logit link to assess the influence of human activities on den site selection patterns by wolves in this semi-arid agroecosystem. For each spatial scale, Pearson correlation coefficients were used to test for multicollinearity among predictors, but no significant correlation between any pair of explanatory variables was detected. At the landscape scale, because of the inherent relationship between topographic contexts with land use, we first examined the possible interactions between elevation and roughness against land use types and length of primary and secondary roads (Table S1), and significant interactions were included in the full model. To do this, we generated a set of additional GLMs containing the pairwise interaction of each land use and type of roads with elevation and roughness (Table S1). We then used the "*anova*" function of the "car" package for R [48] to calculate Likelihood-Ratio χ^2 and Wald χ^2 in order to evaluate the significance level of each interaction. Akaike's Information Criterion corrected for small sample sizes (AICc) [49] was used for model selection and multimodel inference. For each spatial scale, we selected models with

Table 2. Mean (SE) values of fine-scale variables measured in sample plots with and without wolf dens in Hamedan province, Iran.

Variables	Description	Den sites	Random points	P-value
Slope	Measured by a clinometers	15.4 (6)	9 (3.3)	0.000
Hillshade	Measured by a combination of slope and aspect	175.5 (2.11)	177.8 (3.24)	0.234
Herbaceous	Vegetation height less than 25 cm (percentage)	53.3 (20.6)	70.9 (21.4)	0.008
Shrub	Vegetation height between 25 to 200 cm, (percentage)	43.6 (18.3)	25.1 (20.4)	0.003
Tree	Vegetation height above than 200 cm, (percentage)	3.1 (5.3)	3.9 (5.6)	0.406
Soil/Petrology (proportion)	Sable: Particles of clay and sand	35.8 (11.5)	60.6 (7.5)	
	Mm: Soil particle ≤ 1cm	33.8 (7.8)	28.7 (5.3)	
	Cm: Pebbles with size of ≤ 10 cm	15.9 (6.3)	8.4 (3.2)	
	Dc.m: Pebbles with size of ≤ 1 m	9.5 (6.4)	3.2 (2.3)	
	M: Rock with size of ≤ 10 m	5 (6.5)	1.4 (1.5)	
	Dca.m: Rocky materials with size of> 10 m	1.7 (3.5)	0.1 (0.7)	
	PC1 soil: first component of PCA analysis preformed on Soil/Petrology - coarse soil particles -			0.006
	PC2 soil: second component of PCA analysis preformed on Soil/ Petrology -fine soil particles -			0.004
Water availability	Proportion of sites with water bodies within 100 m radius	0.75	0.31	0.001
Farm	Proportion of sites with farmlands within 100 m radius	0.56	0.72	0.283

Comparisons between den sites and random sites were done by Mann-Whitney U-tests excepting for the proportion of presence of water bodies and farmlands within a 100m radius, which were evaluated using Z-proportions tests.

ΔAICc <2, and we calculated Akaike weights (AICc *wi*) [49]. Moreover, for each predictor selected in the set of models with ΔAICc <2, we calculated its estimated importance (or relative evidence weight), computed as the sum of the relative evidence weights of all models in which the variable appears, as well as model-averaged estimates and their unconditional standard errors (SE). Using this approach we reduced model selection bias effects on regression coefficient estimates in all selected subsets [49]. Finally, to verify how well the selected models described our dataset, we performed a Goodness-of-fit test using Hosmer-Lemeshow (HL) procedure [50]. The Area Under the Curve (AUC) of ROC was also calculated as a measure of discrimination capacity of selected candidate models. All analyses were carried out in R version 3.0.1 [51].

Results

Breeding in agroecosystems

Den areas were located in agricultural matrix with a significantly less proportion of irrigated farms (Mann–Whitney U-test, $P < 0.05$; Table 1) and a higher proportion of mosaics of rangelands with scattered dry-farms than random areas (Mann–Whitney U-test, $P < 0.001$; Table 1). We did not find significant differences between den and random areas for the rest of land uses (Table 1). Wolves tended to select elevated and rough areas (where intensive agricultural practices, such as irrigated farms, are less probable; altitude: Mann–Whitney U-test, $P < 0.0001$; roughness: Mann–Whitney U-test, $P = 0.022$; Table 1, Table S1). Finally, as predicted, wolves also avoided areas with abundant primary roads and density of settlements (primary roads: Mann–Whitney U-test, $P = 0.017$; settlements: Mann–Whitney U-test, $P = 0.016$; Table 1). However, location of den sites was not influenced by the development of the network of secondary roads in the den area (Table 1).

We found a significant interaction between irrigated farms and roughness ($\chi^2 = 6.147$, $P = 0.013$; Table S1), and between altitude and secondary roads ($\chi^2 = 3.967$, $P = 0.043$; Table S1). Hence these two interactions were included in the set of predictors for the landscape scale models. Seven candidate models showed ΔAICc < 2 (Table 3), with the best model including rangelands with scattered dry-farms, altitude, roughness, human settlements, primary roads and the interaction between irrigated farms and roughness (Table 3). The probability of a given area being selected as a den area by wolves in this semi-arid agroecosystem raised with an increase in the proportion of rangelands with scattered dry-farms, located at high altitudes and with low human presence (negative estimation for length of primary roads and density of human settlements; Table 4). Model-averaged coefficient estimates indicated that rangelands with scattered dry farms, altitude, roughness, primary roads and human settlements were the most important predictors determining the probability of a given area being selected as a den area by wolves (Table 4). AUC of ROC curve showed good discrimination capacity of selected candidate models and, we did not find evidence of lack of fit in the different models (HL tests, Table S2).

Fine-scale den site selection patterns in agroecosystems

Wolves were prone to excavate dens in rough hillsides with moderate shrub cover (Mann–Whitney U-test, $P < 0.05$; Table 2). At fine-scale, the strongest significant difference between occupied and unoccupied sites was slope (15.4±6.0 vs. 9.0±3.3; Mann–Whitney U-test, $P < 0.0001$; Table 2). In addition, den sites were characterized by significantly lower percentage of open areas (dominated by herbaceous) as well as higher shrub cover (43.6±18.3 vs. 25.1±20.4; Mann–Whitney U-test, $P = 0.003$; Table 2). Water availability was significantly higher in den sites ($Z = 3.276$; $P < 0.001$; Table 2) and wolves tended to locate them in areas with a high proportion of coarse soil particles (Mann–Whitney U-test, $P = 0.0004$; Table 2). As expected, because the study area was dominated by humans, the presence of farm-lands did not differ between occupied/unoccupied sites at fine scale ($Z = 1.073$; $P = 0.283$; Table 2). Also, the difference of the amount of shade received at wolf den and random points was not significant (Table 2).

For den sites, eight candidate models showed ΔAICc <2 (Table 5) and the best model included slope, soil/petrologic terms (PC2soil; fine soil particles) and water availability. These three variables were the most important fine-scale predictors of den site selection by wolves based on their relative importance (Table 6). Averaging the coefficient estimates of the selected candidate models revealed that wolves selected for sites with availability of water bodies, placed in stepper hills and with coarse soil particles (Table 6). Based on AUC, we found a very good discrimination capacity for the selected candidate models ranging from 0.915 to 0.933, and no evidence of lack of fit was detected (HL tests, Table S2).

Discussion

Humans are the main source of disturbance for large carnivores affecting, for example, the composition and security of their habitats [52]. Wolf distribution and habitat suitability is mainly influenced by human-associated factors [32]. Such human influence can be both direct (i.e. mortality; legal hunting, poaching, road kills) [32,53] and indirect (behaviour), for example, wild prey depletion or availability of human-related sources of food [15,16]. However, wolves, as many other large carnivores [2,6,8], do not strictly required areas devoid of humans, showing a high ability adapting to multiple used landscapes. This phenomenon is particularly interesting in agroecosystems where virtually all habitats are agricultural and transformed and wild prey can be rare, with wolves usually feeding on livestock, waste or animal carcasses [7,15,18,54].

In agroecosystems, simply avoiding transformed land cover types is impossible, such as the case of western Iran with the almost complete loss of natural habitats (2%) [39]. As a consequence, wolves are relegated to utilize non-natural land cover types while avoiding negative interactions with humans [2,7,18,19]. So, understanding how wolves adjust the use of space in agricultural lands (one of the most widespread habitats worldwide), adapting to human activities, is therefore a critical step to ensure the persistence and conservation of this species in agroecosystems minimizing human-wolf conflicts. This is particularly important since the occurrence of this contentious species in agroecosystems is beyond anecdotic, with several packs occurring, for example, in our study area, as reflected by the number of wolf dens [35] used here [17,18] (see also [55] for a similar scenario in Spain).

Based on the comparison of human land use between den areas, random areas and the whole study area (Hamedan province) we found that the mixture of rangelands with scattered dry-farms (accounting around 15% of the whole study area; Fig. 2) was preferred by denning wolves, whereas irrigated farms were actively avoided and no patterns were found for extensive and homogeneous dry-farms or rangelands (Fig. 2; Figure S2). The proportion of mixed rangelands with dry-farms was the most predictive variable identifying wolf den areas along with a combined preference for hillsides. Two non-exclusive explanations may be behind of this result. By one hand, dry farming practices requires

Table 3. Selected candidate Generalized Linear Models explaining gray wolf den area selection patterns in Hamedan province, Iran, at the landscape level.

Model	AICc	ΔAICc	AIC w_i
Rng_Dry + Alt + Rough + Setl + Prim + (Irgt × Roug)	125.12	0.02	0.18
Rng_Dry + Alt + Rough + Setl + Prim	125.28	0.18	0.16
Rng_Dry + Alt + Rough + Setl + Prim + (Irgt × Roug) + (Alt × Scond)	126	0.90	0.11
Rng_Dry + Alt + Rough + Setl + Prim + (Alt × Scond) + Dry	126.21	1.11	0.10
Rng_Dry + Alt + Rough + Prim + (Irgt × Roug) + (Alt × Scond)	126.45	1.35	0.09
Rng_Dry + Alt + Rough + Setl + Prim + (Irgt × Roug) + Rng	126.81	1.71	0.07
Rng_Dry + Alt + Rough + Setl + Prim + (Irgt × Roug) + Irgt	127.06	1.96	0.06

Models were ranked according to AICc, and only models with ΔAICc <2 are shown for simplicity. For variables description see Table 1.

low levels of human activity, with human presence not overlapping with the most sensitive period for wolves (denning period) because human activity is limited to only the planting and harvesting seasons. On the other hand, rangelands, which can also show a low intensity of human use depending on livestock practices, can also provide wolves with human-related sources of food (e.g. livestock, carrion, waste). Because of the low abundance of wild prey in the area [56] and the use of human-related food sources by wolves in such ecosystems [54,57], traditional herd roaming in rangelands adjacent to dry-farms by local community may favor food availability (higher density of livestock close to farms), affecting den site selection. On the other hand, this scenario (i.e. the presence of scattered dry-farms) may also increase food availability for scarce wild prey. Further analyses are needed to test these hypotheses.

As we expected, wolf den areas were characterized by lesser density of settlements and primary roads compared with random areas [9,58]. The lack of difference between den and random areas in the length of secondary roads suggests that having lesser disturbance from main surrogates of human activity (primary roads and settlements; areas with an intense human land use), secondary roads may be a less important limiting factor for den site selection by wolves. In fact, because secondary roads generally show a lower human use, wolves may use these linear infrastructures for ease of travel within their territories [2,33].

The lack of refuge - considering the well-established link between the concept of refuge and certain vegetation structures providing safe places to wolves such as forests or scrublands [7,34] - in our study area highlights the importance of rouged terrains with low human use providing good concealment for denning wolves in open areas [7,58,59]. Therefore, although wolves selected for den sites located in places with a higher proportion of shrubs compared to random sites in this agroecosystem (Table 2), on a landscape level, vegetation/habitat types becomes a secondary factor for den selection processes, being strongly modulated by the level of human activities.

On a fine spatial scale, our results indicated that wolves primarily excavated dens into the sides of elevated steep-slope hills (Figure S2), selecting sites with steeper slopes, which is consistent with the selection patterns found in other studies (e.g. similar average values for slope, ca. 15 degrees) [29,60]. The slope in these places will also cause more drainage – in case of torrential rain - than surrounding regions that have gentle slope [29,31,47]. Apart from slope, fine soil particles –PC2 soil- (negative selection) and existence of water bodies (positive selection) were the most important variables affecting den site selection patterns. In an

Table 4. Relative importance (W+), model-averaged coefficient estimates (Estimate), and unconditional standard errors (SE) for the predictors included in the selected candidate models determining the probability of a given area being selected as a den area by wolves in Hamedan province, Iran (models with ΔAICc <2).

Variable	W+	Estimate	SE
Intercept		−4.71	3.71
Rng_Dry	1	0.02	0.01
Alt	1	0.002	0.001
Roug	1	0.02	0.01
Setl	1	−0.01	0.05
Prim	1	−0.003	0.002
Irgt × Roug	0.95	0.001	0.001
Alt × Scond	0.83	0.0004	0.002
Irgt	0.35	−0.001	0.002
Rng	0.27	0.02	0.01
Dry	0.10	−0.02	0.01

For variables description see Table 1.

Table 5. Selected candidate Generalized Linear Models explaining gray wolf den site selection patterns in Hamedan province, Iran, at the fine spatial scale.

Model	AICc	ΔAICc	AICc w_i
Slope + PC2 soil + Water	55.31	0	0.17
Slope + PC2 soil + Water + Hillshade	55.88	0.56	0.13
Slope + PC2 soil + Water + Herbaceous + Shrub	56.33	1.02	0.10
Slope + PC2 soil + Water + Hillshade + PC1 soil	56.36	1.05	0.10
Slope + PC2 soil + Water + PC1 soil	56.45	1.14	0.10
Slope + PC2 soil + Water + Hillshade + Shrub	56.58	1.27	0.09
Slope + PC2 soil + Water + Shrub	56.66	1.35	0.09
Slope + Water + Herbaceous + Shrub	57.25	1. 94	0.06

Models were ranked according to AICc, and only models with ΔAICc <2 are shown for simplicity. For variables description see Table 2.

unusual pattern, we found that the existence of farmlands did not affect selection patterns by denning wolves [18]. High tendency of local communities to place dry-farms in areas with unsuitable topographic conditions for other cultivation practices may also explains why many dens were located in the vicinity of farmlands. We found a significant difference between den and random sites (that were often located within agricultural lands) in terms of soil variables. Most of the areas with a gentle slope and rich soil (PC2 soil) are used for farming by local people. Accordingly, rangelands adjacent to farms are less usable for agriculture and wolves were forced to den in places with coarse soil particles. Finally, we found that the availability of water bodies in the vicinity of den sites is an important factor for denning wolves. As expected, due to high water requirement of lactating females, den sites were selected relatively close to water sources [28,29,30,60]. In semi-arid landscapes, we predict that the dependency of both, denning wolves and humans, to scarce water bodies may have increased human-wolf conflict locally, being an important limiting factor for the persistence of the species.

Our findings at different spatial scales show how wolves can be tolerant to placing their dens in agricultural lands, which demonstrates their resilience to persist in agroecosystems. As agricultural lands dominated this landscape, wolves selected for den areas with low human use irrespective whether such areas were profoundly transformed or not. In our case, this is possible because small dry-farms adjacent to rangelands require minimum human intervention, consequently having a low impact on habitat security and decreasing the risk of mortality for wolves during the breeding period. Thus, spatial and seasonal heterogeneity in human activities become an important factor explaining the persistence of wolves in agroecosystems [61].

As in other regions of the Middle East, agricultural activities in Hamedan province started more than 5000 years ago [62]. Moreover, contrary to European and North American wolf ranges [63] where wolves were exterminated from huge areas during the 19th and 20th centuries [24,32,65], and only began to recolonize some of their former range in recent times [32], such pattern of eradication/re-colonization did not occur in Iran, with wolves persisting in this area continuously over time. Thus, here wolves and human activities have been interacting for a much longer period of time than in other parts of the current and historical wolf range leading to a unique scenario of wolf adaptations to humans.

Effective large carnivore conservation in human-dominated landscape matrix and outside of formally protected areas is of paramount importance in the Anthropocene [64,65]. Successful conservation strategies requires minimizing conflicts between large carnivores and humans, understanding where and when to establish limits of sharing the landscape with these contentious species. Alternatives range from a focus on fencing large carnivores to allowing them to share the landscape with humans (e.g. [66]). However, this debate also requires determining to what extent large carnivores can tolerate living in human-dominated landscapes

Table 6. Relative importance (W+), model-averaged coefficient estimates (Estimate), and unconditional standard errors (SE) for the predictors included in the selected candidate models determining the probability of a given site being selected as a den site by wolves in Hamedan province, Iran (models with ΔAICc<2).

Variable	W+	Estimate	SE
Intercept		−6.68	8.09
Slope	1	0.31	0.09
Water	1	3.10	1.04
PC2 soil	0.94	−0.87	0.45
Hillshade	0.48	−0.03	0.02
Shrub	0.39	0.11	0.10
Herbaceous	0.29	0.14	0.09
PC1 soil	0.20	0.29	0.25

For variables description see Table 2.

considering different spatial and ecological constraints and levels of conflict. Along these lines, our results show how the heterogeneity in human activities emerges as a key factor favoring the persistence of wolves in agroecosystems. Thus, vulnerability of wolves, and other large carnivore species, in human-dominated landscapes could be compensated by the existence of spatial heterogeneity in human activities, favoring a land sharing model of coexistence between large carnivores and people.

However, despite the ability of wolves to persist in agroecosystems, with much of the landscape being devoted to agricultural and livestock activities, human-wolf encounters and conflicts can also increase. As a consequence, because of the high accessibility to wolf dens by people in agroecosystems, lactating wolves and their pups can be very vulnerable to active illegal human persecution [35]. Since wolf core use areas, including den areas, are used by wolf packs more intensively throughout the year and wolves are even prone to use the same den in subsequent years [29,47], there is a pressing need to adopt efficient measures to mitigate human-wolf conflicts in agroecosystems (e.g. discouraging people from destroying wolf dens, changing human behaviors and livestock practices) in order to keep acceptable levels of tolerance and favoring wolf persistence.

Supporting Information

Figure S1 General views of the agroecosystems of Hamedan province, Iran.

Figure S2 Fine-scale pictures showing the environment around den sites in rangelands of Hamedan province, Iran.

Table S1 Results from Generalized Linear Models testing for significant effects of the pairwise interactions between land use and type of roads with elevation and roughness.

Table S2 Results of the assessment of goodness-of-fit and discrimination capacity of selected candidate models explaining the selection of den areas/sites by wolves in Hamedan, Iran.

Acknowledgments

We would like to thank the many local people and the staff of Hamedan DoE who helped us with the field work. Our special thanks go to the wardens who were a great help in finding and surveying dens as well as Elham Nourani for her helps. JVLB was supported by a "Juan de la Cierva" research contract from the Spanish Ministry of Economy and Competitiveness.

Author Contributions

Conceived and designed the experiments: MA JVLB MK. Performed the experiments: MA MK. Analyzed the data: MA JVLB MK. Contributed reagents/materials/analysis tools: MA MK. Wrote the paper: MA JVLB MK.

References

1. Bateman PW, Fleming PA (2012) Big city life: Carnivores in urban environments. J Zool 287: 1–23.
2. Dellinger JA, Proctor C, Steury TD, Kelly MJ, Vaughan MR (2013) Habitat selection of a large carnivore, the red wolf, in a human-altered landscape. Biol Conserv 157: 324–330.
3. Gittleman JL, Funk SM, Macdonald D, Wayne RK (2001) Carnivore conservation. Cambridge University Press, Cambridge, UK.
4. Woodroffe R, Thirgood S, Rabinowitz A (2005) People and wildlife, conflict or co-existence? Cambridge University Press, UK.
5. Treves A, Karanth KU (2003) Human-carnivore conflict and perspectives on carnivore management worldwide. Conserv Biol 17: 1491–1499.
6. Woodroffe R (2011) Ranging behaviour of African wild dog packs in a human-dominated landscape. J Zool 283: 88–97.
7. Llaneza L, Lopez-Bao JV, Sazatornil V (2012) Insights into wolf presence in human dominated landscapes: the relative role of food availability, humans and landscape attributes. Divers Distrib 18: 459–469.
8. Athreya V, Odden M, Linnell JDC, Krishnaswamy J, Karanth U (2013) Big cats in our backyards: persistence of large carnivores in a human dominated landscape in India. PLoS One 8: e57872. doi:10.1371/journal.pone.0057872.
9. Theuerkauf J, Rouys S, Jędrzejewski W (2003) Selection of den, rendezvous, and resting sites by wolves in the Białowieza Forest, Poland. Can J Zool 81: 163–167.
10. Blanco JC, Cortés Y (2007) Dispersal patterns, social structure and mortality of wolves living in agricultural habitats in Spain. J Zool 273: 114–124.
11. Eggermann J, da Costa GF, Guerra AM, Kirchner WH, Petrucci-Fonseca F (2011) Presence of Iberian wolf (*Canis lupus signatus*) in relation to land cover, livestock and human influence in Portugal. Mamm Biol 76: 217–221.
12. Chavez AS, Gese EM (2006) Landscape use and movements of wolves in relation to livestock in a wildland-agriculture matrix. J Wildl Manage 70: 1079–1086.
13. Boitani L (2003) Wolf conservation and recovery. In: Mech LD, Boitani L, editors. Wolves, behavior, ecology and conservation. The University of Chicago Press, Chicago and London. 317–340.
14. Latham ADM, Latham MC, Boyce MS, Boutin S (2011) Movement responses by wolves to industrial linear features and their effect on woodland caribou in northeastern Alberta. Ecol Appl 21: 2854–2865.
15. Meriggi A, Lovari S (1996) A review of wolf predation in southern Europe: does the wolf prefer wild prey to livestock? J Appl Ecol 33: 1561–1571.
16. López-Bao JV, Sazatornil V, Llaneza L, Rodríguez A (2013) Indirect effects on heathland conservation and wolf persistence of contradictory policies that threaten traditional free-ranging horse husbandry. Conserv Lett 6: 448–455.
17. Habib B, Kumar S (2007) Den shifting by wolves in semi-wild landscapes in the Deccan Plateau, Maharashtra, India. J Zool 272: 259–265.
18. Agarwala M, Kumar S (2009) Wolves in agricultural landscapes in Western India. Tropical Resources: Bulletin of the Yale Tropical Resources Institute 28: 48–53.
19. Mech LD (2006) Prediction failure of a wolf landscape model. Wildl Soc Bull 34: 874–877.
20. Conway GR (1987) The properties of agroecosystems. Agricult Sys 24: 95–117.
21. Blanco JC, Cortés Y, Virgós E (2005) Wolf response to two kinds of barriers in an agricultural habitat in Spain. Can J Zool 83: 312–323.
22. Falcucci A, Maiorano L, Tempio G, Boitani L, Ciucci P (2013) Modeling the potential distribution for a range-expanding species: Wolf recolonization of the Alpine range. Biol Conserv 158: 63–72.
23. Tilman D, Balzer C, Hill J, Befort BL (2011) Global food demand and the sustainable intensification of agriculture. PNAS 108: 20260–20264.
24. Boitani L (2000) Action plan for the conservation of wolves (*Canis lupus*) in Europe. Council of Europe Publishing, Strasbourg, France.
25. Harrington FH, Mech LD (1982) Patterns of home-site attendance in two Minnesota wolf packs. In: Harrington FH, Paquet PC, editors. Wolves of the world: perspectives of behavior, ecology, and conservation. Noyes Publications, New Jersey. 81–104.
26. Frame PF, Cluff HD, Hik DS (2007) Response of Wolves to Experimental Disturbance at Homesites. J Wildl Manage 71: 316–320.
27. Frame PF, Cluff HD, Hik DS (2008) Wolf reproduction in response to caribou migration and industrial development on the Central Barrens of mainland Canada. Arctic 61: 134–142.
28. Norris DF, Theberge MT, Theberge JB (2002) Forest composition around wolf (*Canis lupus*) dens in eastern Algonquin Provincial Park, Ontario. Can J Zool 80: 866–872.
29. Trapp JR, Beier P, Mack C, Parsons DR, Paquet PC (2008) Wolf, *Canis lupus*, den site selection in the Rocky Mountains. Can Field Nat 122: 49–56.
30. Person DK, Russell AL (2009) Reproduction and den site selection by wolves in a disturbed landscape. Northwest Sci 83: 211–224.
31. Unger DE, Keenlance PW, Kohn BE, Anderson EM (2009) Factors Influencing Home site Selection by gray wolves in Northwestern Wisconsin and East-Central Minnesota. In: Wydeven AP et al., editors. Recovery of gray wolves in the Great Lakes Region of the United States. Springer Science + Business Media. 175–189, doi: 10.1007/978-0-387-85952-1-11.
32. Mech LD, Boitani L (2003) Wolves: Behavior, Ecology and Conservation. Chicago, University of Chicago Press.
33. Jędrzejewski W, Niedziałkowska M, Nowak S, Jędrzejewska B (2004) Habitat variables associated with wolf (*Canis lupus*) distribution and abundance in northern Poland. Divers Distrib 10: 225–233.

34. Jędrzejewski W, Jędrzejewska B, Zawadzka B, Borowik T, Nowak S, et al. (2008) Habitat suitability model for Polish wolves based on long-term national census. Anim Conserv 11: 377–390.
35. Ahmadi M, Kaboli M, Nourani E, Alizadeh Shabani A, Ashrafi S (2013) A predictive spatial model for gray wolf (*Canis lupus*) denning sites in a human-dominated landscape in western Iran. Ecol Res 28: 513–521.
36. Ziaie H (2008) A field guide to mammals of Iran. 2nd ed. Iranian Wildlife Center, Tehran (in Persian).
37. Behdarvand N, Kaboli M, Ahmadi M, Nourani E, Salman Mahini A, et al. (2014) Spatial risk model and mitigation implications for wolf–human conflict in a highly modified agroecosystem in western Iran. Biol Conserv 177: 156–164.
38. Reyahi-Khoram M, Fotros MH (2011) Land use planning of Hamadan province by means of GIS. International conference on chemical, biological and environment sciences (ICCEBS, 2011) Bangkok.
39. Imani Harsini J (2012) Study on change detection of land use/cover in Hamedan province considering wolves potential habitats during the past 30 years. M.Sc. dissertation. Department of Environmental Sciences, University of Tehran.
40. Safikhani K, Rahiminejhad MR, Kalvandi R (2007) Presentation of flora and life forms of plant species in Kian region (Hamadan province). Watershed Management Research Journal 74: 138–154 (in Persian).
41. McLoughlin PD, Walton LR, Cluff HD, Paquet PC, Ramsay MA (2004) Hierarchical habitat selection by tundra wolves. J Mammal 85: 576–580.
42. Iliopoulos Y, Youlatos D, Sgardelis S (2014) Wolf pack rendezvous site selection in Greece is mainly affected by anthropogenic landscape features. Eur J Wildl Res 60: 23–34.
43. Forest, Range and Watershed Management Organization I.R. of Iran, FRWMO 2010. Iranian Forests, Range and Watershed Management Organization National Land use/Land cover map.
44. Glenz C, Massolo D, Kuonen D, Schlaepfer R (2001) A wolf habitat suitability prediction study in Valais (Switzerland). Landsc Urban Plan 55: 55–65.
45. Ciarniello LM, Boyce MS, Heard DC, Seip DR (2005) Denning behavior and den site selection of grizzly bears along the Parsnip River, British Columbia, Canada. Ursus 16: 47–58.
46. ESRI (2010) ArcGis 9.3. Environmental Systems Research Institute. Redlands, CA.
47. Ballard WB, Dau JR (1983) Characteristics of gray wolf, *Canis lupus*, den and rendezvous sites in Southcentral Alaska. Can Field Nat 97: 299–302.
48. Fox J, Weisberg S (2011) An R Companion to Applied Regression, Second Edition. Thousand Oaks CA: Sage. Available: http://socserv.socsci.mcmaster.ca/jfox/Books/Companion.
49. Burnham KP, Anderson DR (2002) Model selection and inference: a practical information theoretic approach. Springer-Verlag, New York, New York, USA.
50. Hosmer DW, Lemeshow S (2000) Applied logistic regression. Wiley Series in Probability and Statistics. John Wiley and Sons, New York, USA.
51. R Core Team (2013) R: A Language and Environment for Statistical Computing. R Foundation for Statistical Computing, Vienna, Austria. http://www.R-project.org/
52. Weaver JL, Paquet PC, Ruggiero LF (1996) Resilience and conservation of large carnivores in the Rocky Mountains. Conserv Biol 10: 964–976.
53. Liberg O, Chapron G, Wabakken P, Pedersen HC, Hobbs NT, et al. (2012) Shoot, shovel and shut up: cryptic poaching slows restoration of a large carnivore in Europe. Proc Roy Soc Lond B Biol 279: 910–915.
54. Tourani M, Moqanaki EM, Boitani L, Ciucci P (2014) Anthropogenic effects on the feeding habits of wolves in an altered arid landscape of central Iran. Mammalia 78: 117–121.
55. Blanco JC, Cortés Y (2002) Ecología, censos, percepción y evolución del lobo en España: análisis de un conflicto. SECEM, Málaga.176 pp.
56. Darvishsefat AA (2006) Atlas of Protected Areas of Iran (English-Persian), University of Tehran Press, Tehran.
57. Hosseini-Zavarei F, Farhadinia MS, Beheshti-Zavareh M, Abdoli A (2013) Predation by grey wolf on wild ungulates and livestock in central Iran. J Zool 290: 127–134.
58. Capitani C, Mattioli L, Avanzinelli E, Gazzola A, Lamberti P, et al. (2006) Selection of rendezvous sites and reuse of pup raising areas among wolves *Canis lupus* of northeastern Apennines, Italy. Acta Theriol 51: 395–404.
59. Corsi F, Dupre E, Boitani L (1999) A large-scale model of wolf distribution in Italy for conservation planning. Conserv Biol 13: 150–159.
60. Unger DE (1999) A multi-scale analysis of timber wolf den and rendezvous site selection in northwestern Wisconsin and east-central Minnesota. M.Sc. dissertation, University of Wisconsin.
61. Schuette P, Wagner AP, Wagner ME, Creel S (2013) Occupancy patterns and niche partitioning within a diverse carnivore community exposed to anthropogenic pressures. Biol Conserv 158: 301–312.
62. Farshad A, Barrera-Bassols N (2003) Historical anthropogenic land degradation related to agricultural systems: case studies from Iran and Mexico. Geogr Ann A 85: 277–286.
63. Young SP, Goldman EA (1944) The wolves of North America. Dover, New York, USA.
64. Wikramanayake E, McKnight M, Dinerstein E, Joshi A, Gurung B, et al. (2004) Designing a conservation landscape for tigers in human-dominated environments. Conserv Biol 18: 839–844.
65. Muntifering JR, Dickman AJ, Perlow LM, Hruska T, Ryan PG, et al. (2006) Managing the matrix for large carnivores: a novel approach and perspective from cheetah (*Acinonyx jubatus*) habitat suitability modelling. Anim Conserv 9: 103–112.
66. Packer C, Loveridge A, Canney S, Caro T, Garnett ST, et al. (2013) Conserving large carnivores: dollars and fence. Ecol Lett 16: 635–641.

4

Effects of Manure Compost Application on Soil Microbial Community Diversity and Soil Microenvironments in a Temperate Cropland in China

Zhen Zhen[1,2,3], Haitao Liu[1], Na Wang[4], Liyue Guo[1], Jie Meng[1], Na Ding[5], Guanglei Wu[1], Gaoming Jiang[1]*

1 State Key Laboratory of Vegetation and Environmental Change, Institute of Botany, the Chinese Academy of Sciences, Beijing, China, **2** Laboratory of Crop, Guangdong Ocean University, Zhanjiang, China, **3** University of the Chinese Academy of Sciences, Beijing, China, **4** Development and Promotion Center, Shandong Small and Medium-sized Enterprises, Ji'nan, China, **5** State Key Laboratory of Crop Biology, Shandong Agricultural University, Tai'an, China,

Abstract

The long-term application of excessive chemical fertilizers has resulted in the degeneration of soil quality parameters such as soil microbial biomass, communities, and nutrient content, which in turn affects crop health, productivity, and soil sustainable productivity. The objective of this study was to develop a rapid and efficient solution for rehabilitating degraded cropland soils by precisely quantifying soil quality parameters through the application of manure compost and bacteria fertilizers or its combination during maize growth. We investigated dynamic impacts on soil microbial count, biomass, basal respiration, community structure diversity, and enzyme activity using six different treatments [no fertilizer (CK), N fertilizer (N), N fertilizer + bacterial fertilizer (NB), manure compost (M), manure compost + bacterial fertilizer (MB), and bacterial fertilizer (B)] in the plowed layer (0–20 cm) of potted soil during various maize growth stages in a temperate cropland of eastern China. Denaturing gradient electrophoresis (DGGE) fingerprinting analysis showed that the structure and composition of bacterial and fungi communities in the six fertilizer treatments varied at different levels. The Shannon index of bacterial and fungi communities displayed the highest value in the MB treatments and the lowest in the N treatment at the maize mature stage. Changes in soil microorganism community structure and diversity after different fertilizer treatments resulted in different microbial properties. Adding manure compost significantly increased the amount of cultivable microorganisms and microbial biomass, thus enhancing soil respiration and enzyme activities ($p<0.01$), whereas N treatment showed the opposite results ($p<0.01$). However, B and NB treatments minimally increased the amount of cultivable microorganisms and microbial biomass, with no obvious influence on community structure and soil enzymes. Our findings indicate that the application of manure compost plus bacterial fertilizers can immediately improve the microbial community structure and diversity of degraded cropland soils.

Editor: Wen-Xiong Lin, Agroecological Institute, China

Funding: This research was jointly supported by the Key Strategic Project of the Chinese Academy of Sciences (No.KSZD-EW-Z-012-2) and State Key Laboratory of Vegetation and Environmental Change (2014ZDFX04). The funders had no role in study design, data collection and analysis, decision to publish, or preparation of the manuscript.

Competing Interests: The authors have declared that no competing interests exist.

* Email: jianggm@126.com

Introduction

Chemical fertilizers have been extensively applied to sustain global agricultural production since the first Green Revolution [1,2]. Modern high-intensity agricultural ecosystems are characterized by the excessive use of chemical fertilizers, pesticides, and herbicides [1]. Chemical fertilizers directly enhance crop yield because plants directly or indirectly assimilate the nutrients provided by these inorganic fertilizers. However, on one hand, the production and use of these chemicals impart various negative effects on the agricultural ecosystem such as degradation of the soil, loss of crop genetic diversity, reduction in soil microbial diversity, contamination of ground-water resources, and pollution of the atmosphere [3,4]. On the other hand, with the intensive development of animal husbandry, animal dung has become one of the major pollution sources in China. There are a lot of organic matter and nutrients such as nitrogen and phosphorus in animal manure, especially through composting. Application of manure compost may also enhance soil microbial activities that improve the crop growth, and restrain the pests and diseases Compared with chemical fertilizers, manure compost has been comprehensively tested and determined as effective in increasing nutrient availability to crops, thus improving grain yield in a cost-effective and environmentally friendly manner [5,6]. The addition of manure compost can also increase the levels of organic matter and improve soil porosity, structural stability, moisture, and nutrient availability, as well as biological activity [7,8]. Hence, it is becoming a more popular practice to add manure compost to the soils if the degraded cropland is considered to be restored.

Soil microorganisms play important roles in ecological functions such as nutrient cycling and formation of soil aggregates through the decomposition of organic matter [9]. The stability of the microbial community structure has important implications for the rates of soil processes. For example, variations in microbial community structure in the soil influence rates of de-nitrification, nitrification, and nitrogen fixation [10,11]. Organic and inorganic

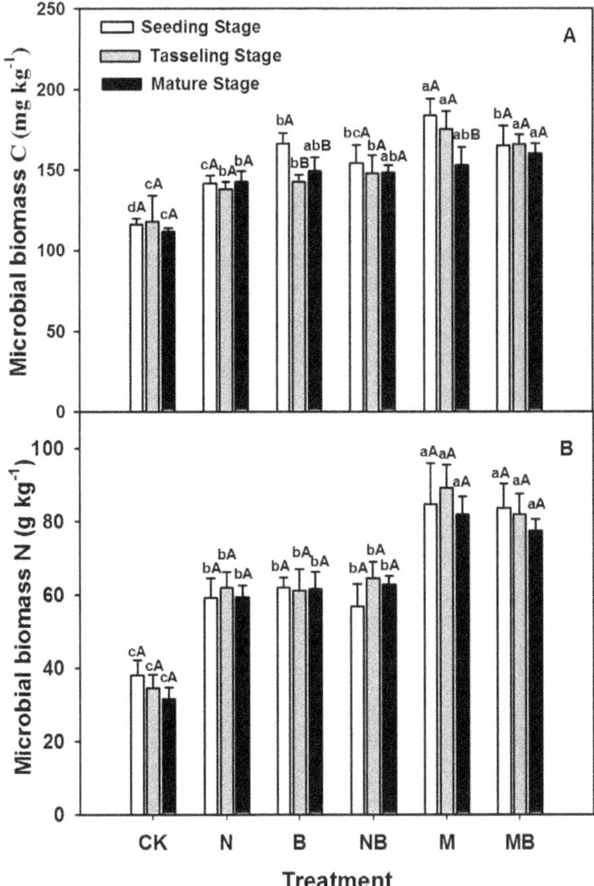

Figure 1. Soil microbial biomass C and soil microbial biomass N in various treatments during maize growth stages. CK: no fertilizer; N: N fertilizer; B: bacteria fertilizer; NB: N fertilizer + bacteria fertilizer; M: manure compost; MB: manure compost + bacteria fertilizer. Data are means ± SD (n = 3). Different lower case letters indicate significant differences (ANOVA, LSD test, $p<0.05$) among fertilizer treatments; different upper case letters indicate significant differences among maize growth stages.

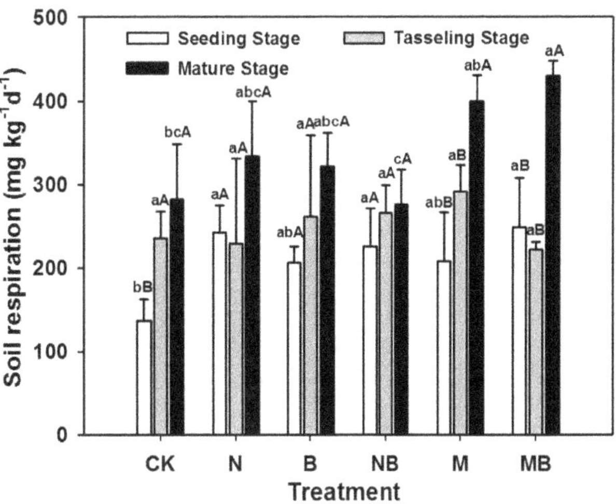

Figure 2. Soil respirations rate in various treatments during maize growth stages. CK: no fertilizer; N: N fertilizer; B: bacteria fertilizer; NB: N fertilizer + bacteria fertilizer; M: manure compost; MB: manure compost + bacteria fertilizer. Data are means ± SD (n = 3). Different lower case letters indicate significant differences (ANOVA, LSD test, $p<0.05$) among fertilizer treatments; different upper case letters indicate significant differences among maize growth stages.

amendments can significantly affect soil microorganisms [12]. Changes in microbial activity and composition, for instance, can influence plant growth by enhancing nutrient turnover and suppressing or mitigating disease incidence [13]. In addition, soil microbial biomass, activity, and community structure are useful indicators of soil quality and health because these parameters are sensitive to changes in cropland management practices [14]. Thus, the adaption of soil microbial structure and functions to the environment are considered essential for sustainable agricultural production [15]. Nevertheless, methods of increasing the soil microbiological diversity and/or restoring the microorganism communities when the soils have been seriously degraded after the long-term application of chemical fertilizers remain largely unknown.

Previous studies have focused on the influence of fertilizers on soil microbial communities, resulting in positive or neutral effects. Some studies have reported that soils in organic farming regimes had higher microbial functional diversity than those in conventional farming systems [16]. According to some investigators, the bacterial diversity was always higher in manure compost-amended farmyard soils regardless of land use patterns or seasons [17]. Although fertilization has resulted in increases in crop yield, this application was not sufficient in triggering a significant improvement in the soil microbial properties [8]. Even some argued [18] that there were no significant differences in bacterial communities between improved and unimproved grassland [18]. While numerous studies have investigated the effects of organic fertilization on microbial communities [19,20], the dynamic patterns of bacteria, fungi, and actinomycetes under different types of fertilization have remained elusive.

Though some studies have recorded the long-term effects of different fertilization practices on soil microbial properties, few attempts have been made to improve the degraded farmland by enhancing soil microbial properties, especially under conditions that do not include cropland resting. Understanding how soil microbial properties respond to manure compost may help to strengthen agricultural ecosystem health practices. In this study, we hypothesized that adding manure compost and bacterial fertilizers to the soils that have been experienced with chemical fertilizers for several decades may rapidly improve soil microbial diversity. Our objectives were: (i) to identify a more effective method in rapidly improving degraded cropland soil quality through fertilizer management; (ii) to analyze the bacterial and fungal community structure and their growth stage-related dynamics in degraded and restored soils.

Materials and Methods

Site description

The study was conducted at the Agricultural Ecosystem Research Station of Shandong Agricultural University in Jiang Jiazhuang Village, Pingyi County, Shandong Province of Eastern China (35°26′21″N, 117°50′11″E). The farm area is in a typical temperate and monsoonal climate, with a mean annual rainfall of 770.2 mm, and an average annual temperature of 13.2°C.

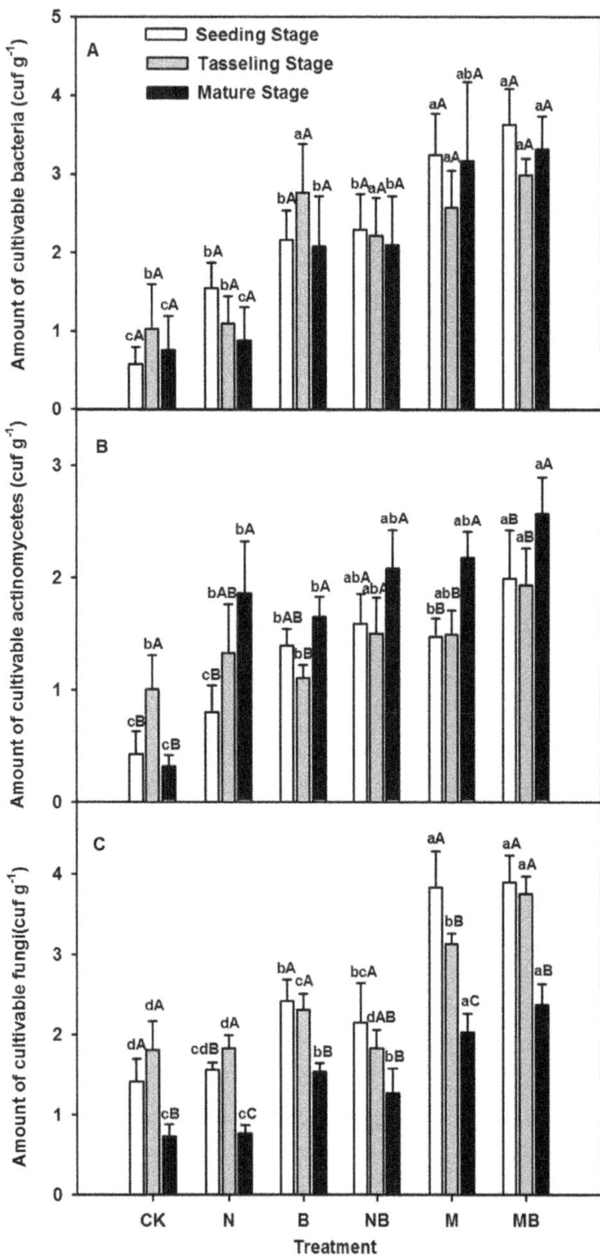

Figure 3. The clony-forming units of bacteria, fungi and actinomyces in various treatments during maize growth stages. CK: no fertilizer; N: N fertilizer; B: bacteria fertilizer; NB: N fertilizer + bacteria fertilizer; M: manure compost; MB: manure compost + bacteria fertilizer. Data are means ± SD (n = 3). Different lower case letters indicate significant differences (ANOVA, LSD test, $p<0.05$) among fertilizer treatments; different upper case letters indicate significant differences among maize growth stages.

Experimental design

Six treatments using different fertilizer applications were designed. Fertilizer treatments were established as follows: no fertilizer (CK), N fertilizer (N), N fertilizer + bacterial fertilizer (NB), manure compost (M), manure compost + bacterial fertilizer (MB), and bacterial fertilizer (B). The experimental pots (24 cm diameter, 45 cm height) were filled with soils collected from the 0–30 cm layer from a maize (*Zea mays* L.) cultivation area.

All the fertilizers were added as basal fertilization before planting maize. Seeds of maize were germinated on moist filter paper for 2 d, and seedlings were then planted in 90 pots (two plants per pot, 15 pots per treatment). The pots were located in a micro-region in the cropland at the study site. Urea (N = 46%) was chosen as the N treatment source. The Technology University of South China provided a bacterial fertilizer (HYSD001, Guangzhou Huayuan Biotechnology Company) that contained a variety of phosphorus-solubilizing and nitrogen-fixing bacteria as exogenous microorganisms. Cattle compost (M), which was fermented for three months (June to September) under high temperatures, was collected from the Hongyi Organic Farm in the village. According to the nitrogen fertilizer requirement (150 $kg{\cdot}hm^{-2}$) for maize growth, equal amounts of N in different treatments except for the B treatment were applied. Based on the 60% water content of cattle compost, the quantity of cattle compost (pH was 7.2; organic matter content was 440.43 $g{\cdot}kg^{-1}$; C/N was 26.7; nitrogen content was 1.6%) in the M and MB treatments was 3.75 $kg{\cdot}m^{-2}$, the quantity of urea (nitrogen content was 46%) in the N and NB treatments was 32.6 $g{\cdot}m^{-2}$, and the quantity of bacterial fertilizer in the B, MB, and NB treatments was 1.56 $g{\cdot}m^{-2}$. A conventional management scheme in China was conducted for pest and weed management during the experiment [21].

Sampling and processing

Soil samples were collected in triplicate at the maize seeding, tasseling, and mature stages from a 0–20 cm depth in each pot. Each soil sample was separated into two parts. One part was air-dried and stored at room temperature for determining soil chemical properties. The other part was passed through a 2-mm sieve, moistened to 60% of their water holding capacity, and immediately stored at 4°C for the measurement of soil microbial properties.

Soil microbial biomass and respiration

Microbial biomass C (MBC) and biomass N (MBN) contents were estimated by using chloroform fumigation extraction [22]. The soil samples were divided into two portions. One 15-g portion (in dry weight) rewetted to 60% was fumigated for 24 h at 25°C with chloroform (ethanol-free). After fumigant removal, the soils were extracted with 60 mL of 0.5 $mol{\cdot}L^{-1}$ K_2SO_4 and placed on a horizontal shaker at a speed of 300 $r{\cdot}min^{-1}$ for 30 min and filtered. The non-fumigated portion was extracted similarly as fumigation commenced. Total dissolved N and C in the K_2SO_4 extracts were determined by using an automated C analyzer (Shimadzu, TOC-VCPH, Japan). Regarding the incomplete extraction, two conversion factors were applied to calculate the biomass, as 0.54 for MBN (k_{EN}) and 0.45for MBC (k_{EC}) [23] in the following equation (1) and (2).

$$MBC = Ec/k_{EC} \tag{1}$$

$$MBN = E_N/k_{EN} \tag{2}$$

where E_C refers to the different organic carbon amount between fumigated and non-fumigated treatment, and E_N represents the difference in total nitrogen [24,25].

Basal soil respiration was determined by using the sealed jar incubation method, which employed a trap of 0.5 mol of NaOH alkali CO_2 [26]. At sampling, the jar lid was opened, the alkali trap was removed, and the solution was back-titrated with 0.5 mol

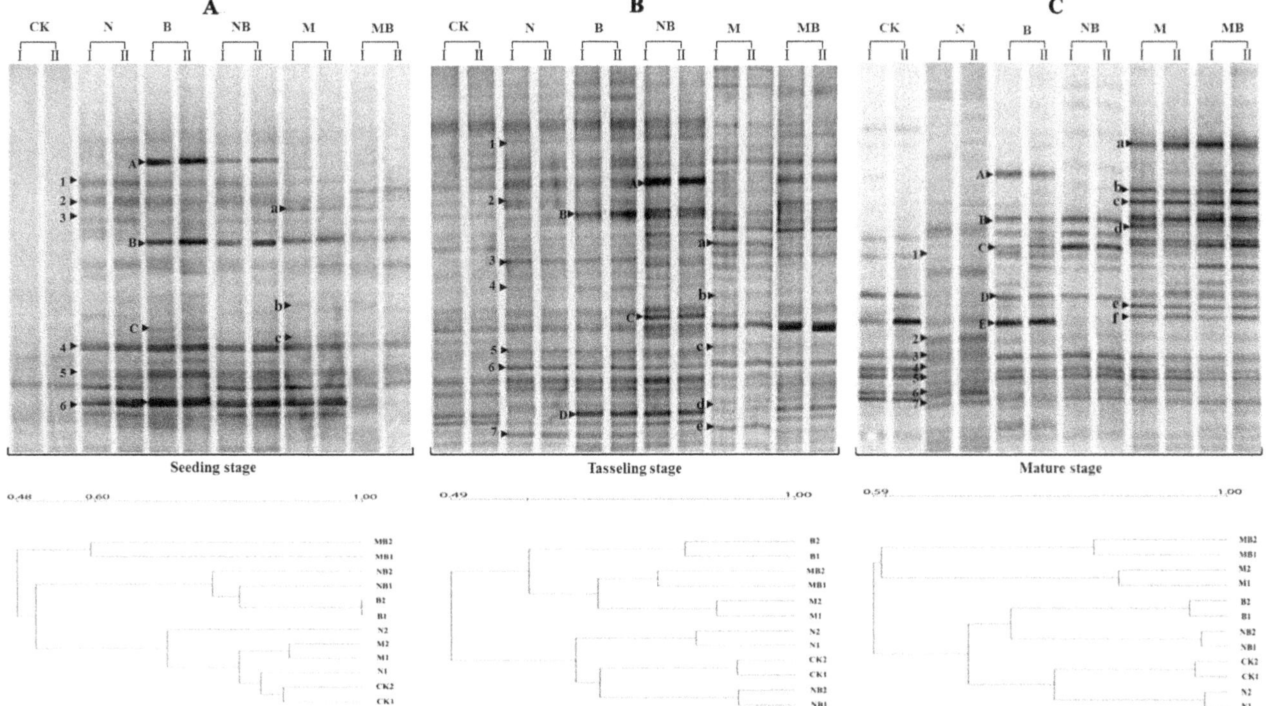

Figure 4. DGGE banding patterns of 16S bacteria fragment and clustering of DGGE profiles in various treatments during maize growth stages. CK: no fertilizer; N: N fertilizer; B: bacteria fertilizer; NB: N fertilizer + bacteria fertilizer; M: manure compost; MB: manure compost + bacteria fertilizer.

HCl in order to assess CO_2 release. The alkali trap was replaced during each measurement.

Plate counts of cultivable microorganisms

The total number of cultivable bacteria, fungi, and actinomyces were counted as colony forming units (CFUs) on agar plates using the dilution plate method. The media used for the enumeration of bacteria, fungi, and actinomyces were beef extract peptone medium, Czapek's medium, and Gause's No. 1 synthetic medium, respectively [27].

DNA extraction and DGGE

To determine soil microbial diversity, we used the molecular biology methods of DGGE. Briefly, soil DNA was extracted by using the Power Soil DNA Extraction Kit (Mobio Laboratories), following the manufacturer's instructions. For bacterial community analysis, the V3 region of 16S rRNA gene were amplified by touchdown polymerase chain reaction (PCR) using two different primer sets, 518R (5′-ATT ACC GCG GCT GCT GG) and GC-338F (5′-CGC CCG CCG CGC GCG GCG GGC GGG GCG GGG GCA CGG GGG GCC TAC GGG AGG CAG CAG), containing the GC clamp on the amplified 16S rDNA template. For fungal community analysis, the fragments of 18S rRNA gene were amplified by nested PCR using the primer sets GC-fung (5′-CGC CCG CCG CGC CCC GCG CCC GGC CCG CCG CCC CCG CCCCATTCCCCGTTACCCGTTG-3′) and NS1 (5′-GTAGTCATATGCTTGTCTC-3′).

PCR mixtures containing 25 μL of Premix Taq (Takara Biotechnologies), 1.5 μL of each primer, and 2 μL of the DNA template were made up to a volume of 50 μL with sterile Milli-Q water. The samples were amplified in a Peltier Thermal Cycler (PTC-200) (Bio-Rad Laboratories, Hercules, CA). The amplification conditions of bacteria and fungi samples were as follows: an initial denaturation of DNA for 5 min at 94°C, followed by 25 cycles of 30 s at 94°C, 30 s at 55°C, and 30 s at 72°C; 10 cycles of 30 s at 92°C, 30 s at 55°C, and 45 s at 72°C; and a final extension at 72°C for 10 min. Blank controls were used through all the steps.

DGGE was performed with a DCode universal mutation detection system (Bio-Rad Laboratories, Hercules, CA). Approximately 1 μg of the bacteria PCR product per lane were loaded onto 8% polyacrylamide (37.5:1 acrylamide: bisacrylamide) gels in a 1× TAE buffer with a bacteria denaturing gradient ranging from 40% to 60%, whereas that for fungi ranged from 15% to 40%. Gel electrophoresis was performed at 60°C for 13 h at 80 V. The gel was subsequently stained in a 1× TAE buffer containing 1:10000 dilution of SYBR Green I nucleic acid staining solution (GenScript, USA) for half an hour before being photographed on a Molecular Imager Gel Doc XR System (Bio-Rad Laboratories).

The soil microbial community diversity indexes were calculated in the following equation (3), (4) and (5):

$$\text{Shannon}-\text{Wiener index } (H): \; H = \sum_{i=1}^{S} Pi \ln Pi \qquad (3)$$

Where: Pi is the ratio of the activity on a particular substrate to the sum of activities on all substrates; S is the sum of band number in one sample.

Evenness index (E):

$$E = H / \ln S \qquad (4)$$

Table 1. Effect of different treatments on soil bacterial community structure diversity as evaluated by Shannon index (*H*), **Evenness** index (*E*) and Simpson index (*C*) in the DGGE-PCR.

Treatments	Shannon-Wiener (*H*)			Evenness index (*E*)			Simpson index (*C*)		
	Seeding	Tasseling	Mature	Seeding	Tasseling	Mature	Seeding	Tasseling	Mature
CK	2.16±0.25cB	2.62±0.01cA	2.53±0.06cA	0.96±0.01bA	0.98±0.01aA	0.95±0.01aA	0.07±0.03aA	0.05±0.01aA	0.06±0.01aA
N	2.84±0.03abA	2.69±0.04cA	2.40±0.05cB	0.96±0.01bA	0.98±0.01aA	0.98±0.01aA	0.05±0.01aA	0.05±0.01aA	0.05±0.01aA
B	2.78±0.02abA	2.82±0.01abA	2.76±0.21bA	0.97±0.01abA	0.98±0.01aA	0.97±0.01aA	0.06±0.01aA	0.05±0.01aA	0.06±0.01aA
NB	2.92±0.04aA	2.86±0.08abA	2.83±0.13bA	0.98±0.01abA	0.96±0.01aA	0.98±0.01aA	0.05±0.01aA	0.05±0.01aA	0.06±0.01aA
M	2.54±0.04bB	3.23±0.05aA	3.44±0.03aA	1.00±0.01aA	0.96±0.01aB	0.96±0.01aB	0.07±0.01aA	0.05±0.01aA	0.05±0.01aA
MB	2.60±0.12bB	3.31±0.06aA	3.41±0.11aA	0.96±0.01bA	0.97±0.01aA	0.96±0.01aA	0.09±0.01aA	0.06±0.01aA	0.05±0.01aA

Data analyzed by two-way ANOVA, LSD test, $p<0.05$. Different lower case letters indicate significant difference among fertilizer treatments; different upper case letters indicate significant difference among maize growth stages.

Where: H – Shannon index; S - band numbers of every substrate.

Simpson index (C):

$$C = \sum Pi^2 \quad (5)$$

Where: Pi is the ratio of the activity on a particular substrate to the sum of activities on all substrates.

Soil enzyme activity

Urease activity was determined by using the phenol-sodium hypochlorite colorimetric method [28], as indicated by mass (mg) of NH_3-N in 1 g soil incubated for 24 h (U). Catalase activity was analyzed by using the titration method [29], and its activity was indicated by depletion (mL) of $KMnO_4$ (0.1 mol L^{-1}) after 20 min of 1 g incubated soil. Invertase activity was determined by using the 3,5-dinitrosalicylate colorimetric method [24], which was indicated by the mass (mg) of glucose of 1 g soil after 24 h. Cellulase activity was measured by using the 3,5-dinitrosalicylate colorimetric method [29], which was indicated by the mass (mg) of decomposed generated glucose (mg) from cellulose of 1 g soil after 72 h.

Statistical analysis

All statistical analyses were carried out using the SPSS 16.0 software. A two-way ANOVA at the 0.05 level was conducted to determine the interactions between fertilizer treatments and maize stages. And one-way ANOVA was used for analysis of significant difference among fertilizer treatments and plant stages.

Results

Soil microbial biomass and soil respiration rate

Soil MBC and MBN over time in different treatments are presented in Fig. 1. For all sampling times, MBC significantly increased in the fertilization treatments against their controls (CK) ($p<0.01$) during the whole maize growth stage, with the largest increment of MBC occurring consistently in the MB treatment. Differences between the MB treatment and other treatments except for the M treatment reached significance at the tasseling stage ($p<0.01$). The ANOVA results showed that MBC in the M and MB treatments were significantly higher than those in the B ($p<0.05$), N ($p<0.01$) and NB treatments at the maize tasseling stage ($p<0.01$). However, no significant differences were noted between MB and M or between N and NB treatments (Fig. 1A). The fertilizers also affected soil MBN, which varied among different treatments. Compared with the CK treatment, all fertilizer treatments showed significantly higher MBN, and this was greater in the M and MB treatments than in the N and NB or B treatments ($p<0.01$). Nevertheless, no statistically significant differences ($p>0.05$) were observed among the B, N, and NB treatments.

Soils with different fertilizers showed higher respiration rates and greater cumulative CO_2 production compared to those observed in the CK treatment (Fig. 2). A significantly enhanced soil respiration rate was observed during the entire period. Compared with other fertilizer treatments, only M and MB treatments showed significant higher respiration rates at mature stage than seeding and tasseling stages. MB treatment exhibited drastically higher soil respiration rate than those of the NB and N treatments at the mature stage ($p<0.01$), whereas M treatment was

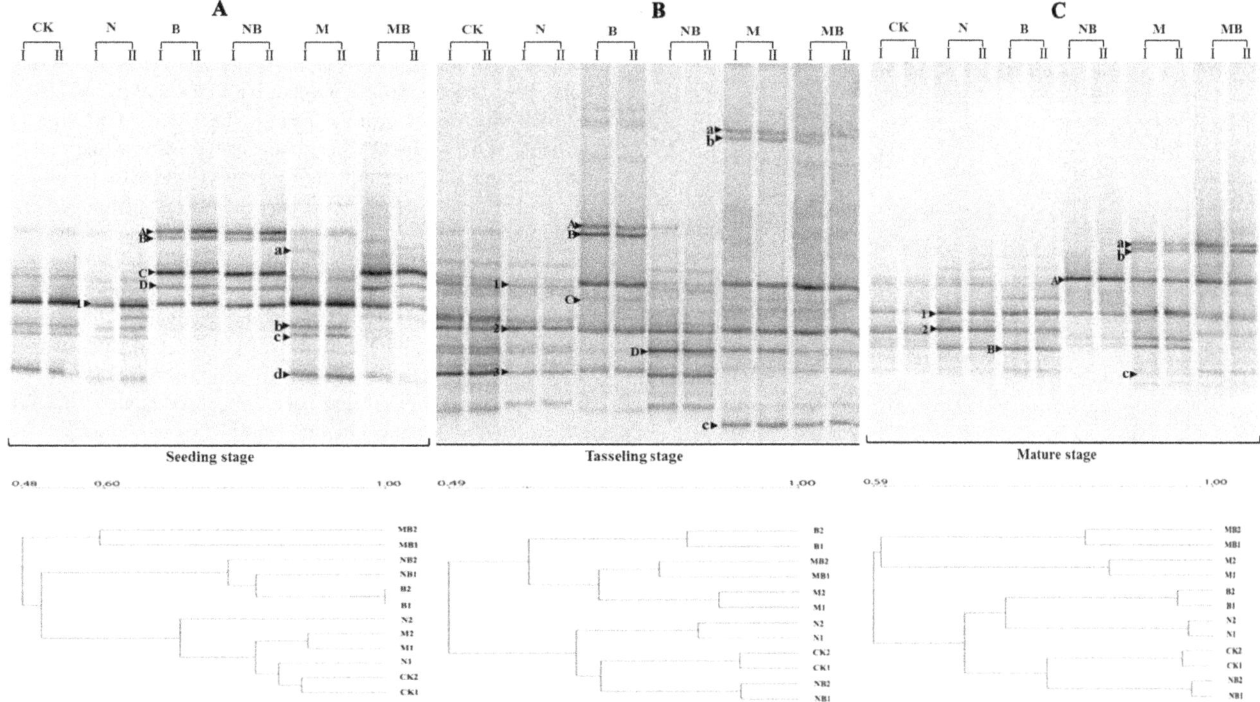

Figure 5. DGGE banding patterns of 18S rDNA fungi fragment and clustering of DGGE profiles in various treatments during maize growth stages. CK: no fertilizer; N: N fertilizer; B: bacteria fertilizer; NB: N fertilizer + bacteria fertilizer; M: manure compost; MB: manure compost + bacteria fertilizer.

only higher than NB treatment. However, soils with N and B treatments only showed higher respiration rates than those of the NB treatment, indicating that fertilizers could significantly enhance the microbial activities, and the highest number and greatest effects were observed from organic amendments.

Plate counts of cultivable microorganisms

There were extremely large variations in the number of bacteria, fungi, and actinomyces among various treatments (Fig. 3). The colony-forming units (CFUs) of bacteria in all fertilizer treatments (except N treatment) were significantly higher than those observed in the CK treatment ($p<0.01$) at the tasseling and mature stages (Fig. 3A). At the mature stage, the number of CFUs of bacteria in the M and MB treatments were significantly higher than that in the N treatment ($p<0.01$), with the bacterial numbers from the MB treatment considerably higher than those observed in the B and NB treatments ($p<0.05$) (Fig. 3A). However, the number of bacteria in the M treatment did not differ from that in the B and NB treatments ($p>0.05$). Fertilizers also largely influenced the amount of soil fungi. For instance, the amount of fungi in all fertilizer treatments was remarkably higher than that in the CK treatment ($p<0.01$) at the mature stage (Fig. 3B). At the same stage, the amount of fungi in the MB treatment was significantly higher than that in the B ($p<0.01$) and N ($p<0.05$) treatments, however this was not the case in the M and NB treatments. And the amount of fungi in all the fertilizer treatments (expect NB treatment) largely increased at mature stage. Similar to bacteria and fungi, the number of cultivable actinomyces was also affected. The amount of cultivable actinomyces in the M and MB treatments was considerably higher than that in the B and NB treatments, both were greatly higher than that observed in the CK and N treatments ($p<0.01$) (Fig. 3C). Meanwhile, the amount of cultivable actinomyces in the CK and N treatments initially increased, but eventually decreased at the end of growth. In the B, NB, M, and MB treatments, cultivable actinomyces decreased during the entire maize growth period.

Soil microbial community composition structure

Changes in the microbial community structure were determined by PCR-DGGE analysis, which targeted domains of bacteria and fungi. As shown in Fig. 4, different fertilizers significantly affected the community structure of soil bacteria and fungi at all sampling times. DGGE bands that could be enhanced by all fertilizers were marked with numbers. Bands enhanced by the N or NB treatments were marked with uppercase alphabet letters, whereas those enhanced by the M or MB treatments were marked with lower case alphabet letters (Fig. 4).

For bacterial community structure, all the fertilizers stimulated the generation of band numbers 1–6. N, B or NB treatments stimulated the occurrence of bands A–D, whereas M or MB enhanced the bands a–c at the seeding stage. The influences from N, B or NB treatment on bacterial community structure were lower than those from M or MB treatment (Fig. 4). The number of DGGE bands was the highest at the tasseling stage, indicating that the formed microbial community was the most diverse at this stage. We therefore confirmed that the largest influences of bacteria community structure occurred after the fertilizer treatments of M and MB. The results at the mature stage were similar to the tasseling stage, *e.g.*, the influences on bacterial community structure from the M or MB treatments were higher than that of the N or NB treatments (Fig. 4B and 4C). Furthermore, statistical analysis showed that manure compost and bacterial fertilizer could enhance the Shannon index, whereas the N and NB treatment decreased the Shannon index, especially from seeding stage to

Table 2. Effect of different treatments on soil fungi community structure diversity as evaluated by Shannon index (*H*), **Evenness** index (*E*) and Simpson index (*C*) in the DGGE-PCR.

Treatments	Shannon-Wiener (*H*)			Evenness index (*E*)			Simpson index (*C*)		
	Seeding	Tasseling	Mature	Seeding	Tasseling	Mature	Seeding	Tasseling	Mature
CK	1.86±0.18cA	2.08±0.01cA	1.04±0.01cB	0.96±0.01aA	0.96±0.01aA	0.95±0.01aA	0.12±0.03cB	0.09±0.01cB	0.18±0.01aA
N	2.03±0.11bA	1.81±0.12dA	0.69±0.01dB	0.93±0.01aA	0.96±0.01aA	0.98±0.01aA	0.15±0.01cA	0.17±0.02bA	0.19±0.01aA
B	1.83±0.16cB	2.55±0.04bA	1.06±0.01cC	0.95±0.02aA	0.98±0.01aA	0.97±0.01aA	0.24±0.02aB	0.29±0.01aA	0.17±0.01aC
NB	1.84±0.17cB	2.26±0.04cA	1.03±0.01cC	0.95±0.01aA	0.96±0.01aA	0.94±0.01aA	0.19±0.03bB	0.25±0.01aA	0.20±0.01aB
M	2.13±0.03abB	2.65±0.11abA	2.71±0.01aA	0.93±0.01aA	0.98±0.01aA	0.96±0.01aA	0.13±0.01cA	0.11±0.01cA	0.15±0.01bA
MB	2.20±0.02aB	2.79±0.08aA	2.55±0.20bA	0.96±0.01aA	0.98±0.02aA	0.96±0.034aA	0.12±0.01cA	0.08±0.005cB	0.14±0.06bA

Data analyzed by two-way ANOVA, LSD test, $p<0.05$. Different lower case letters indicate significant difference among fertilizer treatments; different upper case letters indicate significant difference among maize growth stages.

tasseling stage (Table 1). The Shannon indexes of the M or MB treatments were higher than those of the other treatments, which remained high during the entire experiment except seeding stage (Table 1). At tasseling stage the Shannon indexes of the N, NB, M and MB treatments significantly higher than CK and N treatments. The results at the mature stage were similar to the tasseling stage, the Shannon indexes of M and MB treatments were higher than other fertilizer treatments. Although the Shannon index in the N treatment was the lowest at tasseling and mature stage, a downtrend trend was observed. The evenness indexes (except seeding stage) and the Simpson indexes in all treatments were basically the same, and retained a smooth and steady trend during the entire growth period, and there was no significant difference among maize growth stages (Table 1).

Cluster analysis showed that bacterial community structures in treatments B and NB had a higher degree of similarity, while N, M and CK got closed and MB forms a separate category at the seeding stage (Fig. 4A). M and MB got together and B formed a separate category, whereas CK, NB and N are in another group at the tasseling stage (Fig. 4B). At the mature stage, M and MB had a higher degree of similarity, while other treatments classified to a similar category (Fig. 4C).

In terms of fungal community structure, the number of DGGE bands was highest at tasseling stage, and the largest influences on fungi community structure occurred with the M and MB treatments (Fig. 5). Statistical analysis showed that manure compost could enhance the Shannon index, whereas the N treatment decreased the Shannon index (Table 2). At seeding and tasseling stages, the Shannon index of the MB treatment appreciably higher than other treatments except M treatment, and the Shannon index of the M treatment was the highest at mature stage. However, the Shannon index of the N treatment rapidly decreased during the entire growth period, which was the lowest at tasseling and mature stages. The evenness index in all fertilizer treatments had no significant difference. The simpson index of M and MB treatments were notably lower than those of N, B and NB treatments during the entire experiment (Table 2). Cluster analysis of fungal community structure also showed that fungal community structure in treatments added bacterial fertilizer (B, NB and MB) were similar at the initial seeding stage, while other treatments (CK, N and M) formed another separate category (Fig. 5). M and MB become more and more similar with time, while CK, N and M treatments were in another group, and B forms a separate category at the tasseling and mature stages (Fig. 5).

The results of bacteria and fungi community structure analysis indicated that organic and chemical fertilizers could enhance the bacteria and fungi community by enhancing the microbial diversity at the tasseling and mature stages, while bacterial fertilizer significantly influenced the bacteria and fungi community in initial seeding stage (Table 2).

Soil enzyme activities

Urease activity in M and MB treatments were higher than that in the CK treatment ($p<0.01$), especially in the M treatment (Fig. 6). No significant differences were observed between the N and CK treatments at the seeding, tasseling stages and mature stages. However, urease activity in the N treatment rapidly decreased in the mature stage of maize (Fig. 6A). However, the catalase activities in the M and MB treatments were significantly higher than those observed in the CK treatment at tasseling and mature stages ($p<0.01$), and it was similar for B treatment at the mature stage. On the other hand, the activity of catalase in the N

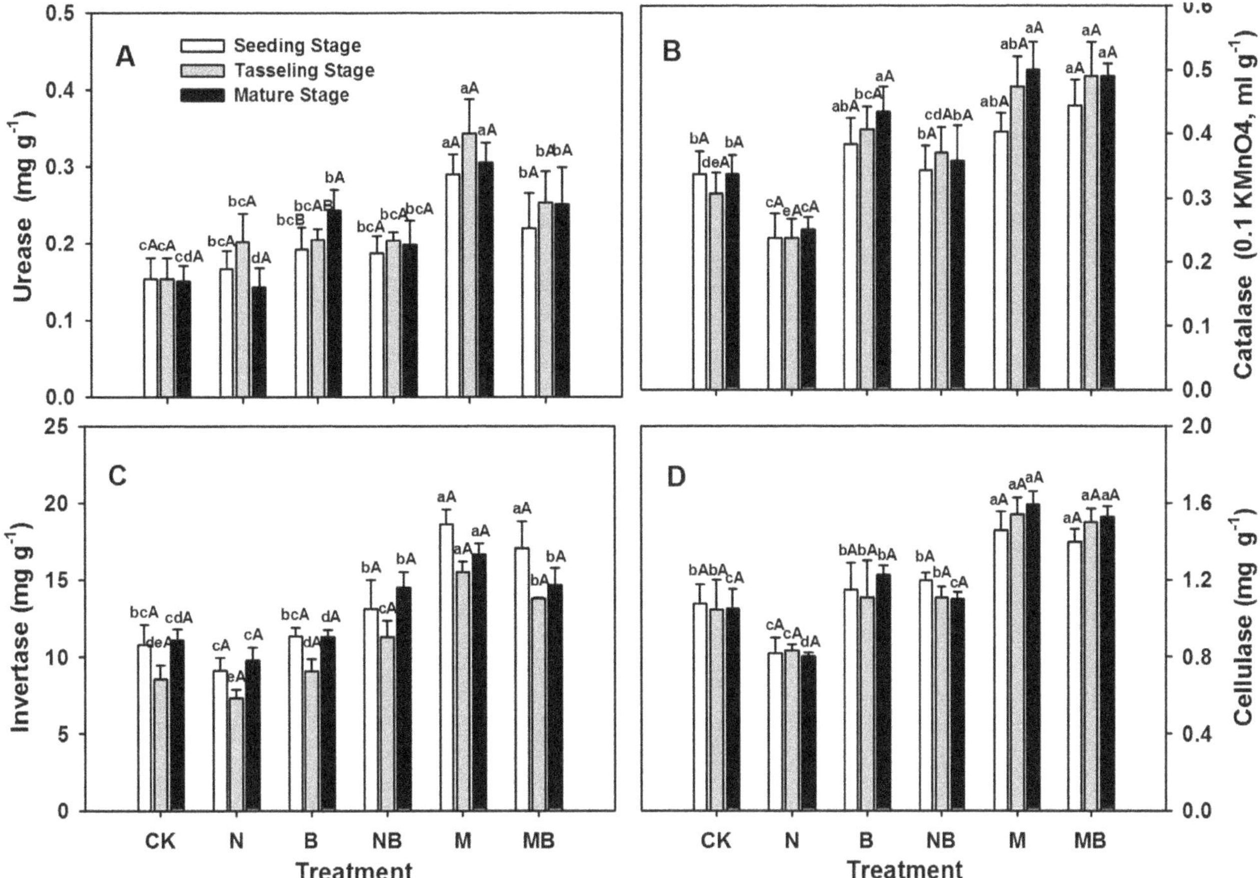

Figure 6. Soil enzyme activity in various treatments during maize growth stages. CK: no fertilizer; N: N fertilizer; B: bacteria fertilizer; NB: N fertilizer + bacteria fertilizer; M: manure compost; MB: manure compost + bacteria fertilizer. Data are means ± SD (n = 3). Different lower case letters indicate significant differences (ANOVA, LSD test, $p<0.05$) among fertilizer treatments; different upper case letters indicate significant differences among maize growth stages.

treatment was significantly lower than that in the CK treatment at all sampling times ($p<0.05$) (Fig. 6B).

Invertase activity in the M and MB treatments was appreciably higher than that in the CK treatment during the entire growth period ($p<0.01$), although no differences were observed between the B and CK treatments. N treatment showed the lowest invertase activity at all times, although this was not statistically different from that observed in the CK treatment (Fig. 6C). The M and MB treatments showed a significantly enhanced cellulase activity ($p<0.01$), whereas that in the N treatment decreased ($p<0.01$). No big differences in cellulase activity were observed among the NB, B, and CK treatments, indicating that the influence of bacterial fertilizer on cellulase activity was not significant (Fig. 6D). These results suggest that manure compost was better than bacteria fertilizer alone in enhancing soil enzyme activities, whereas chemical fertilizers degrade the activities of soil enzymes.

Soil physicochemical and chemical properties

The humus content in the M and MB treatments was remarkably higher than that in the B and CK treatments, whereas humus content in the N treatment was significantly lower than that in the CK treatment. Available N in the M and MB treatments were also considerably higher than that in the CK treatment, whereas there was no difference between the N and CK treatments. Fertilizers showed minimal impact on soil pH, total N, and available P during the entire maize growth period. Positive correlations were observed between soil organic matter (SOM) and microbial biomass (MBC and MBN), humus and microbial biomass (MBC and MBN), SOM and enzymes activities, humus and enzymes activities (urease, catalase, invertase, and cellulase) (Table 3), whereas the correlations among available N, microbial biomass, and enzymes activities were not significant. The addition of fertilizers enriched the soil microbial biomass and soil enzymes by enhancing the soil physicochemical properties of SOM and humus, especially through the addition of manure compost.

PCA combined with a hierarchical clustering analysis

Kaise-Meyer-Olkin Measure showed the KOM value of three phases is 0.798>0.7, and the F value is 0 with Bartlett's Test of Sphericity test. All data obey the normal distribution, and are suitable for principal component analysis (Fig.7).

At seeding stage, three factors were extracted as the main composition factor, and the cumulative variance contribution rate was 93.49 (Fig.7A). The first principal component contribution rate was the largest, accounting for 59.94, and the main relevant parameters were soil microbial indicators (bacteria, fungi, actinomycetes, MBN, MBC), enzyme activities (invertase urease and cellulase) and humus. The main relevant parameters of second and third principal component were SOM and Available N, accounting for 20.06 and 13.49, respectively. According to the selection

Table 3. Correlation of MBC, MBN, Available N, SOM, Humus and enzymes activities.

Variables	MBC	MBN	Available N	SOM	Humus	Urease	Catalase	Invertase
MBN	0.910**							
Available N	0.425	0.518*						
SOM	0.777**	0.820**	0.380					
Humus	0.685**	0.750**	0.621**	0.622**				
Urease	0.721**	0.694**	0.232	0.558*	0.826**			
Catalase	0.585*	0.620**	0.410	0.646**	0.871**	0.835**		
Invertase	0.582*	0.687**	0.238	0.580*	0.776**	0.732**	0.703**	
Cellulase	0.615**	0.663**	0.528*	0.614**	0.948**	0.838**	0.928**	0.790**

*, Pearson correlation is significant at the 0.05 level.
**, Pearson correlation is significant at the 0.01 level.

indexes of principal component, the ahierarchical clustering analysis was used and the result showed that treatments of N, B and NB form a separate category, while M and MB were in another group, and CK formed a separate category (Fig.7A). At tasseling stage, the cumulative variance contribution rate of three principal components factors was 90.12. The first principal component contribution rate was 67.68, with the main relevant parameters of soil microbial indicators (actinomycetes,Shannon, bacteri, MBN and MBC), enzyme activities (Cellulas, Catalase and Invertase) and humus (Fig.7B). The main relevant parameters of second and third principal component were SOM and simpson, accounting for 12.39 and 10.05, respectively. Cluster analysis showed that N, B, NB and CK had a higher degree of similarity, whereas M and MB formed a separate category, respectively (Fig.7B). At mature stage, the cumulative variance contribution rate was 94.45. The variance contribution of the first principal components was 64.154, and the mainly relevant parameters were soil microbial indicators (bacteria, actinomycetes, shannon and MBN), enzyme activities (cellulas and catalase) and humus. The variance contribution of the second principal components was SOM with 20.21. The variance contribution of the third principal components was simpson index, accounting for 10.093. Hierarchical clustering analysis showed that M and MB had a higher degree of similarity and treatments of B and NB forms a category, while CK and N are in another group (Fig.7C).

Discussion

Soil microbial community structure

In cropland soils, the application of fertilizers imparted a stronger effect on microbial communities, as shown by previous studies involving the analysis of PLFA, microbial biomass, and rRNA gene libraries [20]. DGGE profiles revealed that the band numbers in all fertilizer treatments significantly increased at various maize growth stages, with band numbers in M and MB treatments higher than those observed in the other fertilizer treatments (B, NB, and N) (Figs. 4 and 5). Our results suggest that cattle compost changes the soil bacteria and fungi community structure. Previous investigations have also demonstrated that animal compost increased bacterial and fungi diversity by increasing the carbon pool of the soil, thus improving the living conditions for indigenous microbial populations [30,31]. Our study showed that soil organic matter and humus in the M treatment were the highest. The DGGE profiles (Figs. 4C and Figs. 5B) and Shannon index data showed that the bacterial and fungi diversity of the M and MB treatments remained stable, whereas these declined in the N, B, and NB treatments at maize mature stage.

The M treatment resulted in an increase in exogenous microorganisms, which can directly influence the soil microorganism community structure [32]. The specific bands in the DGGE profiles showed that exogenous microorganisms emerged in the M and MB treatments, thus clearly supporting this statement (Figs. 4 and 5). The M treatment greatly improved microbial diversity, indicating that M is conducive to the establishment of a diverse microbial community structure (Tables 1 and 2).

The application of bacterial fertilizer provided a variety of nitrogen-fixing bacteria and phosphorus-solubilizing bacteria that could improve the soil microbial community structure (Figs. 4 and 5). New fungal species, which may act as indicators of exogenous microorganism, were thus introduced (Fig. 5). The Simpson index of soil fungi community in the MB treatment was lowest during the entire stages, and B and NB treatments was the highest

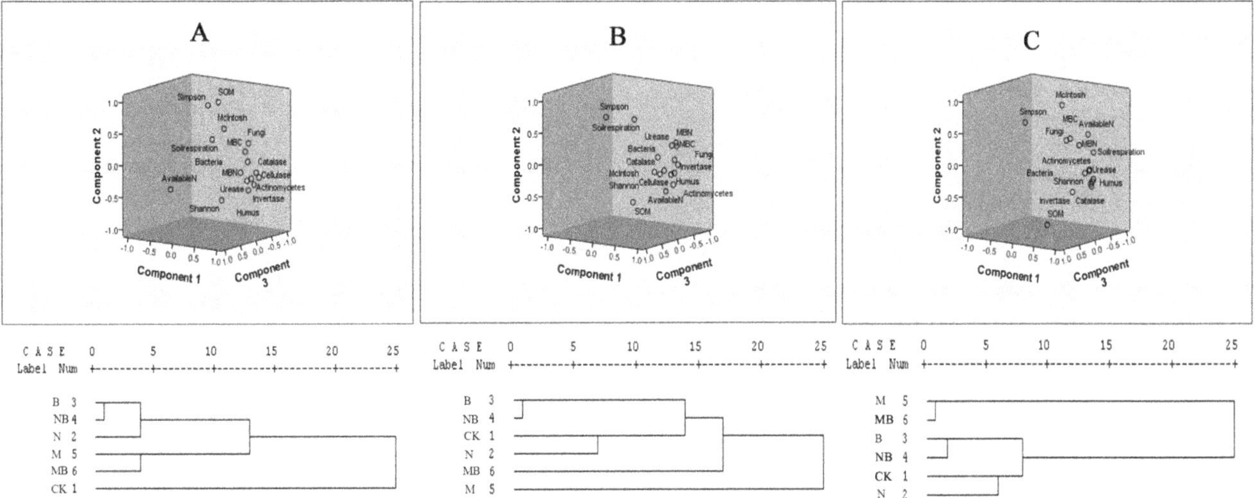

Figure 7. PCA of indexes combined with a hierarchical clustering analysis in various treatments during maize growth stages. CK: no fertilizer; N: N fertilizer; B: bacteria fertilizer; NB: N fertilizer + bacteria fertilizer; M: manure compost; MB: manure compost + bacteria fertilizer. A: Seeding stage; B: Tasseling stage; C: Mature stage.

respectively at seeding stage and last two stages (Table 2), suggesting that exogenous microorganisms played a dominant role in both stages, which combined with compost enhanced soil microbial community uniformity [33],while N fertilizer always displayed less or negative influences compared with the other treatments. For example, the Shannon index of soil bacterial and fungal communities showed that the soil microbial community diversity in the N and NB treatment gradually decreased compared to that in the B treatment, displaying an obvious negative influence from the administration of nitrogen fertilizers (Tables 1 and 2). This may be attributable to the absence of an external carbon source [34]. Nevertheless, only adding N fertilizers may inhibit the soil microorganism propagation rather than increase it, which may further degrade the bacteria and fungi community diversity [35].

Soil microbial biomass and soil activity

Soil microbial biomass is an important source of plant nutrients and is highly correlated with SOC (soil organic carbon) [34]. Enhancement of soil microbial activity was associated with high soil available N for plants [36]. The M and MB treatments significantly increased soil MBC, MBN, and respiration rates, which positively influenced microbial processes and development. These findings also showed that M significantly enhanced the density of soil bacteria, fungi, and actinomycetes. These indexes were significantly positively correlated with humus ($p<0.05$) (Table 3), indicating that the M treatment served as a sufficient carbon source that enhanced the microorganism biomass and activity.

Microbial carbon and nitrogen initially increased and then subsequently decreased during the maize growth period, achieving the highest value at the tasseling stage. This finding might be attributable to the competition of carbon and nitrogen source between soil microorganisms and crops. With the transformation of soil nutrients, the competitiveness of crops is enhanced, thus leading to a decrease in microbial carbon and nitrogen [37].

Effects on soil enzymes

Soil microbial enzymes are mainly driven by metabolic processes, largely reflecting the level of soil microbial activity and the intensity of biochemical reactions [38]. Our investigation demonstrated that different fertilizers have significantly affected soil enzyme activities ($p<0.01$) (Fig. 6). Four kinds of soil enzyme activities consistently displayed the highest levels in the M or MB treatments (Fig. 6). The M treatment can accelerate soil carbon and nitrogen circulation [14] and improve soil quality, with enzyme activities stemming from manure compost. As manure compost contains abundant organic matter, it can drastically increase SOM by providing a rich source of carbon and nutrients for enzyme production microorganisms.

However, the application of N fertilizer had less influence on soil urease activity, whereas it significantly decreased soil catalase, invertase, and cellulase activities ($p<0.01$) (Fig. 6). It is thus possible that the application of chemical fertilizers has inhibited enzyme production in microorganisms. Except for urease activity, the B and NB treatments showed no influence on the activities of the other three enzymes (Fig. 6), suggesting that NB can enhance the soil nitrogen transformation.

Improvement of soil fertility

The N treatment notably reduced microbial biomass and enzyme activities compared to those with the other treatments. This finding could be attributable to the acidifying effect of long-term N fertilizers, as showed by lower pH in previous studies [39]. However, no significant differences in soil pH were observed between most fertilizers and the CK treatment, which maintained a neutral pH during the maize growth period. It is possible that the acidifying effect of the N fertilizer may have not been reflected within a relatively short period. SOC is a pivotal component of soil fertility, and the level of organic matter is influenced by the application of fertilization [40]. In a healthy soil, the level of SOC is considered to be a functional indicator of the net input of organic residues [41]. In our study, manure compost enhanced the accumulation of SOC, which is consistent with the findings of several other previous studies from many countries [42,43]. The improved SOC may be due to a higher humectation rate and a constantly lower decay rate, since SOC is a nutrition pool that maintains microbial activity. SOC, especially humus, was significantly positively correlated with soil microbial biomass and soil enzyme activity (Table 3). However, there was no obvious change in the total N level under manure compost, which may partially be due to a slow release of N from manure compost and smaller losses of N as described by Bhandari et al. [42].

The results of PCA combined with a hierarchical clustering analysis showed that fertilizer treatments had been divided into three groups at the end stage (Fig.7). The application of manure compost treatments (M and MB) had a higher degree of similarity; the application of bacterial fertilizer bacterial fertilizer treatments (B and NB) formed another category, while treatments of CK and N are in another group. Compared to the application of bacterial fertilizer treatments, the distance of application of manure compost treatments was far away with the group of CK and N. These result exactly demonstrated that the application of manure compost has more influences on soil microbial community diversity and soil microenvironments than bacterial fertilizer.

Conclusions

Manure compost, especially manure compost + bacterial fertilizer, consistently resulted in higher levels of soil respiration rate, cultivable microorganisms, and soil enzyme activities, while N fertilizers showed no significant influence or negative results. The number of DGGE bands with bacteria and fungi also indicated that the Shannon index of manure compost treatment was the highest and remained high. Fertilizers, especially manure compost, significantly enhanced soil microbial properties in response to the increase in soil physicochemical properties of soil organic matter and humus. From a soil microbial point of views, manure compost application can be used as an environmentally friendly and rapid measure for restoring degraded cropland.

Author Contributions

Conceived and designed the experiments: GJ ZZ NW. Performed the experiments: ZZ HL. Analyzed the data: ZZ GJ HL NW. Contributed reagents/materials/analysis tools: LG JM ND GW. Wrote the paper: ZZ GJ.

References

1. Tilman D (1998) The greening of the green revolution. Nature 396: 211–212.
2. Leita L, De Nobili M, Mondini C, Muhlbachova G, Marchiol L, et al. (1999) Influence of inorganic and organic fertilization on soil microbial biomass, metabolic quotient and heavy metal bioavailability. Biol Fert Soils 28: 371–376.
3. Kaur T, Brar B, Dhillon N (2008) Soil organic matter dynamics as affected by long-term use of organic and inorganic fertilizers under maize–wheat cropping system. Nutr Cycl Agroecosys 81: 59–69.
4. Chaudhry AN, Jilani G, Khan MA, Iqbal T (2009) Improved processing of poultry litter to reduce nitrate leaching and enhance its fertilizer quality. Asian J Chem 21: 4997–5003.
5. Ahmad R, Jilani G, Arshad M, Zahir ZA, Khalid A (2007) Bio-conversion of organic wastes for their recycling in agriculture: an overview of perspectives and prospects. Ann Microbiol 57: 471–479.
6. Leite LFC, Oliveira FC, Araújo ASF, Galvão SRS, Lemos JO,et al. (2010) Soil organic carbon and biological indicators in an Acrisol under tillage systems and organic management in north-eastern Brazil. Soil Res 48: 258–265.
7. Francis I, Holsters M, Vereecke D (2010) The Gram-positive side of plant–microbe interactions. Environ Microbiol 12: 1–12.
8. Wang W, Niu J, Zhou X, Wang Y (2011) Long-term change in land management from subtropical wetland to paddy field shifts soil microbial community structure as determined by PLFA and T-RFLP. Pol J Eco 59: 37–44.
9. Stark CH, Condron LM, O'Callaghan M, Stewart A, Di HJ (2008) Differences in soil enzyme activities, microbial community structure and short-term nitrogen mineralisation resulting from farm management history and organic matter amendments. Soil Biol Biochem 40: 1352–1363.
10. Hsu SF, Buckley DH (2008) Evidence for the functional significance of diazotroph community structure in soil. ISME J 3: 124–136.
11. Philippot L, Andert J, Jones CM, Bru D, Hallin S (2011) Importance of denitrifiers lacking the genes encoding the nitrous oxide reductase for N_2O emissions from soil. Global Change Biol 17: 1497–1504.
12. Hu J, Lin X, Wang J, Dai J, Chen R, et al. (2011) Microbial functional diversity, metabolic quotient, and invertase activity of a sandy loam soil as affected by long-term application of organic amendment and mineral fertilizer. J Soil Sediment 11: 271–280.
13. Zhang QC, Shamsi IH, Xu DT, Wang GH, Lin XY, et al. (2012) Chemical fertilizer and organic manure inputs in soil exhibit a vice versa pattern of microbial community structure. Appl Soil Ecol 57: 1–8.
14. Bending GD, Turner MK, Jones JE (2002) Interactions between crop residue and soil organic matter quality and the functional diversity of soil microbial communities. Soil Biol Biochem 34: 1073–1082.
15. Wardle DA, Yeates GW, Nicholson KS, Bonner KI, Watson RN (1999) Response of soil microbial biomass dynamics, activity and plant litter decomposition to agricultural intensification over a seven-year period. Soil Biol Biochem 31: 1707–1720.
16. Mäder P, Fliessbach A, Dubois D, Gunstl L, Fried P, et al. (2002) Soil fertility and biodiversity in organic farming. Science 296: 1694–1697.
17. Ge Y, Zhang J-b, Zhang L-m, Yang M, He J-z (2008) Long-term fertilization regimes affect bacterial community structure and diversity of an agricultural soil in northern China. J Soil Sediment 8: 43–50.
18. McCaig AE, Grayston SJ, Prosser JI, Glover LA (2001) Impact of cultivation on characterisation of species composition of soil bacterial communities. FEMS Microbiol Ecolo 35: 37–48.
19. Dinesh R, Srinivasan V, Hamza S, Manjusha A (2010) Short-term incorporation of organic manures and biofertilizers influences biochemical and microbial characteristics of soils under an annual crop. Bioresource technol 101: 4697–4702.
20. Qiu M, Zhang R, Xue C, Zhang S, Li S, et al. (2012) Application of bio-organic fertilizer can control Fusarium wilt of cucumber plants by regulating microbial community of rhizosphere soil. Biolo Fert Soil 48: 807–816.
21. Liu E, Yan C, Mei X, He W, Bing SH, et al. (2010) Long-term effect of chemical fertilizer, straw, and manure on soil chemical and biological properties in northwest China. Geoderma 158: 173–180.
22. Brookes P, Landman A, Pruden G, Jenkinson DS (1985) Chloroform fumigation and the release of soil nitrogen: a rapid direct extraction method to measure microbial biomass nitrogen in soil. Soil Biol Biochem 17: 837–842.
23. Jenkinson DS, Brookes PC, Powlson DS (2004) Measuring soil microbial biomass. Soil Biol Biochem 36: 5–7.
24. Wu J, Joergensen R, Pommerening B, Chaussod R, Brookes PC (1990) Measurement of soil microbial biomass C by fumigation-extraction—an automated procedure. Soil Biol Biochem 22: 1167–1169.
25. Joergensen RG, Mueller T (1996) The fumigation-extraction method to estimate soil microbial biomass: Calibration of the value. Soil Biol Biochem 28: 33–37.
26. Anderson JP (1982) Soil respiration. Methods of soil analysis Part 2 Chemical and microbiological properties: 831–871.
27. Xu G, Zheng H (1986) Handbook of analysis methods of soil microbiology. Agricultural, Beijing, China.
28. Paul EA (2006) Soil microbiology, ecology and biochemistry: Academic press.
29. Guan S, Zhang D, Zhang Z (1986) Soil enzyme and its research methods. Agricultural, Beijing: 274–297.
30. Wu T, Chellemi DO, Graham JH, Martin KJ, Rosskopf EN (2008) Comparison of soil bacterial communities under diverse agricultural land management and crop production practices. Microb Ecol 55: 293–310.
31. Helgason B, Walley F, Germida J (2010) No-till soil management increases microbial biomass and alters community profiles in soil aggregates. Appl Soil Ecol 46: 390–397.
32. Li X, Rui J, Mao Y, Yannarell A, Mackie R (2014) Dynamics of the bacterial community structure in the rhizosphere of a maize cultivar. Soil Biol Biochem 68: 392–401.
33. Hedlund K (2002) Soil microbial community structure in relation to vegetation management on former agricultural land. Soil Biol Biochem 34: 1299–1307.
34. Dijkstra FA, Bader NE, Johnson DW, Cheng W (2009) Does accelerated soil organic matter decomposition in the presence of plants increase plant N availability? Soil Biol Biochem 41: 1080–1087.
35. Pengthamkeerati P, Motavalli P, Kremer R (2011) Soil microbial activity and functional diversity changed by compaction, poultry litter and cropping in a claypan soil. Appl Soil Ecol 48: 71–80.
36. Tu C, Ristaino JB, Hu S (2006) Soil microbial biomass and activity in organic tomato farming systems: Effects of organic inputs and straw mulching. Soil Biol Biochem 38: 247–255.
37. Inselsbacher E, Hinko-Najera Umana N, Stange FC, Gorfer M, Schüller E, et al. (2010) Short-term competition between crop plants and soil microbes for inorganic N fertilizer. Soil Biol Biochem 42: 360–372.
38. Mandal A, Patra AK, Singh D, Swarup A, Ebhin Masto R (2007) Effect of long-term application of manure and fertilizer on biological and biochemical activities in soil during crop development stages. Bioresource Technol 98: 3585–3592.
39. Saha S, Prakash V, Kundu S, Kumar N, Mina BL (2008) Soil enzymatic activity as affected by long term application of farm yard manure and mineral fertilizer under a rainfed soybean–wheat system in NW Himalaya. Eur J Soil Biol 44: 309–315.

40. Drenovsky R, Vo D, Graham K, Scow KM (2004) Soil water content and organic carbon availability are major determinants of soil microbial community composition. Microb Ecol 48: 424–430.
41. Gregorich E, Liang B, Ellert B, Drury CF (1996) Fertilization effects on soil organic matter turnover and corn residue C storage. Soil Sci Soc Am J 60: 472–476.
42. Bhandari A, Ladha J, Pathak H, Padre H, Dawe D, et al. (2002) Yield and soil nutrient changes in a long-term rice-wheat rotation in India. Soil Sci Soc Am J 66: 162–170.
43. Hao X, Liu S, Wu J, Hu R, Tong C, et al. (2008) Effect of long-term application of inorganic fertilizer and organic amendments on soil organic matter and microbial biomass in three subtropical paddy soils. Nutr Cycl Agroecosys81: 17–24.

5

Carbon Sequestration Efficiency of Organic Amendments in a Long-Term Experiment on a Vertisol in Huang-Huai-Hai Plain, China

Keke Hua, Daozhong Wang*, Xisheng Guo*, Zibin Guo

Soil and Fertilizer Research Institute, Anhui Academy of Agricultural Sciences, Hefei, China

Abstract

Soil organic carbon (SOC) sequestration is important for improving soil fertility of cropland and for the mitigation of greenhouse gas emissions to the atmosphere. The efficiency of SOC sequestration depends on the quantity and quality of the organic matter, soil type, and climate. Little is known about the SOC sequestration efficiency of organic amendments in Vertisols. Thus, we conducted the research based on 29 years (1982–2011) of long-term fertilization experiment with a no fertilizer control and five fertilization regimes: CK (control, no fertilizer), NPK (mineral NPK fertilizers alone), NPK+1/2W (mineral NPK fertilizers combined with half the amount of wheat straw), NPK+W (mineral NPK fertilizers combined with full the amount of wheat straw), NPK+PM (mineral NPK fertilizers combined with pig manure) and NPK+CM (mineral NPK fertilizers combined cattle manure). Total mean annual C inputs were 0.45, 1.55, 2.66, 3.71, 4.68 and 6.56 ton/ha/yr for CK, NPK, NPKW1/2, NPKW, NPKPM and NPKCM, respectively. Mean SOC sequestration rate was 0.20 ton/ha/yr in the NPK treatment, and 0.39, 0.50, 0.51 and 0.97 ton/ha/yr in the NPKW1/2, NPKW, NPKPM, and NPKCM treatments, respectively. A linear relationship was observed between annual C input and SOC sequestration rate (SOCsequestration rate = 0.16 Cinput −0.10, R = 0.95, P<0.01), suggesting a C sequestration efficiency of 16%. The Vertisol required an annual C input of 0.63 ton/ha/yr to maintain the initial SOC level. Moreover, the C sequestration efficiencies of wheat straw, pig manure and cattle manure were 17%, 11% and 17%, respectively. The results indicate that the Vertisol has a large potential to sequester SOC with a high efficiency, and applying cattle manure or wheat straw is a recommendable SOC sequestration practice in Vertisols.

Editor: Xiujun Wang, University of Maryland, United States of America

Funding: This work was supported by the Natural Science Foundation of China (Grant No. 41401331), Special Fund for Agro-scientific Research in the Public Interest of China (Grant No. 201203030), the President Distinguished Youth Innovation Fund of Anhui Academic of Agricultural Science (Grants No. 14B1008 and 14B1007), and by the Anhui Province Science and Technology Program (Grant No. 1206c0805033). The funders had no role in study design, data collection and analysis, decision to publish, or preparation of the manuscript.

Competing Interests: The authors have declared that no competing interests exist.

* Email: wdzhong-3@163.com (DZW); 1078681598@qq.com (XSG)

Introduction

Soil organic carbon (SOC) sequestration contributes to the mitigation of greenhouse gas emissions and to the improvement of soil fertility [1]. Net SOC sequestration is the balance of organic C inputs into the soil (via crop residues, organic amendments in compost, animal manure, etc.) and organic C decomposition by soil microbes. SOC sequestration efficiency is commonly expressed by the relationship between annual C input and SOC accumulation rate, which is an indicator of soil C sequestration ability [2]. Therefore, information about the C sequestration efficiency is useful for seeking high efficiency management strategies of enhancing the SOC stock and soil fertility.

C sequestration efficiency is regulated by climate, the quantity, quality of added organic materials, and soil inherent properties [3–5]. Within a climate zone, soil inherent properties (i.e. initial SOC content, soil texture, clay type and aggregates) and cultivation practices have effects on SOC sequestration efficiency. A negative linear relationship has been reported between C sequestration efficiency and initial SOC content [6,7], mainly because SOC tends to increase faster if initial SOC content is far from its saturation level. Feller and Beare [8] reported a positive relationship between SOC sequestration and soil clays. Because clay type (i.e. 2:1 smectite clay minerals versus 1:1 allophanic clay minerals) has influence on the stabilization of SOC [9]. Elliott and Coleman [10] indicated that soil aggregates may physically protect SOC against decomposition by soil microbes. Recently, several studies have examined the SOC sequestration efficiency of different soil types based on long-term fertilization experiment [11–15]. However, no information is available about SOC sequestration efficiency and C sequestration efficiencies of organic amendments in Vertisols [16], which occupy approximately 4.0 million km^2 in Huang-Huai-Hai Plain of China. Vertisols are characterized by a low SOC content (less than 0.6% in the topsoil) a high clay content (more than 35%) and shrinking and swelling properties which contribute to self-mulching [17].

The purpose of this study was to evaluate the SOC sequestration efficiency under long-term fertilization practices in Vertisols, with a winter wheat–soybean double cropping system. Specific objectives were (1) to assess C input and SOC dynamics under various long-term fertilization practices (2) to estimate the SOC sequestration efficiency of different organic amendments.

Materials and Methods

Ethics Statement

The administration of the department of agricultural of Anhui Province gave permission for this research at the study site. We confirm that the field studies did not involve endangered or protected species.

Site description

The experiment is located at the Madian Agro-Ecological Station in Huang-Huai-Hai Plain, Eastern China (N33°13′, E116°37′). The areas has a sub-humid climate, with annual mean, maximum and minimum air temperature of 16.5°C, 36.5°C and −7.4°C, respectively (Fig. 1a). Annual precipitation ranged from about 457 to 1478 mm during the last 24 years (Fig. 1b), about 70% of which occurs from May to September. There are strong inter-annual and seasonal variations in precipitation. For example, a 90-day period of summer drought with a total precipitation of only 138 mm occurred in 1988. Precipitation in the soybean period ranged from 244 mm to 1049 mm with an average of 616 mm, which was much higher than that in the wheat period (259 mm).

The predominant Vertisols have developed in fluvial and lacustrine deposits. They are classified as Calcic Kastanozems, according to the soil classification system of the Food and Agriculture Organization (FAO). Soil pH ranges from 6.0 to 8.6 and SOC content from 5.8 to 7.5 g/kg in the area. The initial (1982) topsoil (0–20 cm) had a total SOC of 5.8±0.08 g/kg, total N 0.96±0.04 g/kg, total P 0.28±0.02 g/kg, bulk density 1.45 g/cm^3 and pH (1:2.5 w/v) 7.4, respectively. The topsoil under NPK treatment in 2013 had coarse sand (2 to 0.2 mm) 8.5±2.0 g/kg, fine sand (0.2 to 0.02 mm) 322.9±21.7 g/kg, silt (0.02 to 0.002 mm) 262.3±24.0 g/kg and clay (<0.002 mm) 406.3±24.0 g/kg, respectively.

Experimental design

The long-term field experiment was initiated in 1982. The experiment had six treatments (Table 1): no fertilizer (CK), mineral nitrogen-phosphorus-potassium fertilizers alone (NPK), mineral NPK fertilizers combined with 2.5 ton/ha/yr (dry base) of wheat straw (NPKW1/2), mineral NPK fertilizers combined with 5.0 ton/ha/yr of wheat straw (NPKW), mineral NPK fertilizers combined with 7.8 ton/ha/yr of pig manure (NPKPM), and mineral NPK fertilizers combined with 12.5 ton/ha/yr of cattle manure (NPKCM). The treatments were laid out in a randomized block design with four replications. The net plot size was 70 m^2 (14.9 m ×4.7 m). Each plot was isolated by 50 cm deep plates. Mineral N, P and K fertilization was urea, calcium superphosphate and potassium chloride, respectively. The amounts applied were similar to the amounts applied by local farmers, i.e., 180 kg N, 90 kg P_2O_5, and 135 kg K_2O/ha/yr. All fertilizers were applied as base fertilizer at the start of the wheat growing season. No fertilizers were applied to soybean growing season because of the high background value of soil N, P and K, which could supply sufficient available nutrients for the growth of soybean. Moreover, extra nitrogen was also obtained by nitrogen fixation by soybean itself. The mean chemical characteristics of pig manure were 360 g C/kg (dry base), 17 g N/kg and 8.9 g P/kg, and those of cattle manure were 370 g C/kg, 7.9 g N/kg and 4.2 g P/kg.

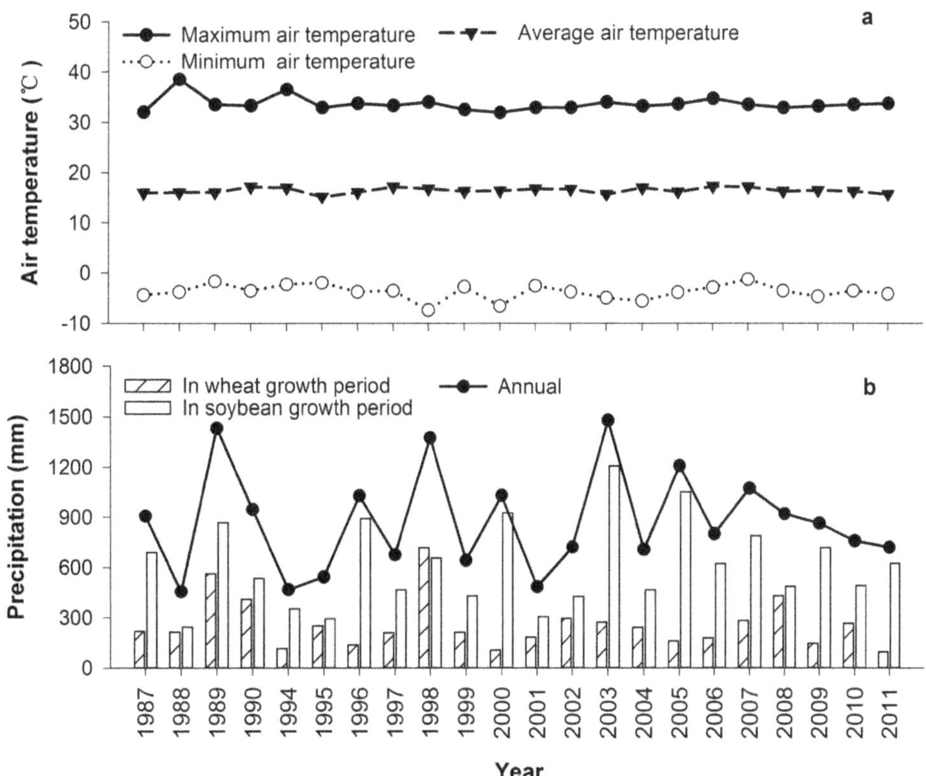

Figure 1. Air temperature and precipitation during the period 1987 to 2011 (no data for 1991 and 1993). (a) Maximum, minimum and average air temperatures. (b) Accumulated precipitation during the wheat growing period (from November to May), the soybean growing period (from June to October), and for the annual total.

Table 1. Mean application rates of mineral and organic fertilizers (kg/ha/yr) per treatment.

Treatment	N	P_2O_5	K_2O	Wheat straw	Pig manure	Cattle manure
CK	0	0	0	0	0	0
NPK	180	90	135	0	0	0
NPKW1/2	180	90	135	2500	0	0
NPKW	180	90	135	5000	0	0
NPKPM	180	90	135	0	7800	0
NPKCM	180	90	135	0	0	12500

Treatments were no fertilization (CK), mineral NPK fertilizers alone (NPK), mineral NPK fertilizers combined with 2.5 ton/ha/yr of wheat straw (NPKW1/2), mineral NPK fertilizers combined with 5.0 ton/ha/yr of wheat straw (NPKW), mineral NPK fertilizers combined with pig manure (NPKPM), and mineral NPK fertilizers combined with cattle manure (NPKCM).

A double cropping system of winter wheat-soybean is the common crop rotation in the region. At the Madian Agro-Ecological Station, plots were plowed to 20 cm depth after crop harvest. Winter wheat (*Triticum aestivum L.*), variety Yedan 13, was grown in rows from late October to May, and soybean (*Glycine max*), variety Zhonghuang 13, from June to September. Herbicides and pesticides were applied during the growth periods when it was needed. Weed residues in all the fields were manually removed after herbicides application. Hence, biomass of weeds left in the field was negligible. Wheat and soybean were harvested manually and all above-ground biomass, except for stubble, were removed from the experimental fields. Wheat grain yields were measured in 1983, 1985, and 1989–2011. Soybean grain yields were measured in 1983, 1985, 1989–1991 and 1998–2011. Grains were air dried, threshed and then weighted separately.

Soil sampling and analyses

Soil samples were collected from the top 20 cm each year after the soybean harvest (in October). Soil samples were randomly taken from three locations in each plot using a soil core sampler (inner diameter 7 cm). The samples were air dried and passed through an 8 mm sieve. Visible pieces of crop residues and roots were removed. Representative sub-samples were then passed through a 2 mm sieve and ground further to pass a 0.25 mm sieve for determination of SOC, total N, total P and other properties. The sieved soil was stored in glass jars until analysis. Soil bulk density was measured using the core method [18]. Soil pH was measured by the potentiometric method in a soil-water extract (1:2.5, w/v water) [18]. SOC content was determined following the vitriol acid-potassium dichromate oxidation method [19]. Total N was determined by using the method described by [19], and total P by Black [20]. Soil texture was measured following the gravitometer methods described by Lu [18]. Four replicates were carried out for each analysis.

Estimation of C inputs, and SOC sequestration rate and efficiency

C inputs into the topsoil (0–20 cm) included materials from crops residues (i.e., mainly roots and stubble), organic amendments (i.e. wheat straw, pig manure, and cattle manure). Annual amount of C inputs by roots from wheat and soybean were estimated at 30% [21] and 19% [22] of above-ground biomass, respectively. Annual C inputs by stubble from wheat and soybean were estimated at 15% of straw biomass. We used a straw to grain ratio of 1.2:1 and 1.6:1 to calculate wheat and soybean straw biomass, respectively [23]. The C contents of wheat and soybean were estimated at 400 g/kg and 453 g/kg (dry base), respectively [23].

The annual C input via crop residues was estimated from

$$C_{input}=((Y_g+Y_s)\times R\times D_r+R_s\times Y_s)\times(1-W)\times C/1000 \quad (1)$$

where Y_g, and Y_s were grain yield and straw yield (kg/ha); R was ratio of root biomass to above ground biomass (AGB); D_γ was ratio of root biomass in topsoil (0–20 cm) to root biomass in the soil profile, which were set at 0.753 for wheat [24] and at 0.984 for soybean [25]; R_s and W represented ratio of stubble to straw biomass and water content of air-dried grain (14% [23]); C was the organic C content of the crop (g/kg).

SOC stock ($SSOC$, ton/ha) in topsoil was calculated using the equation:

$$SSOC=SOC\times BD\times d/10 \quad (2)$$

where SOC was soil organic C content (g/kg), BD was the bulk density (g/cm^3), and d was the thickness of the soil layer (20 cm).

SOC sequestration rate (SSR, ton/ha/yr) was estimated for the topsoil by the following equation [4]:

$$SSR=(SSOC_t-SSOC_0)/t \quad (3)$$

where $SSOC_t$ and $SSOC_0$ were the stock of SOC at time t and in the initial year (1982); t was the duration of experiment. The SOC sequestration efficiency was derived from the slope of the linear regression between annual SOC sequestered and C input [14].

ΔSOC sequestration rates (ΔSSR, ton/ha/yr) via wheat straw, pig manure and cattle manure were estimated for topsoil by the following equation.

$$\Delta SSR=(SSR_{NPK+OM}-SSR_{NPK}) \quad (4)$$

where SSR_{NPK+OM} and SSR_{NPK} were the SOC sequestration rate in treatments with organic amendments combined with mineral NPK fertilizers (NPKW1/2, NPKW, NPKPM and NPKCM) and the treatment with mineral NPK fertilizers alone (NPK), respectively.

C sequestration efficiency of organic amendment (SEO) was calculated by the following equation.

$$SEO(\%)=\Delta SSR/CI\times 100\% \quad (5)$$

where CI was annual C input via organic amendment (ton/ha/yr).

Data analysis

All the statistical analyses were performed using SPSS 13.0 software package (SPSS, Inc., USA). Significant differences were analyzed using *LSD* test at significance level $P = 0.05$ or $P = 0.01$. Graphs were prepared with Sigma plot 10.0 software (Systat Software, Inc., Chicago, IL, USA).

Results

Grain yield and C inputs into the soil

Wheat and soybean yields are shown in Fig. 2. Average annual grain yields followed the order NPKCM> NPKPM> NPKW> NPKW1/2> NPK> CK. Grain yields tended to decrease under the CK treatment, but to increase under all fertilization treatments over time.

Similarly, the CK treatment had a low and decreasing crop C input, while the fertilization treatments had a high and increasing trend of C input via roots and stubble over time (Table 2). Average annual C input via roots and stubble was 2.4 times higher in the NPK treatment than in the CK treatment. Soil amendments with straw and manure slightly increased the C inputs via roots and stubble in the NPKW1/2, NPKW, NPKPM and NPKCW treatments, respectively, comparing to the NPK treatment. Similarly, CK had the lowest annual total C input t (0.45 ton/ha). Compared with CK, the average annual total C inputs in the NPK, NPKW1/2, NPKW, NPKPM and NPKCW treatments were increased by a factor of 2.4, 4.9, 7.2, 9.4, and 13.6, respectively.

SOC dynamics

Changes in SOC contents are shown in Fig. 3. As expected, CK had a low and slightly decreasing SOC content, whereas other treatments had an increasing SOC content. Changer over time in SOC content could be described by linear regression or linear plateau models. In CK, the changes in SOC content were best described by simple linear regression ($y = -0.03x+6.70$, $R = 0.59$, $P<0.01$), suggesting that SOC decreased by 3% per year. Changes in SOC contents of the NPK, NPKW1/2 and NPKW were also best described by simple linear regression, but here significant increases in SOC content were observed. Changes in SOC contents of the NPKPM and NPKCM were best described by linear plateau model. For NPKPM and NPKCW, SOC contents tended to increase rapidly in the first twenty years, but leveled off thereafter. However, a longer duration of soil C sequestration was observed in NPKPM (26 years) than in NPKCM (20 years).

C sequestration rate

Soil bulk density (BD) decreased in all fertilization treatments but not in the CK treatment (Table 3). The decrease was largest in the treatments receiving wheat straw and animal manures, which suggests that long-term application of organic amendments significantly improved soil physical conditions.

The mean SOC sequestration rate over the 29 years experimental period ranged from -0.08 ± 0.03 ton/ha/yr in CK to 0.97 ± 0.04 ton/ha/yr in NPKCM. Mineral fertilizer application (NPK) reversed the SOC decline in the CK treatment into a net SOC sequestration of 0.20 ± 0.02 ton/ha/yr. Mean SOC sequestration rates followed the order: NPKCM> NPKPM> NPKW> NPKW1/2> NPK. These findings are consistent with a large body of evidence indicating that long-term fertilizer, crop residues and manure applications increase the SOC content of arable soils [4,14]. Average ΔSOC sequestration rates (ΔSSR) of wheat straw at low input rate, wheat straw at high input rate, pig manure and cattle manure were 0.19 ± 0.03, 0.30 ± 0.04, 0.31 ± 0.03 and 0.77 ± 0.02 ton/ha, respectively. The result indicated that SOC sequestration rate changes were related to total C inputs.

Relationship between C input and SOC sequestration rate

There was a significantly positive linear correlation between C input and SOC sequestration rate (Fig. 4). The slope of the equation (0.16) indicates that on average 16% of the total C input into the soil was sequestered as SOC; the mean soil C sequestration efficiency was 16%. The results also indicate that 0.63 ton/ha/yr was needed to maintain SOC level constant at the initial level.

The mean relationships between C inputs and ΔSOC sequestration rates through organic amendments are also shown in Fig. 4. Interestingly, C sequestration efficiencies for wheat straw (c

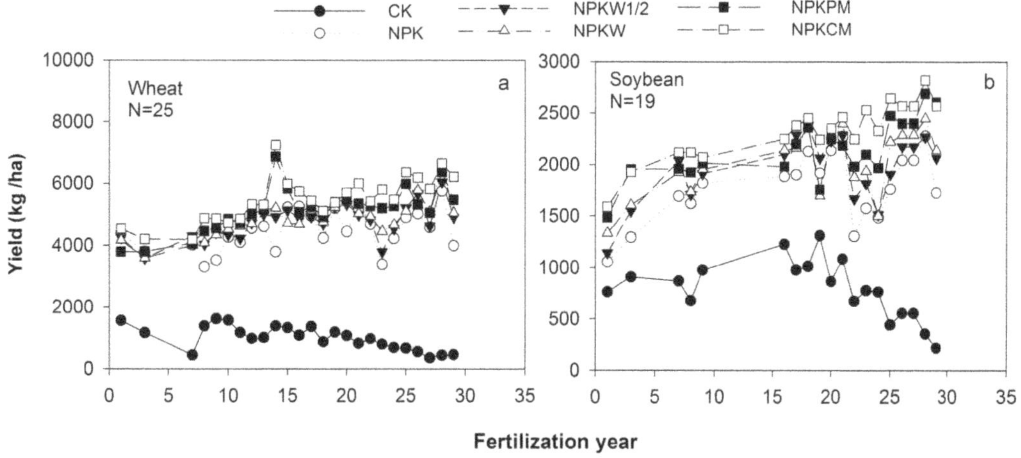

Figure 2. Average annual grain yields in six treatments during the period 1983 to 2011. (a) Wheat yields (b) Soybean yields. Treatments were no fertilization (CK), mineral NPK fertilizers alone (NPK), mineral NPK fertilizers combined with 2.5 ton/ha/yr of wheat straw (NPKW1/2), mineral NPK fertilizers combined with 5.0 ton/ha/yr of wheat straw (NPKW), mineral NPK fertilizers combined with 7.8 ton/ha/yr of pig manure (NPKPM), and mineral NPK fertilizers combined with 12.5 ton/ha/yr of cattle manure (NPKCM). Results are means (n = 4).

Table 2. Average annual crop C input (roots and stubble) and organic amendment C inputs for each treatment during the period 1987 to 2011.

Treatment	Annual crop C input (ton/ha/yr)				Organic amendment C input	Total
	Wheat	CV%	Soybean	CV%	(ton/ha/yr)	(ton/ha/yr)
CK	0.23c	38.3	0.22c	36.2	0	0.45
NPK	1.04b	14.3	0.51b	19.0	0	1.55
NPKW1/2	1.10b	11.8	0.56b	16.7	1.00	2.66
NPKW	1.13b	12.5	0.57b	16.1	2.00	3.71
NPKPM	1.20a	13.7	0.61a	14.0	2.87	4.68
NPKCM	1.28a	13.5	0.66a	12.3	4.62	6.56

Treatments were no fertilization (CK), mineral NPK fertilizers alone (NPK), mineral NPK fertilizers combined with 2.5 ton/ha/yr of wheat straw (NPKW1/2), mineral NPK fertilizers combined with 5.0 ton/ha/yr of wheat straw (NPKW), mineral NPK fertilizers combined with 7.8 ton/ha/yr of pig manure (NPKPM), and mineral NPK fertilizers combined with 12.5 ton/ha/yr of cattle manure (NPKCM). Data are present as means (n = 4). Those with the same letter are not significantly different ($P<0.05$ or 0.01).

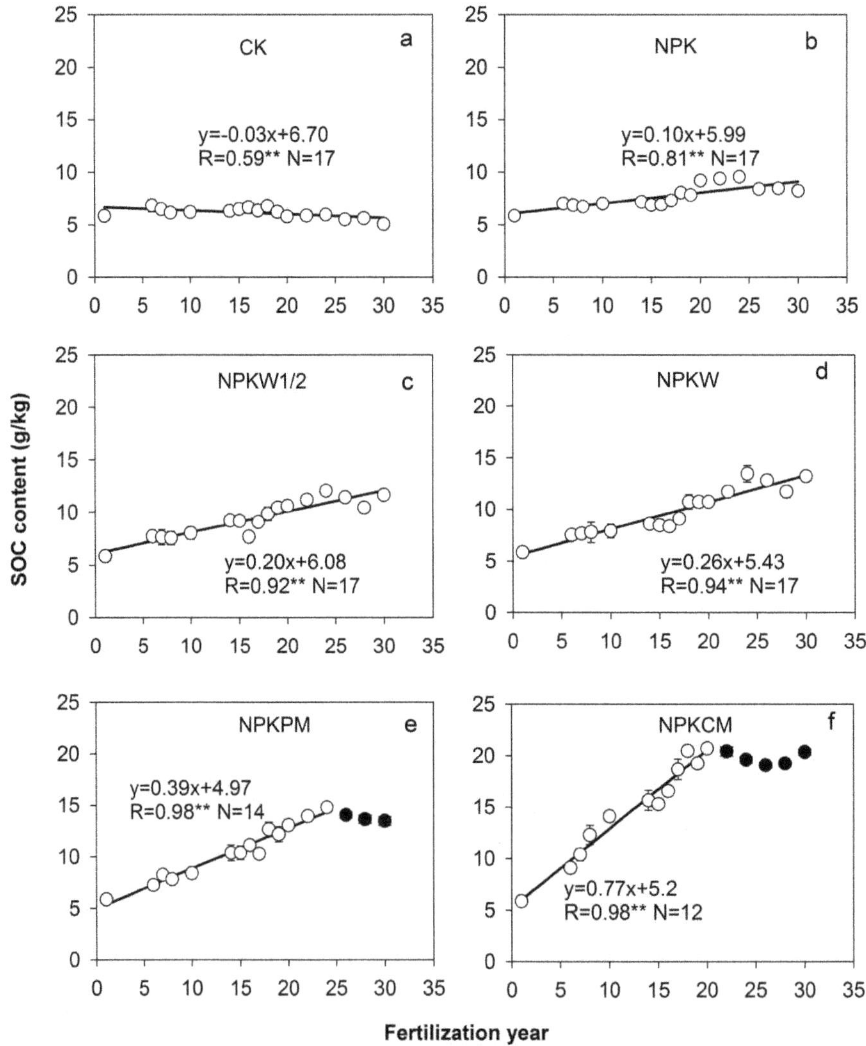

Figure 3. Changes in soil organic carbon (SOC) content over time in the six fertilization treatments during the period 1983 to 2011. CK: no fertilizer (a); NPK: mineral NPK fertilizers alone (b); NPKW1/2: mineral NPK fertilizers combined with 2.5 ton/ha/yr of wheat straw (c); NPKW: mineral NPK fertilizers combined with 5.0 ton/ha/yr of wheat straw (d); NPKPM: mineral NPK fertilizers combined with pig manure (e); NPKCM: mineral NPK fertilizers combined with cattle manure (f); ** indicates significant correlation at $P<0.01$, respectively. Error bars indicated for some years are standard errors (n = 4).

Table 3. Soil bulk density (BD), SOC contents, the estimated mean SOC sequestration rate (SSR) and mean ΔSOC sequestration rate (ΔSSR) during the 29-years period, for all treatments.

Treatment	1982			2011			SSR (ton/ha/yr)	ΔSSR (ton/ha/yr)
	BD (g/cm^3)	SOC (g/kg)	SSOC (ton/ha)	BD (g/cm^3)	SOC (g/kg)	SSOC (t/ha)		
CK	1.45	5.86	16.99	1.44±0.03a	5.08±0.17e	14.63±0.79e	−0.08±0.03e	-
NPK	1.45	5.86	16.99	1.38±0.01a	8.24±0.07d	22.74±0.47d	0.20±0.02d	-
NPKW1/2	1.45	5.86	16.99	1.21±0.02b	11.66±0.46c	28.22±1.58c	0.39±0.05c	0.19±0.03
NPKW	1.45	5.86	16.99	1.19±0.02b	13.23±0.55b	31.49±1.84b	0.50±0.06b	0.30±0.04
NPKPM	1.45	5.86	16.99	1.18±0.02b	13.46±0.42b	31.77±1.53b	0.51±0.05b	0.31±0.03
NPKCM	1.45	5.86	16.99	1.11±0.01c	20.36±0.32a	45.20±1.12a	0.97±0.04a	0.77±0.02

CK: no fertilizer; NPK: mineral fertilization; NPKW1/2: mineral fertilizer combined with 2.5 ton/ha/yr of wheat straw; NPKW: mineral fertilizer combined with 5.0 ton/ha/yr of wheat straw; NPKPM: mineral fertilizer combined with pig manure; NPKCM: mineral fertilizer combined with cattle manure. Data are presented as means (n = 4), and those designated with the same letter are not significantly different ($P<0.01$).

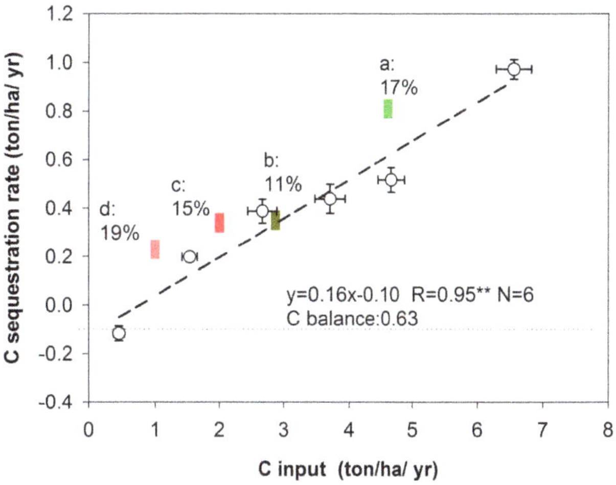

Figure 4. Relationship between C input and SOC sequestration rate. Solid line indicated linear regression curve between C inputs and SOC sequestration rates for the six treatments. a, b, c, and d represent C sequestration efficiencies for wheat straw at low input rate, wheat straw at high input rate, pig manure and cattle manure (calculated by equation (5)), respectively. ** indicated significant at $P<0.01$; error bars indicate standard errors (n = 4).

and d), pig manure (b) and cattle manure (a) differed. The C sequestration efficiencies for wheat straw was 17% (mean value), pig manure 11% and for cattle manure 17%. Hence, application cattle manure or wheat straw has a higher efficiency to enhance SOC.

Discussion

Soil C sequestration duration

Our results clearly show that trends in SOC depend on long-term fertilization regime. The changes of SOC content over time for the three treatments of NPK, NPK1/2W and NPKW are consistent with the observations reported by Zhou et al (2013) [26]. The time of soil C steady state occurring represents soil C sequestration duration [27]. West and Six [28] reported that the duration of C sequestration varies between ecosystems, climate regimes, and fertilization management (e.g., soil organic amendments inputs). West and Post [29] estimated that a change from conventional tillage to no-till results in an average 14% increase in soil C, during a mean experimental period of 15 years. Rui and Zhang [30] found that there was a negative correlation between soil C sequestration rates and duration of soil C sequestration. The C sequestration duration in the NPKW, NPKPM and NPKCM treatments reflect differences in the quantity and quality of C input. Our results indicate that the recommended mean C sequestration period of 20 years for National Greenhouse Gas according to the IPCC Guidelines [31] may not be suitable for all the organic amendments application.

Relation between C input and soil C sequestration

The relationship between C input into the soil and SOC sequestration may present in the SOC sequestration efficiency. Several studies showed that there was no general relationship between C input and SOC sequestration. Some soils might be C saturated [32–34]. Other studies demonstrated that SOC sequestration was linear or logarithmically related to the C input into soil [2,4,15]. In the present study, a clear linear relationship between annual C input and SOC sequestration rate was found, which was similar to findings in upland [4,35], but different from those findings in paddy soils [14,15,36]. The strong linear relationship indicated that the Vertisol had a large capacity to store SOC. Moreover, we did not find indications that the Vertisol had a clear, uniform saturation level for SOC sequestration. The results presented in Fig. 3b, c and d indicate that these maxima (equilibrium levels) have not been reached in 29 years. However, we observed that SOC sequestration rate leveled off after about 20 years in the NPKPM and NPKCM treatments at SOC levels of about 15 and 20 g/kg. The results suggest that SOC saturation level depends on C inputs of organic amendments.

Our results revealed that 0.63 ton/ha/yr (Fig. 4) was needed to maintain the initial SOC level, and that the unfertilized control treatment (CK) did not provide sufficient C input to maintain the initial SOC content. This finding is in agreement with the result from Yan et al. [15], who showed that the C input from unfertilized crops cannot sustain the native SOC content of upland soils. Interestingly, NPK fertilization did provide sufficient C input to contribute to a net SOC sequestration, indicating that the C input provided through the incorporation of only roots and stubble was sufficient to maintain and even enhance the C stock. This contrasts with result reported by Yang [37]. They reported that NPK fertilization was insufficient for maintaining the SOC level under conditions of traditional management with the removal of all aboveground crop biomass. These differences may relate to differences in crop yield levels and in SOC sequestration efficiency. Our study indicates C inputs from wheat and soybean stubbles and roots under conventional NPK fertilization seem sufficient to replenish the C loss through SOC decomposition in Vertisols.

C sequestration efficiency

The initial SOC content and soil texture (i.e. clay content) are important factors regulating SOC sequestration efficiency [9]. The Vertisol in our study had a low initial C content and a high clay content, which suggest that the SOC sequestration might be high. Indeed, we observed a relatively high efficiency irrespective of the type of C input (Fig. 4). Zhang [14] reported a C sequestration efficiency of 6.8% for alluvial soils in Zhengzhou, China, in the same climatic zone as our study. Stewart [38] argued that a greater SOC saturation deficit results in greater sequestration efficiency of added C. We estimated the soil C saturation deficit of the Vertisol at 80%, using the relationship between soil mineral (silt + clay) and C contents (g/kg soil) as proposed by Stewart [33]:

$$\begin{aligned} &\mathit{Csaturation\ deficit} \\ &= (1\text{-}\mathit{initial\ SOC\ content/protective\ capacity}) \times 100\% \\ &\mathit{Protective\ capacity} = 0.21 \times (\mathit{silt} + \mathit{clay\ content}) + 14.76 \end{aligned}$$

Evidently, the high clay+silt content and the low initial SOC content contributed to the relatively high C saturation deficit. The protection of SOC by clay particles was well established [39,40]. Stewart [38] reported a larger enrichment of microbial derived carbohydrates in the clay fraction compared with that of the sand fraction. Hassink [41] reported that C associated with organo-mineral complexes in soil were chemically protected, and the protection increased with an increase in clay content. Six [9] found that soils with a high clay content had a low SOC decomposition rate, likely because SOC in clay soils was chemically stabilized and absorbed onto negatively charged clay minerals [42,43]. Therefore, the clay content has a positive effect on SOC sequestration efficiency [44].

Many factors have been suggested to affect the humification coefficient, including the type of input material, soil type, climate factors and soil nutrients status [45,46]. Maillard and Angers [47] reported that the quality of the organic amendment was the dominant driver determining C sequestration efficiencies of organic amendments, based on a meta-analysis of worldwide published reports. Likely, the quality of the organic amendment in our experiment was also the major reason explaining the differences in C sequestration efficiencies for wheat straw, pig manure and cattle manure. Especially lignin is considered to be one of the most chemical recalcitrant components [48]. In soils, lignins are synthesized from L-phenylalanine and cinnamic acids via various metabolic ways to form lignin precursors such as sinapyl and coniferylalcohols [49]. The lignin structure consists of aromatic rings with side chains and $-OH$ and $-OCH_3$ groups linked by strong covalent bonds. Therefore, lignins are considered as stabilized component of SOC, influencing its pool-size and its turnover [50]. In our study, we observed that C sequestration efficiency of cattle manure was higher than that of pig manure, likely because the cattle manure was composted. Ghosh et al. [51] reported that the lignin content of cattle manure may reach 23.7%, which was significantly higher than that of pig manure. Therefore, application of composted cattle manure seems the preferred strategy for enhancing SOC sequestration in the Vertisol due to its high C sequestration efficiency.

There are a number of uncertainties in our study, which are quite common to many long-term field studies. We estimated the C inputs via roots and stubble from fixed percentages of aboveground biomass. Likely, the C input via roots and stubble depends on the level of fertilization, and varies from year to year due to variations in rainfall. Also, the C content of the pig and cattle manure was not measured each year, while there was quite a big annual variation in composition. Further, there are missing data, especially during the first half of the experiment, due to organizational matters. We obtained C sequestration efficiencies of wheat straw, pig manure and cattle manure based on the two main hypotheses. We assumed that the difference in C input via stubble and roots was negligible small between the treatment NPK and the treatments NPKW1/2, NPKW, NPKPM and NPKCM, and no significant priming effect caused by organic amendments application was existed. These uncertainties in primary data may also lead to uncertainties in the estimated mean SOC sequestration efficiency and C sequestration efficiencies of organic amendments, but do not undermine our main conclusions. Our future studies focus on the accurate quantification of C inputs via roots and stubble, effects of organic amendments application on the C inputs, and the accurate quantification of priming effect by organic amendments application using isotope technique in this ongoing field experiment. Also, SOC fractionation studies are undertaken to establish relationships between C input and SOC sequestration in different SOC pools (i.e. light and heavy C fractions).

Conclusions

The changes in SOC content over time were described by linear equations. However, the SOC sequestration rates clearly decreased in the NPKPM and NPKCM treatments after about 20 years. A significant linear relationship was observed between annual C input and SOC sequestration rate. The Vertisol in our study had a high SOC sequestration potential. Also, the overall mean SOC sequestration efficiency of the wide range of C inputs via roots, stubble, straw and manure was equally high, with an overall mean efficiency of 16% over the 29-years period. The C sequestration efficiencies of wheat straw and cattle manure are higher than pig manure, indicating that cattle manure or wheat straw application is a recommendable SOC sequestration practice in Vertisols.

Acknowledgments

We thank the staff of Madian Agro-Ecological Station and the Ministry of Agriculture, China. We sincerely thank Professor Oene Oenema from Environmental Sciences, Wageningen University, for the useful suggestions and comments. We are also greatly grateful for anonymous reviewers for their very constructive and valuable comments for our manuscript.

Author Contributions

Conceived and designed the experiments: DZW XSG. Performed the experiments: KKH DZW ZBG. Analyzed the data: KKH. Contributed reagents/materials/analysis tools: KKH DZW XSG ZBG. Wrote the paper: KKH.

References

1. Lal R (2004) Soil carbon sequestration impacts on global climate change and food security. Science 304: 1623–1627.
2. McLauchlan KK (2006) Effects of soil texture on soil carbon and nitrogen dynamics after cessation of agriculture, Geoderma 136: 289–299.
3. Tong C, Xiao H, Tang G, Wang H, Huang T, et al. (2009) Long-term fertilizer effects on organic carbon and total nitrogen and coupling relationships of C and N in paddy soils in subtropical China. Soil and Tillage Research 106: 8–14.
4. Zhang WJ, Xu MG, Wang XJ, Huang QH, Nie J, et al. (2012) Effects of organic amendments on soil carbon sequestration in paddy fields of subtropical China. Journal of Soils and Sediments 12: 457–470.
5. Freibauer A, Rounsevell MD, Smith P, Verhagen J (2004) Carbon sequestration in the agricultural soils of Europe. Geoderma 122: 1–23.
6. Li ZP, Lin XX, Che YP (2002) Analysis for the balance of organic carbon pools and their tendency in typical arable soils of Eastern China. Acta Pedologica Sinica 39: 351–360. (In Chinese).
7. Zhou P, Pag GX, Li LQ, Zhang XH (2009) SOC Enhancement in major types of paddy soils in a long-term agro-ecosystem experiment in South China. V. Relationship between carbon Input and soil carbon sequestration. Scientia Agricultura Sinica 42: 4260–4268. (In Chinese).
8. Feller C, Beare MH (1997) Physical control of soil organic matter dynamics in the tropics. Geoderma 79: 69–116.
9. Six J, Conant RT, Paul EA, Paustian K (2002) Stabilization mechanisms of soil organic matter: Implications for C-saturation of soils. Plant Soil 241: 155–176.
10. Elliott ET, Coleman DC (1988) Let the soil work for us. Ecological Bulletins 39: 23–32.
11. Gong W, Yan XY, Wang JY, Hu TX, Gong Y (2009) Long-term manuring and fertilization effects on soil organic carbon pools under a wheat-maize cropping system in North China Plain. Plant Soil 314: 67–76.
12. Zhang JB, Zhu TB, Cai ZC, Qin SW, Muller C (2012) Effects of long-term repeated mineral and organic fertilizer applications on soil nitrogen transformations. European Journal of Soil Science 63: 75–85.
13. Chander K, Goyal S, Mundra MC, Kapoor KK (1997) Organic matter, microbial biomass and enzyme activity of soils under different crop rotations in the tropics. Biology and Fertility of Soils 24: 306–310.
14. Zhang WJ, Wang XJ, Xu MG, Huang SM, Liu H, et al. (2010) Soil organic carbon dynamics under long-term fertilizations in arable land of northern China. Biogeosciences 7: 409–425.
15. Yan X, Zhou H, Zhu QH, Wang XF, Zhang YZ, et al. (2013) Carbon sequestration efficiency in paddy soil and upland soil under long-term fertilization in southern China. Soil and Tillage Research 130: 42–51.
16. Li DC, Zhang GL, Gong ZT (2011) On taxonomy of Shajiang Black Soils in China. Soils 43: 623–629. (In Chinese).
17. Guo ZB, Hua KK, Wang J, Guo XS, He CL, et al. (2014) Effects of different regimes of fertilization on soil organic matter under conventional tillage. Spanish Journal of Agricultural Research 2014 12: 801–808.
18. Lu RK (2000) Analytical methods of soil agricultural chemistry. China Agricultural Science and Technology Press, Beijing. (In Chinese).
19. Walkley A, Black IA (1934) An examination of the degtjareff method for determining soil organic matter and a proposed modification of the chromic acid titration method. Soil science 37: 29–38.
20. Black CA (1962) Methods of soil analysis. Part 2. Chemical and microbiological properties. Madison, Wisc: ASA.
21. Kuzyahov Y, Domenski G (2000) Carbon input by plants into the soil. Journal of Plant Nutrition and Soil science 163: 421–431.
22. IPCC (2006) IPCC guidelines for national greenhouse gas inventories. Prepared by the national greenhouse gas inventories programme. In: Eggleston HS, Buendia L, Miwa K, Ngara T and Tanabe K (Eds), Institute for Global Environmental Strategies. IPCC, Hayama, Japan.
23. NCATS (1994) Chinese organic fertilizer handbook. National Center for Agricultural Technology Service, Chinese Agricultural Publisher. (In Chinese).
24. Miao GY, Zhang YT, Jun Y, Hou YS, Pan XL (1989) A study on the development of root system in winter under inirrigated conditions in semi-arid Loss Plateau. Acta Agronomica Sina 15: 104–115. (In Chinese).
25. Lin WG, Wu JJ, Dong DJ, Zhong P, Wang JS, et al. (2012) Impact of different residue retention system on soybean root distribution in soil profile. Soybean Science 31: 584–588. (In Chinese).
26. Zhou ZC, Gan ZT, Shangguan ZP, Zhang FP (2013) Effects of long-term repeated mineral and organic fertilizer applications on soil organic carbon and total nitrogen in a semi-arid cropland. European Journal of Agronomy 45: 20–26.
27. Six J, Conant RT, Paul EA, Paustian K (2002) Stabilization mechanisms of soil organic matter: implications for C-saturation of soils. Plant Soil 241: 155–176.
28. West TO, Six J (2007) Considering the influence of sequestration duration and carbon saturation on estimates of soil carbon capacity. Climatic Change 80: 25–41.
29. West TO, Post WM (2002) Soil organic carbon sequestration rates by tillage and crop rotation: a global analysis. Soil Science Society of America Journal 66: 1930–1946.
30. Rui WY, Zhang WJ (2010) Effect size and duration of recommended management practices on carbon sequestration in paddy field in Yangtze Delta Plain of China: A meta-analysis. Agriculture, Ecosystems and Environment 135: 199–205.
31. Houghton JT, MeiraFilho LG, Lim B, Tréanton K, Mamaty I, et al. (eds) (1997) Revised 1996 IPCC Guidelines for National Greenhouse Gas Inventories, volumes 1–3. Hadley Centre Meteorological Office, United Kingdom.
32. Gulde S, Chung H, Amelung W, Chang C, Six J (2008) Soil carbon saturation controls labile and stable carbon pool dynamics. Soil Science Society of America Journal 72: 605–612.
33. Stewart CE, Paustian K, Conant RT, Plante AF, Six J (2009) Soil carbon saturation: Implications for measurable carbon pool dynamics in long-term incubations. Soil Biology and Biochemistry 41: 357–366.
34. Chung H, Kathie JN, Plantec A, Six J (2009) Evidence for carbon saturation in a highly structured and organic-matter-rich. Soil Science Society of America Journal 74: 130–138.
35. Kong AY, Six J, Bryant DC, Denison RF, Kessel CA (2005) The relationship between carbon input, aggregation, and soil organic carbon stabilization in sustainable cropping systems. Soil Science Society of America Journal 69: 1078–1085.
36. Cai ZC, Qin SW (2006) Dynamics of crop yields and soil organic carbon in a long-term fertilization experiment in the Huang-Huai-Hai Plain of China. Geoderma 136: 708–715.
37. Yang XM, Zhang XP, Fang HJ, Zhu P, Ren J, et al. (2003) Long-term effects of fertilization on soil organic carbon changes in continuous corn of northeast China: RothC model simulations. Environ Management 32: 459–465.
38. Stewart CE, Paustian K, Conant RT, Plante AF, Six J (2008b) Soil C saturation: evaluation and corroboration by long-term incubations. Soil Biology and Biochemistry 40: 1741–1750.
39. Feller C, Beare MH (1997) Physical control of soil organic matter dynamics in the tropics. Geoderma 79: 69–116.
40. Follett RF, Paul EA, Pruessner EG (2007) Soil carbon dynamics during a long-term incubation study involving ^{13}C and ^{14}C measurements. Soil Science 172: 189–208.
41. Hassink J (1997) The capacity of soils to preserve organic C and N by their association with clay and silt particles. Plant Soil 191: 77–87.
42. Guggenberger G, Christensen BT, Zech W (1994) Land-use effects on the composition of organic matter in particle-size separates of soil: 1. Lignin and carbohydrate signature. European Journal of Soil Science 45: 449–458.
43. Franzluebbers AJ, Haney RL, Hons FM, Zuberer DA (1996) Active fractions of organic matter in soils with different texture. Soil Biology and Biochemistry 28: 1367–1372.
44. Shi XZ, Wang HJ, Yu DS, Weindorf DC, Cheng XF, et al. (2009) Potential for soil carbon sequestration of eroded areas in subtropical China. Soil and Tillage Research 105: 322–327.
45. Lin XX, Wen QX (1987) Utilization of organic wastes in the Taihu region of Jiangsu Province of China. Resources and Conservation 13: 109–116.
46. Galantini JA, Rosell RA, Andriulo A, Miglierina A, Iglesias J (1992) Humification and nitrogen mineralization of crop residues in semi-arid Argentina. Science of the Total Environment 118: 263–270.
47. Maillard E, Angers DA (2014) Animal manure application and soil organic carbon stocks: a meta-analysis. Global Change Biology 20: 666–679.
48. Torres IF, Bastida F, Hernández T, Bombach P, Richnow HH, et al. (2014) The role of lignin and cellulose in the carbon-cycling of degraded soils under semiarid climate and their relation to microbial biomass. Soil Biology and Biochemistry 75: 152–160.
49. Higuchi T (1971) Formation and biological degradation of lignins. Advances in Enzymology and Related Areas of Molecular Biology 34: 207–283.

50. Thévenot M, Dignac MF, Rumpel C (2010) Fate of lignins in soils: a review. Soil Biology and Biochemistry 42: 1200–1211.
51. Ghosh PK, Ramesh P, Bandyopadhyay KK, Tripathi AK, Hati KM, et al. (2004) Comparative effectiveness of cattle manure, poultry manure, phosphor compost and fertilizer-NPK on three cropping systems in Vertisols of semi-arid tropics. I. Crop yields and system performance, Bioresource Technology 95: 77–83.

6

Light Structures Phototroph, Bacterial and Fungal Communities at the Soil Surface

Lawrence O. Davies[1,2]*, Hendrik Schäfer[1], Samantha Marshall[2], Irene Bramke[2], Robin G. Oliver[2], Gary D. Bending[1]

1 School of Life Sciences, Gibbet Hill Campus, University of Warwick, Coventry, United Kingdom, 2 Syngenta, Product Safety, Jealott's Hill International Research Centre, Bracknell, Berkshire, United Kingdom

Abstract

The upper few millimeters of soil harbour photosynthetic microbial communities that are structurally distinct from those of underlying bulk soil due to the presence of light. Previous studies in arid zones have demonstrated functional importance of these communities in reducing soil erosion, and enhancing carbon and nitrogen fixation. Despite being widely distributed, comparative understanding of the biodiversity of the soil surface and underlying soil is lacking, particularly in temperate zones. We investigated the establishment of soil surface communities on pasture soil in microcosms exposed to light or dark conditions, focusing on changes in phototroph, bacterial and fungal communities at the soil surface (0–3 mm) and bulk soil (3–12 mm) using ribosomal marker gene analyses. Microbial community structure changed with time and structurally similar phototrophic communities were found at the soil surface and in bulk soil in the light exposed microcosms suggesting that light can influence phototroph community structure even in the underlying bulk soil. 454 pyrosequencing showed a significant selection for diazotrophic cyanobacteria such as *Nostoc punctiforme* and *Anabaena* spp., in addition to the green alga *Scenedesmus obliquus*. The soil surface also harboured distinct heterotrophic bacterial and fungal communities in the presence of light, in particular, the selection for the phylum Firmicutes. However, these light driven changes in bacterial community structure did not extend to the underlying soil suggesting a discrete zone of influence, analogous to the rhizosphere.

Editor: Stefan Bertilsson, Uppsala University, Sweden

Funding: The authors wish to thank the Biological and Biotechnology Research Council for funding (Training Grant Reference Number: BB/H530970/2). URL: http://www.bbsrc.ac.uk/home/home.aspx. The funders had no role in study design, data collection and analysis, decision to publish, or preparation of the manuscript.

* E-mail: l.o.davies@warwick.ac.uk

Introduction

The upper few millimeters of soil are an area with physico-chemical conditions distinct from those of bulk soil as a result of the surface being exposed to light and other environmental factors such as wind and rain erosion [1]. Soil surface communities are different from those of bulk soil due to the development of photosynthetic communities such as cyanobacteria, algae, mosses, and lichens, which can form biological soil crusts (BSC) with time [2–7]. There has been a dramatic rise in publications reporting on the role of BSCs recently as it has become recognized that this area is a distinct ecosystem with increased nutrient levels [8] and erosion resistance [9] compared to soil without phototroph communities. BSC research to date has focused on arid and semi-arid lands such as the Colorado plateau and Sonoran desert in the USA [2–4,10], Gurbantunggut desert in northwest China [7,9], Negev desert in Israel [11] and Oman [6], where phototroph communities have been estimated to cover up to 70% of the soil surface [12]. BSCs have also been shown to be widespread in temperate soils and under agricultural crops [5,13–15], however, little is known about their community structure and ecological significance.

In arid environments, soil surface communities have several important functions, including the release of exopolysaccharides from fungi and cyanobacteria which bind soil into aggregates, improving soil structure and reducing the impact of wind erosion [16–19]. Another key function of soil surface communities is fixation of N_2 by diazotrophic cyanobacteria such as *Nostoc* spp. [4,20–24], and C fixation by phototrophs [25–27], which may be the reason for higher soil C and N levels in soil with a BSC [8]. The development of phototrophs at the soil surface has also been shown to have a profound impact on plant growth and biomass [28] and result in increased levels of N, K, and Cu in plant tissues [29].

The development of BSC communities in arid environments is characterized by a succession from cyanobacteria dominated to lichen- and moss- dominated crusts [30–32]. Further, a succession within cyanobacteria dominated crusts has also been noted from *Microcoleus vaginatus* to *Nostoc* spp./*Tolypothrix* spp. [4]. However, our understanding of the community structure remains very limited, not least because the majority of studies investigating phototroph diversity in BSCs have used culture dependent methods which are prone to bias [5,7,9,32,33], or molecular methods that target 16S rRNA of bacteria, which ignore the diversity of eukaryotic phototrophs [2,3,6,10,11,34]. Molecular microbial community analysis of bacterial diversity at the soil surface has shown a dominance by cyanobacteria [2,3,6,11], for example, Abed *et al.* [6] found that 77–81% of clones from BSCs of Oman had close homology to cyanobacteria. Consequently, the diversity and community composition of heterotrophic bacteria at the soil surface is not well characterised. Likewise, although fungi have been shown to provide key ecosystem services of BSCs such as structural cohesion provided by hyphal entanglement [19], little is known regarding the fungal community structure at the soil surface [35,36].

In contrast to arid and semi-arid soils, our understanding of the structure and function of soil surface communities in temperate and agricultural soils is limited [5,28]. Phototrophs have been shown to develop under agricultural cropping systems such as wheat, maize and sugar beet between 50 and 80 days after tillage [15]. The presence of these communities reduced soil erosion rates and this reduction increased with the successional age of the crust [15]. However, soil tillage removed this functionally important community for at least 50 days [15]. Phototroph communities may also have other important agricultural functions, for example, several phototrophs have been shown to break down pesticides in pure culture [37] and therefore phototrophs may have a role in pesticide degradation at the soil surface. An understanding of the communities and functions of soil surface communities in temperate environments will inform agricultural management decisions such as the benefits of reduced tillage practices.

In this study, we investigated shifts in phototroph, bacterial and fungal community structure between the soil surface and bulk soil of a pasture soil from a temperate climate throughout the development of phototroph communities at the soil surface. We used universal phototroph primers designed to amplify ribosomal RNA genes of any plastid-containing organisms, 454 pyrosequencing of PCR amplicons, and measured soil pH and nutrient levels with the aim of answering the following questions: (i) How diverse are cyanobacteria and eukaryotic phototrophs at the soil surface? (ii) Does light influence bacterial and fungal community structure and diversity at the soil surface? (iii) Are there successional changes in phototroph, bacterial and fungal communities at the soil surface and underlying bulk soil? (iv) Does the establishment of soil surface communities affect chemical parameters and microbial community structure of underlying bulk soil?

Materials and Methods

Soil

Soil was sourced from Les Barges, Switzerland (CH-1896 Vouvry) in October, 2010. The site did not contain any protected wildlife and it is owned by Syngenta who authorized sampling. Approximately 40 kg was sampled from the top 15 cm of Gartenacker soil (silty loam), which was then sieved to 2 mm and homogenized by mixing to give an average representation of the community structure and chemical properties of the volume of sampled soil. Microbial communities in the upper 15 cm of soil are routinely disturbed and mixed by tillage. Soil was therefore sampled to this depth and homogenized before being setup in microscoms in order to simulate natural mixing of surface communities in agricultural systems. The land had been used for pasture for over 20 years without the application of pesticides. The physico-chemical properties of Gartenacker soil are shown in Table S1.

Test System and Sampling Soil Surface Communities

To follow development of soil surface communities [Figure S1] a modified design was used from Jeffery *et al.* [1] with dimensions of 20 cm×15.5 cm×1.8 cm. Trays were filled with 600 g Gartenacker soil (35% water content) and soil was flattened to minimise soil surface heterogeneity. Trays were covered with either: (i) DS 226 light filter, or (ii) an opaque filter (Lee Filters, Andover, UK). In order to study the impact of light on microbial community development, soil was incubated in a controlled constant environment chamber on a 16 h:8 h light:dark cycle at 200 μmol s^{--1} m^{-1} (Philips Master fluorescent lights (>360 nm) TLD 36 W/840) at a constant temperature of 20°C±2°C. This allowed the development of soil surface communities to be investigated under controlled conditions by removing confounding climatic variables. Trays were setup in triplicate using a randomised design; moisture content was checked weekly by weight and maintained by watering from above using a pipette.

Triplicate trays were destructively sampled at 0, 20, 40, and 80 days. This sampling strategy aimed to follow the development of early-successional phototroph communities based on previous work, which showed development of phototrophs under cropping systems between 50 and 80 days following tillage [15]. At each sampling point, a stainless steel sheet was run under the soil surface at a measured depth of 3 mm to separate the soil surface (upper 3 mm) from the underlying bulk soil (3–12 mm). Surface and bulk soil samples were frozen at −20°C in polyethylene zip bags for 48 h before freeze-drying for 72 h. Freeze-dried soil was homogenised using a mortar and pestle and stored at −20°C.

Soil Chemical Properties

Extractable Mg and K were measured by adding 50 ml 1 M NH_4NO_3 solution to 10 g freeze-dried soil and shaking at 200 rpm for 30 mins. The solution was filtered prior to analysis using an ULTIMA 2 Inductively Coupled Plasma – Atomic Emission Spectroscopy (ICP-AES) (HORIBA Jobin Yvon, Middlesex, UK). Extractable nitrate (NO_3) was measured by adding 50 ml saturated $CaSO_4$ to 20 g freeze-dried soil and shaking at 200 rpm for 30 mins. The solution was filtered prior to colorimetric analysis using a FIAstar 5000 flow injection analyser (FOSS UK Ltd, Warrington, UK). Soil pH was measured by adding 25 ml water to 10 g freeze-dried soil and shaking at 200 rpm for 15 mins prior to pH measurement using an Accumet AR50 electrode (VWR, Leicestershire, UK). Extractable P was measured by adding 100 ml 0.5 M $NaHCO_3$ solution (pH 8.5) to 5 g freeze-dried soil and shaking at 200 rpm for 30 mins. The solution was filtered prior to analysis by ICP-AES [38].

Characterisation of Soil Surface Communities

Chlorophyll *a*. Chlorophyll *a* was extracted according to Ritchie [39]. Briefly, 20 ml 90% (v/v) acetone was added to 5 g freeze-dried soil and shaken at 300 rpm in the dark for 5 hours. Chlorophyll *a* was measured using a Shimadzu UV 1800 spectrophotometer at wavelengths 664 nm and 750 nm before acidifying with 3 M HCl for 90 seconds and re-measuring at 665 nm and 750 nm. Chlorophyll *a* values were calculated from the formulas given in Hansson [40].

Most probable number (MPN) of algae. At day 80, the number of algal cells at the soil surface under light and dark conditions was estimated using MPN. Fresh soil was homogenized and 1 g was transferred aseptically to 10 ml sterile Bold's basal media (BBM) ([41]; method in Supporting Information S1). Serial dilutions were performed at 2, 4, 5, 6, 8, 10, 15, 20, and 25-fold dilutions and 5 replicates of 1 ml aliquots were transferred to a microtitre plate, covered with cling film and incubated for 21 days under a 16 h:8 h light:dark cycle at 200 μmol s^{-1} m^{-1}. Algal growth was recorded by a colour change of BBM from clear to green. Algal abundance was estimated using a MPN calculator according to Blodgett [42].

Microbial Community Structure at the Soil Surface

DNA extraction, PCR amplification of ribosomal RNA markers and Terminal Restriction Fragment Length Polymorphism (TRFLP) to assess phototroph, fungal and bacterial community structure. DNA was extracted using a FastDNA Spin Kit (Qbiogene, Loughborough, UK) according to the manufacturer's handbook. The quantity and quality of DNA in extracts was analysed using a nanodrop ND-1000 spectrophotometer (Labtech International Ltd, Sussex, UK) and by agarose gel electrophoresis, respectively. DNA was extracted from surface and bulk soil samples after 0, 20, 40 and 80 days incubation under light and dark conditions.

The diversity of phototrophs was analysed by PCR targeting 23S rRNA genes of plastids using primers p23SrV_f1 and p23SrV_R1-HEX which produced a product approximately 410 bp in length [43]. Bacterial 16S rRNA genes were amplified using primers 63f and 1087r-VIC giving a 1 kb product [44,45], and for analysis of fungi, PCR targeted the ITS region using primers ITS1f-PET and ITS4r [46,47]. Details of all primer pairs are given in Table S2. PCR was performed using 47 μL MegaMix (Microzone Ltd, Haywards Heath, UK), 1 μL of DNA (10 ng/μL) and 1 μL of either 5 μM (bacteria/phototrophs) or 25 μM (fungi) forward and reverse primers. Samples were run on a GeneAmp 9700 thermocycler (Applied Biosystems, Warrington, UK) using the reaction described in Sherwood & Presting [43] for phototrophs. PCR amplification of 16S rRNA and the ITS region were run in the same reaction using the amplification method described by Marchesi *et al.* [45] with an extension time of 1 min and a final extension time of 10 mins (full methods are in Supporting Information S1).

PCR products were purified using a QIAquick PCR purification kit (Qiagen, Crawley, UK) according to the manufacturer's instructions. Restriction digests were performed at 37°C for 4 hrs followed by 95°C for 15 mins. Digests of 23S rRNA gene fragments of phototrophs used 500 ng PCR product, 2 μL 10X buffer, 0.5 μL 5U *Dde*I (New England Biolabs, Hitchin, UK), made up to 20 μL with Ultra Pure DNase/RNase-free distilled water (Invitrogen, Paisley, UK). *Dde*I was used based on clone libraries using the Restriction Enzyme Mapping Application (REMA, http://bioperl.macaulay.ac.uk). Digests of 16S rRNA gene fragments of bacteria and ITS fragments of fungi used 500 ng and 400 ng of PCR product, respectively, 2 μL 10X buffer, 0.25 μL 5U of either *Msp*I or *Hha*I (New England Biolabs, Hitchin, UK), made up to 20 μL with sterilised distilled water. *Msp*I and *Hha*I were used as they have previously been shown to provide good differentiation between bacterial and fungal taxa [48]. Restriction digests using *Hha*I also contained 0.2 μL (10 mg/ml) bovine serum albumin (New England Biolabs, Hitchin, UK).

All samples were cleaned using Sephadex spin columns and LIZ1200 standard was added prior to electrophoresis using an ABI PRISM 3130×l genetic analyser (Applied Biosystems, Warrington, UK). GeneMarker (Softgenetics, USA) was used to quantify peak area of terminal restriction fragments (TRFs) and values were transformed to relative abundance to standardise data. A constant percentage threshold was selected according to Sait *et al.* [49] to minimise a correlation between total peak area and number of TRFs.

454 Amplicon Pyrosequencing to Determine Diversity of Phototrophs, Fungi and Bacteria at the Soil Surface

Phototroph, bacterial and fungal PCR amplicons from the soil surface incubated under light and dark conditions for 80 days were pyrosequenced by Research and Testing Laboratory (Lubbock, TX, USA) (RTL) using a Roche 454 FLX instrument and Titanium reagents. Bacterial Tag-encoded pyrosequencing was performed as described previously by Dowd *et al.* [50]. Fungal and phototroph pyrosequencing were performed according to the same protocol using the primer pairs described in the previous section. Pyrosequencing gave a total of 67658, 22672 and 77470 reads across six samples for bacteria, fungi and phototrophs, respectively.

Processing of 454 Sequence Data

Sequences were processed using QIIME v. 1.4.0 [51] by selecting sequences with an average quality score >25, containing no ambiguous bases or homopolymers longer than six base pairs, without any primer mismatches, and a sequence length between 250–430 bp (bacteria), 250–390 bp (fungi) and 330–410 bp (phototrophs). Sequences were also denoised using Denoiser [52]. Following denoising, methods of data processing differed for bacteria, fungi and phototrophs. Bacterial OTUs were picked at a 97% similarity threshold using UCLUST [53] and representative sequences were picked using the most abundant method before PyNAST aligning [54] with the 16S rRNA Greengenes database aligned at 97% [55]. Chimeras were identified using ChimeraSlayer [56] and taxonomy was assigned using the RDP classifier and default settings [57]. Processing of fungi and phototrophs used UCHIME [58] for *de novo* chimera identification. Taxonomy was assigned using the RDP classifier for fungi [57] and BLAST [59] for phototrophs. Phylogenetic trees showing sequence abundance data were created using MEGAN 4 [60]. Full details of the number of sequences removed at each processing step are shown in Tables S3–S5. Sequence data have been submitted to the Genbank database under Bioproject Accession No. PRJNA179030.

Statistical Analysis

Parametric tests on non-transformed data were performed where possible. If assumptions were not met, data was log transformed. One-way ANOVA was performed on chlorophyll *a*, pH and soil nutrient data, and t-tests were performed on MPN for algae and phototroph abundance data. All analyses were performed using Minitab version 15. TRF data was analysed using GeneMarker and statistically analysed using non-metric multidimensional scaling (NMDS) analysis, ANOSIM and SIMPER using PRIMER6 (Plymouth, UK). Pyrosequencing data was rarefied at 3317, 6322 and 964 reads for phototrophs, bacteria and fungi, respectively and QIIME v.1.4.0 was used for: ANOVAs to compare taxonomy abundance data and t-tests to compare α diversity. Chao1 was used as a mark-release-recapture assessment of diversity [61] and Observed Species as an assessment of the number of unique OTUs in a sample.

Results

Soil pH and Nutrients

Soil nutrient levels and pH are shown for all sampling points in Figures S2, S3, S4, S5, S6 and after 80 days incubation under light and dark conditions in Table 1. Light had a significant effect on pH, extractable NO_3 and Mg ($p \leq 0.001$) at all sampling points (Figure S2, S3 & S5). At day 80, pH ($p \leq 0.01$) was higher and extractable NO_3 ($p \leq 0.01$) and Mg ($p \leq 0.01$) were lower under light compared to dark incubated samples, however, there was no effect of depth (Table 1). Light did not influence extractable P, however, P was significantly higher at the soil surface compared to underlying bulk soil after 80 days incubation under light conditions ($p \leq 0.01$) (Table 1). Depth also influenced extractable K content with the soil surface having significantly higher extractable K than underlying bulk soil after 80 days incubation under light conditions ($p \leq 0.01$) (Table 1).

Most Probable Number for Algae and Chlorophyll a

MPN assessment of algal abundance estimated a >60-fold greater algal population at the soil surface incubated under light compared to dark conditions for 80 days ($p \leq 0.01$) (Table 1). In addition, light ($p \leq 0.001$) and depth ($p \leq 0.001$) had a significant effect on chlorophyll *a* (Figure 1). Chlorophyll *a* was significantly higher at the soil surface under light at day 20, 40 and 80 ($p \leq 0.001$). Chlorophyll *a* was not detected in bulk soil under light or under dark conditions (Figure 1).

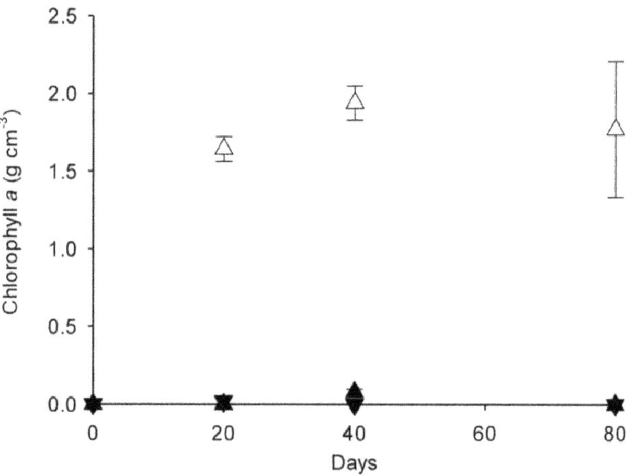

Figure 1. Chlorophyll *a* development in Gartenacker soil. Chlorophyll *a* in the surface (▲) and bulk (▼) of pasture soil after incubation under light (open symbols) or dark (closed symbols) conditions. Errors bars are ±1 standard error.

TRFLP Analysis of Phototroph, Bacterial and Fungal Community Structure at the Soil Surface and Underlying Bulk Soil under Light and Dark Conditions

Phototroph community structure was significantly different at the soil surface ($p \leq 0.01$) and in bulk soil ($p \leq 0.05$) under light conditions compared to dark incubated soil (Figures 2a–2c). There were no significant differences in phototroph community structure between the soil surface and bulk soil incubated under light. NMDS analysis of TRFLP data showed two distinct clusters of samples: Grp I and Grp II (Figure 2a). Dark incubated samples were present in both Grp I and Grp II (Figures 2a–2b), however, all light incubated samples clustered within Grp II (Figure 2c), which suggests that phototroph community structure was more variable under dark compared to light conditions (Figures 2a–2c).

The soil surface incubated under light conditions had significantly different heterotrophic bacterial and fungal communities compared to bulk soil incubated under light and dark incubated samples ($p \leq 0.01$) (Figures 2d & 2e). There was no significant difference in heterotrophic bacterial and fungal community structure between bulk soil incubated under light and dark conditions (Figures 2d & 2e). At day 80, the soil surface harboured distinct bacterial communities under light conditions (Figure 2d).

Microbial Community Structure and Taxonomic Diversity at the Soil Surface and in Bulk Soil after 80 days of Incubation

Phototroph community structure. Pyrosequencing revealed a total of 533 phototrophic OTUs across all samples with an average length of 351 bp, and an average of 71.7 reads assigned to each OTU, out of a total of 38203 processed reads. Chao1 index and Observed Species were both significantly higher at the soil surface incubated in the dark compared to light conditions ($p \leq 0.001$) (Figures 3a & 3b). Moreover, there were an estimated 246 unique phototroph OTUs under dark conditions compared to only 80 under light conditions (Figure 3b). Figures 3a and 3b both show that diversity plateaus under light as sampling depth increased, however, under dark conditions a plateau was not observed. NMDS analysis of phototroph community structure showed a closer clustering of samples under light compared to dark conditions, which suggests that phototroph community structure was less variable under light conditions (Figure 3c).

A wide range of cyanobacteria and eukaryotic phototrophs were detected, including green, red and brown algae, cryptomonads, diatoms, mosses, and angiosperms (Figure 4). Relative composition analysis showed that cyanobacteria, rather than eukaryotic phototrophs, dominated under both treatments, with a relatively greater number of reads assigned to cyanobacteria under light compared to dark conditions ($p < 0.01$) (Table 2). Further, the relative composition of cyanobacteria differed between light treatments e.g. 65.1%±SE 0.96% and 12.6%±SE 2.17% of reads had close homology to *N. punctiforme* PCC 73102 under light and dark conditions, respectively ($p \leq 0.001$), 11.6%±SE 2.02% and 2.4%±SE 0.11% of reads had close homology to *Anabaena variabilis* ATCC 29413 under light and dark, respectively ($p \leq 0.01$), and 2.5%±SE 0.26% and 1.0%±SE 0.29% of reads had close

Table 1. The effect of light and depth on chlorophyll *a*, most probable number (MPN) of algae, pH, and extractable nitrate, phosphorus, potassium and magnesium after 80 days incubation under light and dark conditions (±1 standard error).

Treatment	Depth	MPN (cells $g^{-1} \times 10^3$)	pH	Nitrate (mg kg^{-1})	Phosphorous (mg kg^{-1})	Potassium (mg kg^{-1})	Magnesium (mg kg^{-1})
Light	Surface	69.15±6.5^a	8.1±0.06^a	6.3±0.3^a	78.8±1.8^a	104.2±5.1^a	51.2±1.0^a
	Bulk	Not measured	8.0±0.02^a	4.3±2.2^a	73.0±1.6^b	76.2±3.6^b	50.8±1.3^a
Dark	Surface	1.08±0.15^b	7.5±0.02^b	74.1±1.3^b	75.4±1.7^{ab}	87.4±0.5^b	60.8±1.8^b
	Bulk	Not measured	7.5±0.01^b	70.4±3.1^b	75.6±0.6^{ab}	85.4±2.7^b	59.5±1.9^b

Significant differences between treatments are indicated by different letters ($p \leq 0.01$).

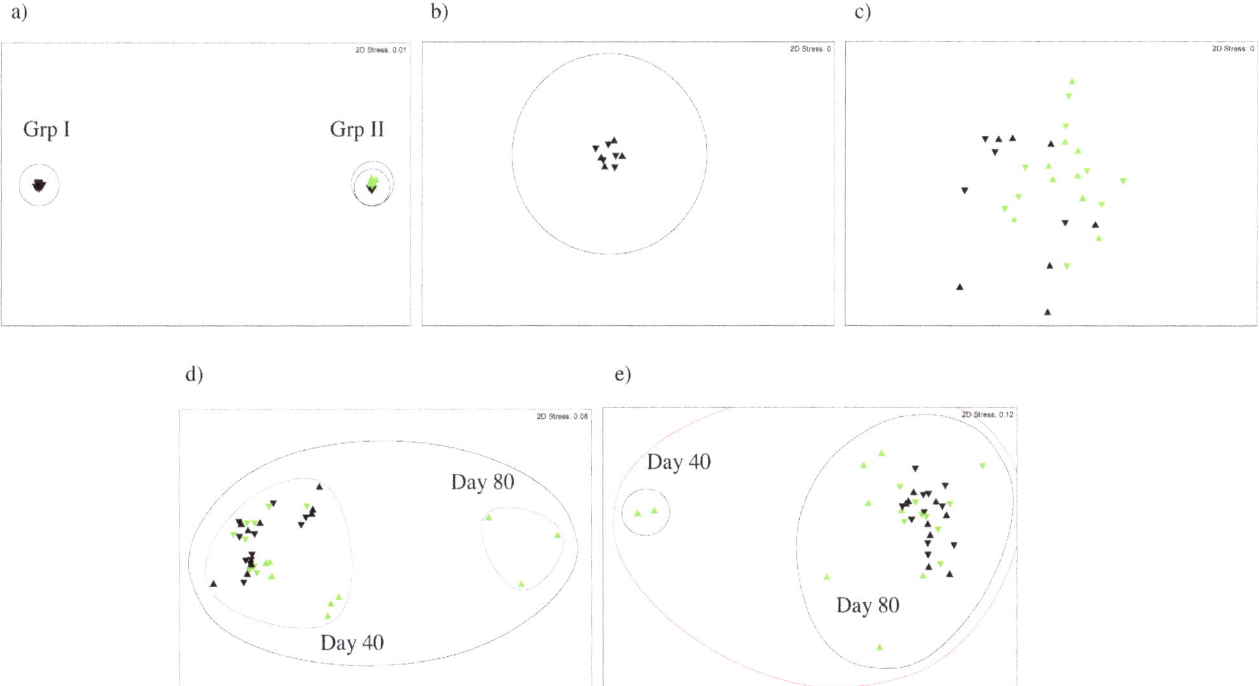

Figure 2. Development of phototroph, bacterial and fungal communities in Gartenacker soil. Phototroph (23S rRNA genes of plastids), bacterial (16S rRNA) and fungal (ITS) community structure at the surface (▲) and bulk (▼) of a pasture soil under light (green) and dark (black) conditions: (a) phototrophs all samples; (b) phototrophs close up of Grp I samples; (c) phototrophs close up of Grp II samples; (d) bacteria all samples (e) fungi all samples. Non-metric dimensional scaling shows clustering based on the similarity of microbial community structure between treatments: 15% (red cluster), 40% (black cluster) and 85% (blue cluster).

homology to *A. cylindrica* PCC 7122 under light and dark, respectively ($p \leq 0.05$) (Figure 4). There were no clearly dominant taxa under dark conditions, rather, seven taxa had a relative read abundance between 6% and 15%, which ranked as follows: *Cyanothece* sp.>*N. punctiforme*>*Thermosynechococcus elongatus*>*Cryptomonas paramecium*>*Ricinus communis*>*Gloeobacter violaceus*>*Scenedesmus obliquus* (Figure 4).

Relative composition analysis showed that a greater proportion of reads were assigned to eukaryotic phototrophs under dark compared to light conditions ($p < 0.001$), in particular cryptomonads, red algae, brown algae, mosses and angiosperms ($p < 0.05$) (Table 2). In contrast, relative composition analysis showed 6.2%±SE 1.25% and 14.8%±SE 1.88% of reads were assigned to *Scenedesmus obliquus* under dark and light conditions, respectively ($p \leq 0.05$). Relative composition analysis also showed a greater number of reads assigned to the green algae *Chlorella variabilis* ($p \leq 0.05$) and *Chlorogonium elongatum* ($p \leq 0.05$), brown alga *Ectocarpus siliculosus* ($p \leq 0.001$), moss *Syntrichia ruralis* ($p \leq 0.05$), angiosperm *Jacobaea vulgaris* ($p \leq 0.001$), diatom *Phaeodactylum tricornutum* ($p \leq 0.05$), and cryptomonads *Rhodomonas salina* ($p \leq 0.001$) and *Cryptomonas curvata* ($p \leq 0.01$) under dark compared to light conditions.

Bacterial community structure. Analysis of pyrosequencing data for bacteria (49766 reads) clustered read data into 6517 bacterial OTUs with an average read length of 340 bp and an average of 7.6 reads assigned to each OTU. Chao1 index and Observed Species were significantly higher at the soil surface under dark compared to light conditions ($p \leq 0.001$) (Figures 5a & 5b). In contrast to phototrophs, NMDS analysis of bacterial community structure showed a closer clustering of dark compared to light incubated samples, which suggests that bacterial community structure was more variable at the soil surface under light conditions (Figure 5c).

At the phylum level, relative composition analysis showed that Proteobacteria dominated the soil surface with 35.1%±SE 0.21% and 36.4%±SE 2.66% of reads assigned under dark and light conditions, respectively (Figure 6). The relative composition of samples showed that 19.3%±SE 4.39% and 5.9%±SE 0.18% of reads had close homology to the phylum Firmicutes under light and dark conditions, respectively ($p \leq 0.05$), and 5.9%±SE 1.21% and 2.0%±SE 0.03% of reads were assigned to the family Bacillaceae under light and dark conditions, respectively ($p \leq 0.05$) (Figure 6). Moreover, relative composition analysis showed that more reads were assigned to the class α-Proteobacteria ($p \leq 0.05$), the order Sphingomonadales ($p \leq 0.001$) and the families Sphingomonadaceae ($p \leq 0.01$) and Rhizobiaceae ($p \leq 0.05$) under light compared to dark conditions (Figure 6).

Relative composition analysis also showed that 5.4%±SE 0.14% and 3.0%±SE 0.04% of reads had close homology to δ-Proteobacteria under dark and light conditions, respectively ($p \leq 0.01$), and 2.5%±SE 0.02% and 1.4%±SE 0.3% of reads had close homology to Syntrophobacteraceae under dark and light conditions, respectively ($p \leq 0.05$) (Figure 6).

Fungal community structure. Pyrosequencing (14577 reads) revealed 472 fungal OTUs with an average length of 316 bp and an average of 30.9 reads assigned to each OTU. However, Observed Species showed a significantly higher number of unique OTUs under dark compared to light conditions ($p \leq 0.001$) (Figures 7a & 7b). NMDS analysis of fungal community structure showed a poor clustering of light incubated samples under light conditions; one sample shared a greater similarity to dark incubated rather than light incubated samples, which suggests

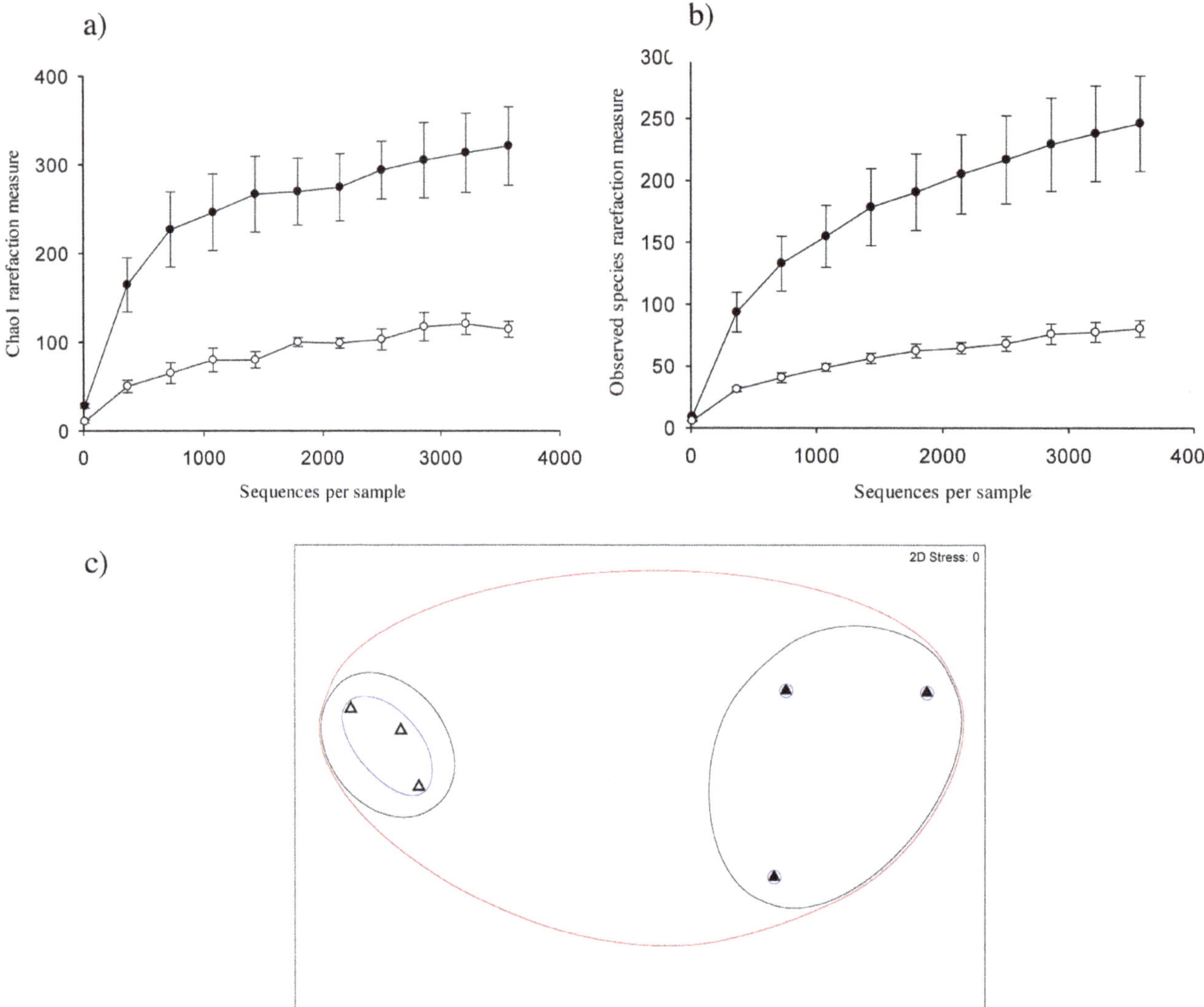

Figure 3. Phototroph diversity at the soil surface under light and dark conditions. α diversity estimates Chao1 (a) and Observed Species (b) and non-metric multidimensional scaling of community structure similarity (c) for phototrophs (23S rRNA genes of plastids) at the soil surface of a pasture soil after 80 days incubation under light (open symbols) or dark (closed symbols) conditions. OTU clustering was performed at the 97% similarity threshold using UCLUST. Error bars are ±1 S.E. Non-metric multidimensional scaling shows clustering based on the similarity of microbial community structure between treatments: 20% (red cluster), 25% (black cluster) and 80% (blue cluster).

that fungal community structure was more variable under light compared to dark conditions (Figure 7c).

Relative composition analysis showed Ascomycota to be the dominant division of fungi at the soil surface with 57.9%±SE 5.96% and 62.4%±SE 2.79% of reads showing close homology under light and dark conditions, respectively (Figure 8). The presence of light produced few shifts in fungal community structure, however, relative composition analysis showed that 2.3%±SE 0.09% and 4.3%±SE 0.53% of reads were assigned to Hypocreales under dark and light conditions, respectively ($p \leq 0.05$) (Figure 8). Relative composition analysis also showed a relatively greater number of reads assigned to both Sordariomycetes incertae sedis and Clavicipitaceae under dark compared to light conditions ($p \leq 0.05$) (Figure 8).

Discussion

Light had a significant effect on phototroph community structure, soil nutrients and pH, and this effect extended to the underlying bulk soil. Light also had a significant, time-dependent impact on heterotrophic bacterial and fungal community structure which was restricted to the soil surface. Soil surface communities are typically defined by the presence of photosynthetic communities in the top 1–3 mm of soil [1], however, we show changes in phototroph community structure at a depth greater than 3 mm, and the presence of distinct heterotrophic microbial communities at the soil surface in the presence of light.

Chlorophyll *a* analysis was used as a broad-scale assessment of phototroph biomass development, and it indicated both the development of phototrophs at the soil surface after 20 days and the restriction of phototrophs to the soil surface under light conditions (Figure 1). The presence of light also significantly increased soil pH and reduced extractable NO_3 and extractable

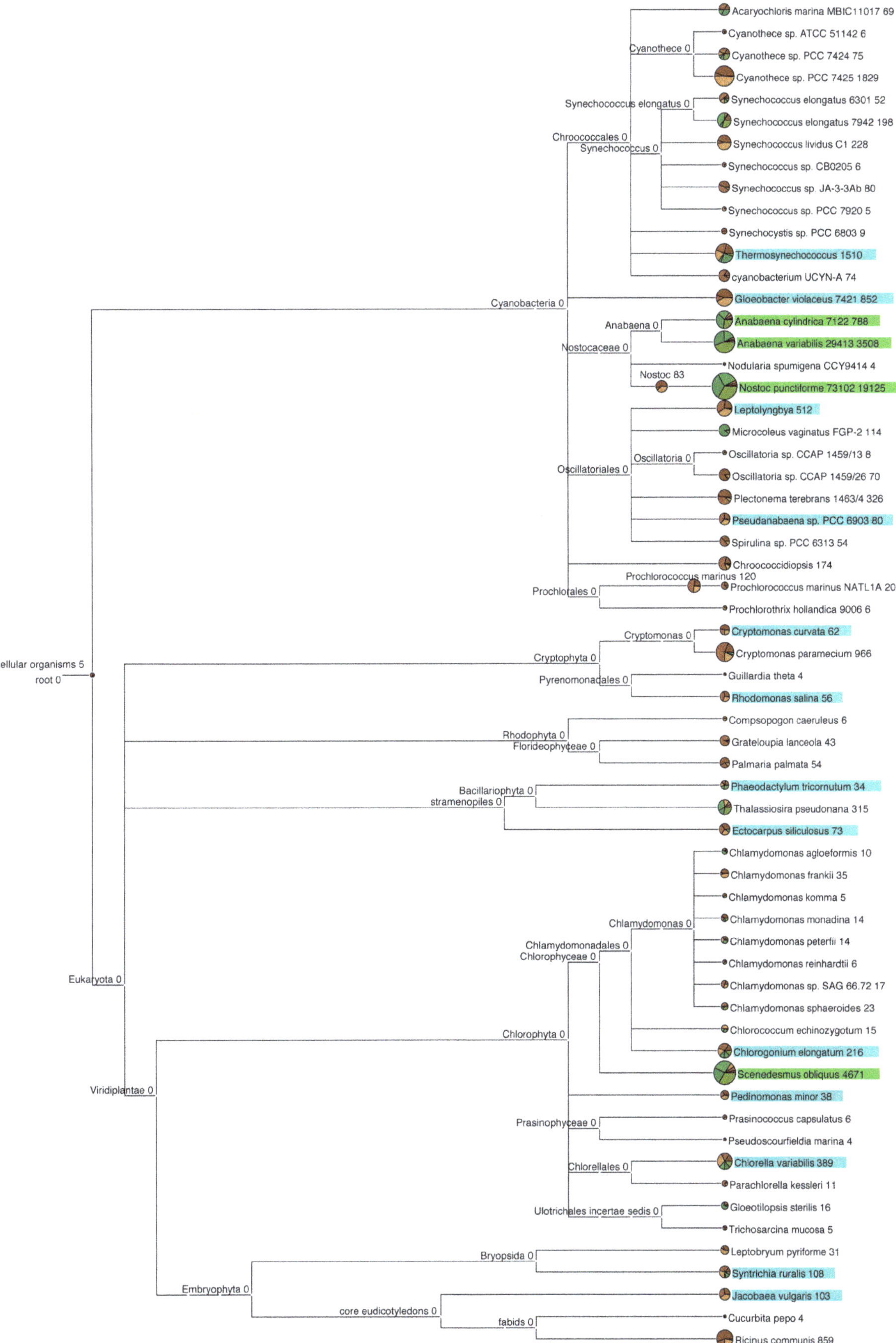

cellular organisms 5
root 0
Cyanobacteria 0
Chroococcales 0
Cyanothece 0
Synechococcus elongatus 0
Synechococcus 0
Acaryochloris marina MBIC11017 69
Cyanothece sp. ATCC 51142 6
Cyanothece sp. PCC 7424 75
Cyanothece sp. PCC 7425 1829
Synechococcus elongatus 6301 52
Synechococcus elongatus 7942 198
Synechococcus lividus C1 228
Synechococcus sp. CB0205 6
Synechococcus sp. JA-3-3Ab 80
Synechococcus sp. PCC 7920 5
Synechocystis sp. PCC 6803 9
Thermosynechococcus 1510
cyanobacterium UCYN-A 74
Gloeobacter violaceus 7421 852
Nostocaceae 0
Anabaena 0
Anabaena cylindrica 7122 788
Anabaena variabilis 29413 3508
Nodularia spumigena CCY9414 4
Nostoc 83
Nostoc punctiforme 73102 19125
Oscillatoriales 0
Oscillatoria 0
Leptolyngbya 512
Microcoleus vaginatus FGP-2 114
Oscillatoria sp. CCAP 1459/13 8
Oscillatoria sp. CCAP 1459/26 70
Plectonema terebrans 1463/4 326
Pseudanabaena sp. PCC 6903 80
Spirulina sp. PCC 6313 54
Chroococcidiopsis 174
Prochlorales 0
Prochlorococcus marinus 120
Prochlorococcus marinus NATL1A 20
Prochlorothrix hollandica 9006 6
Eukaryota 0
Cryptophyta 0
Cryptomonas 0
Pyrenomonadales 0
Cryptomonas curvata 62
Cryptomonas paramecium 966
Guillardia theta 4
Rhodomonas salina 56
Rhodophyta 0
Florideophyceae 0
Compsopogon caeruleus 6
Grateloupia lanceola 43
Palmaria palmata 54
stramenopiles 0
Bacillariophyta 0
Phaeodactylum tricornutum 34
Thalassiosira pseudonana 315
Ectocarpus siliculosus 73
Viridiplantae 0
Chlorophyta 0
Chlorophyceae 0
Chlamydomonadales 0
Chlamydomonas 0
Chlamydomonas agloeformis 10
Chlamydomonas frankii 35
Chlamydomonas komma 5
Chlamydomonas monadina 14
Chlamydomonas peterfii 14
Chlamydomonas reinhardtii 6
Chlamydomonas sp. SAG 66.72 17
Chlamydomonas sphaeroides 23
Chlorococcum echinozygotum 15
Chlorogonium elongatum 216
Scenedesmus obliquus 4671
Pedinomonas minor 38
Prasinophyceae 0
Prasinococcus capsulatus 6
Pseudoscourfieldia marina 4
Chlorellales 0
Chlorella variabilis 389
Parachlorella kessleri 11
Ulotrichales incertae sedis 0
Gloeotilopsis sterilis 16
Trichosarcina mucosa 5
Embryophyta 0
Bryopsida 0
Leptobryum pyriforme 31
Syntrichia ruralis 108
core eudicotyledons 0
Jacobaea vulgaris 103
fabids 0
Cucurbita pepo 4
Ricinus communis 859

Figure 4. Phototroph community structure at the soil surface under light and dark conditions. The diversity and abundance of phototrophs (23S rRNA genes of plastids) at the soil surface of a pasture soil after 80 days incubation under light or dark conditions. Data is presented in MEGAN as an OTU table created in QIIME at a 97% similarity threshold (uclust). The number of reads that can be assigned to each taxon are shown at the end of each node. Pie charts show the proportion of reads assigned to each sample incubated under light (green) and dark (brown) conditions with replicates displayed as shades of these colours. Taxonomic assignments with only one read were removed. Significant differences in the read abundance of sequences between light and dark samples are highlighted in green when abundance is significantly higher under light conditions and in blue when abundance is significantly higher under dark conditions ($p<0.05$).

Mg at both the soil surface and underlying bulk soil under light compared to dark conditions (Table 1). Therefore, although phototrophs appeared to be restricted to the soil surface, the influence of light extended to bulk soil (Figure 1; Table 1).

TRFLP analysis of phototrophs was used as a fine-scale assessment of community structure, and it showed development of distinct communities at the soil surface and bulk soil under light compared to dark incubated soil (Figures 2a–2c). In contrast to chlorophyll *a* data, TRFLP analysis showed no difference in phototroph community structure between the soil surface and underlying bulk soil under light conditions (Figures 2a–2c). Therefore, fine-scale molecular analysis has shown a new depth of influence of light on phototroph community structure that previous broad-scale assessments have missed [1]. It has previously been shown that approximately 0.3% of light is transmitted beyond the top 2 mm of soils with the highest transmittance of light [62]. Therefore, these shifts in phototroph community structure in bulk soil may be driven by attenuated light penetrating small cracks present at the soil surface. Alternatively, penetration of filamentous cyanobacteria into underlying soil may be a consequence of primary production at the soil surface under light conditions. These hypotheses require further testing, particularly in cracking clay soils where light penetration through soil cracks could result in significant shifts in phototroph community structure at even greater depths.

Distinct bacterial and fungal communities developed at the soil surface under light conditions compared to bulk soil, and dark incubated soil (Figures 2d & 2e). Although chlorophyll *a* data showed the development of phototrophs after only 20 days (Figure 1), shifts in bacterial and fungal communities were only evident after 40 days (Figures 2d & 2e). This time lag may be controlled by the time taken for light to indirectly affect soil pH and/or nutrient availability. The influence of light on bacterial and fungal communities was restricted to the soil surface which suggests they are either directly responding to light which is attenuated at lower depths, and/or indirectly responding to nutrients that are only altered at the soil surface, presumably as a result of the growth of phototrophs, such as extractable P or extractable K. Alternatively, heterotrophic bacterial and fungal communities may have an indirect response to availability of C fixed by phototrophs at the soil surface.

Light may also exert an additional indirect effect on community structure by elevating temperature and therefore accelerating the frequency of drying-rewetting cycles at the soil surface. It has previously been shown that drying-rewetting regimes can influence bacterial composition [63,64] and fungal PLFA [65]. Placella *et al.* (2012) showed significant declines in the relative abundances of Actinobacteria and Acidobacteria, significant increases in the relative abundances of β- and γ- proteobacteria, and specific α-proteobacteria such as Sphingomonadales, and a bell-shaped response for Bacilli after soil re-wetting [64]. Relative composition analysis showed a similar effect of light on Bacilli and Sphingomonadales in the current study, which could be a consequence of more pronounced wet-dry cycles under light compared to dark conditions (Figure 6). However, it is important to note that Placella *et al.* (2012) investigated shifts in active communities over a short time-period (72-hour) after total soil water content was increased by ~30% [64]. In contrast, weekly monitoring of soil moisture content in the current study showed water content did not differ by >1% between light and dark incubated samples.

Studies of the soil surface have typically focused on how bacterial and fungal communities differ based on geographical location, desert type, or aridity level; a direct impact of light on heterotrophic communities, however, has not been reported previously [6,11,36]. Moreover, we show community shifts between 40 and 80 days following a simulated tillage event, which adds to studies conducted under agricultural cropping systems, which have shown phototroph development between 50 and 80 days after tillage [15].

Phototroph diversity has been investigated using cultivation-dependent techniques [7,9,34,32,33] or molecular analysis targeting bacterial diversity in arid lands [2,3,6,10,11,34], however, we reveal the diversity of both cyanobacteria and eukaryotic phototrophs at the soil surface of a temperate soil using 454 pyrosequencing. Using relative composition analysis, we show specific cyanobacterial taxa being selected for by light, namely *N. punctiforme*, *A. cylindrica* and *A. variabilis* (Figures 3–4; Table 2). The fact that relative composition analysis showed that significantly more reads were assigned to cryptomonads, red algae, brown algae, mosses, and angiosperms in the dark reflects that these proliferated less than cyanobacteria in the light but are nonetheless present in the seed bank of phototrophs (Figure 4; Table 2). The dominant cyanobacteria of BSCs has been shown to be influenced by several factors, including the type of BSC [3], successional stage [4], underlying soil substrata [2], and the level of aridity [11]. We show a selection for the diazotrophic cyanobacterium *N. punctiforme* at the surface of temperate soil, consistent with results documented in mature, or late-successional BSCs from arid lands

Table 2. Relative read abundance of sequences with close homology to cyanobacteria and eukaryotic phototrophs from the soil surface of a pasture soil after incubation under light or dark conditions for 80 days (±1 standard error).

Taxonomy	Light (%)	Dark (%)
Cyanobacteria	63.8±3.36	82.7±2.03**
Eukaryotes	36.2±3.36	17.3±2.03**
- Green algae	12.8±1.76	15.8±1.89
- Red algae	1.1±0.036	0.01±0.00*
- Brown algae	0.67±0.08	0.03±0.02**
- Diatoms	1.1±0.21	0.93±0.24
- Cryptomonads	10.6±3.91	0.24±0.21*
- Mosses	1.1±0.18	0.10±0.03**
- Angiosperms	9.5±3.15	0.11±0.07*

Significant differences between light and dark treatments is indicated by a *($p\leq0.05$) or **($p\leq0.01$).

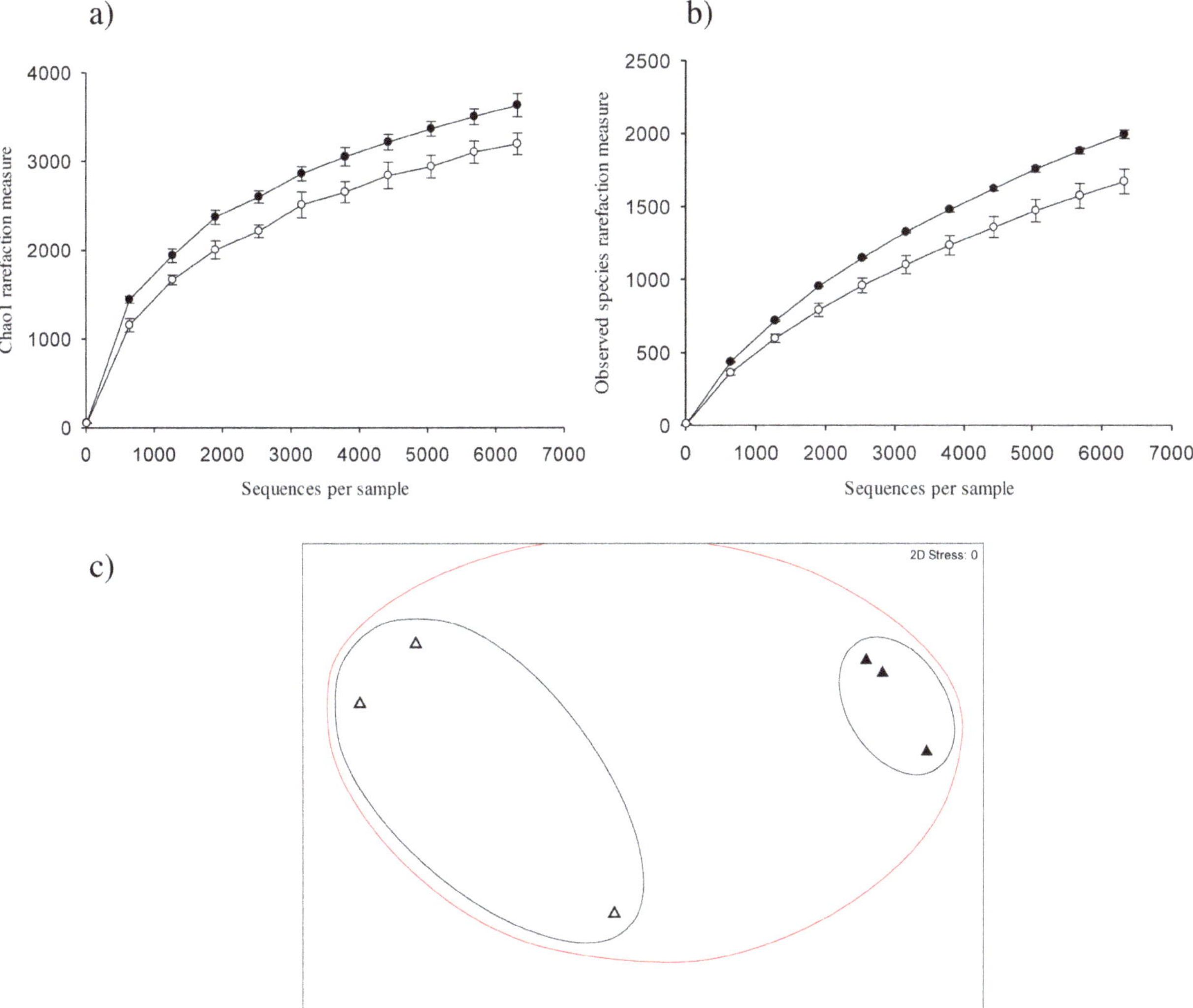

Figure 5. Bacterial diversity at the soil surface under light and dark conditions. α diversity estimates Chao1 (a) and Observed Species (b) and non-metric multidimensional scaling of community structure similarity (c) for bacteria (16S rRNA) at the soil surface of a pasture soil after 80 days incubation under light (open symbols) or dark (closed symbols) conditions. OTU clustering was performed at the 97% similarity threshold using uclust. Error bars are ±1 S.E. Non-metric multidimensional scaling shows clustering based on the similarity of microbial community structure between treatments: 45% (red cluster) and 55% (black cluster).

[2,4,6,7,10,34] (Figure 4). This suggests that diazotrophic cyanobacteria may also be important ecosystem engineers in temperate environments, in addition to arid zones [4,20–24]. However, the contribution of surface communities to N_2 fixation in temperate soils or agricultural systems remains to be elucidated. Such data could be beneficial for informing agricultural management decisions, for example, the realization that diazotrophs were able to fix an agriculturally significant proportion of N_2 could influence decisions relating to soil tillage and the amount, frequency and timing of N fertiliser application under cropping systems.

454 pyrosequencing revealed that light also selected for heterotrophic bacteria at the soil surface (Figure 6). We found that in contrast to the desert soils studied to date [2,3,6,11], few bacterial sequences (<4%) had close homology to cyanobacteria, allowing shifts in heterotrophic bacteria to be assessed (Figure 5–6). The comparative reduction in bacterial diversity under light conditions was not due to a selection for cyanobacteria (Figure 6; Table 2) as α diversity was still significantly lower under light conditions ($p \leq 0.01$) after the removal of photosynthetic bacterial OTUs from analysis (results not shown). The differences in diversity may be due to an input of C through photosynthesis and or N by N_2 fixation, which could indirectly select for specific heterotrophic bacteria. This is analogous to the 'rhizosphere effect.' The rhizosphere is the area of soil under the influence of roots. Studies have shown that the rhizosphere can select for particular microbial communities and that this selection is plant-specific [66]. A similar effect may be occurring at the soil surface under light conditions. Moreover, taken with TRFLP results which show that the impact of light on bacterial community structure is restricted to the upper 3 mm of the soil surface (Figure 2c), a new research area of microbial influence may be emerging, which we term the 'crustosphere.'

TRFLP and 454 pyrosequencing revealed that light also significantly impacted fungal community structure at the soil surface (Figure 2e & 8). The relatively few shifts in fungal communities could be due to the development stage of phototroph communities. BSCs typically undergo a succession from cyano-bacteria- to lichen- to moss- dominated crusts in arid zones [32].

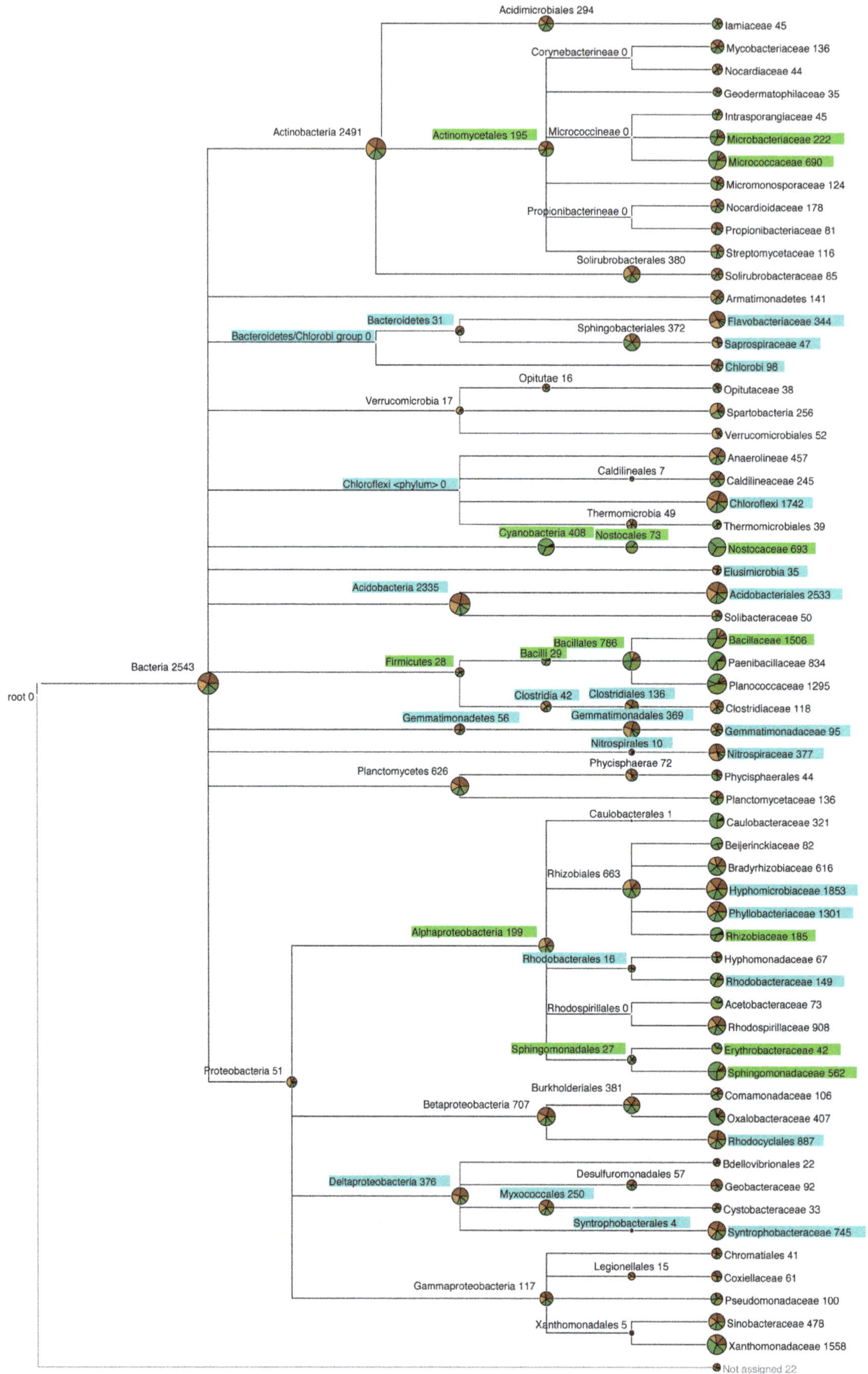

Acidimicrobiales 294
Iamiaceae 45
Corynebacterineae 0
Mycobacteriaceae 136
Nocardiaceae 44
Geodermatophilaceae 35
Intrasporangiaceae 45
Actinobacteria 2491
Actinomycetales 195
Micrococcineae 0
Microbacteriaceae 222
Micrococcaceae 690
Micromonosporaceae 124
Propionibacterineae 0
Nocardioidaceae 178
Propionibacteriaceae 81
Streptomycetaceae 116
Solirubrobacterales 380
Solirubrobacteraceae 85
Armatimonadetes 141
Bacteroidetes 31
Flavobacteriaceae 344
Bacteroidetes/Chlorobi group 0
Sphingobacteriales 372
Saprospiraceae 47
Chlorobi 98
Opitutae 16
Opitutaceae 38
Verrucomicrobia 17
Spartobacteria 256
Verrucomicrobiales 52
Anaerolineae 457
Caldilineales 7
Caldilineaceae 245
Chloroflexi <phylum> 0
Chloroflexi 1742
Thermomicrobia 49
Thermomicrobiales 39
Cyanobacteria 408
Nostocales 73
Nostocaceae 693
Elusimicrobia 35
Acidobacteria 2335
Acidobacteriales 2533
Solibacteraceae 50
Bacillales 786
Bacillaceae 1506
Bacilli 29
Firmicutes 28
Paenibacillaceae 834
Bacteria 2543
Planococcaceae 1295
root 0
Clostridia 42
Clostridiales 136
Clostridiaceae 118
Gemmatimonadetes 56
Gemmatimonadales 369
Gemmatimonadaceae 95
Nitrospirales 10
Nitrospiraceae 377
Phycisphaerae 72
Planctomycetes 626
Phycisphaerales 44
Planctomycetaceae 136
Caulobacterales 1
Caulobacteraceae 321
Beijerinckiaceae 82
Bradyrhizobiaceae 616
Rhizobiales 663
Hyphomicrobiaceae 1853
Phyllobacteriaceae 1301
Alphaproteobacteria 199
Rhizobiaceae 185
Rhodobacterales 16
Hyphomonadaceae 67
Rhodobacteraceae 149
Rhodospirillales 0
Acetobacteraceae 73
Rhodospirillaceae 908
Sphingomonadales 27
Erythrobacteraceae 42
Proteobacteria 51
Sphingomonadaceae 562
Burkholderiales 381
Comamonadaceae 106
Betaproteobacteria 707
Oxalobacteraceae 407
Rhodocyclales 887
Bdellovibrionales 22
Desulfuromonadales 57
Deltaproteobacteria 376
Geobacteraceae 92
Myxococcales 250
Cystobacteraceae 33
Syntrophobacterales 4
Syntrophobacteraceae 745
Chromatiales 41
Legionellales 15
Coxiellaceae 61
Gammaproteobacteria 117
Pseudomonadaceae 100
Xanthomonadales 5
Sinobacteraceae 478
Xanthomonadaceae 1558
Not assigned 22

Figure 6. Bacterial community structure at the soil surface under light and dark conditions. The diversity and abundance of bacteria (16S rRNA gene) at the soil surface of a pasture soil after 80 days incubation under light or dark conditions. Data is presented in MEGAN as an OTU table created in QIIME at a 97% similarity threshold (uclust). The OTU table is presented at the taxonomic level of family. The number of reads that can be assigned using the RDP classifier at a confidence level of 80% are shown at the end of each node. Pie charts show the proportion of reads assigned to each sample incubated under light (green) and dark (brown) conditions with replicates shown as shades of these colours. Taxonomic assignments accounting for $<0.5\%$ total sequence abundance were removed. Significant differences in the read abundance of sequences between light and dark samples are highlighted in green when abundance is significantly higher under light conditions and in blue when abundance is significantly higher under dark conditions ($p<0.05$).

In the present study, the soil surface was dominated by cyanobacteria (Table 2). However, if the surface was left to develop to a lichen dominated community, more significant shifts in fungal community structure may be evident as lichen symbioses develop. However, parallels can still be drawn between soil surface fungal communities of temperate and arid lands, for example, relative composition analysis showed that Ascomycota were the dominant fungi in the present study in addition to surveys in the Colorado plateau, Chihuahuan desert and Sonoran deserts, USA [35–36].

In conclusion, the application of fine-scale molecular analysis gave new insights into soil surface community structure. We show differences in phototroph community structure in bulk soil in the presence of light, which have not previously been detected. We

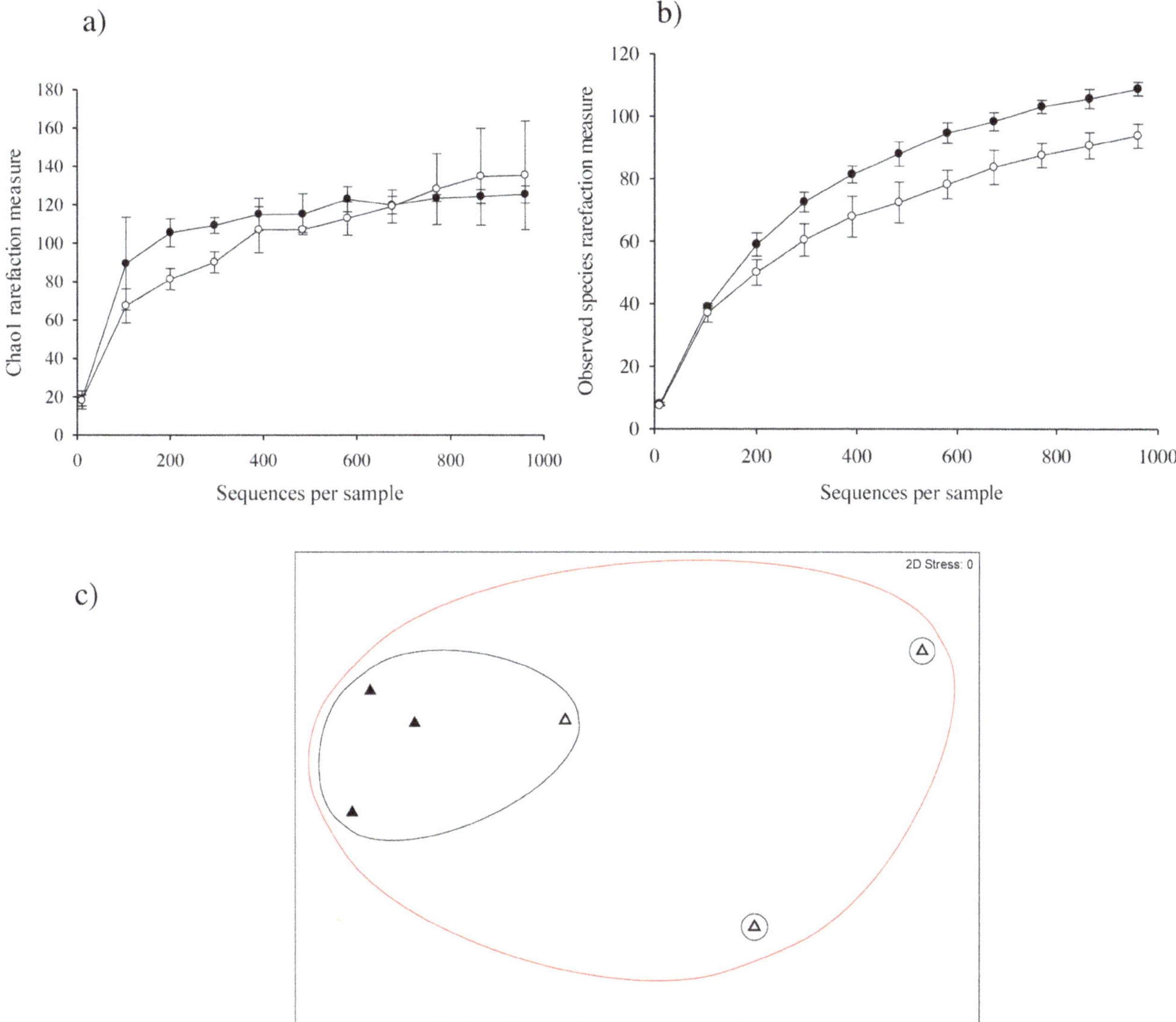

Figure 7. Fungal diversity at the soil surface under light and dark conditions. α diversity estimates Chao1 (a) and Observed Species (b) and non-metric multidimensional scaling of community structure similarity (c) for fungi (ITS region) at the soil surface of a pasture soil after 80 incubation under light (open symbols) or dark (closed symbols) conditions. OTU clustering was performed at the 97% similarity threshold using UCLUST. Error bars are ± 1 S.E. Non-metric multidimensional scaling shows clustering based on the similarity of microbial community structure between treatments: 55% (red cluster) and 70% (black cluster).

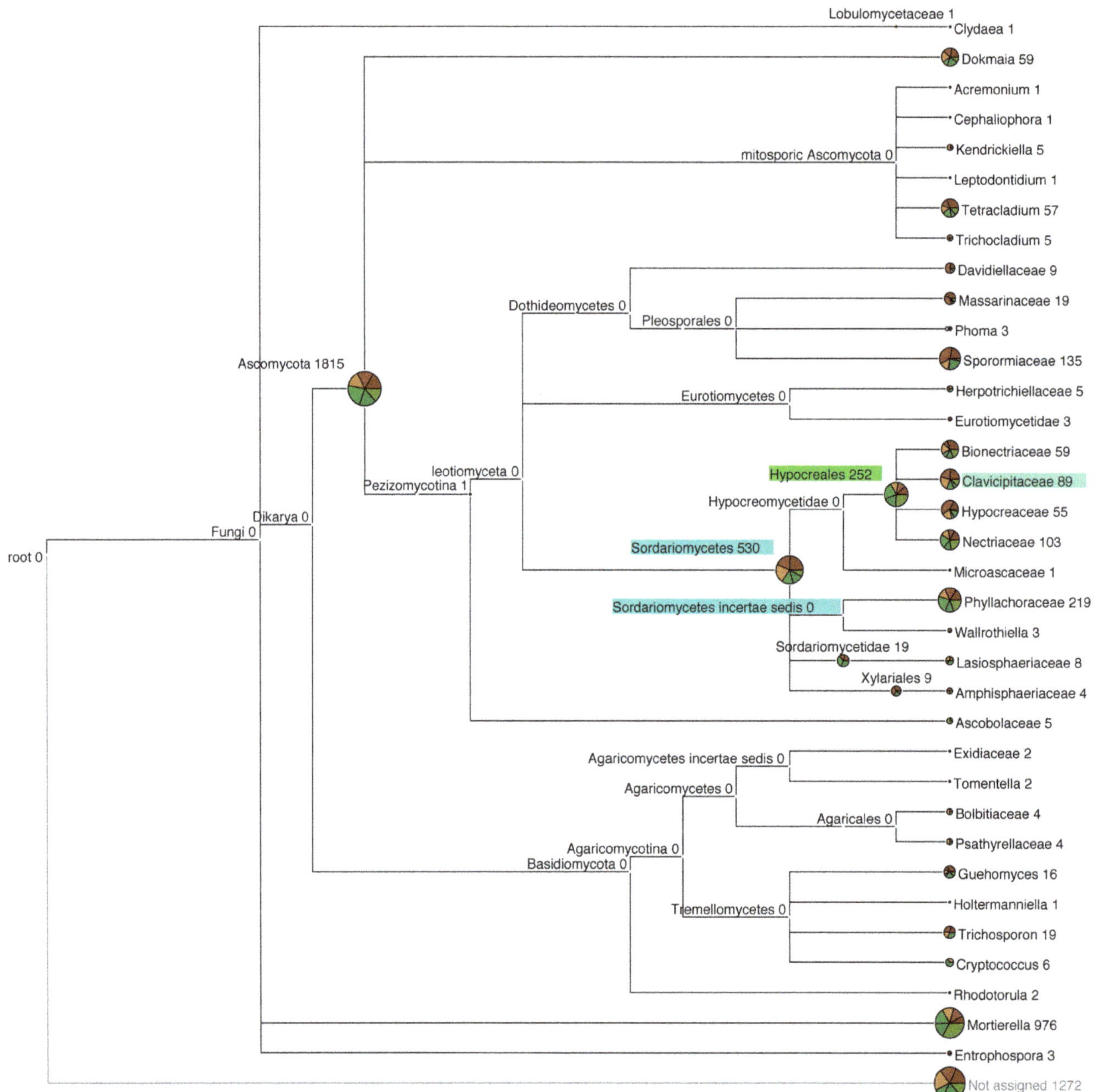

Figure 8. Fungal community structure at the soil surface under light and dark conditions. The diversity and abundance of fungi (ITS region) at the soil surface of a pasture soil after 80 days incubation under light or dark conditions. Data is presented in MEGAN as an OTU table created in QIIME at a 97% similarity threshold (uclust). The number of reads that can be assigned using the RDP classifier at a confidence level of 80% to each taxon are shown at the end of each node. Pie charts show the proportion of reads assigned to each sample incubated under light (green) and dark (brown) conditions with replicates shown as shades of these colours. Significant differences in the read abundance of sequences between light and dark samples are highlighted in green when abundance is significantly higher under light conditions and in blue when abundance is significantly higher under dark conditions ($p<0.05$).

also show that the soil surface harbours distinct heterotrophic bacterial and fungal communities. Future work should focus on the ecological significance of both phototrophic and heterotrophic communities, particularly in temperate zones, including their functional importance in agro-ecosystems.

Supporting Information

Figure S1 Phototroph development at the soil surface. Development of phototrophs at the surface of a pasture soil; (a) 9 days incubation under light conditions; (b) 14 days incubation under light conditions, and; (c) Comparison of dark and light incubated soil after 40 days incubation.

Figure S2 Soil pH. pH at the surface (▲) and bulk (▼) of pasture soil after incubation under light (open symbols) or dark (closed symbols) conditions. Errors bars are ±1 standard error.

Figure S3 Soil nitrate. Nitrate at the surface (▲) and bulk (▼) of pasture soil after incubation under light (open symbols) or dark (closed symbols) conditions. Errors bars are ±1 standard error.

Figure S4 Soil potassium. Potassium at the surface (▲) and bulk (▼) of pasture soil after incubation under light (open symbols) or dark (closed symbols) conditions. Errors bars are ±1 standard error.

Figure S5 Soil magnesium. Magnesium at the surface (▲) and bulk (▼) of pasture soil after incubation under light (open symbols) or dark (closed symbols) conditions. Errors bars are ±1 standard error.

Figure S6 Soil phosphorous. Phosphorous at the surface (▲) and bulk (▼) of pasture soil after incubation under light (open symbols) or dark (closed symbols) conditions. Errors bars are ±1 standard error.

Table S1 Soil properties of Gartenacker topsoil (10–20 cm) taken from Switzerland.

Table S2 Primer pairs used to investigate bacterial, fungal and phototroph community structure in Gartenacker soil incubated under light and dark conditions.

Table S3 Number of phototroph sequences removed at each processing step.

Table S4 Number of bacterial sequences removed at each processing step.

Table S5 Number of fungal sequences removed at each processing step.

Acknowledgments

We wish to thank Dr. Sally Hilton for help with TRFLP and Mr. Matthew Mitchell for help with soil chemical analysis.

Author Contributions

Conceived and designed the experiments: LOD HS SM IB RGO GDB. Performed the experiments: LOD. Analyzed the data: LOD HS SM IB RGO GDB. Contributed reagents/materials/analysis tools: LOD HS SM IB RGO GDB. Wrote the paper: LOD HS SM IB RGO GDB.

References

1. Jeffery S, Harris JA, Rickson RJ, Ritz K (2009) The spectral quality of light influences the temporal development of the microbial phenotype at the arable soil surface. Soil Biol Biochem 41: 553–560.
2. Garcia-Pichel F, Lopez-Cortes A, Nubel U (2001) Phylogenetic and morphological diversity of cyanobacteria in soil desert crusts from the Colorado Plateau. Appl Environ Microbiol 67: 1902–1910.
3. Redfield E, Barns SM, Belnap J, Daane LL, Kuske CR (2002) Comparative diversity and composition of cyanobacteria in three predominant soil crusts of the Colorado Plateau. Fems Microbiol Ecol 40: 55–63.
4. Yeager CM, Kornosky JL, Housman DC, Grote EE, Belnap J, et al. (2004) Diazotrophic community structure and function in two successional stages of biological soil crusts from the Colorado plateau and Chihuahuan desert. Appl Environ Microbiol 70: 973–983.
5. Langhans TM, Storm C, Schwabe A (2009a) Community assembly of biological soil crusts of different successional stages in a temperate sand ecosystem, as assessed by direct determination and enrichment techniques. Microb Ecol 58: 394–407.
6. Abed RMM, Al Kharusi S, Schramm A, Robinson MD (2010) Bacterial diversity, pigments and nitrogen fixation of biological desert crusts from the Sultanate of Oman. FEMS Microbiol Ecol 72: 418–428.
7. Zhang B, Zhang Y, Downing A, Niu Y (2011) Distribution and composition of cyanobacteria and microalgae associated with biological soil crusts in the Gurbantunggut Desert, China. Arid Land Res Manag 25: 275–293.
8. Beraldi-Campesi H, Hartnett HE, Anbar A, Gordon GW, Garcia-Pichel F (2009) Effect of biological soil crusts on soil elemental concentrations: implications for biogeochemistry and as traceable biosignatures of ancient life on land. Geobiology 7: 348–359.
9. Zhang J, Zhang YM, Downing A, Cheng JH, Zhou XB, et al. (2009) The influence of biological soil crusts on dew deposition in Gurbantunggut Desert, Northwestern China. J Hydrol 379: 220–228.
10. Nagy ML, Perez A, Garcia-Pichel F (2005) The prokaryotic diversity of biological soil crusts in the Sonoran Desert (Organ Pipe Cactus National Monument, AZ). Fems Microbiol Ecol 54: 233–245.
11. Zaady E, Ben-David EA, Sher Y, Tzirkin R, Nejidat A (2010) Inferring biological soil crust successional stage using combined PLFA, DGGE, physical and biophysiological analyses. Soil Biol Biochem 42: 842–849.
12. Pointing SB, Belnap J (2012) Microbial colonization and controls in dryland systems. Nat Rev Microbiol 10: 551–562.
13. Li XR, Jia XH, Long LQ, Zerbe S (2005) Effects of biological soil crusts on seed bank, germination and establishment of two annual plant species in the Tengger Desert (N China). Plant and Soil 277: 375–385.
14. Veluci RM, Neher DA, Weicht TR (2006) Nitrogen fixation and leaching of biological soil crust communities in mesic temperate soils. Microb Ecol 51: 189–196.
15. Knapen A, Poesen J, Galindo-Morales P, De Baets S, Pals A (2007) Effects of microbiotic crusts under cropland in temperate environments on soil erodibility during concentrated flow. Earth Surf Proc Land 32: 1884–1901.
16. Belnap J, Gillette DA (1997) Disturbance of biological soil crusts: Impacts on potential wind erodibility of sandy desert soils in southeastern Utah. Land Degrad Dev 8: 355–362.
17. Eldridge DJ, Leys JF (2003) Exploring some relationships between biological soil crusts, soil aggregation and wind erosion. J Arid Environ 53: 457–466.
18. Chamizo S, Canton Y, Miralles I, Domingo F (2012) Biological soil crust development affects physicochemical characteristics of soil surface in semiarid ecosystems. Soil Biol Biochem 49: 96–105.
19. Tisdall JM, Nelson SE, Wilkinson KG, Smith SE, McKenzie BM (2012) Stabilisation of soil against wind erosion by six saprotrophic fungi. Soil Biol Biochem 50: 134–141.
20. Belnap J (2002) Nitrogen fixation in biological soil crusts from southeast Utah, USA. Biol Fert Soils 35: 128–135.
21. Johnson SL, Budinoff CR, Belnap J, Garcia-Pichel F (2005) Relevance of ammonium oxidation within biological soil crust communities. Environ Microbiol 7: 1–12.
22. Zhao Y, Xu M, Belnap J (2010) Potential nitrogen fixation activity of different aged biological soil crusts from rehabilitated grasslands of the hilly Loess Plateau, China. J Arid Environ 74: 1186–1191.
23. Stewart KJ, Coxson D, Siciliano SD (2011) Small-scale spatial patterns in N_2-fixation and nutrient availability in an arctic hummock-hollow ecosystem. Soil Biol Biochem 43: 133–140.
24. Li XR, He MZ, Zerbe S, Li XJ, Liu LC (2010) Micro-geomorphology determines community structure of biological soil crusts at small scales. Earth Surf Proc Land 35: 932–940.
25. Housman DC, Powers HH, Collins AD, Belnap J (2006) Carbon and nitrogen fixation differ between successional stages of biological soil crusts in the Colorado Plateau and Chihuahuan Desert. J Arid Environ 66: 620–634.
26. Yoshitake S, Uchida M, Koizumi H, Kanda H, Nakatsubo T (2010) Production of biological soil crusts in the early stage of primary succession on a high arctic glacier foreland. New Phytol 186: 451–460.
27. Castillo-Monroy AP, Maestre FT, Rey A, Soliveres S, Garcia-Palacios P (2011) Biological soil crust microsites are the main contributor to soil respiration in a semiarid ecosystem. Ecosystems 14: 835–847.
28. Langhans TM, Storm C, Schwabe A (2009b) Biological soil crusts and their microenvironment: impact on emergence, survival and establishment of seedlings. Flora 204: 157–168.
29. Harper KT, Belnap J (2001) The influence of biological soil crusts on mineral uptake by associated vascular plants. J Arid Environ 47: 347–357.

30. Lange OL, Kidron GJ, Budel B, Meyer A, Kilian E, et al. (1992) Taxonomic composition and photosynthetic characteristics of the biological soil crusts covering sand dunes in the western Negev desert. Funct Ecol 6: 519–527.
31. Belnap J (1993) Recovery rates of cryptobiotic crusts - inoculant use and assessment methods. Great Basin Nat 53: 89–95.
32. Li XR, Wang XP, Li T, Zhang JG (2002) Microbiotic soil crust and its effect on vegetation and habitat on artificially stabilized desert dunes in Tengger Desert, North China. Biol Fert Soils 35: 147–154.
33. Hawkes CV, Flechtner VR (2002) Biological soil crusts in a xeric Florida shrubland: composition, abundance, and spatial heterogeneity of crusts with different disturbance histories. Microb Ecol 43: 1–12.
34. Steven B, Gallegos-Graves LV, Starkenburg SR, Chain PS, Kuske CR (2012) Targeted and shotgun metagenomic approaches provide different descriptions of dryland soil microbial communities in a manipulated field study. Environ Microbiol Rep 4: 248–256.
35. Bates ST, Garcia-Pichel F (2009) A culture-independent study of free-living fungi in biological soil crusts of the Colorado Plateau: their diversity and relative contribution to microbial biomass. Environ Microbiol 11: 56–67.
36. Bates ST, Nash TH, Garcia-Pichel F (2012) Patterns of diversity for fungal assemblages of biological soil crusts from the southwestern United States. Mycologia 104: 353–361.
37. Thomas KA, Hand LH (2012) Assessing the metabolic potential of phototrophic communities in surface water environments: Fludioxonil as a model compound. Environ Toxicol Chem 31: 2138–2146.
38. Faithful NT (2002) Methods in agricultural chemical analysis, reference book 427. Wallingford, UK: CABI Publishing. 266 p.
39. Ritchie RJ (2006) Consistent sets of spectrophotometric chlorophyll equations for acetone, methanol and ethanol solvents. Photosynth Res 89: 27–41.
40. Hansson LA (1988) chlorophyll-a determination of periphyton on sediments - identification of problems and recommendation of method. Freshwater Biol 20: 347–352.
41. Bold HC (1949) The morphology of *Chlamydomonas chlamydogama* sp. nov. Bull Torrey Bot Club 76: 101–108.
42. Blodgett R (2006) Appendix 2: most probable number from serial dilutions, In: Hammack, T (ed). Bacteriological analytical manual 8th edition. AOAC International: Maryland, USA, 135.
43. Sherwood AR, Presting GG (2007) Universal primers amplify a 23S rDNA plastid marker in eukaryotic algae and cyanobacteria. J Phycol 43: 605–608.
44. Hauben L, Vauterin L, Swings J, Moore ERB (1997) Comparison of 16S ribosomal DNA sequences of all *Xanthomonas* species. Int J Syst Bacteriol 47: 328–335.
45. Marchesi JR, Sato T, Weightman AJ, Martin TA, Fry JC, et al. (1998) Design and evaluation of useful bacterium-specific PCR primers that amplify genes coding for bacterial 16S rRNA. Appl Environ Microbiol 64: 795–799.
46. White TJ, Bruns TD, Lee S, Taylor J (1990) Analysis of phylogenetic relationship by amplification and direct sequencing of ribosomal RNA genes. In: M.A. Innis, D.H. Gelfond, J.J. Sainsky 1 and T.J. White (ed.), PCR protocol: a guide to method and applications. Academic Press, New York, N.Y.
47. Gardes M, Bruns TD (1993) ITS primers with enhanced specificity for basidiomycetes - application to the identification of mycorrhizae and rusts. Mol Ecol 2: 113–118.
48. Hilton S, Bennett AJ, Keane G, Bending GD, Chandler D, et al. (2013) Impact of shortened crop rotation of oilseed rape on soil and rhizosphere microbial diversity in relation to yield decline. PLoS ONE 8. Doi: e59859. doi:10.1371/journal.pone.0059859.
49. Sait L, Galic M, Strugnell RA, Janssen PH (2003) Secretory antibodies do not affect the composition of the bacterial microbiota in the terminal ileum of 10-week-old mice. Appl Env Micro 69: 2100–2109.
50. Dowd SE, Callaway TR, Wolcott RD, Sun Y, McKeehan T, et al. (2008) Evaluation of the bacterial diversity in the feces of cattle using 16S rDNA bacterial tag-encoded FLX amplicon pyrosequencing (bTEFAP). BMC Microbiol 8: 125.
51. Caporaso JG, Kuczynski J, Stombaugh J, Bittinger K, Bushman FD, et al. (2010a) QIIME allows analysis of high-throughput community sequencing data. Nat Methods 7: 335–336.
52. Reeder J, Knight R (2010) Rapidly denoising pyrosequencing amplicon reads by exploiting rank-abundance distributions. Nat Methods 7: 668–669.
53. Edgar RC (2010) Search and clustering orders of magnitude faster than BLAST. BMC Bioinformatics 26: 2460–2461.
54. Caporaso JG, Bittinger K, Bushman FD, DeSantis TZ, Andersen GL, et al. (2010b) PyNAST: a flexible tool for aligning sequences to a template alignment. BMC Bioinformatics 26: 266–267.
55. DeSantis TZ, Hugenholtz P, Larsen N, Rojas M, Brodie EL, et al. (2006) Greengenes, a chimera-checked 16S rRNA gene database and workbench compatible with ARB. Appl Environ Microbiol 72: 5069–5072.
56. Haas BJ, Gevers D, Earl AM, Feldgarden M, Giannoukos G, et al. (2011) Chimeric 16S rRNA sequence formation and detection in Sanger and 454-pyrosequenced PCR amplicons. Genome Res 21: 494–504.
57. Wang Q, Garrity GM, Tiedje JM, Cole JR (2007) Naive Bayesian classifier for rapid assignment of rRNA sequences into the new bacterial taxonomy. Appl Environ Microbiol 73: 5261–5267.
58. Edgar RC, Haas BJ, Clemente JC, Quince C, Knight R (2011) UCHIME improves sensitivity and speed of chimera detection. BMC Bioinformatics 27: 2194–2200.
59. Altschul SF, Gish W, Miller W, Myers EW, Lipman DJ (1990) Basic local alignment search tool. J Mol Biol 215: 403–410.
60. Huson DH, Mitra S, Ruscheweyh H-J, Weber N, Schuster SC (2011) Integrative analysis of environmental sequences using MEGAN4. Genome Res 21: 1552–1560.
61. Chao A (1984) Non-parametric estimation of the number of classes in a population. Scand J Stat 11: 265–270.
62. Benvenuti S (1995) Soil light penetration and dormancy of jimsonweed (Datura-stramonium) seeds. Weed Sci 43: 389–393.
63. Fierer N, Schimel JP, Holden PA (2003) Influence of drying-rewetting frequency on soil bacterial community structure. Microb Ecol 45: 63–71.
64. Placella SA, Brodie EL, Firestone MK (2012) Rainfall-induced carbon dioxide pulses result from sequential resuscitation of phylogenetically clustered microbial groups. Proc Natl Acad Sci U S A 109: 10931–10936.
65. Gordon H, Haygarth PM, Bardgett RD (2008) Drying and rewetting effects on soil microbial community composition and nutrient leaching. Soil Biol Biochem 40: 302–311.
66. Morgan JAW, Bending GD, White PJ (2005) Biological costs and benefits to plant-microbe interactions in the rhizosphere. J Exp Bot 56: 1729–1739.

7

Vertical Profiles of Soil Water Content as Influenced by Environmental Factors in a Small Catchment on the Hilly-Gully Loess Plateau

Bing Wang[1], Fenxiang Wen[1], Jiangtao Wu[1], Xiaojun Wang[1]*, Yani Hu[2]

1 College of Environmental Science and Resources, Shanxi University, Taiyuan, China, **2** Library, Hebei University of Science and Technology, Shijiazhuang, China

Abstract

Characterization of soil water content (SWC) profiles at catchment scale has profound implications for understanding hydrological processes of the terrestrial water cycle, thereby contributing to sustainable water management and ecological restoration in arid and semi-arid regions. This study described the vertical profiles of SWC at the small catchment scale on the hilly and gully Loess Plateau in Northeast China, and evaluated the influences of selected environmental factors (land-use type, topography and landform) on average SWC within 300 cm depth. Soils were sampled from 101 points across a small catchment before and after the rainy season. Cluster analysis showed that soil profiles with high-level SWC in a stable trend (from top to bottom) were most commonly present in the catchment, especially in the gully related to terrace. Woodland soil profiles had low-level SWC with vertical variations in a descending or stable trend. Most abandoned farmland and grassland soil profiles had medium-level SWC with vertical variations in varying trends. No soil profiles had low-level SWC with vertical variations in an ascending trend. Multi-regression analysis showed that average SWC was significantly affected by land-use type in different soil layers (0–20, 20–160, and 160–300 cm), generally in descending order of terrace, abandoned farmland, grassland, and woodland. There was a significant negative correlation between average SWC and gradient along the whole profile ($P<0.05$). Landform significantly affected SWC in the surface soil layer (0–20 cm) before the rainy season but throughout the whole profile after the rainy season, with lower levels on the ridge than in the gully. Altitude only strongly affected SWC after the rainy season. The results indicated that land-use type, gradient, landform, and altitude should be considered in spatial SWC estimation and sustainable water management in these small catchments on the Loess Plateau as well as in other complex terrains with similar settings.

Editor: Andrew C. Singer, NERC Centre for Ecology & Hydrology, United Kingdom

Funding: This work was supported by the Natural Science Foundation of China (No. 41201277 and 41101025) and the Natural Science Foundation of Shanxi Province of China (No. 2014011034-2). The funders had no role in study design, data collection and analysis, decision to publish, or preparation of the manuscript.

Competing Interests: The authors have declared that no competing interests exist.

* Email: xjwang@sxu.edu.cn

Introduction

Soil water content (SWC) is a critical factor for plant growth and a determinant of plant distribution in arid and semiarid areas such as China's Loess Plateau [1, 2]. Vertical distribution of SWC can greatly affect soil water movement [3], thereby greatly affecting the biomass production and water use efficiency of plants (e.g., switchgrass) under water stress [4]. Plant-available water stored in the soil profile has a buffering capacity, which, in deep layers, prolongs or alleviates the effects of seasonal or inter-annual drought on plant growth and soil water flux to the atmosphere [5–7]. Research has provided strong evidence that deep soil water depletion plays a key role in sustainable agriculture, ecological restoration, and terrestrial water cycling on the Loess Plateau [8–10]. However, measurement of SWC profiles has been frequently conducted at different spatial scales. The results thus need to be converted before comparison analysis or practical uses. The SWC profile in small catchment is considered to be at a moderate scale for data exchanging. In particular, small catchment is thought to be the basic unit for integrated soil and water loss management in complicated terrain of the Loess Plateau [11, 12]. Characterization of SWC profiles and evaluation of relevant influencing factors at the small catchment scale have implications for hydrological modeling of soil water dynamics and sustainable management of soil water resources in similar areas.

Classical statistics is frequently used to analyze the variability of SWC profiles at the small catchment scale, which involves the estimation of descriptive parameters such as average (mean), variance, standard deviation (STD), and coefficient of variation (CV). Average SWC at individual soil depth intervals or across the whole soil profile is extensively determined. The CV of SWC is also routinely calculated as the temporal variable in a certain period of time or the spatial variable across a specific area. The SWC profile can be divided into distinct intervals by considering its average and CV which exhibit complex spatial-temporal relationships in several plots or watersheds [13–16]. Additionally, ranking method, clustering method, and semivariogram model have been applied for the division of SWC profile [3, 17–20]. However, the above-mentioned methods cannot clearly reflect the variation trend in SWC profiles. Thus, great effort has been made

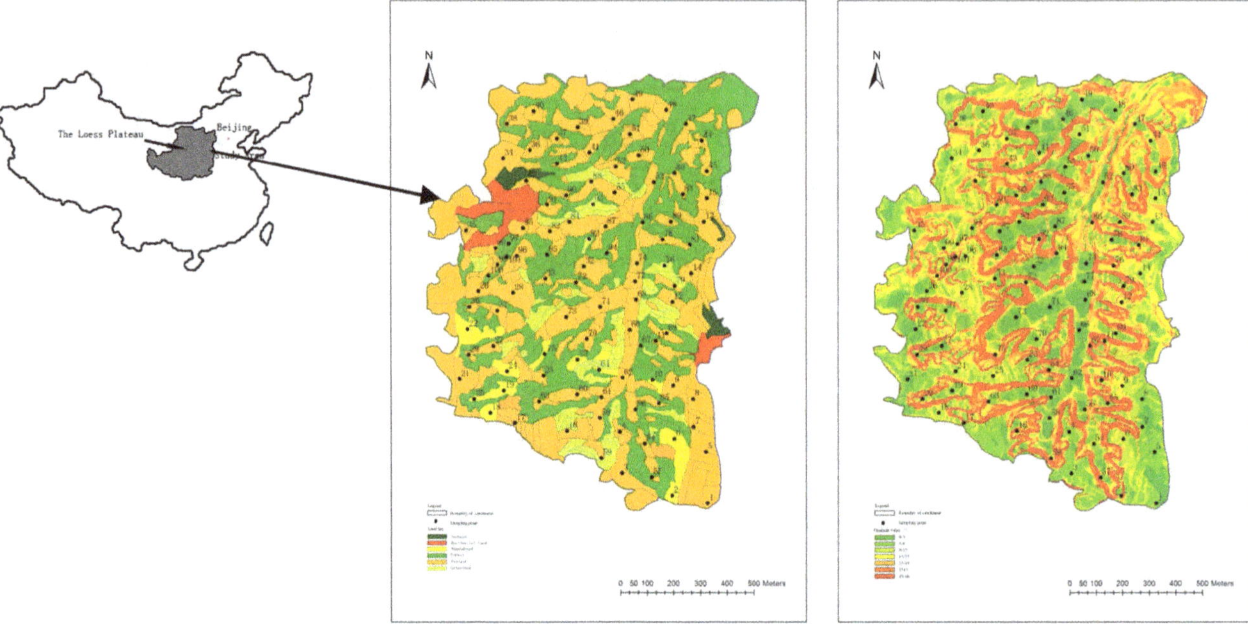

Figure 1. Location of Sanyanjing catchment and distribution of 101 sampling points in the Sanyanjing catchment.

to describe the vertical profiles of SWC through comparing variation curves or variation ranges, in small watersheds related to different land-use types, vegetation species, and/or terrain factors [3, 17, 18, 20–22]. If a massive sample size is involved, however, it becomes difficult to distinguish the vertical profiles and major influencing factors of SWC by direct comparisons.

In recent decades, a great number of studies have been conducted on the spatiotemporal variability of SWC and related influencing factors worldwide. Canton [23] pointed out that wasteland-scale spatial variability of SWC is mainly controlled by surface cover and soil properties in a semi-arid region of Spain, where surface cover counteracts the influence of terrain factors (including gradient, aspect, topographic wetness index, and distance from the river) on SWC distribution. Burnt [24] reported a topographic index which can simulate changes in high-level SWC in a humid climate zone of Devin County, UK. O'loughlin [25] estimated the spatial pattern of SWC distribution in a small catchment using humidity index model based on digital terrain dataset. Hawley [15] discovered that topography is the major factor responsible for the spatial distribution of SWC in an agricultural region, where resultant SWC variation is diminished by vegetation in a moist climate zone in Chickasha, USA. In arid and semi-arid areas, catchment-scale distribution of SWC is strongly affected by land-use/vegetation and topographic indices, e.g., land-use type, soil organic matter content, tillage, soil physical properties, gradient, and aspects [6, 17–19, 21, 26].

Many researchers have focused on the quantification of environmental parameters such as topographic factor, vegetation type, soil texture, and land-use type, in attempt to evaluate their impacts on the variability of SWC. At the small catchment scale, little information is available on the major factors affecting vertical profiles of SWC in cinnamon soil (Haplic Lixisols, FAO) zone on the hilly and gully Loess Plateau [3, 19]. As a regional water reservoir experiencing depletion, the plateau region requires measurements and characterization of deep SWC profiles for the thick soil layer. However, soil sampling for SWC profile analysis at the catchment scale has been commonly conducted at <200 cm depth [18–20, 23]. Wang et al. [21] exceptionally examined SWC along the 0–21 m soil profile on the Loess Plateau, but the reliability of their tests might be affected by a small sample size (11 sites). Deep soil sampling at a larger number of sites and statistical analysis of parameters involving soil depth information will contribute to better understanding of the vertical profiles and influencing factors of SWC.

In the present study, we characterized the vertical profiles of SWC in a small catchment on the Loess Plateau by cluster analysis of two descriptive parameters (mean and regression gradient). Sampling was carried out in the 0–300 cm profile at 101 points throughout the catchment before and after the rainy season, to meet the demand for deep depth, spatial representativeness and temporal comparability. The influences of selected environmental factors (land-use type, topographic factors, and landform) on average SWC were examined by multi-linear regression [6, 26–32]. The results were discussed in order to provide new insights to the vertical profiles and influencing factors of SWC on the Loess Plateau, further providing reference data for sustainable management of water resource in small catchment areas in the semi-arid region with complex terrain.

Material and Methods

1 Site description

This study was conducted in Sanyanjing catchment (112°2′13″, 37°46′23″), which is located on the east margin of the Loess Plateau in Shouyang county, mid-east Shanxi province, China (Figure 1). The catchment has a total area of 1.32 km^2 and the elevation ranges from 1001 to 1160 m. It is a hilly and gully area with mostly deep gully erosion slopes. The landform consists of ridge and gully.

The catchment area has a semi-arid continental climate (Cwa by Koppen Climate Classification) with an average annual precipitation of 474.2 mm (1967–1999). Snow in the winter accounts for ~8% and rainfall in July to September for ~73% of annual precipitation. Monthly average precipitation, potential evapotranspiration and precipitation in 2013 are shown in Figure 2. Annual mean temperature in this area is 8.1°C, with a maximum of 34.7°C and a minimum −20.6°C. The soil type is cinnamon soil (Haplic lixisols, FAO), which consists of 54–62% silt and 10.95–30.15% sand with the bulk density of 1.3–1.4 g/cm^3. Soil texture was measured using a particle size analyzer (SEDIMAT 4–12, UGT, Germany). Soil bulk density was determined through sampling with cutting rings (inner diameter 5.0 cm, volume 100 cm^3) and drying in an oven (105°C, 24 h). The profile of soil texture related to different landforms is listed in Table 1. Maximum soil depth is mostly down to 300 cm on the ridge and bare rock could be rarely seen only on the northern margin of the gully area.

The distribution of land-use types across the catchment is shown in Figure 1. Terrace is the dominant land-use type, accounting for about 60% of the study area. Few terraces had been abandoned for natural restoration of vegetation because of the Grain for Green project since 2000. Grassland is mainly covered with herbs and semi-shrubs, which had never been reclaimed for several decades. About 80% of the woodland is covered with semi-shrubs at steep slopes unsuitable for sampling.

2 Soil sampling

Ethics statement. Sampling activities at the farmland were allowed by the owners. No specific permissions were required at other locations because they were not privately-owned or protected in any way and the field activities did not involve any endangered or protected species.

A total of 101 sampling points were designed in a 150 m×150 m grid throughout the catchment area by considering major land-use types, including terrace (83), abandoned farmland (9), grassland (3), and woodland (6). Soil sampling was carried out during two periods in 2013, from April 29 to May 4 (before the rainy season) and from October 28 to November 1 (after the rainy season). No precipitation occurred during the two sampling periods or a week before sampling. Each sample was taken at 20 cm intervals along the 0–300 cm soil profile using an auger (inner diameter 5.0 cm). The samples were kept in capped aluminum boxes for transportation. Measurement of SWC was conducted using an oven-drying method (105°C, 24 h). At the majority of the sampling points, soils were collected along a vertical profile over 300 cm. A few exceptions were in the north of the gully at the lowest altitude where weathered rock was occasionally encountered. Background information of the 101 sampling points is summarized in Table 2.

3 Data analysis

To identify the variability of SWC profiles at the catchment scale, descriptive parameters were calculated for each profile. Further, we calculated the linear regression coefficient (K value) between SWC and soil depth to represent the variation trend of SWC vertical profiles and the mean value to describe the average level of SWC along the 0–300 cm profile.

The SWC profiles were classified using a combined cluster analysis of the K and mean values. Cluster analysis is the process of grouping a set to data objects into multiple groups (or clusters), so that objects within a cluster share high similarity but are dissimilar to those in other clusters. In this approach, dissimilarities and similarities are assessed based on the attribute values describing the objects and often involve distance measures. Cluster analysis is a statistical classification method for discovering whether the individuals of a population gall into different groups by making quantitative comparisons of multiple characteristics [34]. Here a combined cluster analysis was conducted in three steps: 1) cluster of the mean to three groups, which present the average level of SWC along the vertical soil profile, 2) cluster of K to three groups, which reflect the variation trend of SWC profiles (top to bottom), and 3) combination of the two sets of groups into nine new groups using the between-groups linkage method with squared Euclidean distance criteria [34].

For group comparisons, SWC profiles (0–300 cm) of the same group were averaged and re-plotted. The average curves of SWC were compared between groups to identify the major factors influencing SWC in individual soil layers (0–20, 20–160, and 160–300 cm). On the basis of cluster analysis, the influences of land-use type, topography and landform on average SWC in individual soil layers were examined by multi-regression analysis. The independent variables were land-use type, landform type, Sin(gradient), Sin(aspect), flow accumulation (calculated cell numbers to a grid cell from surrounding cells with the ArcGIS hydrology analysis

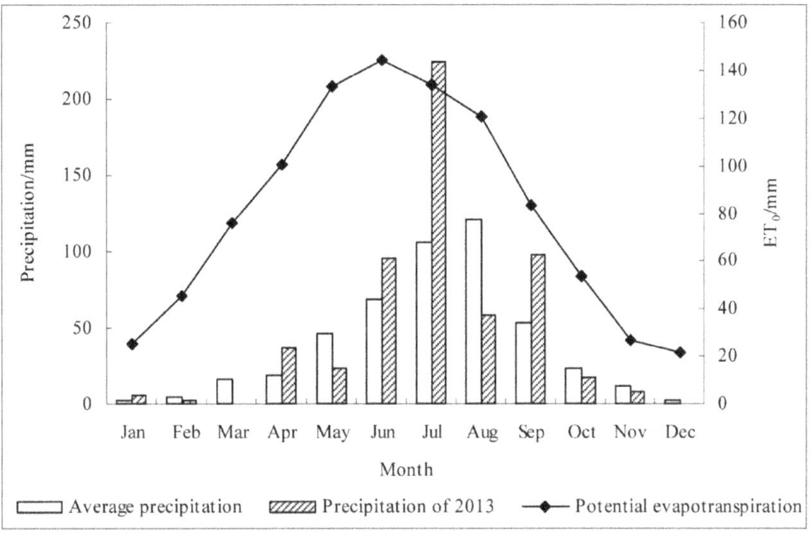

Figure 2. Average annual precipitation and potential evapotranspiration of 1967–1999 and precipitation in 2013 in the Sanyanjing catchment, Shanxi province, China.

Table 1. Soil texture in vertical profiles related to different landforms in the Sanyanjing catchment in Shanxi province, China.

Point description	Soil depth (cm)	Sand (%)		Silt (%)		Clay (%)
		>0.05 mm (%)	0.05–0.02 mm (%)	0.02–0.0063 mm (%)	0.0063–0.002 mm (%)	<0.002 mm (%)
	0–20	26.05	30.40	20.60	7.00	15.95
	20–40	21.35	34.80	20.80	6.90	16.15
	40–60	30.25	25.90	21.60	4.50	17.75
	60–80	28.25	28.80	21.70	2.80	18.45
	80–100	18.45	38.00	20.80	6.70	16.05
	100–120	26.45	30.10	20.40	5.20	17.85
Terrace at	120–140	26.95	27.90	22.00	4.40	18.75
ridge	140–160	16.35	32.60	24.20	8.10	18.75
(Point 33)	160–180	27.75	22.40	22.70	7.90	19.25
	180–200	36.55	17.80	20.60	8.00	17.05
	200–220	18.45	29.00	24.30	9.00	19.25
	220–240	20.45	28.50	23.20	8.60	19.25
	240–260	22.05	27.30	23.90	8.80	17.95
	260–280	31.35	20.30	22.20	9.40	16.75
	280–300	24.65	29.80	20.70	5.90	18.95
	0–20	30.15	20.00	23.20	6.40	20.25
	20–40	27.75	21.80	21.90	6.10	22.45
	40–60	25.95	26.80	20.50	4.50	22.25
	60–80	23.85	27.09	21.20	1.20	26.65
	80–100	0.95	31.00	20.10	20.10	27.85
	100–120	10.95	23.40	18.40	18.40	28.85
Terrace at	120–140	7.45	27.30	19.20	19.20	26.85
ridge	140–160	17.45	19.50	18.90	18.90	25.25
(Point 39)	160–180	10.45	25.80	18.20	18.20	27.35
	180–200	15.75	24.70	17.00	17.00	25.55
	200–220	13.35	21.40	18.60	18.60	28.05
	220–240	20.85	8.80	22.00	22.00	26.35
	240–260	20.25	12.40	20.00	20.00	27.35
	260–280	12.15	23.70	19.10	19.10	25.95
	280–300	24.95	29.00	20.50	0.50	25.05

module), and elevation. The former two factors were categorical variables converted into dummy variables before introduced into the regression analysis; and the latter four factors were continuous variables produced using digital elevation model at 1-m resolution.

SWC data were statistically analyzed in SPSS13.0 (SPSS Inc., Chicago, IL, USA), and topographic features were analyzed in ArcGIS 10.0 (ESRI, Redlands, CA, USA). SWC profiles were drawn in Microsoft Excel 2010 (Microsoft Corp., Redmond, WA, USA) and then clustered in SPSS 13.0 by considering descriptive parameters (maximum, minimum, mean, CV, STD, and K). Multi-regression analysis was performed in SPSS 13.0, with a P-value less than 0.05 considered statistically significant.

Results

1 Vertical profiles and descriptive parameters of SWC

The vertical profiles (0–300 cm) of SWC at 101 sampling points before the rainy season were drawn (Figure 3). These SWC profiles showed dynamic variations across the catchment study area, with substantial differences in the soil layers below 100 cm. At a few sampling points, there were obvious soil water depletion (e.g., 67, 75, 85, and 94) and an increasing trend (top to bottom) of SWC (e.g., 12 and 36). High degrees of soil desiccation were rarely detected in the lower soil layers, and low SWC was mainly found in the lower soil layers of woodland.

Descriptive parameters such as maximum, minimum, mean, and CV, STD are commonly used to reveal the spatial-temporal variability of SWC. However, these parameters cannot reflect the variation trend of SWC vertical profiles. To this end, the K value of SWC to profile depth was introduced for quantification of variation trend of SWC vertical profiles (Figure 4). Results showed that before the rainy season, SWC substantially varied between 5.87% and 34.72%, whereas the mean, STD, CV, and K values respectively ranged from 10.57% to 21.76%, 0.47 to 4.53, 3% to 24%, and −0.0405 to 0.0274 along the vertical soil profile (0–

Table 2. Background information of 101 soil sampling points in the Sanyanjing catchment study area in Shanxi province, China.

Land-use type	Vegetation	Landform type	Soil profile/cm	Sampling points
Terrace (n=83)	Maize	Ridge	300	1, 3–5, 7–9, 12–18, 21–22, 25–26, 28, 32–35, 37, 51, 53, 91, 95, 96, 98
			260	99
		Gully	300	29, 30, 38–46, 48–50, 54–60, 62, 63, 65, 66, 68–74, 77–84, 87–90
			280	47
			260	31, 61
			220	52
			160	86
	Millet	Ridge	300	11
	Maize +five-year-walnut	Gully	300	27
Abandoned farmland (n=9)	Subshrubs + herbs	Ridge	300	6, 10, 23
			240	2
	Subshrubs + herbs + few ulmus pumila	Gully	220	19
	Robinia peseudoacacia + subshrubs + herbs	Ridge	300	24
	Robinia peseudoacacia+ subshrubs + herbs	Gully	300	64
	Herbs + few almond-apricot	Gully	300	20
	Poplar + subshrubs +herbs	Gully	300	75
Grassland (n=3)	Subshrubs +herbs	Ridge	300	92, 93
		Gully	300	76
Woodland (n=6)	Poplar	Gully	300	97
			280	85
	Poplar + herbs	Gully	140	101
	Poplar + subshrubs	Gully	280	100
	Poplar + subshrubs + herbs	Gully	300	67, 94

300 cm). The ranges of the parameters after the rainy season were generally similar with those before the rainy season.

According to the division criteria of Nielson [24], CV in the range of 10–100% indicates moderate variability. Thus, the vertical variability of SWC at all sampling points in Sanyanjing catchment (Figure 4) can be classified to the medium degree. K is the linear regression coefficient between SWC and soil depth. A positive value of K indicates that SWC increases with increasing soil depth. Inversely, a negative value of K indicates that SWC decreases with increasing soil depth. The positive and negative K values of SWC data (Figure 4) are indicative of different variation trends of SWC vertical profiles in the catchment.

2 Clustering of SWC profiles

The vertical profile of SWC across the catchment can be described more clearly using cluster analysis. The 101 SWC profiles before the rainy season were classified into the first three groups by considering the mean value of SWC (Figure 5a), and the second three groups by considering the K value of SWC to soil depth (Figure 5b). The mean value of SWC ranged from 10.57% to 13.13% (low level), 14.15% to 16.86% (medium level), and 17.13% to 21.76% (high level) in the first three groups, whereas the K value of SWC to soil depth ranged from −0.0405 to 0.0106 (decreasing trend), 0.0144 to 0.0163 (stable trend), and 0.0194 to 0.0274 (increasing trend) in the second three groups. By combining the two cluster series, we obtained nine groups of SWC profiles (Table 3).

Before the rainy season, vertical profiles of SWC in groups 1–3 featured low-level SWC (Table 3). In group 1, SWC decreased along the vertical profile (0–300 cm) in woodland (2) and abandoned farmland (1) located in the gully area. In group 2, SWC remained stable along the vertical profile in terrace (2) and grassland (1) located on the ridge as well as woodland (1) located in the gully (Table 4). No sampling points were classified into group 3 with increasing SWC along the vertical profile.

Vertical profiles of SWC in groups 4–6 featured medium-level SWC (Table 3). In group 4, SWC decreased along the vertical profile on the ridge related to terrace (4) as well as in the gully related to abandoned farmland (1) and woodland (1) located in the gully. In group 5, SWC remained stable along the vertical profile in terrace (7) mostly located in the gully, few terrace (3) and abandoned farmland (1) located on the ridge, and grassland (1) and woodland (1) located in the gully. In group 6, SWC increased along the vertical profile in terrace (2) located on the ridge (Table 4).

Vertical profiles of SWC in groups 7–9 featured high-level SWC (Table 3). In group 7, SWC decreased along the vertical profile in terrace (5), grassland (1), and abandoned farmland (1) located on the ridge, and terrace (2), abandoned farmland (1) and woodland (1) located in the gully. In group 8, SWC remained stable along

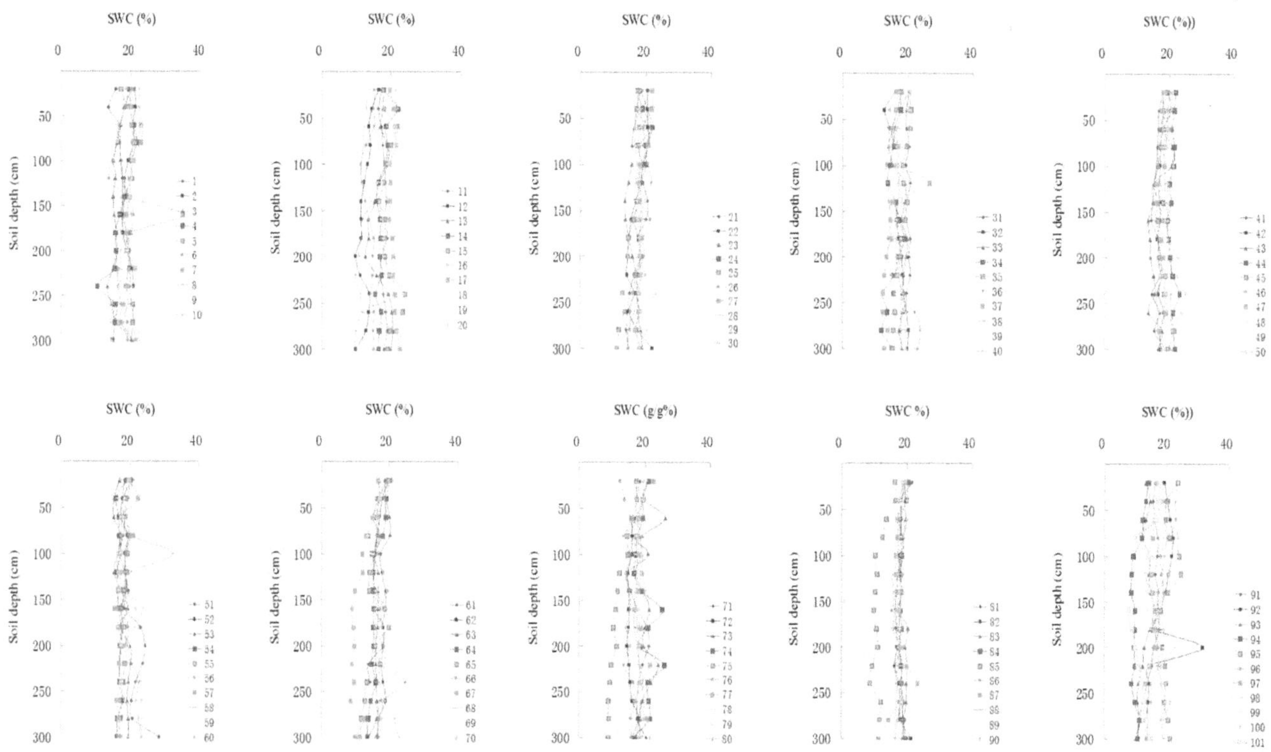

Figure 3. Vertical profiles of soil water at 101 sampling points in the Sanyanjing catchment.

the vertical profile at up to 58 of 101 sampling points, far more than other groups. Most sampling points of group 2 were located in the gully related to terrace (41), and few were on the ridge related to terrace (14) and abandoned farmland (3). In group 9, SWC increased along the vertical profile in terrace located on the ridge (3) and in the gully (1) (Table 4).

Similar grouping of SWC profiles was obtained with data collected after the rainy season. Overall, soil profiles of group 8 with high-level SWC in a stable trend were most commonly present in the catchment, more after the rainy season than before the rainy season. Group 3 of SWC profiles with low level and increasing trend was absent in the study area.

3 The relationships between average SWC and selected environmental factors

According to cluster analysis, there were nine combinations of SWC profiles in terms of average level and variation trend. We averaged SWC profiles of the same group and plotted the average curves (Figure 6), to examine differences of SWC profiles among various types. From Figure 4, we divided the whole soil profile (0–30 cm) into three layers (0–20, 20–160, and 160–300 cm) for multiple linear regression analysis. The results showed that selected environmental factors had significant linear correlations with average SWC at individual layers of 0–20 cm ($P<0.001$, $R^2=0.30$; $P<0.001$, $R^2=0.37$), 20–160 cm ($P=0.01$, $R^2=0.19$; $P<0.001$, $R^2=0.39$), and 160–300 cm ($P<0.001$, $R^2=0.32$; $P<0.001$, $R^2=0.43$; Table 5).

Before the rainy season, average SWC in the lower soil layer (10–20 cm) was significantly lower in grassland and woodland than in terrace, with no significant difference between abandoned farmland and terrace ($P_{D51}=0.109$, $P_{D52}=0.003$, $P_{D53}=0.047$, $P_{X1}=0.013$, and $P_{D61}=0.005$; Table 5). Additionally, average SWC decreased with increasing gradient, with higher levels on the ridge than in the gully.

In the lower soil layer (20–160 cm), average SWC decreased significantly with increasing gradient ($P=0.026$; Table 5), and was significantly lower in woodland than in the other three types of land-use types, with no significant differences among the latter three types. Other environmental factors had no significant linear correlation with average SWC ($P>0.05$).

In the deeper soil layer (160–300 cm), there also existed a significantly negative correlation between Sin(gradient) and average SWC ($P=0.001$; Table 5). Average SWC obviously increased with increasing gradient and was significantly higher in terrace than in abandoned farmland, grassland, and woodland (in descending order).

Similar results can be seen in the data collected after the rainy season. That is, land-use type was the major environmental factor affecting average SWC, whereas landform and altitude strongly affected average SWC only in specific periods and soil layers (Table 5). In the whole vertical profile (0–300 cm), SWC occurred at high levels from upper to deeper layers in terrace, with the lowest level in woodland. Compared with data of terrace, average SWC was relatively low in grassland and woodland in the top (0–20 cm) and deeper soil layers (160–300 cm), with significantly low levels in abandoned farmland soils in the deeper layer only. In the deeper soil layer (160–300 cm), average SWC varied with different land-use types in descending order of terrace > abandoned farmland > grassland > woodland.

Discussion

1 Vertical profiles of SWC at the catchment scale

According to Wang [21], the variability of SWC (as indicated by the CV) varies notably across the whole Loess Plateau, i.e., 15% in

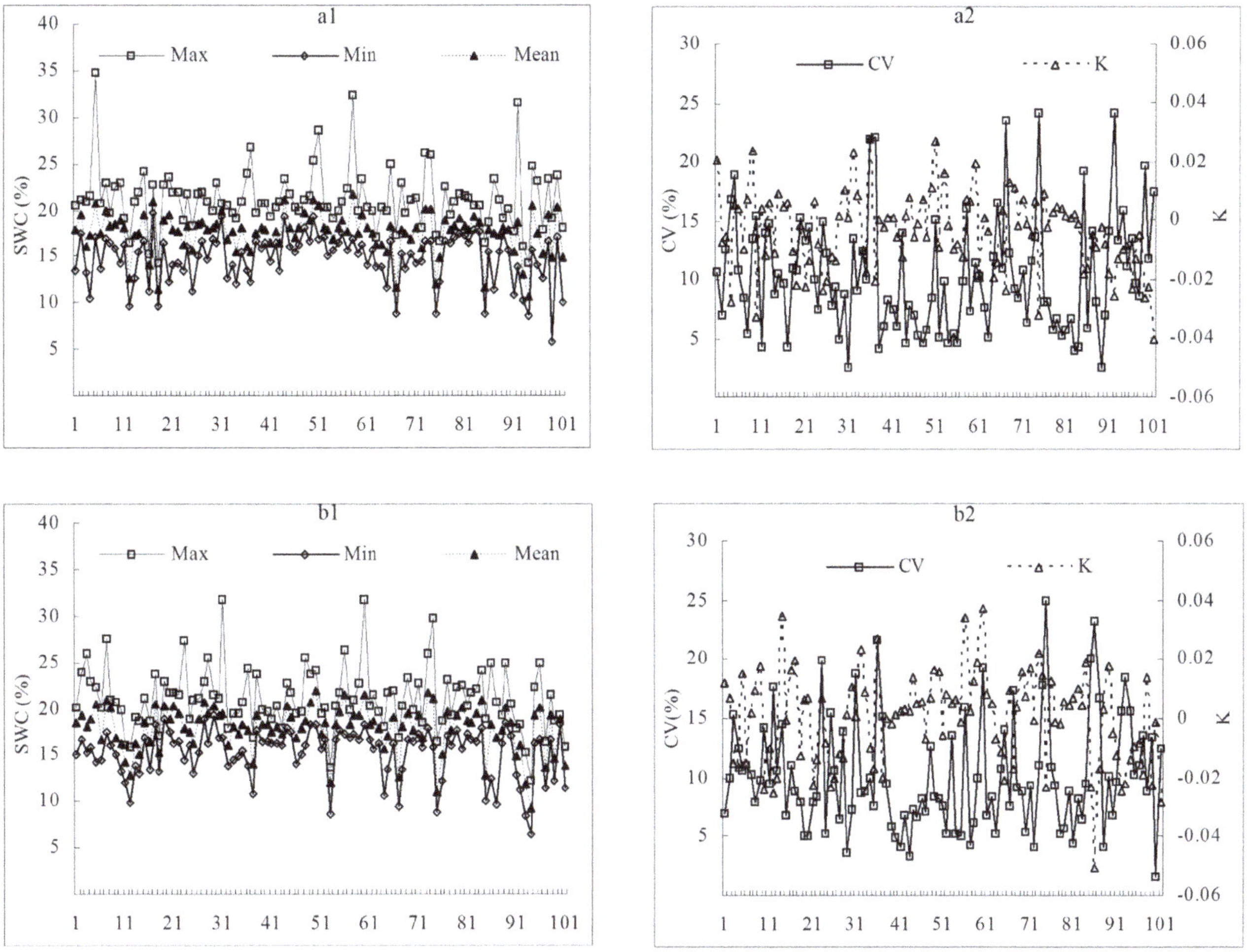

Figure 4. Statistical parameters of soil water content at 101 sampling points across the Sanyanjing catchment. (a. before the rainy season; and b. after the rainy season.)

Changwu and 55% in Shenmu. In the small catchment of Sanyanjing, SWC profiles exhibited weak and medium degrees of variability at 0–300 cm depth [33], with CV in the range of 3–24% (Figure 2). The lower variability of SWC profiles in our study area may be related to the higher SWC levels across the catchment (Pearson correlation coefficient between average SWC and CV,

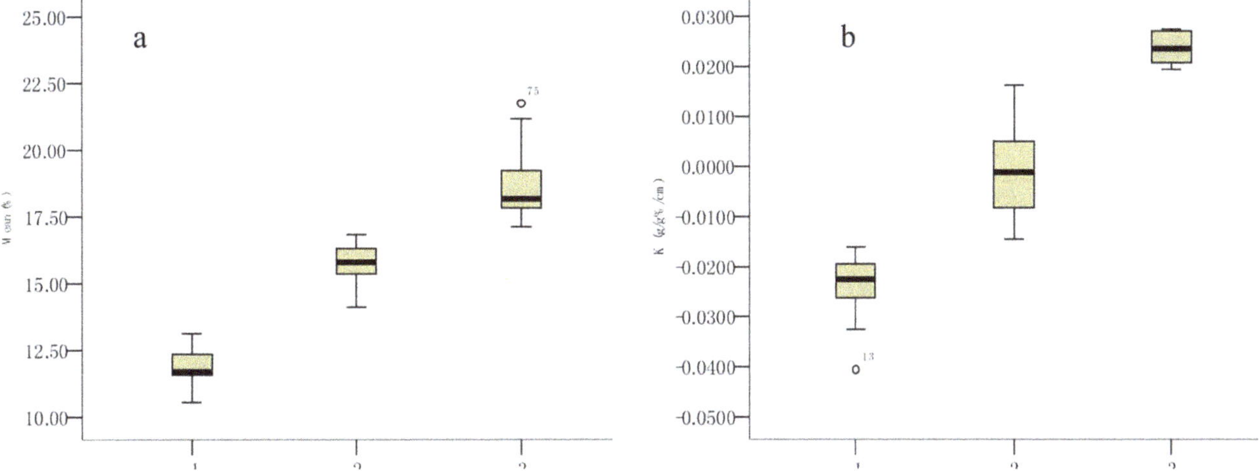

Figure 5. Grouping of 101 vertical soil water profiles in the Sanyanjing catchment before the rainy season by cluster analysis of the mean value (a) and regression gradient (K, b).

Table 3. Combined grouping of 101 vertical profiles of soil water content (0–300 cm) in Sanyanjing catchment by cluster analysis of the mean value and regression gradient.

Cluster by mean	Cluster by K	Combined grouping	Quantity of points		Point Nos.	
			Before the rainy season	After the rainy season	Before the rainy season	After the rainy season
1	1	1	3	0	65,67, 85	-
	2	2	4	1	12,18,93,94	94
	3	3	0	0	-	-
2	1	4	6	15	25,37,64,91,99,101	11,12,18,37,53,64,67,75,76,85,91,93,97,99,101
	2	5	13	0	3,16,23,34,41,43,56,63,65,70,72,76,97	3,16,23,34,41, 43,56,63,65,70,72,76,97
	3	6	2	1	32,36	14
3	1	7	11	1	4,10,19,21,26,35,61,86,92,96,100	86
	2	8	58	80	2,5–8,11,13–15,17, 20,22,24,27–31,33, 38–40,42,44–50,52, 53,55,56–59,62,66, 68,69,71,73,74, 77–84,87–90,95,98	2–10,13,15,16, 17,19–36,38–52,54,55,57–59, 61–63,65,66,68,69–74, 77–84, 87–90,92,95,96,98,100
	3	9	4	3	1,9,51,60	1, 51,60

Table 4. Grouping richness of 101 vertical profiles of SWC (0–300 cm) in relation to different land-use types in Sanyanjing catchment before the rainy season.

Land-use type	Grouping								
	1	2	3	4	5	6	7	8	9
Terrace	0	2	0	4	10	2	7	54	4
Abandoned	1	0	0	1	1	0	2	0	0
Grassland	0	1	0	0	1	0	1	0	0
Woodland	2	1	0	1	1	0	1	0	0

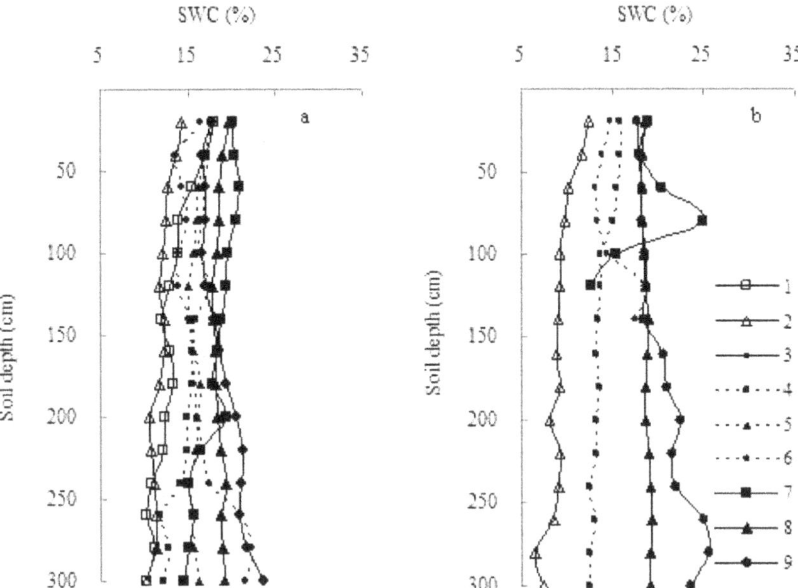

Figure 6. Vertical soil water profiles in relation to different groups in the Sanyanjing catchment study area before and after the rainy season (a. before the rainy season; and b. after the rainy season).

−0.40; $P<0.01$). Qiu [17] found that wetter soil with greater vertical variations in an increasing trend along the SWC profile (mean 13.03%; and STD, 2.3%) is representative in a dry year in Danangou catchment on the Loess Plateau, where the land-use pattern (including slope farmland, terrace, and orchard) differs from that in our study area.

Cluster analysis of the mean and K values provides a clear description for the overall variability of SWC in the vertical profile. Based on combined grouping, the 101 vertical SWC profiles were classified into nine groups with high, medium, and low levels associated with increasing, stable, and decreasing trends (Table 3). More than half of the SWC profiles were obtained from terrace soils in the gully and classified into group 8 (58/101 before the rainy season and 80/101 after the rainy season) with high-level SWC in a stable trend (Tables 3, 4). Despite that all sampling points of woodland were also located in the gully, their average SWC remained the lowest among different land-use types and mostly descended along the vertical profile (Table 4). The above differences can be attributed to the lower soil water consumption by maize crop in the terrace, which generally has shallower root distribution and less above-ground biomass than trees in the woodland. Our observations coincide with previous findings on the Loess Plateau that soil water conditions of terrace, gully farmland, and dam land are better than that of artificial woodland. The latter land-use type is associated with soil desiccation, especially in deep soil layers [2, 22, 35, 36].

Although the cluster analysis divided vertical SWC profiles into nine groups, only eight types were present in the Sanyanjing catchment and no sampling points were classified into group 3 (i.e., low-level SWC with an increasing trend from top to bottom). According to previous research in semi-arid regions, if SWC occurs at low level in the upper soil layer, deep-root crops, shrubs, and trees will consume more soil water in the deeper soil layers through root extraction [37–40]. Additionally, it is hard to achieve soil water recharge in the deeper soil layers by precipitation infiltration because the depth of soil water infiltration is shallow. Therefore, soil desiccation exists in the lower soil layer in case of no groundwater recharge [8–10]. These mechanisms explain the absence of high-level SWC with an increasing trend along the vertical profile in the small catchment of Sanyanjing (Table 3).

2 Effects of environmental factors on average SWC at the catchment scale

Consistent with cluster analysis (Table 3), multiple regression analysis showed that land-use type had a significant effect on soil water status in the small catchment of Sanyanjing (Table 4, 5). This result coincides with the data previously reported in small catchments on the Loess Plateau [17, 18, 22, 23, 29]. For example, Zhang [23] concluded that average SWC (20–200 cm) descends with different land-use types (farmland > grassland > shrub land > and woodland, n = 80) in the small catchment of Zhifanggou. Bai [42] found that average SWC (0–500 cm) ranges from of 9% to 16% in orchard, gradient farmland, terrace, and grassland, but remains less than 10% in shrub land and most woodland (n = 91) in Nangou catchment in the central area of Loess Plateau, Ansai, Shaanxi. The consistency of the data demonstrates that cluster analysis is a reliable method for characterization of SWC profiles.

The effect of land-use type on SWC can be related to the differences existing in anthropogenic activity and vegetation type [22]. Average SWC was found significantly higher in terrace and abandoned farmland than in grassland and woodland along the 0–300 cm profile (Table 5). Abandoned farmland and terrace are associated with artificial tillage in the surface soil layer, which improves soil porosity and loosens soil structure, further enhancing soil water infiltration [22, 43]. Additionally, soil water consumption by crops is less than that in grassland and woodland due to lower leaf area index [21], contributing to the accumulation of SWC. The above mechanisms account for the greater average of SWC profiles with a stable trend to soil depth in terrace and abandoned farmland.

Difference in root distribution is another factor contributing the effect of land–use type on SWC [40]. In the Sanyanjing catchment, average SWC of woodland was higher in the 0–20 cm soil layer but lower in the 20–160 and 160–300 cm soil

Table 5. Multi-linear regression analysis of soil water content and selected environmental factors in three layers (0–20, 20–160, and 160–300 cm) of the vertical soil profile in the Sanyanjing catchment study area.

				Before the	rainy season				
Model		Y_1			Y_2			Y_3	
	Unstandardized coefficients B	Standardized coefficients (Beta)	Sig.	Unstandardized coefficients B	Standardized coefficients (Beta)	Sig.	Unstandardized coefficients B	Standardized coefficients (Beta)	Sig.
Constant	14.714		0.066	6.031		0.498	18.514		0.069
X_1	−2.743	−0.23	0.013	−2.751	−0.22	0.026	−4.471	−0.296	0.001
X_2	−0.112	−0.037	0.68	0.042	0.013	0.889	−0.261	−0.068	0.449
X_3	4.86E-06	0.025	0.783	−7.80E-06	−0.038	0.694	1.20E-05	−0.05	0.578
X_4	0.005	0.081	0.507	0.011	0.177	0.175	0	0.005	0.97
D_{51}	1.09	0.146	0.109	0.019	0.002	0.98	−0.788	−0.084	0.354
D_{52}	−3.472	−0.278	0.003	−0.845	−0.064	0.503	−2.553	−0.162	0.073
D_{53}	−1.715	−0.191	0.047	−3.216	−0.342	0.001	−4.745	−0.384	0
D_{61}	−1.512	−0.348	0.005	−0.607	−0.134	0.307	−0.833	−0.151	0.212
	$R^2 = 0.30$	($P < 0.001$)		$R^2 = 0.19$	($P = 0.001$)		$R^2 = 0.32$	($P < 0.001$)	
				After the	**rainy season**				
Model		Y_1			Y_2			Y_3	
	Unstandardized coefficients B	Standardized coefficients (Beta)	Sig.	Unstandardized coefficients B	Standardized coefficients (Beta)	Sig.	Unstandardized coefficients B	Standardized coefficients (Beta)	Sig.
Constant	11.426		0.111	−5.811		0.452	−9.045		0.41
X_1	−2.01	−0.178	0.041	−2.763	−0.221	0.01	−3.688	−0.206	0.015
X_2	0.313	0.11	0.201	−0.132	−0.042	0.618	0.294	0.063	0.44
X_3	6.30E-06	−0.034	0.692	1.80E-05	−0.089	0.285	1.60E-05	−0.054	0.508
X_4	0.007	0.125	0.278	0.023	0.359	0.002	0.027	0.291	0.009
D_{51}	−0.622	−0.088	0.307	−1.161	−0.149	0.08	−1.81	0.162	0.053
D_{52}	−2.374	−0.2	0.021	−2.372	−0.181	0.032	−4.166	−0.222	0.008
D_{53}	−2.599	−0.305	0.001	−4.807	−0.51	0	−6.862	−0.467	0
D_{61}	−2.082	−0.506	0	−1.512	−0.332	0.004	−3.14	−0.476	0
	$R^2 = 0.37$	($P < 0.001$)		$R^2 = 0.39$	($P = 0.001$)		$R^2 = 0.43$	($P < 0.001$)	

Dependent Variable: Y_1 (soil water content of 0–20 cm layer).
Y_2 (average soil water content of 20–160 cm layer).
Y_3 (average soil water content of 160–300 cm layer).
Independent Variables: $X_1 = \text{Sin(gradient)}$, $X_2 = \text{Sin(aspect)}$, $X_3 = \text{flowaccu}$, $X_4 = \text{elevation}$.
Dummy Variables: $X_5 = \text{terrace}$, $(D_{51}, D_{52}, D_{53}) = (0,0,0)$; $X_5 = \text{abandoned farmland}$, $(D_{51}, D_{52}, D_{53}) = (1,0,0)$.
$X_5 = \text{grassland}$, $(D_{51}, D_{52}, D_{53}) = (0,1,0)$; $X_5 = \text{woodland}$, $(D_{51}, D_{52}, D_{53}) = (0,0,1)$.
$X_6 = \text{ridge}$, $(D_{61}) = 1$; $X_6 = \text{gully}$, $(D_{61}) = 0$.
D represents sub-variable; binary variables 0 and 1 for the absence and presence of some land-use type or landform, respectively.

layers than data of grassland (Table 5). The varying trends of SWC profiles between grassland and woodland can be related to different distribution of root system in individual soil layers and stratified root extraction of soil water. The rooting depth of maize crop is reported to be approximately 100 cm and most maize roots are distributed in the soil layer of 0–20 cm, shallower than average rooting depths in grassland (20–60 cm) and woodland (20–100 cm) [40]. Diverse root distribution patterns can lead to different levels of soil water consumption by plants, contributing to great variability of SWC level.

In addition to land-use type, topographic factors strongly affected SWC in the study area (Table 4). This is because the distribution of wind and solar radiation varies with different topographic conditions, leading to different levels of soil evaporation, runoff on gradient, and soil water infiltration [41]. Gradient negatively affected SWC in the soil layers of 0–20, 20–160, and 160–300 cm (Table 5), possibly due to the increased runoff with increasing gradient and resultant reduction of precipitation infiltration [17, 27, 29, 41]. Other topographic factors including aspect and flow accumulation had no significant effects on average SWC in the three soil layers (Table 5). Similarly, Gómez [29] referred that aspect has no obvious influence on SWC in burned and unburned areas. Shi [19] suggested that aspect and catchment area significantly affect SWC during the wet period only, whereas elevation has a significant effect on SWC in arid and humid periods but not in semi-arid and semi-humid periods. In the present study, we found the effect of elevation on SWC of the three layers varying with the period of time and being significant after the rainy season only.

As for the landform type, location of sampling points significantly affected SWC only in the surface layer (0–20 cm) before the rainy season and throughout all the three layers (0–20, 20–160, and 160–300 cm) after the rainy season, with greater values in the gully than on the ridge (Table 5). The effect of landform type on SWC can be related to different levels of soil evaporation as affected by wind strength and solar radiation and soil physical properties. Similarly, Zhang [22] suggested that average SWC descends with different landforms as gully > terrace > slop land > hill top.

Overall, land-use type is the most significant factor affecting SWC while topographic factors and landform type are interacting jointly at the catchment-scale. Because the impact of environmental factors on SWC varies in different periods, it is necessary to increase the observation frequency, in order to better understand the spatiotemporal distribution and influencing factors of SWC in the small catchment. Such work will provide reference data for selecting reasonable environmental parameters in catchment scale SWC simulation over different periods of time.

Conclusions

In this study, cluster analysis enables catchment-scale characterization of soil water profiles in terms of average level and variation trend along the vertical profile, allowing for simple and clear interpretation of the results. A total of nine groups of soil water profiles are recognized but those with low-level soil water content and a decreasing trend are not present in the Sanyanjing catchment. Land-use type, gradient, landform type, and altitude are the major environmental factors significantly influencing average soil water content in the hilly and gully catchment with complex terrain. The former two factors strongly affect soil water content along the 0–300 cm soil profile, whereas effects exerted by the latter two factors vary by soil layer and season.

Understanding the vertical profile of soil water content and evaluation of related major influencing factors in individual soil layers can help with sustainable land use and water management in catchment areas on the hilly and gully Loess Plateau as well as in arid and semi-arid areas with complex terrain. For better estimation of soil water profiles in small catchments, other factors such as fertilization, coverage, and soil physical properties may be considered with respect to specific soil layers.

Acknowledgments

We give our thanks to Prof./Dr. Wenzhao Liu (Institute of Soil and Water Conservation, Northwest A & F University, Yangling) for suggestions on experimental arrangement.

Author Contributions

Conceived and designed the experiments: BW FW JW XW YH. Performed the experiments: BW FW JW XW YH. Analyzed the data: BW FW JW XW YH. Contributed reagents/materials/analysis tools: BW FW JW XW YH. Wrote the paper: BW FW JW XW YH.

References

1. Engelbrecht BMJ, Comita LS, Condit R, Kursar TA, Tyree MT, et al. (2007) Drought sensitivity shapes species distribution patterns in tropical forests. Nature 447: 80–82.
2. Yang L, Wei W, Mo BR, Chen LD (2011) Soil water under different artificial vegetation restoration in the semi-hilly region of the Loess Plateau. Acta Ecologica Sinica 31: 3060–3068 (in Chinese with English abstract).
3. Xing G, Zhang XM, Fei XL, Wu YX (2012) Study on soil moisture content under different land use types in Sunjiacha basin. Agricultural Research in the Arid Areas 30: 225–229 (in Chinese with English abstract).
4. Li JW, Zuo HT, Li QF, Fan XF, Hou XC (2011) Effect of soil water spatial distribution pattern on switchgrass during first growing season. Acta Agrista Sinica 19: 43–50 (in Chinese with English abstract).
5. Jipp PH, Nepstad DC, Cassel DK, Carvalho C (1998) Deep soil moisture storage and transpiration in forests and pastures of seasonally-dry Amazonia. Climatic Change 39: 395–412.
6. Grassini P, You JS, Hubbard KG, Cassman KG (2010) Soil water recharge in a semi-arid temperate climate in the central US Great Plains. Agricultural Water Management 97: 1063–1069.
7. Markewitz D, Devine S, Davidson EA, Brando P, Nepstad DC (2010) Soil moisture depletion under simulated drought in the Amazon: impacts on deep root uptake. New Phytologist 187: 592–607.
8. Li YS (2001) Fluctuation of yield on high-yield field and desiccation of the soil on dryland. Acta Pedologica Sinica 38: 353–356 (in Chinese with English abstract).
9. Huang MB, Dang TH, Gallichand J, Goulet M (2003) Effect of increased fertilizer applications to wheat crop on soil-water depletion in the Loess Plateau, China. Agricultural Water Management 58: 267–278.
10. Liu WZ, Zhang XC, Dang TH, Zhu OY, Li Z, et al. (2010) Soil water dynamics and deep soil recharge in a record wet year in the southern Loess Plateau of China. Agricultural Water Management 97: 1133–1138.
11. Beldring S, Gottschalk L, Seibert J, Tallaksen LM (1999) Distribution of soil moisture and groundwater levels at patch and catchment scales. Agricultural and Forest Meteorology 98–99: 305–324.
12. Li B, Rodell M (2013) Spatial variability and its scale dependency of observed and modeled soil moisture over different climate regions. Hydrology and Earth System Sciences 17: 1177–1188.
13. Henninger DL, Petersen GW, Engman ET (1976) Surface soil moisture within a watershed: Variations, factors influencing, and relationship to surface runoff. Soil Science Society of American Journal 40: 773–776.
14. Jones EB, Owe M, Schmugge TJ (1982) Soil moisture variation patterns observed in Hand county, South Dakota. Water Recources Bulletin 18: 949–954.
15. Hawley ME, Jackson TJ, Mccuen RH (1983) Surface soil moisture variation on small agricultural watersheds. Journal of Hydrology 62: 179–200.
16. Robinson M, Dean TJ (1993) Measurement of near surface soil water content using a capacitance probe. Hydrological Processes 7: 77–86.
17. Qiu Y, Fu BJ, Wang J, Chen LD (2000) Quantitative analysis of relationships between spatial and temporal variation of soil moisture content and

environmental factors at a gully catchment. Acta Ecologica Sinica 20: 741–747 (in Chinese with English abstract).

18. Zeng C, Shao MA, Wang QJ, Zhang J (2011) Effects of land use on temporal-spatial variability of soil water and soil-water conservation. Acta Agriculturae Scandinavica Section B-Soil and Plant Science 61: 1–13.
19. Shi ZH, Zhu HD, Chen J, Fang NF, Ai L (2012) Spatial heterogeneity of soil moisture and its relationships with environmental factors at small catchment level. Chinese Journal of Applied Ecology 23: 889–895 (in Chinese with English abstract).
20. Chen LD, Huang ZL, Gong J, Fu BJ, Huang YL (2007) The effect of land cover/vegetation on soil water dynamic in the hilly area of the loess plateau, China. Catena 70: 200–208.
21. Wang YQ, Shao MA, Liu ZP, Orton R (2013) Regional-scale variation and distribution patterns of soil saturated hydraulic conductivities in surface and subsurface layers in the loessial soils of China. Journal of Hydrology 487: 13–23.
22. Zhang R, Cao H, Wang YQ, Huang CQ, Tan WF (2012) spatial variability of soil moisture and its influence factors in watershed of gully region on the loess plateau. Research of Soil and Water Conservation 19: 52–58 (in Chinese with English abstract).
23. Canton Y, Sole-benet A, Domingo F (2004) Temporal and spatial patterns of soil moisture in semiarid badlands of SE Spain. Journal of Hydrology 285: 199–214.
24. Burnt TP, Butcher DP (1985) Topographic controls of soil moisture distributions. Journal of Soil Science 36: 469–486.
25. O'loughlin EM (1981) Saturation regions in catchments and their relations to soil and topographic properties. Journal of Hydrology 53: 229–246.
26. Huang J, WU P, Zhao XN (2012) Effects of rainfall intensity, underlying surface and slope gradient on soil infiltration under simulated rainfall experiments. Catena 104: 93–102
27. Qiu Y, FU BJ, Wang J, Chen LD (2003) Spatiotemporal prediction of soil moisture content using multiple-linear regression in a small catchment of the Loess Plateau, China. Catena 54: 173–195.
28. Qiu Y, Fu B, Wang J, Chen L, Meng Q, Zhang Y (2010) Spatial prediction of soil moisture content using multiple-linear regressions in a gully catchment of the Loess Plateau, China. Journal of Arid Environments 74: 208–220.
29. Gómez-Plaza A, Martínez-Mena M, Albaladejo J, Castillo VM (2001) Factors regulating spatial distribution of soil water content in small semiarid catchments. Journal of Hydrology 253: 211–226.
30. Dripps WR, Bradbury KR (2007) A simple daily soil-water balance model for estimating the spatial and temporal distribution of groundwater recharge in temperate humid areas. Hydrogeology Journal 15: 433–444.
31. Yao XL, Fu BJ, Lu YH, Sun FX, Wang S, et al. (2013) Comparison of four spatial interpolation methods for estimating soil moisture in a complex terrain catchment. PLoS One 8(1): e54660.
32. Wang MB, Li HJ (1995) Quantitative study on the soil water dynamics of various forest plantations in the loess plateau region in northwestern Shanxi. Acta Ecologica Sinica 15: 172–184 (in Chinese with English abstract).
33. Wang YQ, Zhang XC, Han FP (2008) Profile variability of soil properties in check dam on the Loess Plateau and its functions. Environmental Science 29: 1020–1026 (in Chinese with English abstract).
34. Jain AK (2010) Data clustering: 50 years beyond K-means. Pattern Recognition Letters 31: 651–666.
35. Huang YL, Chen LD, Fu BJ, Wang YL (2005) Spatial pattern of soil water and its influencing factors in gully catchment of the Loess Plateau. Journal of Natural Resources 20: 483–492 (in Chinese with English abstract).
36. Zou JL, Shao MA, Gong SH (2011) Effects of different vegetation and soil types on profile variability of soil moisture. Research of Soil and Water Conservation 18: 12–17 (in Chinese with English abstract).
37. Kizito F, Dragila M,Se'ne M, Lufafa A, Diedhiou I, et al. (2006) Seasonal soil water variation and root patterns between two semi-arid shrubs co-existing with Pearl millet in Senegal, West Africa. Journal of Arid Environments 67: 436–455.
38. Li J, Chen B, Li XF, Zhao YJ, Ciren YJ, et al. (2008) Effects of deep soil desiccation on artificial forestlands in different vegetation zones on the Loess Plateau, China. Acta Ecologica Sinica 28: 1429–1445 (in Chinese with English abstract).
39. Cheng LP, Liu WZ (2013) Long term effects of farming system on soil water content and dry soil layer in deep loess profile of Loess Tableland in China. Journal of Integrative Agriculture. 13(6): 1382–1392.
40. Wang XZ, Jiao F (2011) Partition of soil moisture profiles based on sequential clustering method. Journal of Northwest A &F University 39: 191–201,196 (in Chinese with English abstract).
41. Fu XL, Shao MA, Wei XR, Wang HM, Zeng C (2013) Effects of monovegetation restoration types on soil water distribution and balance on a hillslope in northern Loess Plateau of China. Journal of Hydrologic Engineering 18: 413–421.
42. Bai TL, Yang QK, Shen J (2009) Soil variability of soil moisture vertical distribution and related affecting factors in hilly and gully watershed region of Loess Plateau. Chinese Journal of Ecology 28: 2508–2514 (in Chinese with English abstract).
43. Lian G, Guo XD, Fu BJ, Hu CX (2006) Spatial variability of bulk density and soil water in a small catchment of the Loess Plateau. Acta Ecologica Sinica 26: 647–654 (in Chinese with English abstract).

8

Transfer of Cadmium from Soil to Vegetable in the Pearl River Delta area, South China

Huihua Zhang[1]*, Junjian Chen[1]*, Li Zhu[2], Guoyi Yang[1], Dingqiang Li[1]*

1 Guangdong Institute of Eco-environmental and Soil Sciences, Guangzhou, China, **2** Management School, Jinan University, Guangzhou, China

Abstract

The purpose of this study was to investigate the regional Cadmium (Cd) concentration levels in soils and in leaf vegetables across the Pearl River Delta (PRD) area; and reveal the transfer characteristics of Cadmium (Cd) from soils to leaf vegetable species on a regional scale. 170 paired vegetables and corresponding surface soil samples in the study area were collected for calculating the transfer factors of Cadmium (Cd) from soils to vegetables. This investigation revealed that in the study area Cd concentration in soils was lower (mean value 0.158 mg kg^{-1}) compared with other countries or regions. The Cd-contaminated areas are mainly located in west areas of the Pearl River Delta. Cd concentrations in all vegetables were lower than the national standard of Safe vegetables (0.2 mg kg^{-1}). 88% of vegetable samples met the standard of No-Polluted vegetables (0.05 mg kg^{-1}). The Cd concentration in vegetables was mainly influenced by the interactions of total Cd concentration in soils, soil pH and vegetable species. The fit lines of soil-to-plant transfer factors and total Cd concentration in soils for various vegetable species were best described by the exponential equation ($y = ax^b$), and these fit lines can be divided into two parts, including the sharply decrease part with a large error range, and the slowly decrease part with a low error range, according to the gradual increasing of total Cd concentrations in soils.

Editor: Manuel Reigosa, University of Vigo, Spain

Funding: Funding was provided by Natural Science Foundation of China (NSFC No. 41171387 and 31270516), the Natural Science Foundation of Guangdong Province, China (No.S2012030006144), the Science and Technology Planning Project of Guangdong Province, China (No.2012A020100003 and 2011B030900013), and Science and Technology Planning Project of Guangzhou (No.2013J2200003). The funders had no role in study design, data collection and analysis, decision to publish, or preparation of the manuscript.

Competing Interests: The authors have declared that no competing interests exist.

* Email: hhzhang@soil.cn (HZ); jjchen@soil.gd.cn (JC); dqli@soil.gd.cn (DL)

Introduction

Cadmium is non-essential element to biota. It is known to be toxic for plants as well as animals to a much higher extent and at lower concentrations than e.g. Zn, Pb, or Cu. It especially affects humans because of their longevity and the accumulation of Cd in their organs by eating Cd-contaminated food [1–3].

Soil Cd is naturally derived from parent materials because of chemical weathering, and as a contaminant in many areas of anthropogenic activities such as mining, smelting, composts, phosphate fertilizer application, waste disposal, and vehicle exhausts [4–7]. The main forms of cadmium in soils and sediments are the exchangeable fraction, followed by the Fe-Mn oxides and residual fractions. Several studies indicated the Cd in soils contaminated by anthropogenic activates such as mining and smelting, seem to be more bioavailable than Cd from unimpacted soils [5]. To protect the safe of soil environment and prevent the soil Cd contamination, various soil quality standards were established in many countries or regions, such as 0.3 mg kg^{-1} in China [8], 0.8 mg kg^{-1} in the Netherlands and Switzerland, 0.5–1 mg kg^{-1} in Austria [9], and 5 mg kg^{-1} in Taiwan [10].

Because soil Cd is easily accumulated by plants, soil-plant-human transfer of Cd has been considered as a major pathway of human exposure to soil Cd [11,12]. Many studies have been conducted in China such as Cd in rice [10,13–15], in orchard [16,17], and in vegetable [18–21]. And the maximum permissible concentrations of Cd of 0.05 mg kg^{-1} for non-environmental pollution vegetable – so called "No-Polluted vegetable" [22] and maximium concentration of 0.2 mg kg^{-1} in leaf vegetables for food security – so called "Safe vegetable" [23] are used in China.

Although undoubtedly soil Cd is the primary source of vegetable Cd, and total soil Cd concentration was commonly used for soil environmental quality estimation, its usefulness to predict soil-to-plant transfer was often questioned since the phytoavailability of Cd in soils [12,21,24]. Yang et al [25] proposed that various vegetable species showed significantly different accumulation capacity for Cd in the same soil sample sites; Soil-to-vegetable transfer factor (TF) of Cd was much higher in a pot experiment than that in a field trial. Previous studies have showed that the TF is decided especially by the soil properties and plant species [26]. Wang et al [12] pointed out that TFs can be considered as a useful index of metals potentially transfer abilities from soil to plant, and the TFs of Cd for leafy vegetables were higher than those for non-leafy vegetables.TF has been widely used in the evaluation of potential health risk of human exposure to metals from soil [3,5,12,15,16].

In the PRD area, the vegetable planting area is about 4830 km^2. The vegetable production was 1129.2×10^4 tons in 2011. The leaf vegetable is the dominant consumption for Guangdong people, the planting area is about 33% of the total planting area [27]. As a report of the Statistics Bureau of Guangdong Province, the export of fresh vegetable was up to 78.3 $\times 10^4$ tons [28], mainly selled to Japan, Korea, Malaysia, Russia, and America. Hence, the quality safety of vegetable is not only

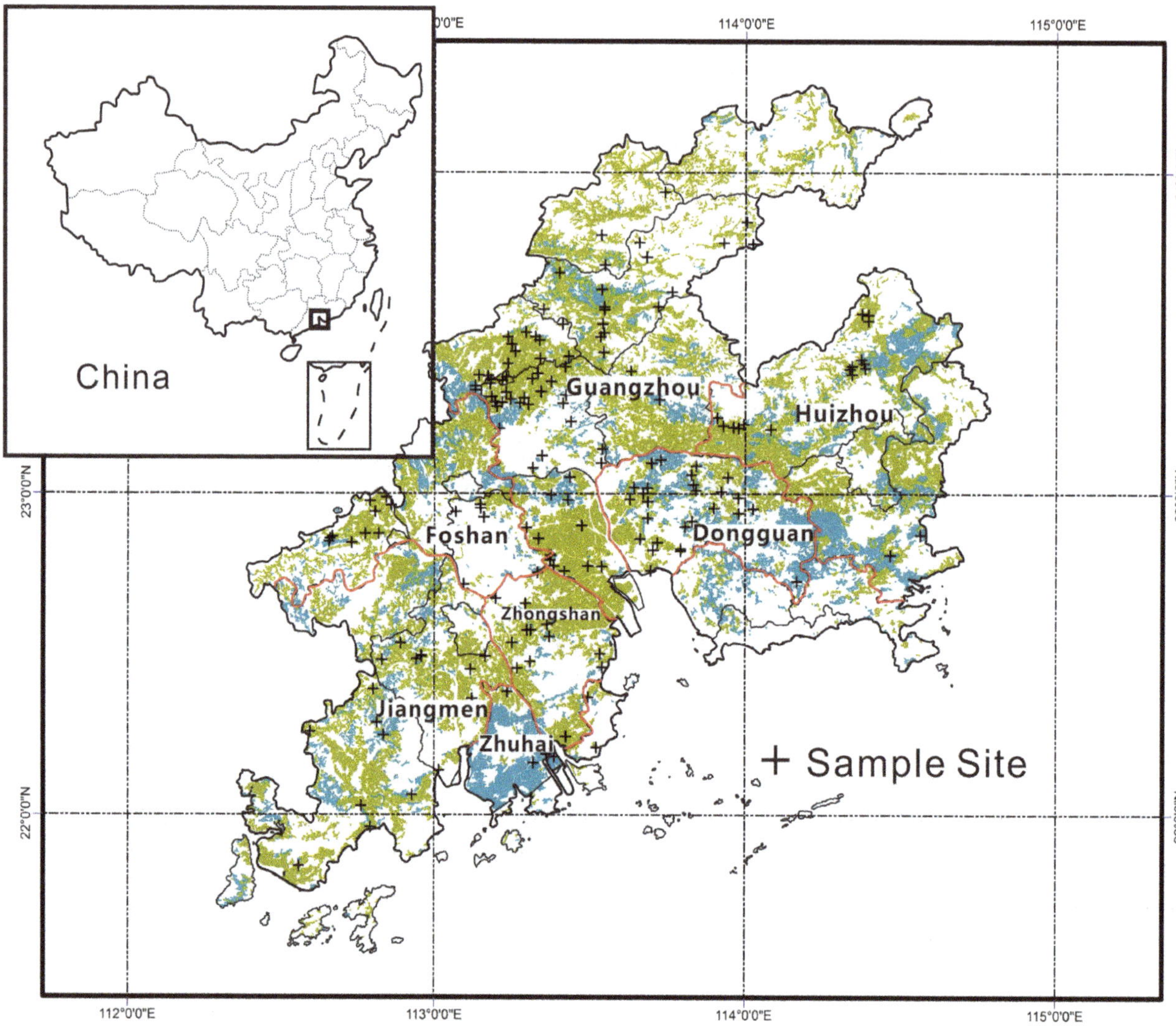

Figure 1. Locations of soil and vegetable samples in the Pearl River Delta area, Guangdong, China. (The yellow and blue areas are the dry lands and the paddy fields, respectively; the red line is the boundary of the city).

concerned by the local government, but it should receive much more attention from the international community for the increasing export and import trade.

In this study, we conducted a systemic environmental quality survey of the vegetable soil and various vegetables in the PRD area. The aim of this study was to: (1) obtain the information on concentrations and spatial patterns of Cd in soils and leaf vegetables across the PRD area; (2) investigate the relationship between soil Cd and vegetable Cd on a regional scale; (3) reveal the transfer characteristic of Cd from soil to various leaf vegetable species.

Materials and Methods

Ethics Statement

This study was carried out on collective-owned lands, and the owners of the lands gave us permission to conduct the study on these sites. The field studies did not involve endangered or protected species.

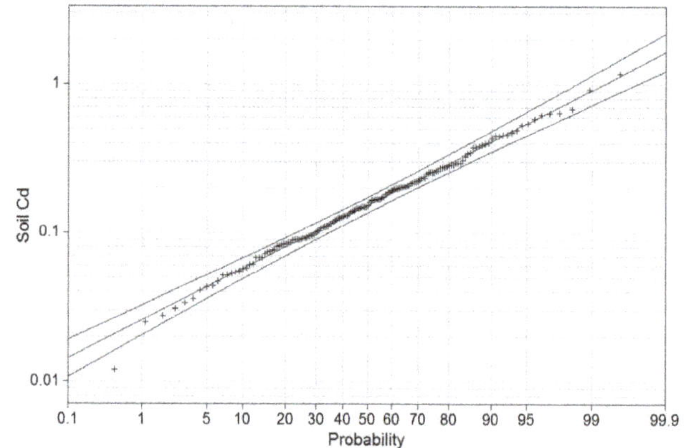

Figure 2. Lognormal probability plot for total Cd concentration in vegetable soils in the Pearl River Delta area, Guangdong, China (mg kg^{-1}, DW).

Table 1. Descriptive statistics of soil Cd concentrations (mg kg^{-1}, DW) and vegetable Cd concentration (mg kg^{-1}, FW) in the PRD area.

Sample Site	n	Soil Cd concentration			Vegetable Cd concentration			pH (H_2O)	OM[e] (%)
		Mean ± SD	Range	C.V. (%)	Mean ± SD	Range	C.V. (%)	Mean ± SD	Mean ± SD
Dongguan	27	0.138±0.079	0.041–0.433	57.2	0.024±0.012	0.004–0.057	48.4	5.76±0.87	2.58±0.70
Foshan	16	0.291±0.200	0.025–0.685	69.0	0.017±0.011	0.002–0.037	62.5	5.84±1.12	2.61±1.01
Guangzhou	73	0.226±0.190	0.027–1.180	84.3	0.027±0.024	0.002–0.082	86.8	6.34±0.96	2.44±0.66
Jiangmen	17	0.188±0.157	0.012–0.644	83.9	0.014±0.016	0.002–0.068	117.8	5.49±0.67	2.28±0.90
Zhuhai	6	0.204±0.160	0.044–0.474	78.5	0.008±0.005	0.001–0.014	58.1	6.11±0.99	2.34±0.63
Zhongshan	13	0.316±0.150	0.083–0.530	47.3	0.022±0.021	0.001–0.067	92.7	6.02±0.97	2.45±0.90
Huizhou	18	0.107±0.052	0.052–0.200	39.6	0.026±0.018	0.010–0.082	69.1	6.69±0.43	2.41±0.62
Total	170	0.208±0.169 0.158 (1.39)[d]	0.012–1.180	81.2	0.024±0.019 0.015 (1.57)[d]	0.002–0.082	84.3	6.12±0.96	2.42±0.74
Threshold		0.3[a]			0.05[b]; 0.2[c]				

[a]The maximum permissible concentrations of Cadmium for agriculture soils (SEPAC, 1995).
[b]the maximum permissible concentrations of Cadmium for non-environmental pollution vegetable (AQSIQ, 2001).
[c]The maximum level of Cadmium for leaf vegetables (MOH, 2012).
[d]Geometric Mean (Geometric standard deviation).
[e]Orgnic matter.

Study area

The PRD area is located in the south of Guangdong province occupying 41698 km^2 of land area (Figure 1). The area has a subtropical – tropical monsoon climate with an average annual temperature of 21–22°C and average annual rainfall of 1600–2000 mm. The main soil types in the PRD area are Ultisol, mostly developed on the granite parent materials in the local hills, and paddy soils developed on the fluvial sediments. In the study area, 10–15 crops of vegetables are planted annually, and the rotation of vegetable and rice is applied in mostly paddy soils.

Over the last 30 yr, rapid urbanization and industrialization has taken place in this area. Heavy metal contents in soils and sediments were elevated compared with historical monitoring results [29]. As a result, heavy metal accumulation in agricultural soils has also become increasingly serious in this area because of increasing reliance on fertilizers and agrochemicals [30].

Field sampling and preparation

Both surface soil samples and vegetable samples used in this study were collected from locations shown in Figure 1. All sample sites were far away at least 100 m from the obviously polluted area such as industries, feedlot, wastewater and highway for avoiding the directly anthropogenic influence. The planting area of each site was larger than 1000 m^2 for.a certain vegetable species. In this study the vegetable species included Pakchoi (*Brassica rapa chinensis*) (n = 31), Chinese flowering cabbage (*Brassica campestris L. ssp. chinensis*) (n = 37), Leaf mustard (*Brassica juncea Coss*) (n = 16), Romaine lettuce (*Lactuca sativa L. var. longifolia*) (n = 31), Chinese lettuce (*Lactuca sativa L. var. asparagina*) (n = 15), Cauliflower (*Brassica oleracea L.var. botrytis L.*) (n = 7), Water spinach (*Ipomaea aquatic Forssk*) (n = 7), Celery (*Apium graveolens*) (n = 6), Chinese chives (*Allium tuberosum*) (n = 5), Spinach (*Spinacia oleracea*) (n = 5), Amaranth (*Amaranthus mangostanus L.*) (n = 5), watercress (*Nasturtium officinale*) (n = 5).

Vegetables and field soils at a soil depth of 0–15 cm which was the root concentrated layer in this study were collected when the vegetable were suitable for harvest. Each vegetable and soil sample consisted of five subsamples, and combined and mixed well. The fresh vegetable samples were put in clean plastic bags and immediately transported to the laboratory for sample treatment.

The vegetable samples were cleaned with tap water and Milli-Q water, and then the edible parts (including leaves and steams) were separated, weighed (Fresh weight). The washed samples were then dried in an oven at 60°C, and their dry weights (DW) were recorded. The dry vegetable samples were ground to pass through a 250 μm sieve in a steel grinder. Soil samples were air-dried at room temperature (25°C) and ground to pass through a 2 mm nylon sieve. The fine vegetable and soil sample powders were stored in polythene zip bags.

Chemical analyses

The sieved soil samples were ground further to pass through a 150 μm nylon sieve. The prepared soil samples were digested to dryness using an acid mixture of 10 ml HF, 5 ml $HClO_4$, 2.5 ml HCl, and 2.5 ml HNO_3. Total Cd concentrations in soils were determined by Inductively Coupled Plasma-Atomic Emission Spectrometry (ICP-AES) (Model PS 1000 AT, USA). The recovered soil Cd concentration of the National Research Center for GeoAnalysis soil standard reference materials (SRM) (ESS-4, Beijing, China; Standard value: 0.083 mg kg^{-1}) was 0.081 mg kg^{-1} (n = 32). The analytic precision was 2.4% for soil Cd.

Seventy five soil samples were analyzed for exchangeable Cd using the first step of Tessier sequential extraction[31]. The extraction procedure is 2.00 g of air-dried soil (<2 mm) were mixed with 16 ml 1 M $MgCl_2$ solution (pH = 7.0), the mixture was shaken for one hour on 25°C. the suspension was filtered immediately after shaking. The Cd concentration in the filtrate solutions were measured by Inductively Coupled Plasma-Atomic Emission Spectrometry (ICP-AES) (Model PS 1000 AT, USA).

The ground vegetable samples were ashed in a muffle furnace for 16 h at 500°C, dissolved in 0.5 M HNO_3, and diluted to 25 mL with deionized water. Cd concentrations in vegetables were determined by Inductively Coupled Plasma-Atomic Emission Spectrometry (ICP-AES) (Model PS 1000 AT, USA). A plant standard reference material (GSV-4, Beijing, China; Standard value:0.057 mg kg^{-1}) was used in order to control the determination quality. The recovered Cd concentration was 0.059 mg kg^{-1} (n = 10). The analytic precision 3.5% for vegetable Cd. The Cd concentration in vegetable samples was expressed on a fresh weight basis.

The pH of soil was measured by taking 10 g of sample into 25 ml of deionized water [32]. The soil organic matter content was measured using potassium bichromate oxidation process [33].

Data analysis

The transfer factors (TFs) of Cd from soil to vegetable (edible part) were calculated using the following equation [12,15–17,20,34]:

$$TF = \frac{C_{vegetable}}{C_{soil}}$$

Where $C_{vegetable}$ is Cd concentration (FW) in the edible parts of vegetables, and C_{soil} is the total Cd concentration (DW) in the soil where the vegetable was grown.

The data statistical analysis was performed using the Minitab 16 statistical software (Minitab Inc., USA). Spatial interpolation was performed using ordinary kriging. For the low-density sampling, the ordinary kriging estimate can be thought of simply as an optimally weighted average of the data [35]. It provides a best linear unbiased prediction of spatial distribution. The spatial interpolation and contour maps displaying the spatial distribution of Cd concentrations in soils and vegetables, and transfer factors were produced based on the geostatistical analysis by using the software of ArcGIS 9.0.

Results and Discussions

Cd concentrations in vegetable soils

Total Cd concentrations in soils range from 0.012 to 1.18 mg kg^{-1} in the PRD area (Table 1). These values fit a log normal distribution (Figure 2), therefore the geometric mean (GM) value of 0.158 mg kg^{-1} and geometric standard deviations (GSD) of 1.39 were used to represent the central tendency and variations of the data. The present GM value was much higher than its background concentration (0.04 mg kg^{-1}) of soil in Guangdong province [35] and significantly lower than Cd mean concentrations of 0.858 mg kg^{-1} in the reclaimed tidal flat soil and 1.4–1.8 mg kg^{-1} in some Cd-contaminated vegetable soils in the PRD area [20,21]. The present results reflect the overall level of soil Cd concentration on the region al scale, not is responsible for local point-source contamination.

According to the maximum permissible concentrations of cadmium for agriculture soils of 0.3 mg kg^{-1} [8], thirty vegetable soils were contaminated with Cd in the study area, mainly located in the Guangzhou, Zhongshan, and Foshan cities. The soil Cd mean concentration was up to 0.316 mg kg^{-1} in the Zhongshan city, followed by 0.291 mg kg-1 in the Foshan city, 0.226 mg kg-1

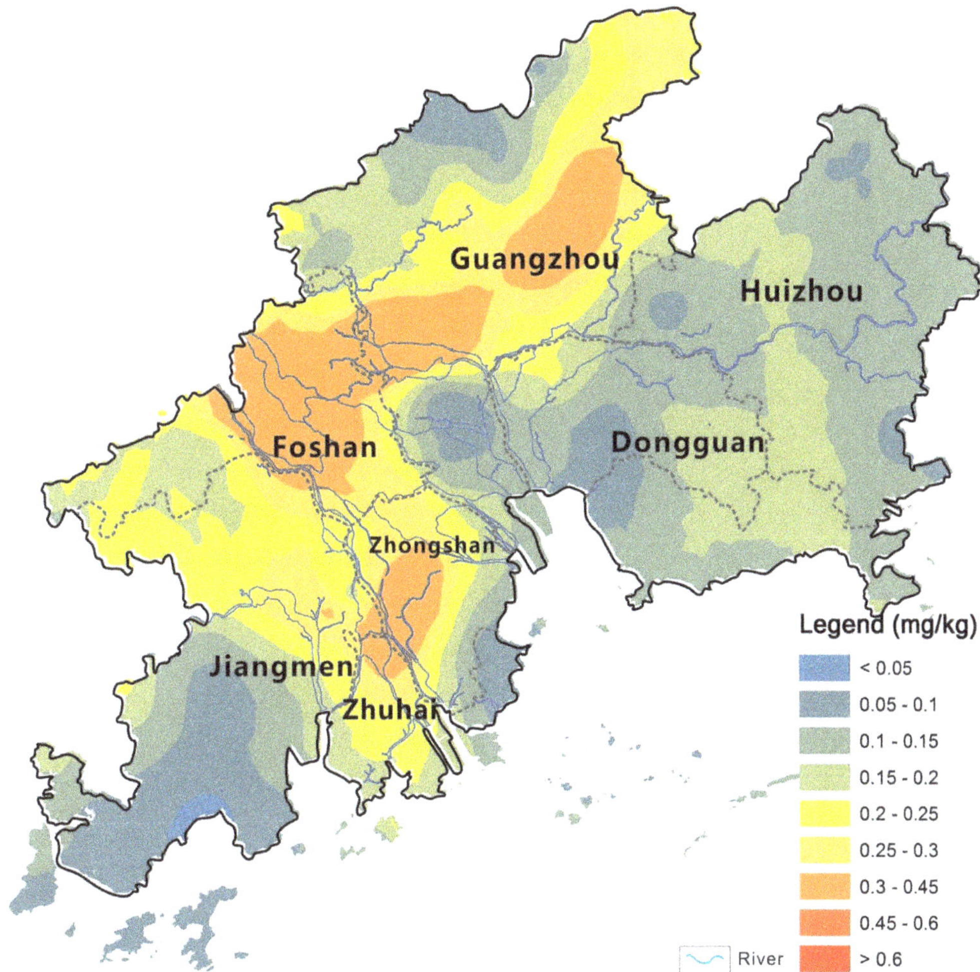

Figure 3. Spatial pattern of total Cd concentration in the soils in the PRD area (mg kg^{-1}, DW).

in the Guangzhou City, and 0.204 mg kg-1 in the Zhuhai City (Table 1). Generally, spatial characteres of soil Cd show that the western areas (including Foshan, Guangzhou, Zhongshan, Zhuhai) influenced by the West River and North River has the higher Cd concentration than the eastern areas (including Dongguan and Huizhou) influenced by the East River (Figure 3). The results coincided with spatial characters of other heavy metal concentrations (Pb, Zn) in sediments in the PRD area, that is, there were higher heavy metal concentrations in sediments of the West River and North River than that of the East River [36]. The spatial correlation between soil Cd cocnetrtiaons and regional rivers suggested that the cadmium produced by the mining and smelting activities in the upper reaches of the West River and North River was the main source of Cd in soils of the PRD area, and the West River and North River was the important transfer approach for Cd entering into surrounding soils through river irrigation.

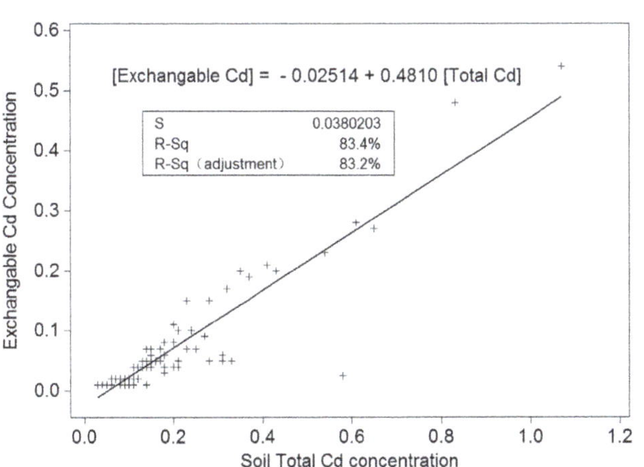

Figure 4. Relationship between exchangeable Cd concentration and total Cd concentration in vegetable soils (mg kg^{-1}, DW).

Although many literatures pointed out that metal availability in soil is of main concern, because the available concentration is an indication of the amount available for plant uptake, and provided some methods to determine plant availability of heavy metals [12,21,24]. Our results reveal that the significant positive correlation between soil total Cd and exchangeable Cd was shown in Figure 4, which can be described by a linear equation:

$$y = 0.4810x - 0.02514 \ (r^2 = 0.834; p < 0.001)$$

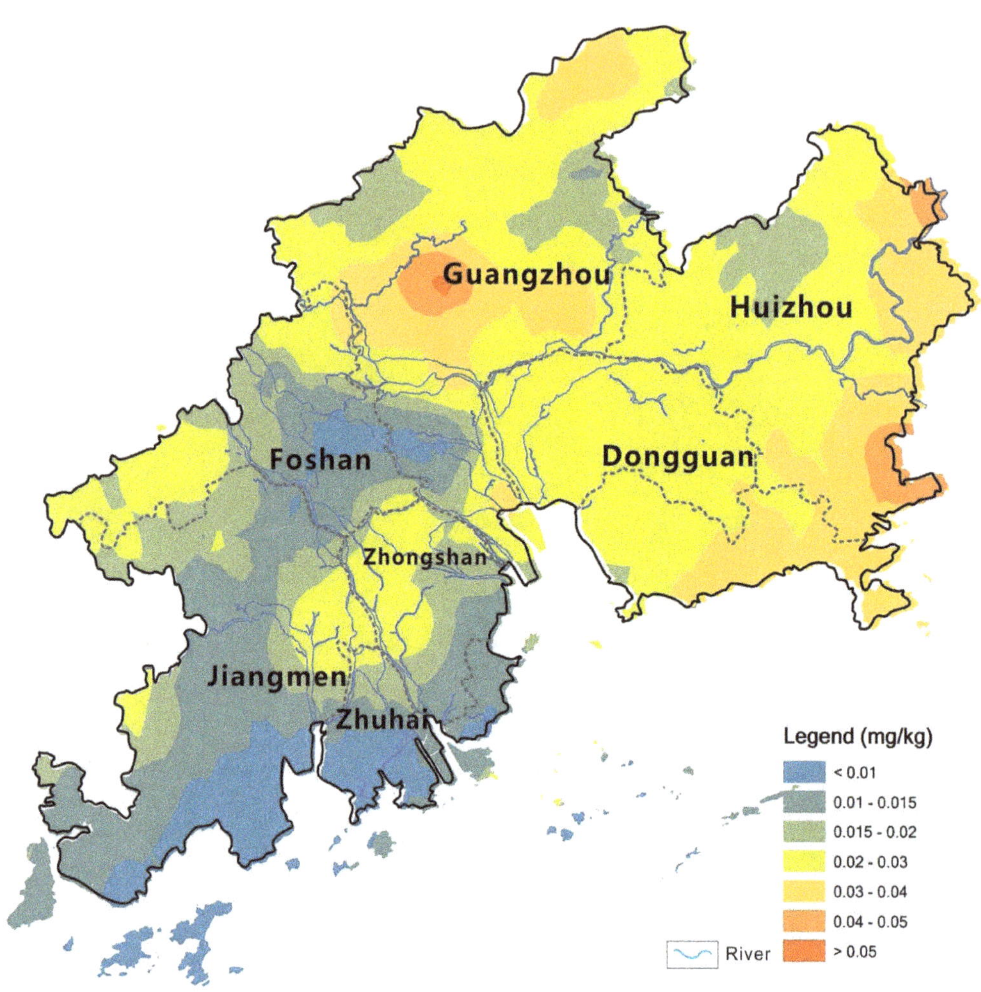

Figure 5. Spatial pattern of Cd concentrations in the vegetables in the PRD area (mg kg^{-1}, FW).

Table 2. Cd concentration (mg kg^{-1}, FW) in different vegetable species and the TF values in the PRD area.

Vegetable species	n	Cd		TF	
		Mean ± SD	Range	Mean ± SD	Range
Pakchoi (*Brassica rapa chinensis*)	31	0.023±0.015	0.002–0.074	0.222±0.220	0.010–0.920
Chinese flowering cabbage (*Brassica campestris L. ssp. chinensis*)	37	0.022±0.015	0.001–0.068	0.160±0.142	0.025–0.774
Leaf mustard (*Brassica juncea Coss*)	16	0.023±0.014	0.001–0.054	0.532±1.087	0.022–4.50
Romaine lettuce (*Lactuca sativa L. var. longifolia*)	31	0.026±0.021	0.003–0.082	0.148±0.111	0.001–0.371
Chinese lettuce (*Lactuca sativa L. var. asparagina*)	15	0.031±0.026	0.005–0.078	0.188±0.219	0.017–0.821
Cauliflower (*Brassica oleracea L.var. botrytis L.*)	7	0.010±0.013	0.002–0.032	0.048±0.042	0.014–0.103
Water spinach (*Ipomaea aquatic Forssk*)	7	0.011±0.019	0.001–0.053	0.045±0.052	0.003–0.117
Celery (*Apium graveolens*)	6	0.043±0.020	0.019–0.064	0.234±0.115	0.110–0.376
Chinese chives (*Allium tuberosum*)	5	0.027±0.032	0.001–0.074	0.229±0.241	0.001–0.563
Spinach (*Spinacia oleracea*)	5	0.037±0.027	0.006–0.067	0.213±0.034	0.182–0.262
Amaranth (*Amaranthus mangostanus L.*)	5	0.060±0.017	0.030–0.082	0.753±0.442	0.258–1.577
watercress (*Nasturtium officinale*)	5	0.021±0.020	0.002–0.039	0.087±0.046	0.003–0.129
Total	170	0.024±0.019	0.001–0.082	0.189±0.211	0.001–4.50

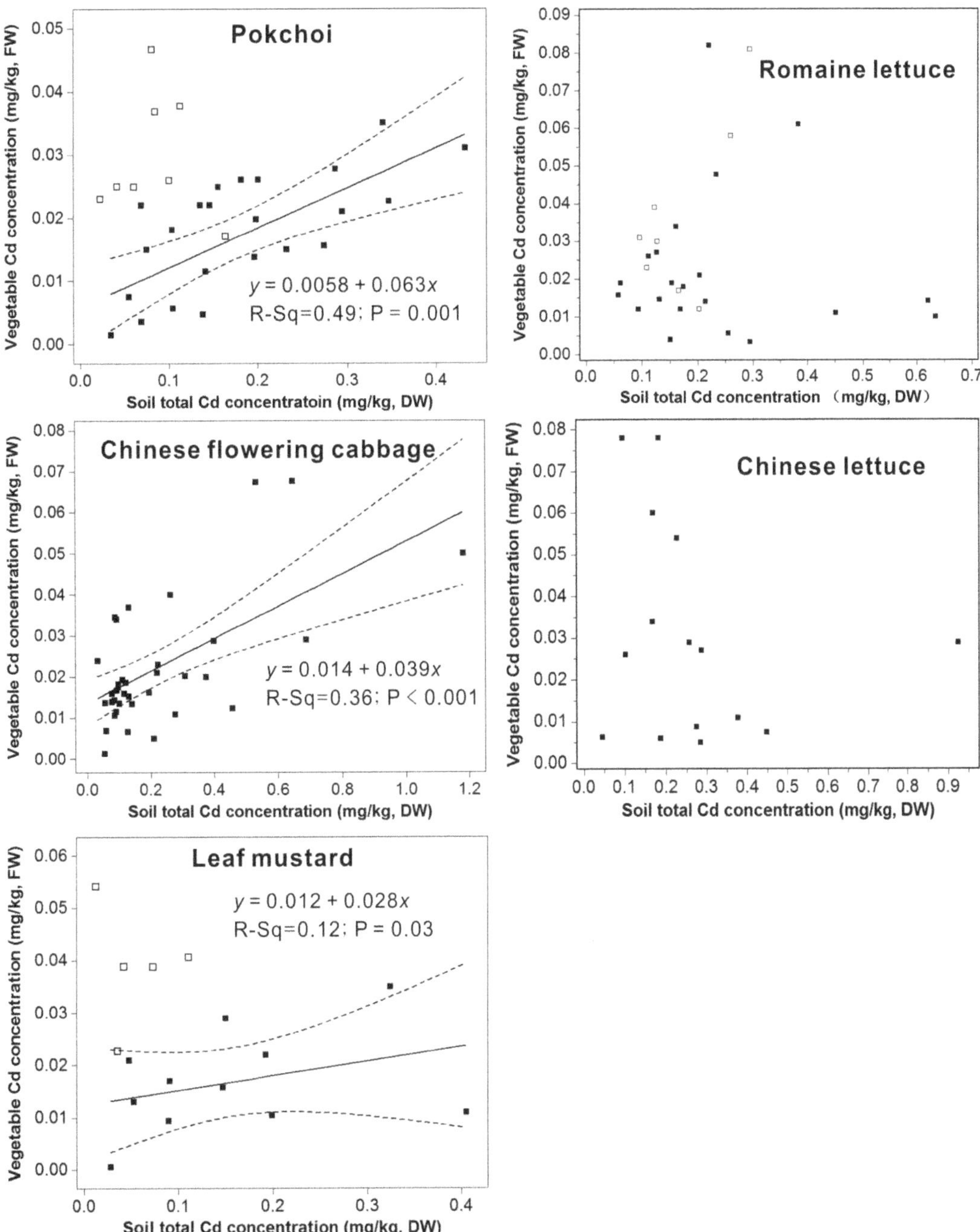

Figure 6. Relationships between soil total Cd concentrations (mg kg^{-1}, DW) and vegetable Cd concentrations (mg kg^{-1}, FW) for five species of main vegetables (soil pH: 3.79–7.72) (hollow diamonds represented samples with lower pH (3.79−5.00) removed for the regression analysis; the solid line is the sample regression line; dotted lines indicated the 95% confidence interval).

Where y is the soil exchangeable Cd concentration, x refers to soil total Cd concentration.

This indicated that exchangeable Cd concentration in soils in the PRD area was mainly controlled by the soil total Cd concentration.

Cd concentrations in vegetables

The present results show that in the PRD area the range of Cd concentration in all vegetable samples was 0.002–0.082 mg kg^{-1} with a geometric mean value of 0.015 mg kg^{-1} (Table 1). In this study Cd concentrations in all vegetable samples met the national standard of Safe vegetables (0.2 mg kg^{-1}), 150 vegetable samples met the standard of No-Polluted vegetables (0.05 mg kg^{-1}). Twenty vegetable samples were in between Safe vegetables and No-polluted vegetables, of which14 vegetable sample sites were in the Guangzhou.

According to the different cities, the mean vegetable Cd concentration, was in the order of Guangzhou (0.027 mg kg^{-1})> Huizhou (0.026 mg kg^{-1})> Dongguan (0.024 mg kg^{-1})> Zhongshan (0.022 mg kg^{-1})> Foshan (0.017 mg kg^{-1})> Jiangmen (0.014 mg kg^{-1})> Zhuhai (0.008 mg kg^{-1}) (Table 1). This order is significantly different from Cd concentrations in soils.

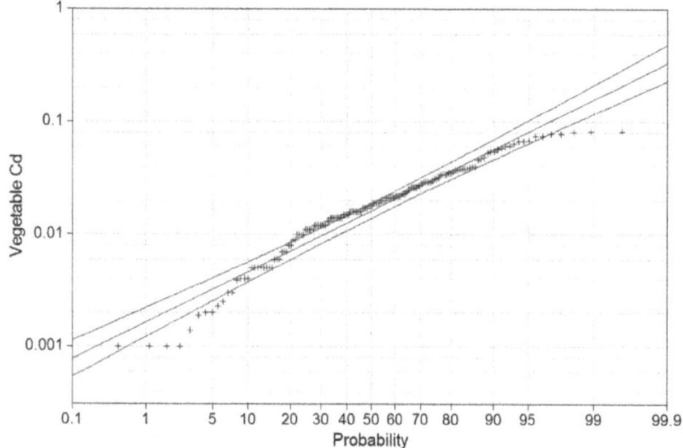

Figure 7. Lognormal probability plot for vegetable Cd concentrations in the Pearl River Delta area, Guangdong, China (mg kg^{-1}, FW).

The spatial pattern of Cd concentrations in vegetables showed that no spatial correlation with that of total Cd cocentrtaions in soils on the regional scale (Figure 3 and Figure 5).

The mean and range of Cd concentrations in various vegetable species were listed in Table 2. In the study area the dominant vegetable species are the pakchoi, Chinese flowering cabbage, leaf mustard, Romaine lettuce and Chinese lettuce for the local residents. The mean Cd concentrations in these vegetables range from 0.022 mg kg^{-1} to 0.031 mg kg^{-1} (Table 2). The amaranth vegetables had the highest Cd concentration of 0.082 mg kg^{-1} with mean value of 0.06 mg kg^{-1}. This value was closed to 0.078 mg kg^{-1} in amaranth on the reclaimed tidal flat soil reported by Li et al [20]. Compared with the previous researches on the Cd-contaminated soil [20,21], the present results revealed that there were low Cd concentrations of various vegetables in the study area.

Relationships between Cd in soils and in vegetables

Many studies have shown that the concentration of heavy metals in vegetables is influenced by many factors, such as total concentration of heavy metals in soils, vegetable species, soil pH value, soil organic matter, climate, atmospheric depositions and temperature [19,21,37–39]. McBride [40] suggested that combination of soil pH and soil total Cd concentration was reasonably predictive of Cd concentration in the above-ground plant tissue.

In this study, we discussed the correlation between Cd concentrations in five species of dominant leaf vegetables and Cd concentrations in soils. The results revealed that in the condition of soil pH>5, Cd concentrations in both pakchoi and Chinese flowering vegetables had obvious positive correlations with Cd concentrations in soils; and Cd concentrations in leaf mustard vegetables and Cd concentrations in soils had a weak positive correlation (Figure 6). These can be described by linear equations, respectively:

$$y = 0.063x + 0.0058\ (Pakchoi : r^2 = 0.49; p = 0.001)$$

$$y = 0.039x + 0.014\ (Chinese\ flowering\ cabbage : r^2 = 0.36; p < 0.001)$$

$$y = 0.028x + 0.012\ (Leaf\ mustard : r^2 = 0.12; p = 0.03)$$

Where y is the Cd concentration in the various vegetable, x is the soil total Cd concentration.

The results of the line regression indicated that in the condition of soil pH>5, soil total Cd concentrations explained 49%, 36% and 12% of the variability of Cd concentration in Pakchoi and Chinese flowering cabbage and leaf mustard, respectively. But Cd concentrations in the Romaine lettuce and Chinese lettuce had no correlations with total Cd concentrations in soils (Figure 6). This phenomenon should be related with various vegetable families. Pakchoi, Chinese flowering cabbage and leaf mustard belong to the brassica of cruciferous plants; Romaine lettuce and Chinese lettuce belong to the Lactuca of Compositae plants. The difference of correlations between Cd concentrations in vegetables and total Cd concentrations in soils should be the result of different absorption capacity of various vegetable families to Cd. And the result showed that the influence of soil total Cd concentrations on the Cd in various vegetable species is obviously different [39,41].

When soils pH is less than 5, total Cd concentrations in soils were low because the environmental capacity of soil Cd was small in acid soils due to the strong eluviation, but Cd concentrations in the pakchoi and leaf mustard vegetables were much higher than the reference line (Figure 6). Obviously, low soil pH can raise the available Cd content for pakchoi and leaf mustard vegetables even in the soils with low Cd concentrations.

The analysis of Pearson correlations among soil Cd, vegetable Cd, soil pH, organic matter and transfer factor showed that there was no obvious correlation among soil Cd, vegetable Cd, organic matter and soil pH; significant correlations were be found between transfer factor and soil Cd ($r^2 = -0.295$) and between transfer factor and vegetable Cd ($r^2 = 0.435$) (Table 3). Our results confirmed that soil pH played important role to Cd absorption of vegetables, but there was no linear correlation between soil pH and Cd concentration in vegetables in this study. Moreover, The lognormal probability distribution graph shows that some vegetable samples of upper tail had less high value than would be

Table 3. Pearson correlations of soil Cd, vegetable Cd, soil pH, organic matter and Transfer factor.

	soil Cd	vegetable Cd	pH	Organic matter
vegetable Cd	0.205(0.088)			
pH	0.033(0.784)	0.137(0.258)		
Organic matter	0.065(0.590)	−0.007(0.952)	0.076(0.532)	
TF	−0.295(0.013)	0.435(0.000)	−0.143(0.239)	−0.103(0.396)

Cell content: Pearson correlation (p-value).

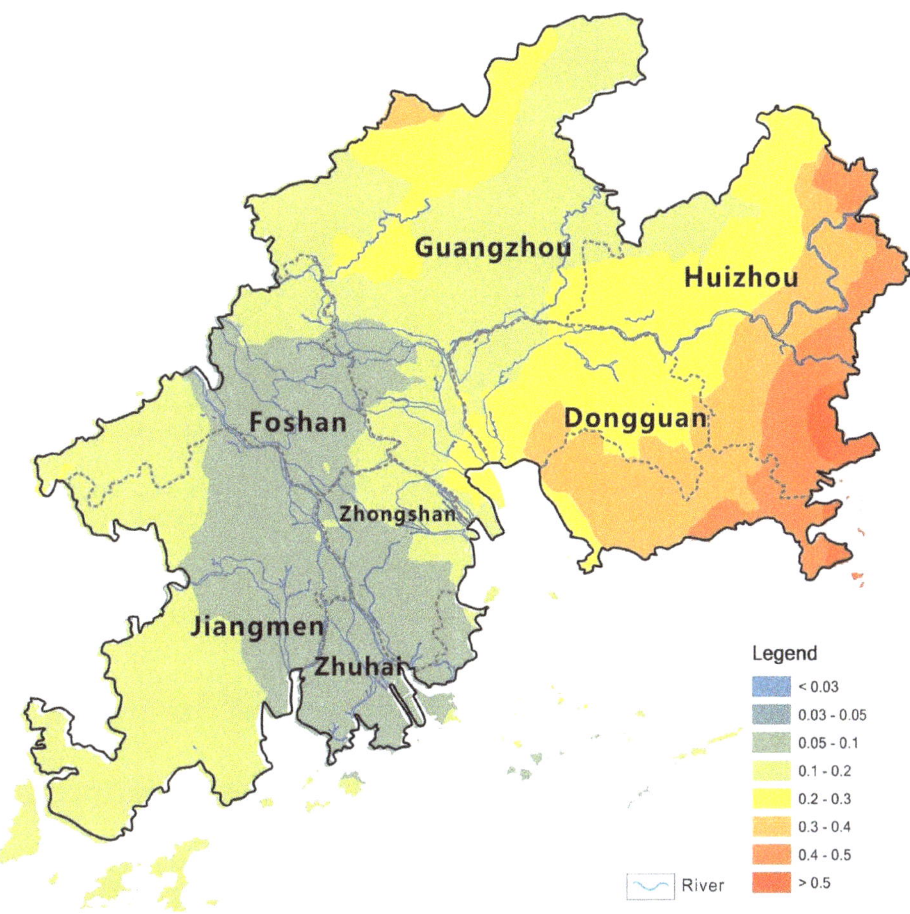

Figure 8. Spatial pattern of soil-to-vegetable transfer factor of Cd in the PRD area.

expected (Figure 7), suggesting the vegetable's ability to absorb Cd would be decreasing with the increasing the Cd content in the vegetables.

Transfer factors of Cd from soil to vegetable

The soil-to-vegetable transfer factors (TFs) reflected the ability of vegetables to take up soil metals. TFs varied significantly with plant species [12,19,21,38,39], and were commonly viewed as a "constant" for a given plant species and a given metal. TFs were used as an important character index for establishing the soil environmental quality criteria or assessing the health risk of soil contamination.

In this study, the TFs of Cd from soil to vegetable for the 12 vegetable species varied from 0.045 in Water spinach to 0.753 in Amaranth (Table 2). The TF mean values in the main five species of leaf vegetables in the study area were 0.532 for leaf mustard, 0.222 for Pakchoi, 0.188 for Chinese lettuce, 0.160 for Chinese flowering cabbage, and 0.148 for Romaine lettuce. Compared with TFs of Cd reported by Wang et al [12], and Hu et al [21], these values were much higher because of the lower Cd concentration in soils. The spatial pattern of TFs also showed that the sample sites with high TFs were always corresponding to the lower Cd concentrations in soils in the study area (Figure 3 and Figure 8).

The regression equations between the TF values and total Cd concentrations in soils can be described by power model ($y = ax^b$) for five kinds of main vegetables:

$$y = 0.0233x^{-0.883}\ (Pakchoi: r^2 = 0.433; p < 0.001)$$

$$y = 0.0392x^{-0.582}$$
$$(Chinese\ flowering\ cabbage: r^2 = 0.354; p < 0.001)$$

$$y = 0.0214x^{-0.915}\ (Leaf\ mustard: r^2 = 0.406; p = 0.008)$$

$$y = 0.0172x^{-1.047}\ (Romiane\ lettuce: r^2 = 0.358; p < 0.001)$$

$$y = 0.0173x^{-1.116}\ (Chinese\ lettuce: r^2 = 0.403; p = 0.011)$$

Where y is the soil-to-vegetable transfer factor of Cd, x is the total Cd concentration in the soils.

For main leaf vegetables, the transfer factors of Cd decreased with increasing total Cd concentrations (Figure 9), suggesting that the ability of vegetable to take up soil Cd decreased with the total soil Cd increasing. Figure 9 also illustrated that when soil Cd concentration was low (about 0.1–0.2 mg kg^{-1}), the TFs decreased sharply; with the gradual increasing of soil total Cd concentration, TFs decreased only slowly, till to a stable value.

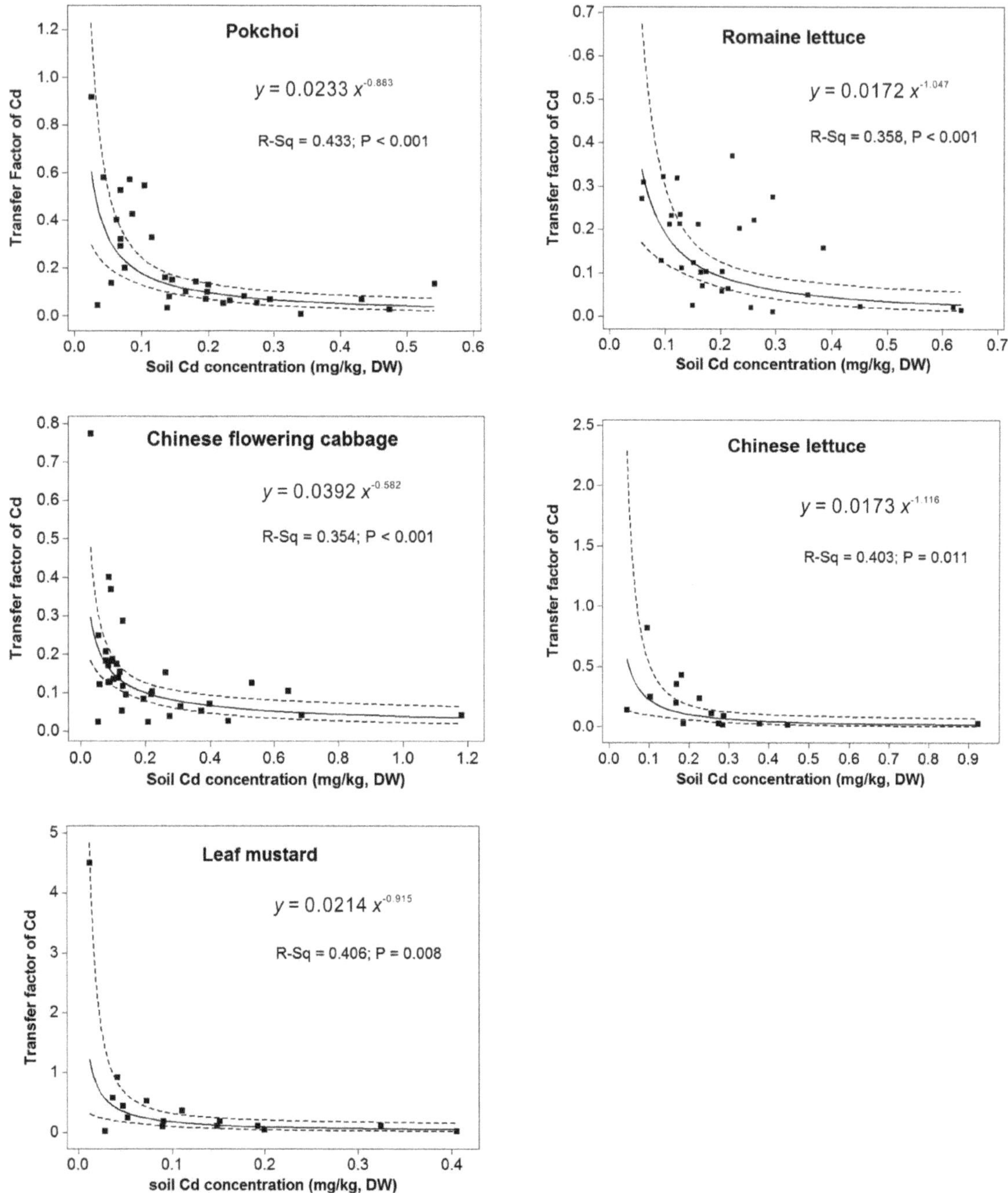

Figure 9. Relationships between soil-to-vegetable transfer factors of Cd and total Cd concentrations in soils for five species of main vegetables (the solid line is the sample regression curve; dotted lines indicated the 95% confidence interval).

Moreover, the 95% confidence interval of regression lines showed that there were a larger error range at the low Cd concentration interval of soils (0–0.1 or 0.2 mg kg^{-1}) for the various leaf vegetable species, and when soil Cd concentrations were greater than 0.1 or 0.2 mg kg^{-1}, the TFs of various vegetables would be reasonable with a lower error range for the assessment of environmental risk (Figure 9). As a result, for carrying out rational environmental health assessment in different Cd concentrations of soils, the present results indicated that the fit line of TFs for vegetable species should be discussed to divide into two parts, including the sharply decrease part and the slowly decrease part, according to the gradual increasing of soil total Cd concentration. This work is agree with the suggestion proposed by Wang et al [12] that the transfer factors of a given crop-metal system estimated from the regression model between the TF values and the corresponding soil metal concentration, are much more reasonable than using the arithmetic means or geometric means at a given soil metal concentration.

Conclusions

The present results showed that the total Cd concentration in soils is safe for planting leaf vegetables in the study area. All vegetable samples met the national standard of safe vegetables, and 88% of which met the standard of no-polluted vegetables. Total Cd concentrations in soils and soil pH played an important role to Cd absorption of vegetables, but there were no obviously linear correlations. There was no spatial correlation between total Cd concentrations in soils and Cd concentrations in vegetables. For a givenl vegetable species the regression model between the TF value and the corresponding total Cd concentrations in soils could be used to estimate the transfer abilities of Cd from soil to vegetable at a given total Cd concentration in soils.

Acknowledgments

We thank four reviewers and editors for their very useful comments that improved the quality of this manuscript.

Author Contributions

Performed the experiments: HZ JC. Analyzed the data: LZ GY. Contributed reagents/materials/analysis tools: HZ DL. Wrote the paper: HZ.

References

1. Singh BR, McLaughlin MJ (1999) Cadmium in Soils and Plants. In: Mclaughlin M., Singh BR, editors. Cadmium in soils and plants. Boston: Kluwer Academic. 357–367.
2. Tudoreanu L, Phillips CJC (2004) Modeling cadmium uptake and accumulation in plants. Adv Agron 84: 121–157. doi:10.1016/S0065-2113(04)84003-3.
3. Singh A, Sharma RK, Agrawal M, Marshall FM (2010) Risk assessment of heavy metal toxicity through contaminated vegetables from waste water irrigated area of Varanasi, India. Trop Ecol 51: 375–387.
4. WHO (1992) Cadmium: Environmental Health Criteria, 134. Geneva.
5. Chlopecka A, Bacon JR, Wilson MJ, Kay J (1996) Forms of Cadmium, Lead, and Zinc in Contaminated Soils from Southwest Poland. J Environ Qual 25: 69. doi:10.2134/jeq1996.00472425002500010009x.
6. Kirkham MB (2006) Cadmium in plants on polluted soils: Effects of soil factors, hyperaccumulation, and amendments. Geoderma 137: 19–32. doi:10.1016/j.geoderma.2006.08.024.
7. Wu L, Tan C, Liu L, Zhu P, Peng C, et al. (2012) Cadmium bioavailability in surface soils receiving long-term applications of inorganic fertilizers and pig manure. Geoderma 173-174: 224–230. doi:10.1016/j.geoderma.2011.12.003.
8. SEPAC (1995) Environmental quality standard for soil (GB 15618–1995). State Environmental Protection Administration of China, Beijing, China
9. Desaules A (2012) Critical evaluation of soil contamination assessment methods for trace metals. Sci Total Environ 426: 120–131. doi:10.1016/j.scitotenv.2012.03.035.
10. Römkens PFAM, Guo HY, Chu CL, Liu TS, Chiang CF, et al. (2009) Prediction of Cadmium uptake by brown rice and derivation of soil-plant transfer models to improve soil protection guidelines. Environ Pollut 157: 2435–2444. doi:10.1016/j.envpol.2009.03.009.
11. Sun W, Sang L, Jiang B (2012) Trace metals in sediments and aquatic plants from the Xiangjiang River, China. J Soils Sediments 12: 1649–1657. doi:10.1007/s11368-012-0596-8.
12. Wang G, Su M, Chen Y, Lin F, Luo D, et al. (2006) Transfer characteristics of cadmium and lead from soil to the edible parts of six vegetable species in southeastern China. Environ Pollut 144: 127–135. doi:10.1016/j.envpol.2005.12.023.
13. Wang Q, Dong Y, Cui Y, Liu X (2001) Instances of soil and crop heavy metal contamination in China. Soil Sediment Contam 10: 497–510.
14. Xiong X, Allinson G, Stagnitti F, Li P, Wang X, et al. (2004) Cadmium contamination of soils of the Shenyang Zhangshi Irrigation Area, China: an historical perspective. Bull Environ Contam Toxicol 73: 270–275. doi:10.1007/s00128-004-0423-z.
15. Williams PN, Lei M, Sun G, Huang Q, Lu Y, et al. (2009) Occurrence and partitioning of cadmium, arsenic and lead in mine impacted paddy rice: Hunan, China. Environ Sci Technol 43: 637–642.
16. Li JT, Qiu JW, Wang XW, Zhong Y, Lan CY, et al. (2006) Cadmium contamination in orchard soils and fruit trees and its potential health risk in Guangzhou, China. Environ Pollut 143: 159–165. doi:10.1016/j.envpol.2005.10.016.
17. Bi X, Ren L, Gong M, He Y, Wang L, et al. (2010) Transfer of cadmium and lead from soil to mangoes in an uncontaminated area, Hainan Island, China. Geoderma 155: 115–120. doi:10.1016/j.geoderma.2009.12.004.
18. Zhou Z., Fan Y., Wang M (2000) Heavy metal contamination in vegetables and their control in China. Food Rev Int 16: 239–255. doi:10.1081/FRI-100100288.
19. Cui Y, Zhu Y, Zhai R, Chen D, Huang Y, et al. (2004) Transfer of metals from soil to vegetables in an area near a smelter in Nanning, China. Environ Int 30: 785–791. doi:10.1016/j.envint.2004.01.003.
20. Li Q, Chen Y, Fu H, Cui Z, Shi L, et al. (2012) Health risk of heavy metals in food crops grown on reclaimed tidal flat soil in the Pearl River Estuary, China. J Hazard Mater 227–228: 148–154. doi:10.1016/j.jhazmat.2012.05.023.
21. Hu J, Wu F, Wu S, Sun X, Lin X, et al. (2013) Phytoavailability and phytovariety codetermine the bioaccumulation risk of heavy metal from soils, focusing on Cd-contaminated vegetable farms around the Pearl River Delta, China. Ecotoxicol Environ Saf 91: 18–24. doi:10.1016/j.ecoenv.2013.01.001.
22. AQSIQ (2001) Safety qualification for agricultural product – Safety requirements for non-environmental pollution vegetable (GB 18406.1–2001). General Administration of Quality Supervision, Inspection and quarantine of China, Beijing, China.
23. MOH (2012) Maximum levels of contaminants in Foods (GB 2762–2012). Ministry of Health of the People's Republic of China.
24. Hart JJ, Welch RM, Norvell WA, Kochian LV (2002) Transport interactions between cadmium and zinc in roots of bread and durum wheat seedlings. Physiol Plant 116: 73–78. doi:10.1034/j.1399-3054.2002.1160109.x.
25. Yang Y, Zhang F-S, Li H-F, Jiang R-F (2009) Accumulation of cadmium in the edible parts of six vegetable species grown in Cd-contaminated soils. J Environ Manage 90: 1117–1122. doi:10.1016/j.jenvman.2008.05.004.
26. McLaughlin M, Parker D, Clarke J (1999) Metals and micronutrients – food safety issues. F Crop Res 60: 143–163. doi:10.1016/S0378-4290(98)00137-3.
27. Wan Z (2013) The development report of Guangdong vegetables industry in 2012. Institute of Agricultural Economics and Rural Development.
28. SBG (2012) Guangdong Statistical Yearbook –2012. Beijing: Statistics Bureau of Guangdong Province, China Statistic Press.
29. Li X, Wai OWH, Li YS, Coles BJ, Ramsey MH, et al. (2000) Heavy metal distribution in sediment profiles of the Pearl River estuary, South China. Appl Geochemistry 15: 567–581. doi:10.1016/S0883-2927(99)00072-4.
30. Wong S, Liu X, Zhang G, Qi S, Min Y (2002) Heavy metals in agricultural soils of the Pearl River Delta, South China. Environ Pollut 119: 33–44. doi:10.1016/S0269-7491(01)00325-6.
31. Tessier A, Campbell PGC, Bisson M (1979) Sequential extraction procedure for the speciation of particulate trace metals. Anal Chem 51: 844–851. doi:10.1021/ac50043a017.
32. Chinese National Standard Agency (1988) Determination of pH value in forest soil. GB 7859–87, UDC 634.0.114:631.422: 171–173.
33. Yu T, Wang Z (1988) Soil Analytical Chemistry. Beijing: Science Press.
34. Chojnacka K, Chojnacki A, Górecka H, Górecki H (2005) Bioavailability of heavy metals from polluted soils to plants. Sci Total Environ 337: 175–182. doi:10.1016/j.scitotenv.2004.06.009.
35. Zhang HH, Chen JJ, Zhu L, Li FB, Wu ZF, et al. (2011) Spatial patterns and variation of soil cadmium in Guangdong Province, China. J Geochemical Explor 109: 86–91. doi:10.1016/j.gexplo.2010.10.014.
36. Zhang C, Wang L (2001) Multi-element geochemistry of sediments from the Pearl River system, China. Appl Geochemistry 16: 1251–1259. doi:10.1016/S0883-2927(01)00007-5.
37. Voutsa D, Grimanis A, Samara C (1996) Trace elements in vegetables grown in an industrial area in relation to soil and air particulate matter. Environ Pollut 94: 325–335. doi:10.1016/S0269-7491(96)00088-7.
38. Peris M, Micó C, Recatalá L, Sánchez R, Sánchez J (2007) Heavy metal contents in horticultural crops of a representative area of the European Mediterranean region. Sci Total Environ 378: 42–48. doi:10.1016/j.scitotenv.2007.01.030.
39. Li Y, Li L, Zhang Q, Yang Y, Wang H, et al. (2013) Influence of temperature on the heavy metals accumulation of five vegetable species in semiarid area of northwest China. Chem Ecol 29: 353–365. doi:10.1080/02757540.2013.769970.
40. McBride M, Murray B (2002) Cadmium uptake by crops estimated from soil total Cd and pH. Soil Sci 167.
41. Hooda PS, McNulty D, Alloway BJ, Aitken MN (1997) Plant Availability of Heavy Metals in Soils Previously Amended with Heavy Applications of Sewage Sludge. J Sci Food Agric 73: 446–454. doi:10.1002/(SICI)1097-0010(199704)73:4<446::AID-JSFA749>3.0.CO;2-2.

9

The Kill Date as a Management Tool for Cover Cropping Success

María Alonso-Ayuso, José Luis Gabriel, Miguel Quemada*

School of Agriculture Engineering, Technical University of Madrid, Madrid, Spain

Abstract

Integrating cover crops (CC) in rotations provides multiple ecological services, but it must be ensured that management does not increase pre-emptive competition with the subsequent crop. This experiment was conducted to study the effect of kill date on: (i) CC growth and N content; (ii) the chemical composition of residues; (iii) soil inorganic N and potentially mineralizable N; and (iv) soil water content. Treatments were fallow and a CC mixture of barley (*Hordeum vulgare* L.) and vetch (*Vicia sativa* L.) sown in October and killed on two different dates in spring. Above-ground biomass and chemical composition of CC were determined at harvest, and ground cover was monitored based on digital image analysis. Soil mineral N was determined before sowing and after killing the CC, and potentially mineralizable N was measured by aerobic incubation at the end of the experiment. Soil water content was monitored daily to a depth of 1.1 m using capacitance sensors. Under the present conditions of high N availability, delaying kill date increased barley above-ground biomass and N uptake from deep soil layers; little differences were observed in vetch. Postponing kill date increased the C/N ratio and the fiber content of plant residues. Ground cover reached >80% by the first kill date (~1250°C days). Kill date was a means to control soil inorganic N by balancing the N retained in the residue and soil, and showed promise for mitigating N losses. The early kill date decreased the risk of water and N pre-emptive competition by reducing soil depletion, preserving rain harvested between kill dates and allowing more time for N release in spring. The soil potentially mineralizable N was enhanced by the CC and kill date delay. Therefore kill date is a crucial management variable for maximizing the CC benefits in agricultural systems.

Editor: Anil Shrestha, California State University, Fresno, United States of America

Funding: This work was funded by the Spanish Comisión Interministerial de Ciencia y Tecnología (project AGL 2011-24732) and the Regional Government of Madrid (Project AGRISOST, S2009/AGR-1630). The funders had no role in study design, data collection and analysis, decision to publish, or preparation of the manuscript.

Competing Interests: The authors have declared that no competing interests exist.

* Email: miguel.quemada@upm.es

Introduction

The potential for cover crops (CC) to provide ecological services has been documented in diverse cover cropping systems and environments [1]. Replacing fallow periods with CC may enhance soil aggregate stability [2], water retention capacity [3], nutrient supply [4] and disease suppression [5]. Moreover, plant residues covering the soil following crop kill can improve soil protection over time and help control weeds, preserve soil moisture, ameliorate compacted soils, and reduce soil erosion [6]. However, improperly managed CC may have a detrimental effect on the cash crop either by competing for water and nutrients, building up diseases or retarding seed germination [6,7]. Overall, there are two crucial management factors that determine CC success: species selection and CC termination [8]. In this manuscript, we focus on the termination of CC incorporated into field crop rotations, particularly on kill date. Previous experiments have shown that changing kill date by a few weeks in spring may have a large effect on the water and N pre-emptive competition with the subsequent cash crop, nitrate leaching control, and soil moisture conservation [7,9–12], but there is no agreement on the advantages and limitations of postponing CC termination. Therefore, a better understanding of the effect of the living CC and its dead mulch on the dynamics of N and water might improve our ability to choose a kill date that maximizes the benefits of CC in the system.

The effect of kill date on soil water availability is a balance between the water extracted by the living CC and the evaporation prevented by the residue mulch [9]. Integrating CC into the cropping system increases above-ground biomass and resistance to decomposition, which enhances the development of a dead mulch covering the soil surface [13]. Some authors have found that, despite the soil water depletion, moisture conservation was improved when the CC was killed later [9,11]. However, other experiments have shown that an earlier CC kill reduced water pre-emptive competition, preserved top soil moisture, and increased water availability for the subsequent crop [7,12]. Similarly, postponing kill date enhanced N pre-emptive competition as soil available N is depleted by the CC uptake and only slowly replenished by the mineralization of a high C/N residue [6]. Legume species may compensate for this effect by fixing more N_2 when killed late [10]. The water and N balance is driven by various factors that are all affected by kill date, such as above-ground biomass accumulation, ground cover (GC), and N uptake

[6]. The chemical composition of crop residues, such as lignin content and the C/N ratio, are important characteristics governing the decomposition process and determining the persistence of the residue layer covering the soil [14]. However, there are few experiments that combine multiple approaches to study these variables, and to our knowledge, none that involves continuous monitoring of soil water dynamics.

We hypothesize that a combined study of CC growth, GC and water and N dynamics would contribute to understanding the effect of kill date and improve management practices. This experiment was conducted over two years to study the effect of kill date on the following factors: (i) the growth and N content of the CC; (ii) the chemical composition and residue quality of the CC; (iii) soil inorganic N dynamics and its potentially mineralizable N; and (iv) the soil water content. A CC mixture of barley and vetch was selected because mixing legumes and non-legumes is known to be an efficient technique to merge N and water management benefits of the individual species [9,15].

Materials and Methods

1. Field experiment

The experiment was conducted from October 2011 to November 2013 at La Chimenea field station (40°03′N, 03°31′W, 550 m a.s.l.), which is located in the central Taxus river basin in Aranjuez (Madrid, Spain). The soil is a silty clay loam (Typic Calcixerept) [16] with a high content of organic matter and carbonates, pH ~8 and low stone content throughout the profile. The most relevant soil characteristics are presented in Table 1. Total organic C was measured by the Walkley-Black method [17]. Calcium carbonate was determined by a volumetric calcimeter method [18]. The climate is semiarid Mediterranean [19] with high interannual variability (Figure 1). Annual rainfall averages ~350 mm with less rain in the summer and more in the autumn and a mean annual temperature of 14.2°C. Weather data were recorded throughout the experimental period by a data logger (CR10X. Campbell Scientific Ltd, Shepshed, UK) located ~100 m from the field site.

The experiment consisted of twelve plots (15 m×12 m) randomly assigned to three treatments (fallow and two different CC kill dates) with four replications. Cover crop treatments were sown with a barley (*Hordeum vulgare* L., cv. Vanessa) + vetch (*Vicia sativa* L., cv. Aitana) mixture in early October (October 6, 2011; October 8, 2012). The mixture, 30% (54 kg ha^{-1}) barley seeds and 70% (45 kg ha^{-1}) vetch, was broadcast by hand and buried (~5 cm) with a shallow cultivator. All treatments received ~18 mm irrigation on the sowing date with a sprinkle irrigation system (9.5 mm h^{-1}) to ensure uniform CC establishment. Plots from the first kill treatment (FK) were killed in mid-March (March 13, 2012; March 14, 2013) when barley was at the end of the booting growth stage and vetch at the stem elongation stage. Plots from the second kill treatment (SK) were killed in mid-April (April 9, 2012; April 13, 2013) when barley was at the emerging inflorescence growth stage and vetch was at the stem elongation stage. Cover crops were killed by application of glyphosate (N-phosphonomethyl glycine, 0.7 kg a.e. ha^{-1}) followed by a shredder. The last day of April was chosen as a hypothetical planting date (HPD) for both years, as it is a typical date for planting summer cash crops in the area. The HPD represented the beginning of the mulch period during which the CC residue remained on the ground. To compare the residue mulch effect on water conservation, SWC was homogenized by adding water to the drier plots with a drip irrigation system (30.5 mm to FK treatment and 35.5 mm to SK) in April 2012. As rainfall was abundant in March 2013, no supplemental irrigation was applied. Fallow plots were kept free of weeds by hand weeding during the trial period. The experiment was conducted in a field that was cultivated with triticale during the two previous years and had not received organic amendments or N fertilizer for four years prior to the beginning of the trial.

2. Cover crop analysis

To measure the above-ground biomass and GC in each plot, four 0.5 m×0.5 m squares were marked after sowing the CC. Ground cover was monitored in each square based on digital photos taken from a nadir perspective at 1.2 m height every other week [20]. The images were taken with a Nikon S210 Coolpix camera that had a lens resolution of 5 Mpixels. The images were saved at a resolution of 1200×800 pixels and the GC percentage was determined with the software SigmaScan Pro (Systat software, Chicago, IL, USA) by means of the "Turf analysis" macro developed by Karcher and Richardson [21]. The process consisted of creating a pixel layer by selecting the hue and saturation ranges that identified the surface covered by the crop. The GC was determined to be the number of pixels of the layer divided by the total number of pixels constituting the image of the marked squares [20]. The GC evolution was adjusted to the Gompertz function [22], and the thermal times (°C days) until the GC reached 30% (t_{30}) and 80% (t_{80}) were calculated.

Just before killing the CC, the above-ground biomass of the marked squares was hand harvested by cordless grass shear at the ground level, separated by species, oven dried for 48 h at 65°C,

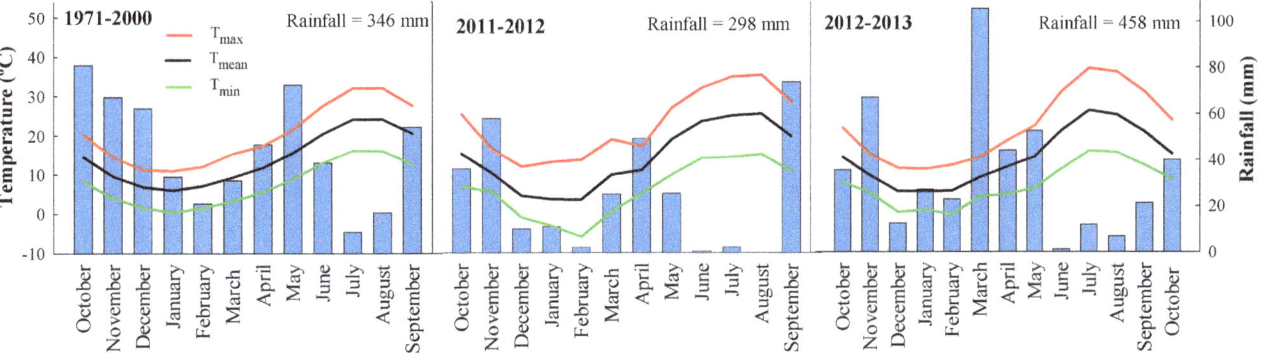

Figure 1. Monthly rainfall (bars) and average maximum, mean and minimum temperatures from 1971 to 2000 and during the two experimental seasons at Aranjuez (Madrid, Spain).

Table 1. Soil properties at the beginning of the experiment.

Depth (cm)	0–23	23–40	40–70	70–120
pH (1:2.5)	8.16	8.06	8.02	7.84
Organic Matter (g kg^{-1})	31.8	29.2	21.9	22.3
CO_3 (g CO_3^{2-} kg^{-1})	198.0	201.3	159.0	181.0
Sand (g kg^{-1})	260	250	250	250
Silt (g kg^{-1})	490	510	520	460
Clay (g kg^{-1})	250	240	230	290

and weighed to determine biomass of barley, vetch, and mixture in kg ha^{-1}. The C and N concentrations in the above-ground biomass were determined for a subsample of each species from each plot using the Dumas combustion method (LECO CHNS-932 Analyzer, St. Joseph, MI, USA). Total N content was calculated for each specie as the product of above-ground biomass times N concentration. The atmospheric N_2 fixation by the legume was estimated by the natural abundance method [23] based on the $\delta^{15}N$ (‰) determination (Europe Scientific 20–20 IRMS Analyzer, Crewe, UK) for subsamples from the vetch and a barley reference cultivated as a sole crop in an adjacent field. Cover crop residue quality was assessed by measuring neutral detergent fiber (NDF), acid detergent fiber (ADF), and lignin (L) with the Goering and Van Soest method [24] in subsamples of each crop species from each plot.

3. Soil inorganic nitrogen content (N_{min})

Four soil cores were taken from each plot to a depth of 1.2 m in 0.2 m intervals with an Eijkelkamp helicoidal auger (Eijkelkamp Agrisearch Equipment, Geisbeek, Netherlands) just before sowing the CC, after the second kill date each year, and at the end of the experiment. Soil cores were combined by depth to provide a composite profile of six samples. For each plot, soil N_{min} was calculated for each layer. Soil samples were placed in a plastic box and immediately firmly closed then transported, and refrigerated (4–6°C). Within five consecutive days, a soil subsample from each box was extracted with 1 M KCl (~30 g of soil: 150 ml of KCl), centrifuged, decanted, and a subsample of the supernatant volume was stored in a freezer until later analysis. Nitrate concentration in the extracts was determined by the Griess-Ilosvay method [25] after reduction of NO_3^- to NO_2^- with a Cd column. Ammonium in the soil extracts was determined by the salicylate-hypochlorite method [26].

4. Soil nitrogen mineralization potential (N_0)

Soil N mineralization potential was estimated by adapting the procedure proposed by Stanford and Smith [27]. A subsample of the top soil layer (0–0.2 m) free from residue contamination was collected from each plot after the second kill date of the second year, air-dried, and sieved (<6 mm). A homogeneous mixture of 30 g of that soil and an equal weight of sand was packed to a depth of 5 cm in plastic syringes (3.5 cm diameter, 10 cm long) to achieve a bulk density of 1.25 g cm^{-3}. Thin glass wool layers were placed between the soil and the bottom of the syringe and over the soil sample to avoid soil loss during leaching and to minimize moisture loss. Initial soil inorganic N was removed by leaching

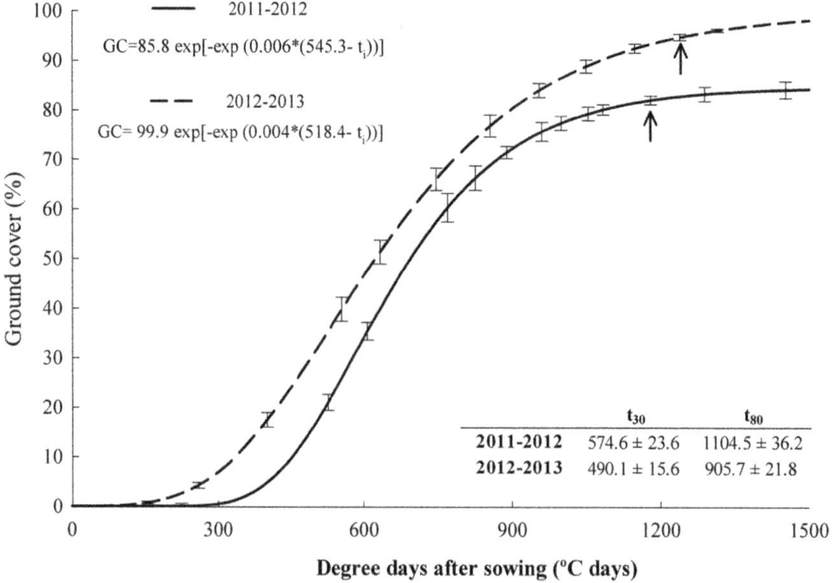

Figure 2. Ground cover (GC) development during the cover crop growth period in both experimental seasons. Arrows show the first kill date. Lines represent the Gompertz model adjusted to the observed values. Fitted models and the thermal time until the ground cover reaches 30 and 80% (t_{30} and t_{80}) are shown.

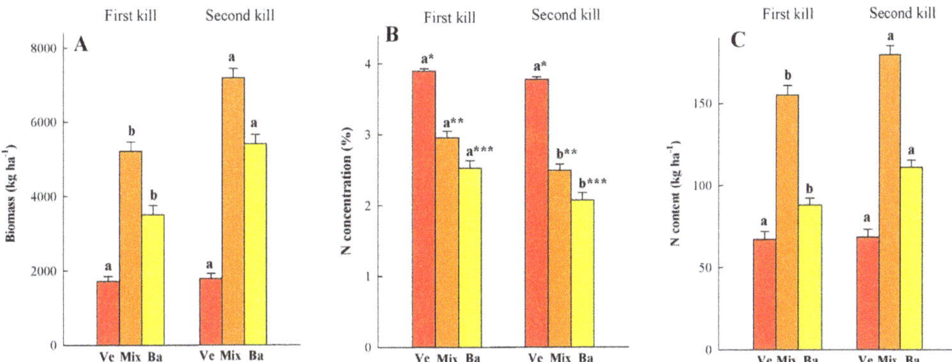

Figure 3. Biomass (A), N concentration (B) and N content (C) for vetch (Ve), barley (Ba) and the mixture (Mix) measured for the first and second kill dates. Values are the mean of two cropping seasons. Letters above bars show significant differences between kill dates for each species, and asterisks between species for each kill date. Small bars represent the standard error.

with 100 ml of 0.01 M $CaCl_2$ followed by 20 ml of a N-free nutrient solution (0.0095 M $CaSO_4$, 0.000047 M KH_2PO_4, 0.00138 M K_2SO_4, 0.0003 M $MgSO_4$). The excess water was removed using a vacuum pump until a weight within 2 g of that measured before leaching was reached. The syringes were covered with a porous parafilm and incubated aerobically at 35°C. Syringes were removed from the incubator at 14, 28, 42, 56, and 60 d after preparation and were leached with 100 ml of 0.01 M $CaCl_2$ solution followed by 20 ml of N-free solution. Leachates were made up to 100 ml with $CaCl_2$, and subsamples were stored in a freezer at −25°C until later analysis. After the leaching procedure, the cores were allowed to drain under vacuum until a weight within 2 g of that measured before leaching was reached. Nitrate concentrations in the leachates were determined by the Griess-Ilosvay method, ammonium by the salicylate-hypochlorite method and total N by the Dumas combustion method, as described above. The N_0 and the mineralization rate constant (k) were estimated after fitting a non-linear regression model ($Nt = N_0 \exp(-k\ t)$) for describing the cumulative N mineralized (Nt) with time (t) in each soil sample.

5. Soil volumetric water content (SWC)

The SWC was monitored daily during the field trial using the EnviroScan capacitance probes (Sentek Pty Ltd, Stepney, Australia) that has been described in detail elsewhere [28]. Seventy-two capacitance sensors were mounted on twelve plastic extrusions (four repetitions per treatment), introduced into access tubes located in the middle of each plot and connected to three data loggers. Sensor readings were automated and stored in the data loggers and downloaded weekly. To ensure the measurement reliability, a normalization procedure was conducted that obtained reference readings by exposing each sensor to air and water (~20°C). The sensors were centered at 10, 30, 50, 70, 90, and 110 cm below the soil surface in each plastic extrusion, and normalized readings were registered every 6 h. A daily average of the four readings from the 0–20 (10), 20–40 (30), 40–60 (50), 60–80 (70), 80–100 (90), 100–120 (110) cm-deep soil layers was transformed into SWC using a calibration equation that was obtained at the experimental site [29]. The SWC data set was comprised of two CC seasons, which started before sowing in October and lasted until August each year, and used to study the effect of the living CC and the dead mulch on soil moisture.

6. Statistical analysis

Analyses of variance (ANOVA) and t-test were performed in order to determine differences between kill dates for each variable. Year was considered a random effect and, interaction between kill date and year was also tested. Means were separated by Tukey's test at the 0.05 probability level ($P \leq 0.05$). Least significant differences (LSD) were calculated for SWC ($P \leq 0.05$). Prior to conducting the ANOVA, tests were conducted to verify if the assumptions of ANOVA were met. The Gompertz model was

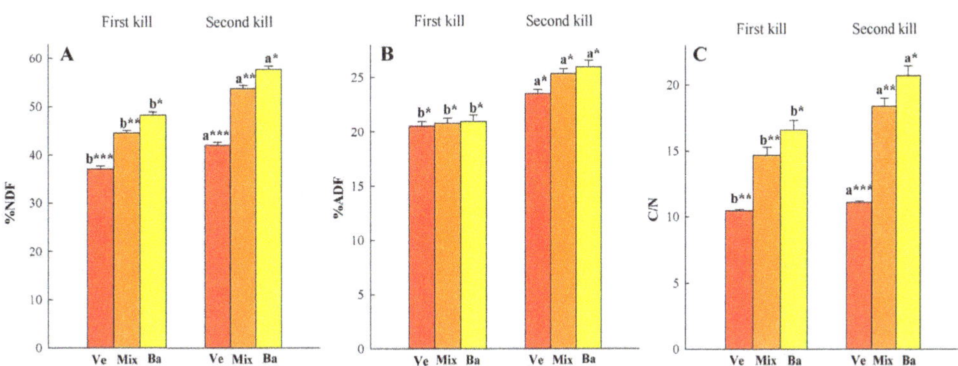

Figure 4. Neutral detergent fiber-NDF- (A) and acid detergent fiber -ADF- (B) fractions, and C/N ratio (C) in vetch (Ve), barley (Ba) and the mixture (Mix), measured for the first and second kill dates. Values are the mean of two cropping seasons. Letters above bars show significant differences between kill dates for each species, and asterisks between species for each kill date. Small bars represent the standard error.

Table 2. Soil inorganic N content (kg N ha^{-1}) in the different soil layers and the entire profile at different sampling dates for the fallow (Fa), first kill (FK) and second kill (SK) treatments.

	Oct 2011	Apr 2012	Oct 2012	Apr 2013	Oct 2013
0–40 cm					
Fa	94 (13.7)	126.8 (7.9) a	135.3 (15.4)	29.7 (3.8) b	71.54 (5.38) c
FK	94 (13.7)	46.9 (7.1) b	121.5 (16.6)	56.9 (3) a	119.37 (28.48) b
SK	94 (13.7)	37 (8.2) b	154 (8)	24.8 (4.5) b	171.72 (2.36) a
40–80 cm					
Fa	24.7 (9.5)	71.6 (10.6) a	86.1 (4.6) a	40.1 (1.8) a	72.22 (7.43)
FK	24.7 (9.5)	30.4 (5.3) b	48.3 (4.4) b	42.6 (3.5) a	65.38 (19.89)
SK	24.7 (9.5)	25.4 (5.1) b	48.1 (7.1) b	27.1 (4.1) b	39.39 (2.68)
80–120 cm					
Fa	26.1 (11.1)	60.6 (3.5)	66.7 (11)	68.8 (3.3) a	100.37 (11.74) a
FK	26.1 (11.1)	56.1 (24.5)	61.5 (29.3)	47 (7.4) b	61.59 (17.91) ab
SK	26.1 (11.1)	34.5 (3.7)	33.5 (6.3)	32.3 (8.3) b	39.24 (5.83) b
Soil profile (0–120 cm)					
Fa	144.8 (22.8)	259 (8.0) a	288.1 (25.4)	138.6 (5.1) a	244.13 (19.05)
FK	144.8 (22.8)	133.32 (27.5) b	231.3 (31.2)	146.6 (10.7) a	246.34 (52.87)
SK	144.8 (22.8)	96.9 (15.4) b	235.6 (14.2)	84.2 (13.3) b	247.34 (7.9)

Means with standard error in parentheses.
Within a date and depth, means with the same letter are not significantly different between kill dates at $P<0.05$.

fitted to the GC and the N mineralization potential model was fitted to the cumulative N mineralized using a non-linear regression procedure. The models were evaluated for their ability to simulate the observed data by comparing the mean of the lack of fit to the mean square due to pure error by using the variance ratio, or F-test. When the lack of fit was significantly smaller than the pure error, the model fitted the data. Further discussion regarding this evaluation procedure can be found elsewhere [30,31]. Statistical analyses were accomplished using the PASW Statistics Software (SPSS, Chicago, IL, USA).

Results

1. Weather conditions

The first experimental year was drier than the second (Figure 1). In the first year, rainfall was substantially lower (298 mm) than the 30-year average although in the second, it was greater (408 mm). Differences mainly occurred in spring; from March to May, the rainfall was 202 mm in 2013 compared with 101 mm in 2012. The rest of the season was rather similar for both years except for intense precipitation in late September 2012. Air temperature followed a typical Mediterranean seasonal distribution over the two years. Winter was cooler in the first year with an average minimum temperature of −3 and −5°C in January and February, respectively.

2. Cover crop: GC, above-ground biomass and N content

The GC followed a classical Gompertz model (Figure 2). The F-test comparing the mean squares due to pure error and lack of fit was not significant at 0.01 level, therefore, the model was adjusted to the GC observations. The first kill occurred between 157 and 159 d after sowing (~1200°C days) whereas the second year occurred between 184 and 186 d after sowing (~1500°C days). The GC>30%, a crucial threshold for erosion control [32], was attained between 37 and 41 d after sowing (>490°C days). The GC>80%, considered to be full cover for erosion control and direct soil evaporation [32], was reached between 95 and 130 d after sowing (>900°C days). Differences were observed in the maximum GC attained, 85% in 2012 and 100% in 2013. In both

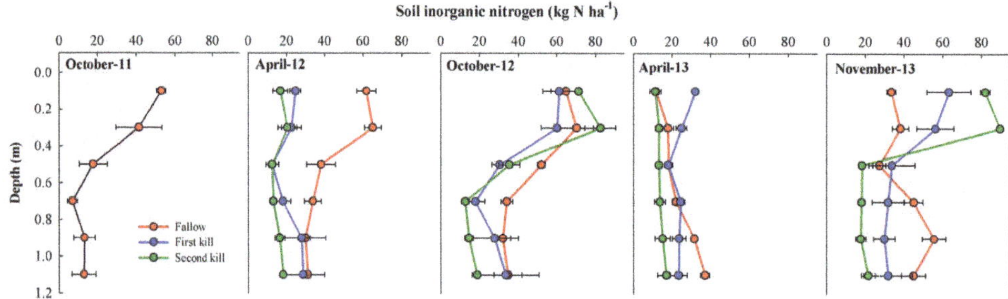

Figure 5. Soil inorganic nitrogen (kg N ha^{-1}) in the upper 1.2 m of the soil profile for the fallow (Fa), first kill (FK) and second kill (SK) treatments at different sampling times.

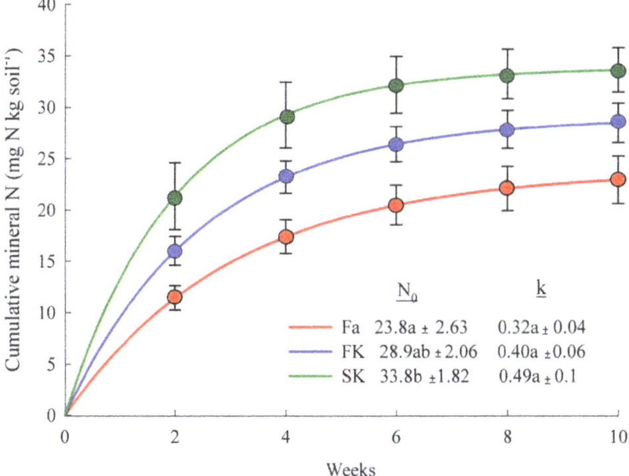

Figure 6. Cumulative N mineralization in soils from the fallow (Fa), first kill (FK) and second kill (SK) treatments during a 10-week aerobic incubation. Soil N mineralization potential (N_0) and N mineralization rate (k) were calculated by fitting a non-linear regression model ($Nt = N_0 \exp(-k\ t)$). Values followed by different letters are significantly different between treatments at ≤0.05 by Tukey's test. Bars represent the standard error.

years, maximum GC was already attained by the first kill date. The ground continued to be covered by the CC residue mulch more than six months after CC killing. Above-ground biomass increased from the FK to the SK by ~2000 kg ha^{-1}, and most of the increase was due to barley (Figure 3). Although barley biomass was greater at the SK than at the FK, no differences were found in vetch biomass between kill dates. Vetch above-ground biomass was double in 2013 than in 2012 whereas barley above-ground biomass did not vary by year. At the end of the mulch period, residues left on the soil surface were greater in the SK (2214 kg ha^{-1} in 2012, and 4014 in 2013) than in the FK treatment (966 kg ha^{-1} in 2012, 2151 in 2013. The ground was fully covered by the residues in both treatments.

Nitrogen concentration in the above-ground biomass was higher in the FK than in the SK. Nitrogen concentration in the mixture decreased from 2.7 to 2.35% in 2012 and from 3.2 to 2.6% in 2013. As expected, vetch had a higher N concentration than barley whereas the N concentration of the mixture was intermediate (Figure 3). Differences in N content between the FK and SK, calculated as the product of above-ground biomass and N concentration, were not significant for the mixture or vetch (Figure 3). However, barley N content in the SK was higher than in the FK; N uptake increased from 79 to 107 kg N ha^{-1} in 2012 and from 97 to 115 kg N ha^{-1} in 2013. The vetch reached high N_2 atmospheric fixation rates during the experiment. In the first season, N_2 atmospheric fixation was >80% of N content and ~100% in the second. No differences in N_2 atmospheric fixation were found between kill dates.

Residue quality fractions varied between kill dates (Figure 4). The NDF and ADF fractions increased when the CC was killed later. An interaction effect between treatment and year occurred for lignin content as differences in the lignin fraction between kill dates were only significant in barley and in the mixture in 2012. Lignin fraction in the mixture increased from 2.2 to 3.6% and in the barley from 1.2 to 2.9% in 2012. The C/N of barley, vetch and mixture residues was always lower for the FK than for the SK.

3. Soil inorganic nitrogen content

At the beginning of the experiment, average N_{min} of the upper layers of the soil profile was 145 kg N ha^{-1} (Figure 5, Table 2). During the first CC growing period, soil N_{min} increased in the fallow treatment, particularly in the upper 0.8 m, while it decreased in the remaining depths. During the mulch period, soil N_{min} in the fallow treatment increased slightly whereas, in the CC treatments, N_{min} increased largely in the upper soil layers, but there was no difference between the treatments by October in 2012.

During the second CC growing season, soil N_{min} was depleted in all treatments. After the CC kill in April 2013, the N_{min} was 84 kg N ha^{-1} in the SK treatment and ~140 kg N ha^{-1} in the others. The FK treatment accumulated more N_{min} in the upper layers, most likely due to the early mineralization that occurred between the CC kill and soil sampling (~4 weeks). At the end of the mulch period in November 2013, a large increase in soil N_{min} occurred in all treatments. Although no difference was observed for average N_{min} (~245 kg N ha^{-1}) over the entire profile, differences between treatments were observed at some soil depths. In the SK, most of the N_{min} (~70%) was in the top layer (0–40 cm) whereas it was ~29% in the fallow treatment. Similarly, while >40% of soil N_{min} was in the bottom layer in the fallow, < 16% was in the bottom layer in the SK. The N_{min} distribution in the FK treatment was intermediate between the others.

4. Soil nitrogen mineralization potential

The one-pool exponential model fit the cumulative mineral N from the aerobic incubation, as the F-test comparing the mean squares due to pure error and lack of fit was not significant at 0.01 level. The N_0 was higher for soils from the SK treatment (34 mg N kg soil^{-1}) than those from the fallow (24 mg N kg soil^{-1}). The N_0 from the FK treatment was intermediate. No differences between treatments were observed in the N mineralization (Figure 6).

5. Soil water content

Soil water content was affected by the presence of CC and by the kill date with differences observed in the upper layers and the whole soil profile (Figure 7, Table 3). During both CC growth periods, SWC followed a similar pattern. At sowing, the three treatments started with low SWC in the entire profile (~220 mm) in both seasons, and only small differences appeared in the top layers. The precipitation during the three months following CC sowing recharged the soil profile, reaching SWC>300 mm in 2013 due to abundant rainfall (Figure 7a). No differences were observed between treatments during this period in any season. However, during the next three months, CC extracted water from the upper layers, and by the first kill date, SWC was higher in the fallow than in the CC treatments. After the first kill date, the SWC in the FK treatment varied compared to the SK depending on the annual weather conditions. During the first season, no differences were observed between the FK and SK treatments, most likely due to low precipitation, and by the second kill date, both had similar SWC and were lower than the fallow (Table 3). During the second season, the high precipitation between both kill dates recharged the soil profile, and the FK treatment SWC was similar to the fallow except in the deeper soil layer (100–120 cm, Figure 7e) where the fallow treatment remained wetter. For the second kill date of 2013, the SK treatment SWC was lower, and differences with the FK were obvious down to 80 cm and down to 120 cm with the fallow (Table 3).

Although initial differences in SWC were observed between the two years of the study during the mulch period, such differences were not evident toward the end of the season. After water

Table 3. Soil water content (mm) in each of the soil layers and the entire profile at different sampling dates for the fallow (Fa), first kill (FK) and second kill (SK) treatments.

		0–20		20–40		40–60		60–80		80–100		100–120		0–120	
		mm													
2011–2012															
07/10/2011	Fa	23.9	a	40.3	a	38.1	a	40.8	a	37.4	a	40.4	a	221	a
CC sowing	FK	18.7	a	30.8	b	39.4	a	41.7	a	44.2	a	43.4	a	218	a
	SK	18.5	a	35.1	ab	42.6	a	41.0	a	37.0	a	41.0	a	215	a
13/03/2012	Fa	25.2	a	42.7	a	40.5	a	37.5	a	34.1	a	34.9	a	215	a
First kill date	FK	12.4	b	28.7	b	33.3	a	35.5	a	35.5	a	36.5	a	182	b
	SK	11.4	b	31.7	b	35.5	a	32.2	a	33.1	a	35.5	a	179	b
09/04/2012	Fa	33.9	a	42.2	a	40.1	a	38.3	a	35.4	ab	36.3	a	226	a
Second kill date	FK	18.3	b	28.1	b	33.2	a	36.0	a	36.3	a	37.5	a	189	b
	SK	12.6	c	28.6	b	33.8	a	29.8	a	30.8	b	36.4	a	172	c
30/04/2012	Fa	9.4	c	27.6	c	32.6	a	36.2	a	34.1	ab	36.4	a	251	a
HPD	FK	10.6	b	27.4	b	36.9	a	42.7	a	41.5	a	44.0	a	266	a
	SK	12.8	a	32.0	a	38.7	a	37.3	a	37.4	b	43.6	a	269	a
12/09/2012	Fa	9.4	a	27.6	a	32.6	a	36.2	a	34.1	b	36.4	a	176	a
Mulch period end	FK	10.6	a	27.4	a	36.9	a	42.7	a	41.5	a	44.0	a	203	a
	SK	12.8	a	32.0	a	38.7	a	37.3	a	37.4	ab	43.6	a	202	a
2012–2013															
08/10/2012	Fa	43.9	b	55.3	a	31.3	a	33.6	a	32.0	b	34.4	a	231	a
CC sowing	FK	48.5	ab	55.7	a	36.3	a	39.7	a	38.5	a	40.5	a	259	a
	SK	52.7	a	55.2	a	39.0	a	34.7	a	35.3	ab	41.0	a	258	a
15/03/2013	Fa	50.3	a	56.0	a	58.8	a	56.4	a	44.7	a	39.1	a	305	a
First kill date	FK	43.4	b	47.3	b	41.1	b	43.8	b	40.8	a	38.3	a	255	b
	SK	41.2	b	42.2	c	39.3	b	39.3	b	37.4	a	39.2	a	239	b
11/04/2013	Fa	49.4	b	56.1	a	57.7	a	60.0	a	55.9	a	47.7	a	327	a
Second kill date	FK	53.9	a	59.1	a	61.2	a	58.2	a	43.8	b	39.8	b	316	a
	SK	39.1	c	50.4	b	41.0	b	40.1	b	37.8	b	39.3	b	248	b
30/04/2013	Fa	44.2	b	52.6	a	54.6	a	57.6	a	55.6	a	51.8	a	311	a
HPD	FK	52.7	a	55.4	a	57.6	a	59.5	a	50.0	a	42.4	b	307	a
	SK	40.3	c	41.7	b	42.2	b	40.8	b	38.8	b	40.5	b	244	b
01/09/2013	Fa	7.8	a	37.7	a	47.0	a	56.2	a	56.5	a	54.9	a	315	a
Mulch period end	FK	8.5	a	32.5	ab	45.3	a	50.9	a	50.1	ab	49.2	a	318	a
	SK	8.8	a	27.5	b	41.4	a	44.0	a	43.3	b	47.2	a	244	b

Within a date and depth, means with the same letter are not significantly different at $P<0.05$.

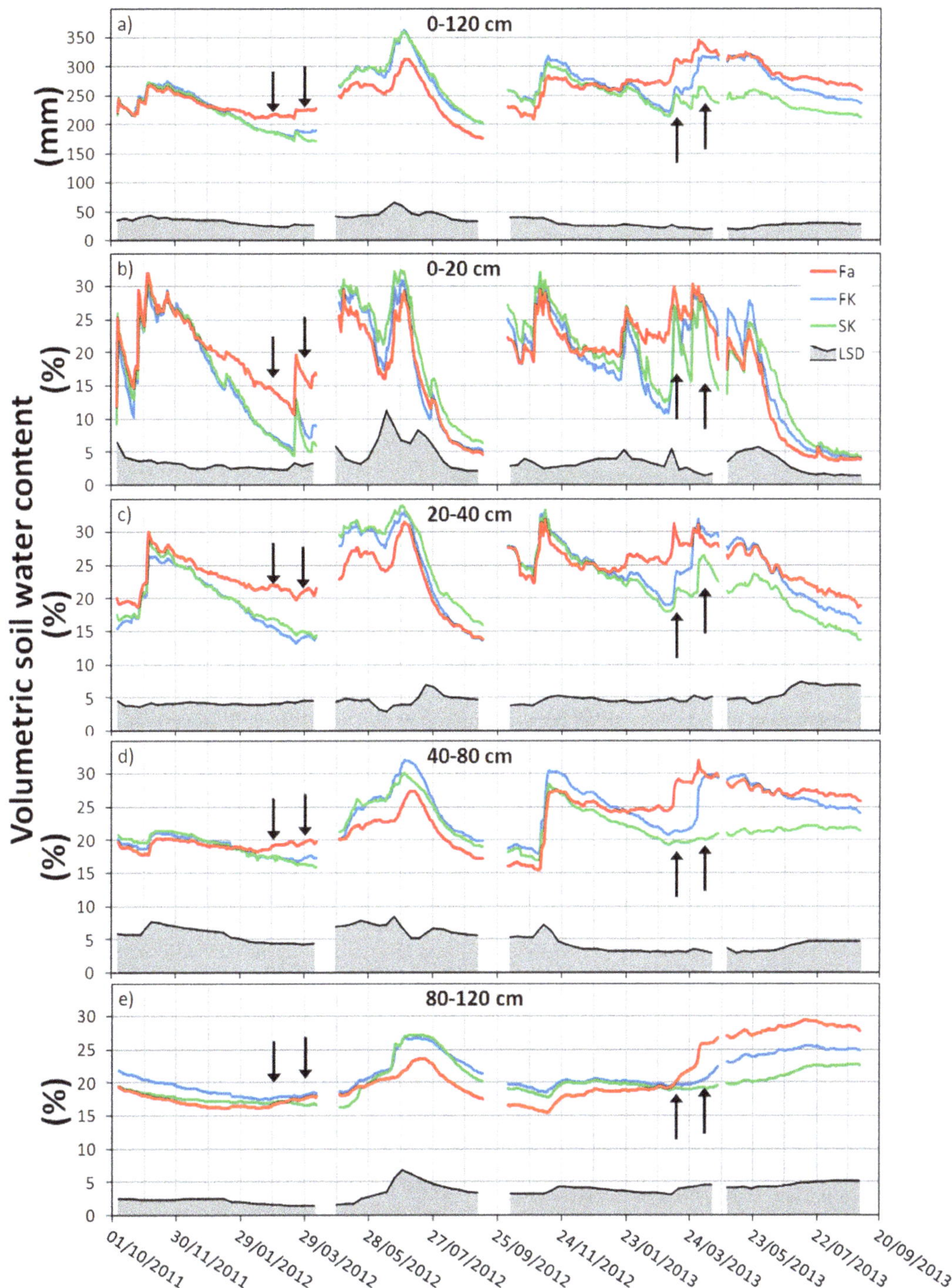

Figure 7. Soil water content monitored with the capacitance sensors at different depths for the fallow (Fa), first kill (FK) and second kill (SK) treatments during the experimental period. Arrows represent the first and second kill dates. The shadowed area in the bottom represents the LSD at 0.05 probability level.

application in April 2012, the SWC in the entire profile at the HPD was similar in all treatments (~262 mm; Table 3). During the wet/dry cycles that occurred in spring and summer, the CC treatments always retained more moisture than the fallow which showed the ability of the residue mulch to reduce soil water evaporation losses (Figure 7). After mid-June, when the soil was left to dry under the summer heat, differences in SWC between treatments disappeared, first in the upper layers and later in the entire soil profile. Even though slower drying was observed in the CC treatments, no differences were observed between treatments by the end of September (Table 3). In 2013, no water was applied, and the SWC in the entire soil profile was lower for the SK than for the FK and fallow treatments at the HPD (Table 3). At this time, no differences in SWC were observed between the FK and fallow treatments in the entire soil profile. However, the FK upper layer was wetter than the fallow because of the reduction in soil

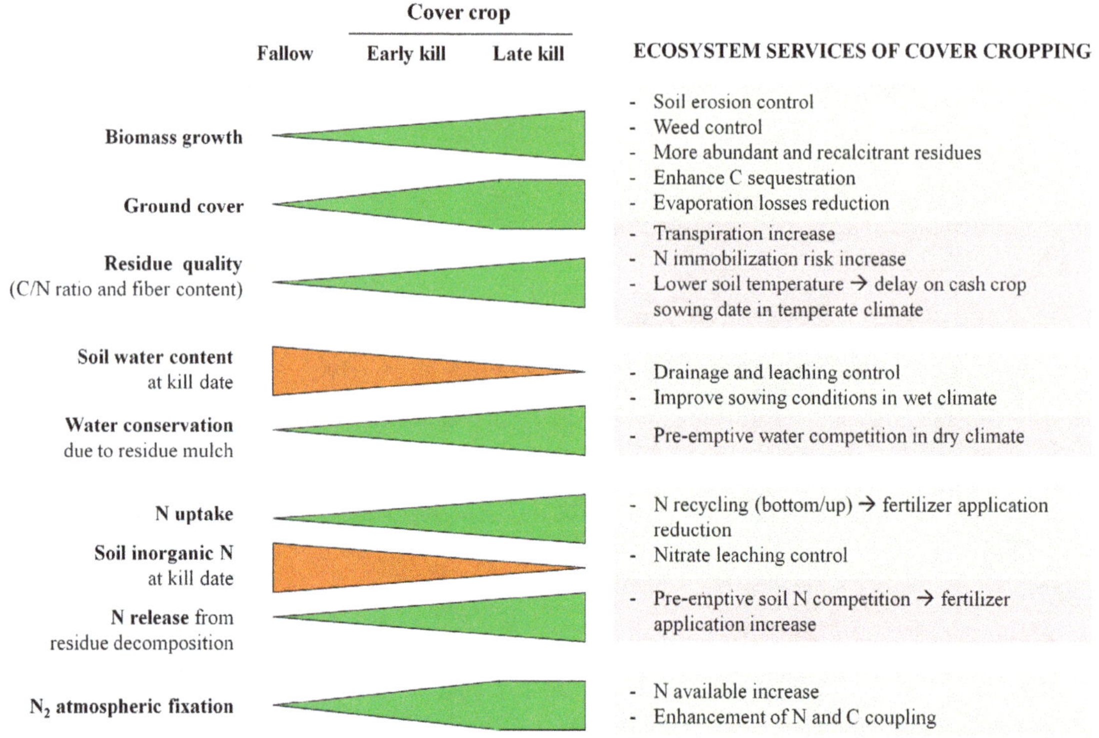

Figure 8. Schematic representation of the effect of kill date on environmental variables and ecosystem services with fallow as a reference condition.

water evaporation caused by the residue mulch (Table 3, Figure 7b). During summer, water losses in the top layer were greater in the fallow than the CC treatments, showing the effect of the mulch. By the end of the experiment, no differences in the SWC of the entire profile were observed between treatments, but some SK subsoil layers remained drier as the result of water depletion caused by the living CC.

Discussion

This study confirmed that CC kill date is an important management strategy as it affected key variables of the soil-plant system, influencing the success of the subsequent cash crop. Previous studies have focused mainly on the effect of N on the succeeding crop [6] and on the CC residue quality [14], but there is a lack of information on the combined effects of CC growth, N dynamics, and water dynamics. This information is necessary to optimize CC benefits and for development of more accurate simulation models.

Delaying kill date did not increase GC whereas the year had an effect on the maximum cover attained. We could not find data to compare the effect of kill date on GC. Nevertheless, our results agreed with those of previous researchers that showed the capacity of the asymptotic Gompertz function to capture the characteristics of the GC evolution of several CC [22,33]. Ground cover is an important variable used in studies related to soil erosion [32], evaporation [34], weed control [35,36] or radiation interception [22]. In addition, the evolution of digital technologies allows for non-destructive monitoring using reliable and efficient techniques [20]. For all of these reasons, we propose that GC should become a common variable for the characterization of CC that facilitates comparisons between species, varieties and management strategies in different regions. In the present study, GC differed with seasonal weather conditions; the dry 2011–2012 season was less favorable for CC growth than the wet and mild 2012–2013 season. In both seasons, GC>80% was already attained by the first kill date, so no additional benefits in terms of soil erosion or weed control were expected by the second kill date [32,35].

As expected, delaying kill date increased the above-ground biomass of the barley/vetch mixture. This increase was mainly due to barley as there was no difference in vetch biomass between kill dates. The barley/vetch ratio, expressed as seed weight at sowing, was 3:7, but barley was a stronger competitor due to faster establishment and growth rate [20]. The observed increase in biomass is common in many other studies [8,13], but the relationship between the legume and the non-legume differed. In Central Italy, a barley/vetch mixture with a 1:1 sowing rate yielded between 48 and 66% vetch biomass 160 d after planting which increased vetch in the sowing mixture up to 75% and produced between 53 and 76% vetch biomass at harvest [37]. A possible explanation for the large barley dominance in our study is that soil available N was not a limiting factor. Other authors reported similar superior dry matter values for the non-legume compared with the legume in mixtures, and this dominance became stronger with higher levels of N [38]. This hypothesis is corroborated by the plant and soil available N results in our study. Most of the N content in the vetch/barley mixture accumulated before the FK date. Nitrogen content of barley was greater at SK than FK (Figure 3), and in the SK treatment, the inorganic N in the soil was depleted in April, particularly from the deeper layers (Table 2, Figure 5). The high competition with barley for soil N forced the vetch to rely mainly on N_2 fixation for its N requirement, which explains the large degree of N_2 fixation observed in the legume. We were expecting the vetch to fix a significant amount of N_2 between both kill days, as observed in other experiments [10,13], but no increase in vetch N content was observed as in the study of Benincasa et al. [8]. The strong competition with barley for light explains the low vetch growth between kill dates and reinforces the need for sowing mixtures with a large proportion of vetch seeds (>50%) to favor legume activity. Most authors have found that the N content in CC and mixtures increased as kill date was delayed [10,13], but our results show that the main effect of delaying kill date increased above-ground biomass rather than N content.

The effect of postponing kill date on chemical residue composition was in agreement with previous findings in the literature [8,12,39]. Nitrogen concentration decreased from the first to the second kill date, being greater for vetch than for barley (Figure 3). These differences were reinforced in the C/N ratio as a slight increase in C content was observed in the SK samples compared with the FK (Figure 4). Vetch N concentration was similar to that found by other authors, but barley values were greater than those reported in other CC experiments [10,40,41]. In the FK samples, barley N concentration was ~2.9% and ~2.2% in the SK which agrees with data from barley cultivated as a cash crop using synthetic N fertilizer [42]. These data confirm that barley did not suffer from lack of N. The increase in fiber fractions, which is related to a decrease in the labile fraction, from delaying kill date was consistent with other studies and was expected to lead to slower C and N mineralization [43]. Increases in lignin fraction were only significant in barley in the first season, but this is not surprising as there were only three weeks between kill dates, and lignin content is less sensitive to change [39]. In the drier season (2011–2012), barley residue and the mixture had significantly lower fiber and lignin content than in the wetter season for both kill dates. A reduction in NDF, ADF and lignin cereal crop straw in dry years was previously noted in a study conducted over several years and locations in Washington State [44]. Our results confirmed that delaying kill date increased the amount of CC residue and the residues were more recalcitrant to decomposition. This is particularly interesting when the goal is preserving soil moisture [11], controlling weeds [35], or increasing the slow release of the N pool [45]. In the present experiment, a detailed study of CC residue decomposition was not conducted primarily because the lowest part of the residue layer stuck to the soil when sampling, and we decided that it was reliable for determining the amount of the remaining residue but not its quality.

The topsoil N_0 was also affected by the kill date two years after starting the experiment. Soil N mineralization potential assess the capacity of a soil to mineralize N and is a good indicator of land use as it is very sensitive to small changes in labile C and N fractions [46]. The increase of N_0 in the CC treatments proves that CC contribute to the enhancement of labile soil organic pools. The low contribution of CC to total organic C is a known fact as residues contain low quantities of recalcitrant organic matter [2]. However, CC add substrate for soil microorganisms during their growth period (root exudation, root turnover), and residue decomposition increases the organic matter labile fractions [6]. Many studies of N mineralization from CC residues have been published, but there is a lack of studies that focus on the effects on the soil in which they were cropped. In our study, no differences were observed in total soil C, but delaying kill date increased the C supplied by CC and, as a consequence, topsoil N_0. Nitrogen mineralization rate, a parameter that mainly depends on environmental conditions, was not different between treatments and was within the range found in the literature [47].

Consistent with other studies, CC absorbed most of the available soil inorganic N during their growth period, modifying the degree of pre-emptive competition and the risk of N loss with respect to the fallow (Table 2, Figure 5). This study also showed that kill date affected these parameters and could be a means to adapt soil N availability to the requirements of a particular agroecosystem. Delaying kill date increased the risk of pre-emptive competition, a negative effect that was reflected not only in the deeper layers as is common [4,48], but in the upper layers during the second season as well. The early kill date allowed more time for N release from CC residues, and by the HPD, more N was available in the topsoil. This positive effect of an early kill date on reducing pre-emptive competition has not been shown before and was evident when comparing FK with the SK and the fallow in the wet year, in which N_{min} was washed out of the soil during the fall and winter. The well-known CC effect of recycling N in the soil-plant system by uptaking inorganic N from deeper layers and subsequently releasing it to the top layers was clear in this study [6]. In the first season, the soil N_{min} depleted in April by the CC uptake was replenished by October by the N released from the residue mineralization. In the second season, CC prevented most of the available N from being lost from the soil, as happened in the fallow, and released the N by November. The difference in November was not in the amount of N_{min} but its distribution in the soil profile. In the fallow, N_{min} was mostly in the deeper layers and had greater potential for leaching during fall and winter. However, in the SK treatment, N_{min} was in the top layers making it readily available for uptake by the subsequent crop. The N_{min} profile of the FK soil was intermediate. Organic N remained in the residue layer at the end of the experiment, but as mentioned before, we could not obtain a reliable measure of N concentration.

Even if N uptake by CC depleted the soil by April in the first season and seemed to increase the risk of pre-emptive competition with respect to the fallow, we propose that this may be a positive effect in many agricultural systems as it may help control N losses. In a tomato drip irrigation study, it was observed that high soil N_{min} levels at spring planting were linked to large leaching losses as crops are often overwatered to ensure establishment [49]. Excessive irrigation during the crop establishment period may represent up to 80% of total nitrate leaching losses [34,50]. Therefore, keeping soil N_{min} at low levels and correcting fertilizer needs based on crop nutritional status during the cropping season holds promise for controlling N losses and increasing the efficiency of N use [51].

The larger above-ground biomass observed in the SK affected SWC. The CC extracted soil water by transpiration, increasing evapotranspiration losses when compared with the fallow [34]. Spring growth of the vetch/barley mixture was vigorous and soil water depletion was enhanced between the first and the second kill date. The differences in the whole profile SWC at the first kill date between CC treatments and the fallow were >35 mm in the first season and >55 in the second. At the second kill date, the SWC in the FK treatment was similar to the fallow. By the HPD and during the rest of the mulch period, the treatment effect on water availability was variable. While in the first season there was no difference, most likely because the soil was replenished by irrigation, the SWC was 60 mm lower in the SK than in the other treatments during the second season. Similar results were found in California [52,53], where reductions in spring SWC were up to 80 mm due to CC extraction. In Minnesota, the difference in SWC in the upper 0.6 m caused by delaying the rye kill date by three weeks (from April to May) was 27 mm [12]. Our results confirm that late killing of CC increases pre-emptive water competition compared with early killing, a risk that might be mitigated in rainy years or by irrigation if water is available.

A relevant aspect of the increased water uptake by CC is the effect on the upper layers' SWC. Sufficient top layer moisture at planting is crucial to ensure crop establishment and plant survival in many crops [49]. In the present study, the difference in SWC between the fallow and the CC treatments at the first kill date due to water depletion ranged between 13–19 mm in the upper 0.2 m of the soil and was enhanced between the fallow and the SK by the second kill date. However, the effect of CC residue mulching led to a higher SWC in the upper layers of the FK treatment than in the fallow at the HPD. When comparing CC treatments, the results are less clear; more water in the upper layers was retained in the SK than in the FK in the first season, but the SK was drier in the second. The FK treatment always increased water retention in the upper layers with respect to the fallow, but the result varied by year in the SK treatment. The top layer SWC was influenced by CC water uptake during winter, the residue mulch effect, and the amount of rain between the CC kill date and cash crop planting [9]. To clarify the discrepancy between the soil water depletion by CC uptake and conservation by the mulched CC residue, an experiment comparing sheltered and un-sheltered plots in Maryland showed that delaying kill date of a vetch CC produced more biomass and resulted in better soil moisture conservation in the upper 0.3 m [11]. However, recent experiments in Minnesota reported that delaying the rye kill date from the tillering to the booting stage came at the cost of SWC depletion in the top 0.3 m even if the residues continued to cover the soil after termination [12]. In Indiana, early kill of the CC resulted in better water conservation in the topsoil (0.1 m) than late kill. In drought years, early kill date produced an increase in maize yield but had a negative impact in normal precipitation years due to a lowering of soil temperature and workability at the maize planting date [7]. Our results indicate that early kill of the CC enhanced spring topsoil water moisture with respect to late kill, particularly by preserving soil water replenished by precipitation between kill dates.

A limitation found when comparing our study with others that focus on CC killing was that kill dates were always expressed as days after sowing. Given the variety of climatic regions in which cover cropping is practiced, the number of days after sowing is not a valid variable. We propose using degree-days accumulated and growth stage as more suitable variables for comparison between different regions. Soil type also has a major influence on the water balance and N cycling. Simulation models that can take multiple environmental factors into account might allow to overcome these limitations and increase the applicability of this and others studies to different locations and years.

One criticism of this study might be that a crop was not planted after the CC to obtain real data on crop water uptake. We are aware of this situation, but we think that this type of dataset is necessary to advance our understanding of the complex effect of CC kill date on crop resources. Separating the mulching and the effect of transpiration might allow for the calibration of key parameters, proper simulation of both processes, and finally, of the whole system. Replacing fallows by CC is known to provide many benefits, but the agronomic application needs to account for the requirements of specific agroecosystems. Kill date is a management tool to regulate the effects of CC on many environmental variables as presented in Figure 8. In dry environments, residue moisture conservation is the main goal of CC [9] whereas the risk of delaying sowing because of soils that are too wet is a major concern in cooler climates [7]. Depleting soil N enhances pre-emptive competition with the subsequent crop [6] but also controls N losses and creates opportunities for rational fertilizer application [34]. Late termination of CC maximizes their biomass production and the formation of a thick mulch, enhancing C sequestration, water conservation and weed control [36]. Our work does not close the debate on kill date but contributes to more rational decision-making for the successful use of CC.

Conclusions

Kill date is a major management variable with CC that may lead to increasing water and N use efficiency and controlling N losses and pre-emptive competition with the subsequent crop. The ground cover of CC was >80% by the first kill date so no more benefits in terms of soil erosion or weed control derived from the living CC were expected by the second kill date. However, delaying kill date increased above-ground biomass, providing more CC residues with a higher fiber content and C/N ratio that were more recalcitrant to decomposition and, therefore, more suitable to protect the soil and enhance the slow release of the N pool.

The early kill date decreased the risk of pre-emptive competition by diminishing the N uptake of CC, and allowing more time for N release from CC residues. Cover crops generally uptake inorganic N from deeper layers to be released on the topsoil through residue mineralization. Delaying kill date enhanced this recycling effect that prevented losses of available N by keeping soil inorganic N at low levels. Kill date was a means of controlling this process and showed promise in mitigating N losses and increasing N use efficiency. The N mineralization potential in topsoil was enhanced by the presence of CC and by delaying kill date, proving that CC contribute to the enhancement of labile organic pools.

Delaying kill date increased pre-emptive competition of water, as the water extracted by CC from the first to the second kill date was greater than the water conservation due to the extra residue generated by the SK treatment. The soil water content in the upper layers at the time of planting the subsequent crop increased by the first kill date with respect to the fallow by preserving rain harvested between kill dates, but the soil water content in the late kill could be even lower than in the fallow depending on the climatic conditions of the year.

Supporting Information

Table S1 Biomass, N content and residue left. Above-ground biomass and N content of cover crops at the first kill (FK) and second kill (SK) dates and residue covering the soil at the end of the mulch period (~7 months after cover crop killing). Means with standard error in parentheses. Within a row, means with the same letter are not significantly different between kill dates at $P<0.05$.

Table S2 Chemical composition of cover crops. Cover crop residue neutral detergent fiber (NDF), acid detergent fiber (ADF) and lignin (L) fractions and C/N ratio for the first kill (FK) and second kill (SK) dates. Means with standard error in parentheses. Within a row, means with the same letter are not significantly different between kill dates at $P<0.05$.

Image S1 Aerial view of the experimental site. Plots are marked as fallow (FA), first kill (FK) nd second kill (SK) plus the replication number. Image taken between kill dates (March 29, 2012).

Image S2 Ground level view of the experimental site. Image taken between kill dates (April 4, 2012).

Acknowledgments

We would like to thank the staff of La Chimenea field station (IMIDRA) for their helpful assistance.

Author Contributions

Conceived and designed the experiments: MQ. Performed the experiments: MA JLG MQ. Analyzed the data: MA JLG MQ. Contributed to the writing of the manuscript: MA JLG MQ.

References

1. Tonitto C, David MB, Drinkwater LE (2006) Replacing bare fallows with cover crops in fertilizer-intensive cropping systems: A meta-analysis of crop yield and N dynamics. Agriculture, Ecosystems & Environment 112: 58–72.
2. Kuo S, Sainju UM, Jellum EJ (1997) Winter cover crop effects on soil organic carbon and carbohydrate in soil. Soil Science Society of America Journal 61: 145–152.
3. Quemada M, Cabrera ML (2002) Characteristic moisture curves and maximum water content of two crop residues. Plant and Soil 238: 295–299.
4. Gabriel JL, Quemada M (2011) Replacing bare fallow with cover crops in a maize cropping system: Yield, N uptake and fertiliser fate. European Journal of Agronomy 34: 133–143.
5. Abawi GS, Widmer TL (2000) Impact of soil health management practices on soilborne pathogens, nematodes and root diseases of vegetable crops. Applied Soil Ecology 15: 37–47.
6. Thorup-Kristensen K, Magid J, Jensen LS (2003) Catch crops and green manures as biological tools in nitrogen management in temperate zones. Advances in Agronomy 79: 227–302.
7. Stipesevic B, Kladivko EJ (2005) Effects of winter wheat cover crop desiccation times on soil moisture, temperature and early maize growth. Plant Soil and Environment 51: 255–261.
8. Benincasa P, Tosti G, Tei F, Guiducci M (2010) Actual N Availability from Winter Catch Crops Used for Green Manuring in Maize Cultivation. Journal of Sustainable Agriculture 34: 705–723.
9. Clark AJ, Meisinger JJ, Decker AM, Mulford FR (2007) Effects of a grass-selective herbicide in a vetch-rye cover crop system on corn grain yield and soil moisture. Agronomy Journal 99: 43–48.
10. Tosti G, Benincasa P, Farneselli M, Pace R, Tei F, et al. (2012) Green manuring effect of pure and mixed barley-hairy vetch winter cover crops on maize and processing tomato N nutrition. European Journal of Agronomy 43: 136–146.
11. Clark AJ, Decker AM, Meisinger JJ, McIntosh MS (1997) Kill date of vetch, rye, and a vetch-rye mixture: 2. Soil moisture and corn yield. Agronomy Journal 89: 434–441.
12. Krueger ES, Ochsner TE, Porter PM, Baker JM (2011) Winter Rye Cover Crop Management Influences on Soil Water, Soil Nitrate, and Corn Development. Agronomy Journal 103: 316–323.
13. Clark AJ, Decker AM, Meisinger JJ, McIntosh MS (1997) Kill date of vetch, rye, and a vetch-rye mixture: 1. Cover crop and corn nitrogen. Agronomy Journal 89: 427–434.
14. Wagger MG, Cabrera ML, Ranells NN (1998) Nitrogen and carbon cycling in relation to cover crop residue quality. Journal of Soil and Water Conservation 53: 214–218.
15. Tosti G, Benincasa P, Farneselli M, Tei F, Guiducci M (2014) Barley-hairy vetch mixture as cover crop for green manuring and the mitigation of N leaching risk. European Journal of Agronomy 54: 34–39.
16. Soil Survey Staff (2014) Keys to Soil Taxonomy. Washington, DC, USA: USDA, Natural Resources Conservation Service.
17. Nelson DW, Sommers LE (1996) Total carbon, organic carbon and organic matter. In: Sparks D, editor. Methods of soil analysis, part 3: chemical methods. Madison, WI, USA: ASSA and SSSA. 961–1010.
18. Loeppert RD, Suarez DL (1996) Carbonate and Gypsum. In: Sparks D, editor. Methods of soil analysis, part 3: chemical methods. Madison, WI, USA: ASA and SSSA. 437–474.
19. Papadakis J (1966) Climates of the world and their agricultural potentialities. Rome, Italy: DAPCO.
20. Ramirez-Garcia J, Almendros P, Quemada M (2012) Ground cover and leaf area index relationship in a grass, legume and crucifer crop. Plant Soil and Environment 58: 385–390.
21. Richardson MD, Karcher DE, Purcell LC (2001) Quantifying turfgrass cover using digital image analysis. Crop Science 41: 1884–1888.
22. Bodner G, Himmelbauer M, Loiskandl W, Kaul HP (2010) Improved evaluation of cover crop species by growth and root factors. Agronomy for Sustainable Development 30: 455–464.
23. Unkovich M, Herridge D, Peoples M, Cadisch G, Boddey B, et al. (2008) Measuring plant-associated nitrogen fixation in agricultural systems. Camberra, Australia: Australian Centre for International Agricultural Research. 258 p.
24. Goering HK, Van Soest PJ (1970) Forage fiber analysis: apparatus, reagents, procedures, and some applications. Washington, DC: Agricultural Research Service. Handbook no. 379. US Government Printing Office.
25. Keeney DR, Nelson DW (1982) Nitrogen–inorganic forms. In: Page AL, editor. Methods of soil analysis Part 2: Chemical and microbiological properties. Madison, WI, USA: ASA and SSSA. 643–698.
26. Crooke WM, Simpson WE (1971) Determination of ammonium in Kjeldahl digests of crops by an automated procedure. Journal of the Science of Food and Agriculture 22: 9–10.
27. Stanford G, Smith SJ (1972) Nitrogen mineralization potentials of soils. Soil Science Society of America Proceedings 36: 465–472.
28. Paltineanu IC, Starr JL (1997) Real-time soil water dynamics using multisensor capacitance probes: Laboratory calibration. Soil Science Society of America Journal 61: 1576–1585.
29. Gabriel JL, Lizaso JI, Quemada M (2010) Laboratory versus field calibration of capacitance probes. Soil Science Society of America Journal 74: 593–601.
30. Quemada M, Cabrera ML (1995) CERES-N model predictions of nitrogen mineralized from cover crop residues. Soil Science Society of American Journal 59: 1059–1065.
31. Whitmore AP (1991) Method for assessing the goodness of computer simulation of soil processes. Journal of Soil Science 42: 289–299.
32. Quinton JN, Edwards GM, Morgan RPC (1997) The influence of vegetation species and plant properties on runoff and soil erosion: results from a rainfall simulation study in south east Spain. Soil Use and Management 13: 143–148.
33. Ramirez-Garcia J, Gabriel JL, Alonso-Ayuso M, Quemada M (2014) Quantitative characterization of five cover crop species. Journal of Agricultural Science. doi:10.1017/S0021859614000811.
34. Gabriel JL, Muñoz-Carpena R, Quemada M (2012) The role of cover crops in irrigated systems: Water balance, nitrate leaching and soil mineral nitrogen accumulation. Agriculture Ecosystems & Environment 155: 50–61.
35. Teasdale JR, Mohler CL (1993) Light transmittance, soil-temperature, and soil-moisture under residue of hairy vetch and rye. Agronomy Journal 85: 673–680.
36. Saini M, Price AJ, Van Santen E, Arriaga FJ, Balkcom KS, Raper RL (2008) Planting and termination dates affect winter cover crop biomass in a conservation-tillage corn-cotton rotation: Implications for weed control and yield. In: Endale DM editor. Tifton, GA, USA. Southern Conservation Agricultural Systems. 137–141.
37. Tosti G, Benincasa P, Guiducci M (2010) Competition and Facilitation in Hairy Vetch-Barley Intercrops. Italian Journal of Agronomy 5: 239–248.
38. Ofori F, Stern WR (1987) The combined effects of nitrogen-fertilizer and density of the legume component on production efficiency in a maize cowpea intercrop system. Field Crops Research 16: 43–52.
39. Wagger MG (1989) Time of desiccation effects on plant composition and subsequent Nitrogen release from several winter annual cover crops. Agronomy Journal 81: 236–241.
40. Clark AJ, Decker AM, Meisinger JJ (1994) Seeding rate and kill date effects on hairy vetch cereal rye cover crop mixtures for corn production. Agronomy Journal 86: 1065–1070.
41. Benincasa P, Guiducci M, Tei F (2011) The nitrogen use efficiency: Meaning and sources of variation-case studies on three vegetable crops in Central Italy. HortTechnology 21: 266–273.
42. Arregui LM, Quemada M (2008) Strategies to improve nitrogen use efficiency in winter cereal crops under rainfed conditions. Agronomy Journal 100: 277–284.
43. Quemada M, Cabrera ML (1995) Carbon and nitrogen mineralized from leaves and stems of four cover crops. Soil Science Society of America Journal 59: 471–477.
44. Stubbs TL, Kennedy AC, Reisenauer PE, Burns JW (2009) Chemical composition of residue from cereal crops and cultivars in dryland ecosystems. Agronomy Journal 101: 538–545.
45. Quemada M, Cabrera ML, McCracken DV (1997) Nitrogen release from surface-applied cover crop residues: Evaluating the CERES-N submodel. Agronomy Journal 89: 723–729.
46. Cookson WR, Abaye DA, Marschner P, Murphy DV, Stockdale EA, et al. (2005) The contribution of soil organic matter fractions to carbon and nitrogen mineralization and microbial community size and structure. Soil Biology and Biochemistry 37: 1726–1737.
47. Quemada M, Diez JA (2007) Available nitrogen for corn and winter cereal in Spanish soils measured by electro-ultrafiltration, calcium chloride, and incubation methods. Communications in Soil Science and Plant Analysis 38: 2061–2075.
48. Thorup-Kristensen K (1994) The effect of nitrogen catch crop species on the nitrogen nutrition of succeeding crops. Fertilizer Research 37: 227–234.
49. Vázquez N, Pardo A, Suso ML, Quemada M (2005) A methodology for measuring drainage and nitrate leaching in unevenly irrigated vegetable crops. Plant and Soil 269: 297–308.
50. Salmerón M, Isla R, Cavero J (2011) Effect of winter cover crop species and planting methods on maize yield and N availability under irrigated Mediterranean conditions. Field Crops Research 123: 89–99.
51. Vázquez N, Pardo A, Suso ML, Quemada M (2006) Drainage and nitrate leaching under processing tomato growth with drip irrigation and plastic mulching. Agriculture Ecosystems & Environment 112: 313–323.
52. McGuire AM, Bryant DC, Denison RF (1998) Wheat yields, nitrogen uptake, and soil moisture following winter legume cover crop vs. fallow. Agronomy Journal 90: 404–410.
53. Mitchell JP, Peters DW, Shennan C (1999) Changes in soil water storage in winter fallowed and cover cropped soils. Journal of Sustainable Agriculture 15: 19–31.

10

Soil Charcoal to Assess the Impacts of Past Human Disturbances on Tropical Forests

Jason Vleminckx[1]*, Julie Morin-Rivat[2,3], Achille B. Biwolé[2], Kasso Daïnou[2,4], Jean-François Gillet[2], Jean-Louis Doucet[2], Thomas Drouet[5], Olivier J. Hardy[1]

1 Service d'Évolution Biologique et Écologie, Faculté des Sciences, Université Libre de Bruxelles, Brussels, Belgium, **2** Laboratoire de Foresterie des Régions tropicales et subtropicales, Unité de Gestion des Ressources forestières et des Milieux naturels, Gembloux Agro-Bio Tech, Université de Liège, Gembloux, Belgium, **3** Laboratory of Wood Biology & Xylarium, Royal Museum for Central Africa, Tervuren, Belgium, **4** Laboratory of Applied Ecology, University of Abomey-Calavi, Cotonou, Benin, **5** Laboratoire d'Écologie Végétale et Biogéochimie, Faculté des Sciences, Université Libre de Bruxelles, Brussels, Belgium

Abstract

The canopy of many central African forests is dominated by light-demanding tree species that do not regenerate well under themselves. The prevalence of these species might result from ancient slash-and-burn agricultural activities that created large openings, while a decline of these activities since the colonial period could explain their deficit of regeneration. To verify this hypothesis, we compared soil charcoal abundance, used as a proxy for past slash-and-burn agriculture, and tree species composition assessed on 208 rainforest 0.2 ha plots located in three areas from Southern Cameroon. Species were classified in regeneration guilds (pioneer, non-pioneer light-demanding, shade-bearer) and characterized by their wood-specific gravity, assumed to reflect light requirement. We tested the correlation between soil charcoal abundance and: (i) the relative abundance of each guild, (ii) each species and family abundance and (iii) mean wood-specific gravity. Charcoal was found in 83% of the plots, indicating frequent past forest fires. Radiocarbon dating revealed two periods of fires: "recent" charcoal were on average 300 years old (up to 860 BP, n = 16) and occurred in the uppermost 20 cm soil layer, while "ancient" charcoal were on average 1900 years old (range: 1500 to 2800 BP, n = 43, excluding one sample dated 9400 BP), and found in all soil layers. While we expected a positive correlation between the relative abundance of light-demanding species and charcoal abundance in the upper soil layer, overall there was no evidence that the current heterogeneity in tree species composition can be explained by charcoal abundance in any soil layer. The absence of signal supporting our hypothesis might result from (i) a relatively uniform impact of past slash-and-burn activities, (ii) pedoturbation processes bringing ancient charcoal to the upper soil layer, blurring the signal of centuries-old Human disturbances, or (iii) the prevalence of other environmental factors on species composition.

Editor: Bruno Hérault, Cirad, France

Funding: This study is part of the FRFC project (« Fonds pour la Recherche Fondamentale et Collective», convention n ° 2.4577.10) entitled "Dynamics of tree and herbaceous communities of the tropical forests of Central Africa, in relation with past anthropogenic and climatic disturbance". J. Vleminckx and J. Morin-Rivat are funded by the "Fonds pour la formation à la Recherche dans l'Industrie et dans l'Agriculture (FRIA)". The authors gratefully acknowledge the CoForChange project funded through the ERA-net BiodivERsA (http://www.coforchange.eu/fr/). The authors declare that the funders only provided financial support for the field inventories and had no role in the study design, data collection and analysis, decision to publish, or preparation of the manuscript.

Competing Interests: The authors have declared that no competing interests exist.

* Email: jason.vleminckx@ulb.ac.be

Introduction

For a long time, many tropical forests have been viewed as "virgin" or "primary" ecosystems, undisturbed by anthropogenic activities. However, increasing evidence from different continents has suggested that these forests may actually hide influences of past Human disturbances [1]. In South-East Asia for example, it has been suspected that agricultural activities developed as early as 7000 BP in Papua New Guinea [2] and 8000 BP in Thailand [3], in areas covered today by lowland uninhabited rainforests. In Northwest Belize, Ross [4] demonstrated that modern tree species composition appeared to display important differences between areas of high and low settlement of ancient Maya, because of different intensities of past forest gardening (recruitment of useful species). Similarly, in central Amazonia, "terra preta" soils covering an area of about 500 km^2, on which stands apparent "pristine" rainforests, have resulted from intense burning and agricultural activities occurring about 2500 BP, which have considerably enhanced the fertility of these soils and may thus have impacted floristic diversity [5].

In central African moist forests, accumulating evidences show that Human has had a profound impact on the vegetation dynamics during the three last millennia. While only sparse evidence is recorded for Human presence in central Africa during the early and middle Holocene [6,7], archaeological surveys have suggested a dramatic expansion of an ancestral Bantu population coming from the southern part of the actual Cameroon-Nigeria border during the third millennium BP [8]. Archaebotanical data have indicated that this expansion coincided with an increasing seasonality in the precipitation regime that would have generated a disruption of the forest cover and replaced it by savannas or open forest formations [9,10], and therefore facilitated the Human colonization of central Africa [11,12]. Direct evidence from pollen and diatom frequency diagrams have demonstrated increasing aridity in the region [13,14] during the second half of the third millennium BP (especially between 2400 and 2100 BP), while charred botanical remains from this period have indicated that Bantus introduced the culture of pearl millet (*Pennisetum glaucum*, Poaceae), which requires a prolonged dry season [15]. Moreover, charcoal dating and identification have revealed higher abundance of pioneer species characterizing open forest formations [16]. The dryer areas of Central Africa were then more subject to fires during the dry season, but even when more humid conditions returned, the forest did not regenerate immediately as fires continued to maintain an open vegetation. According to archeological data, these resilient fires may not have been naturally caused only but also prompted by slash-and burn agricultural activities introduced by Bantu farmers [16,17]. For some unclear reasons, a dramatic decline of Human occupation in Western Central Africa occurred about 1400 BP [7,18], and it is only around the late middle age (about 600 BP) that evidence for Human activities are found again in the region [6,19].

Nowadays, while natural openings are scarce in Central African moist forests, the upper canopy is often dominated by long-lived light-demanding tree species which seem to suffer from low regeneration rate underneath their own shade [20]. This has been explained by a decrease of forest clearing activities since Human settlements from deep forests zones have been forced to move along roadsides during the colonial period [21]. Considering these facts and the massive amounts of evidence (potsherds, charcoals, iron objects) recorded in the literature for ancient Human occupation in vast uninhabited regions of Central African forest [7,15], it has been suggested that until a recent past, the widespread practice of slash-and-burn agriculture by farmers may have favored the competitiveness of light-demanding species and profoundly influenced the dynamics of tree diversity organization.

The pioneer works of Jones [22,23] in the mahogany forest of Okomu (Nigeria) were among the firsts to link vegetation and past Human disturbances in African rainforests. Jones suggested that the abundance of old light-demanding species he observed resulted from intense past Human activities [19,22]. Mahogany forests found in Cameroon [24], Central African Republic [25,26], Republic of Congo [26] and Democratic Republic of the Congo [27] have been suspected to hide a similar history to those of Okomu [28]. More recently, Biwolé et al. [29] have emphasized coincidence between populations of light-demanding trees species and the age of the last Human disturbances in a forest of Southern Cameroon.

Therefore, if past slash-and-burn agricultural practices favored the establishment of light-demanding species and if their intensity was spatially heterogeneous, we expect a correlation between forest stand composition and signs of past Human disturbances, such as soil charcoal abundance. To our knowledge, this expectation has never been tested using an adequate statistical sampling design in tropical forests. To bridge this gap, we aimed to test the hypothesis that Human disturbance intensity varied in time and space and favored the establishment of long-lived light-demanding tree species in Central Africa. To do so, we have carried out tree inventories and soil charcoal analyses at a landscape scale in three rainforest areas located in Southern Cameroon, using soil charcoal abundance as a proxy for past slash-and-burn activities. We addressed the following specific questions: (1) Is charcoal abundance spatially structured throughout the landscape and according to soil depth? (2) What is the age distribution of charcoal and is age related to soil depth? (3) What is the spatial structure of floristic and functional properties of local tree assemblages, considering: (i) species abundance, (ii) relative abundance of three regeneration guilds (pioneer, non-pioneer light-demanding, shade-bearer), and (iii) wood-specific gravity (i.e. $g.cm^{-3}$ of wood, used as a proxy for shade-tolerance)? (4) Does soil charcoal abundance explain variation in local tree assembly properties?

Materials and Methods

Study area

Study areas corresponded to three forest logging concessions. Two of them, hereafter named Area 1 and Area 2, were located in South-East Cameroon while another, Area 3, was located in the South-Western part of the country (Figure 1). The vegetation of these areas consisted of transition formations between lowland evergreen and semi-evergreen moist tropical forest [24], where the upper canopy was dominated by typical long-lived light-demanding species like *Pericopsis elata* (Fabaceae) in Area 1, *Triplochiton scelroxylon* (Malvaceae) in Area 2 and *Lophira alata* (Ochnaceae) in Area 3. Topography was characterized by hilly and highly weathered plateaus displaying dense hydrographic networks, situated at an elevation of ca. 700 m a.s.l. in Areas 1 and 2, and ca. 500 m a.s.l. in Area 3. The climate within all study areas is equatorial, with mean annual temperatures of ca. 24–25°C and mean annual rainfall reaching ca. 1600–1700 mm (www.climatedata.eu). Rainfall displays a bimodal distribution, with a rainy season stretching from September to November and another from March to June, separated by two relatively less humid periods. Ferralsols constituted the most dominant type of soil, with substantial occurrence of Acrisols in Area 3 [30]. Permission to carry out our field inventories were provided by logging companies (contact details are given in Appendix S1).

Inventory sites

A total of twelve sites were inventoried, of which six were located in Area 1, three in Area 2 and three in Area 3 (Figure 1). Each site corresponded to a linear transect along which 11 to 20 rectangular plots (40×50 m) were set up every 250 m and geo-referenced. This linear disposition of the plots within sites was designed to be able to apply torus-translation tests [31] (see details below). In Areas 1 and 2, sites were disposed along virtual transects extending over 101.5 and 66.5 km, respectively, while in Area 3, sites were irregularly disposed. Sites coordinates and the number of plots per site are given in Table S1 in File S1.

Charcoal abundance

At each corner and at the centre of each plot, we excavated a soil pit with an auger (5 cm diameter), collecting soil samples at the following depth layers (in cm): 0–10, 10–20, 20–40, 40–60, 60–80 and 80–100. The two uppermost layers were smaller in order to analyze charcoal abundance at finer resolution, as we suspected that charcoal reflecting recent Human disturbance were mostly located in these superficial soil layers. Soil charcoal abundance is classically quantified by determining its mass after sieving the soil volume in water [32,33]. This procedure being too time-consuming in the field for the large number of measures needed, we opted for a faster yet less precise method. Therefore, for each soil volume (280 cm^3 for 20 cm of vertical profile), we searched manually for charcoals remains and visually attributed a "charcoal abundance index" (CAI) defined as follows: 0 = absence of charcoal; 1 = charcoal traces (scattered pieces or powder) and/or

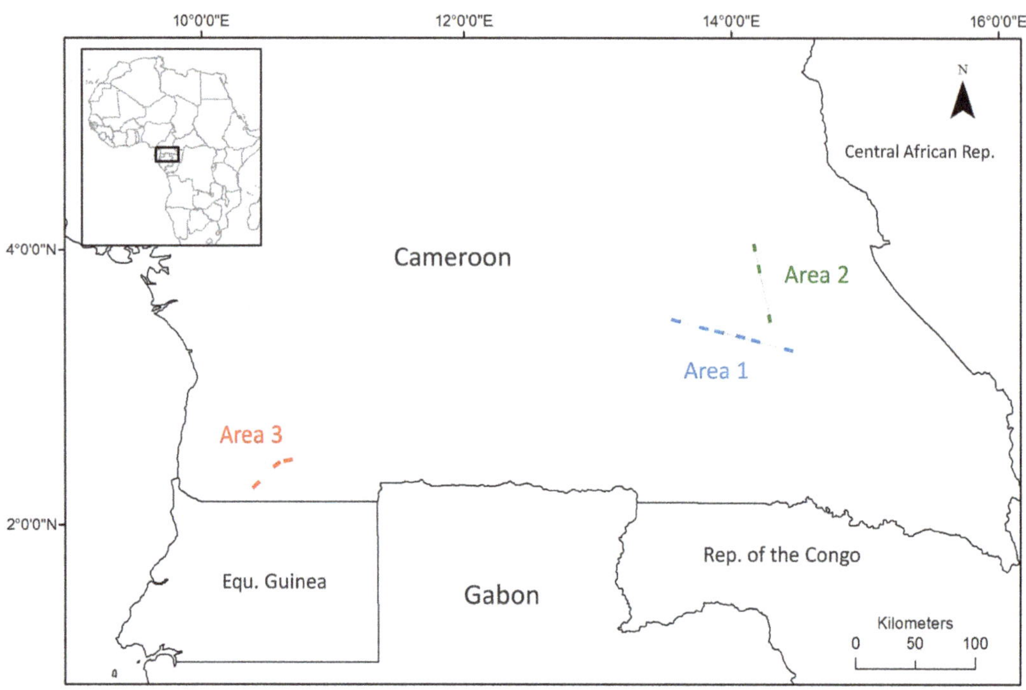

Figure 1. Geographical location of the three study areas. Each site is represented by a rectangle. Sites in Areas 1 and 2 are linearly disposed along a virtual transect represented by a dashed line.

≤5 pieces with a length <5 mm or 1 piece ≥5 mm; 2 = more than 5 pieces with a length <5 mm and/or more than 1 piece ≥ 5 mm. CAI could not be estimated for layers deeper than 20 cm in site 4, due to logistic constraints. Within each of the three investigated areas, CAI was estimated by the same person (JV, JMR and ABB in Area 1, 2 and 3, respectively).

We validated our CAI method in two ways. First, we tested the correlation between our CAI and the real charcoal mass for 36 soil volumes (each of 140 cm^3) excavated with an auger in a plot where previous digging revealed important charcoal presence (plot nr 2 of site 1, Area 1). For each volume, we first estimated the CAI, and then used a sieve of 2 mm mesh to collect all the charcoal pieces contributing significantly to the charcoal mass. We observed a significant correlation between CAI and ln(1+charcoal mass, in mg) (r-Pearson = 0.91, $P<0.001$) and no overlap of charcoal mass distributions between CAI values (see Appendix S2). Second, as Chabal [34] showed that the total number of charcoal particles found in a soil volume is highly correlated to the total charcoal mass, we tested the correlation between our CAI and the number of charcoal particles counted for the same soil volume on (i) 1510 soil volumes sampled in Area 2 where charcoals collected by hand were systematically counted, and (ii) 900 soil samples collected for another project (in Area 1) for which charcoals were collected using a sieve of 2 mm mesh. Pearson correlations between CAI and ln(1+charcoal particles number) reached 0.92 and 0.93 in both datasets (P<0.001, Appendix S2).

Radiocarbon dating

We selected 60 charcoal samples found at depth ranging from 5 to 150 cm collected in 37 random pits. In eight soil pits, charcoal samples from 2 to 5 different soil layers were dated to better assess the age-depth relationship. ^{14}C dating was performed at the Poznań Radiocarbon Laboratory (Poland). Calibration was performed under the OxCal v4.1.7 program [35–37] with the IntCal09 atmospheric calibration curve [38] and expressed in BP.

Species functional traits

Within each plot, we inventoried all the trees displaying a diameter at breast height (dbh) ≥20 cm. Over the three study areas, we found on average 32 individuals, 19 species, 18 genera and 13 families per plot. Table 1 summarizes abundance and diversity data for each study area. Species diversity was computed as the effective number of species expected in a random sample of $k = 2$ or $k = 100$ individuals [39], in order to give more weight to abundant and rare species, respectively. For all tree species identified in our plots, we compiled information on two key functional traits reflecting their shade-tolerance: regeneration guild (RG) and wood–specific gravity (WSG). Although well-defined classification of species in RG does not exist, three guilds could be defined, based on Hawthorne [40] and field observations in Central Africa (J. L. Doucet, A. Fayolle & J-F Gillet, pers. obs.; www.coforchange.eu): (i) pioneer species (P) require large gaps for establishment, (ii) non-pioneer light-demanding species (NPLD) can establish in shade but need a gap in the forest to grow to their full height, while (iii) shade-bearer species (SB) can be found in shade both as young and older plants. The relative abundance of each guild was computed for each of the 208 plots, after discarding unclassified species. WSG (g/cm^3) was used because this variable reflects the diameter growth rate (see references in Slik [41]), as fast-growing (light-demanding) species are more likely to have lighter wood than slow-growing (shade-bearer) species [41,42]. The WSG of most species could be assessed using a database [43]. When WSG data was missing for a species we assigned it the average WSG of its congeneric species, based on previous evidence showing that WSG is phylogenetically conserved [44]. For each plot, we calculated the mean WSG weighted by species relative abundances. The number of species for which the RG and

Table 1. Abundance and diversity data for each study area.

	Area 1	Area 2	Area 3
Nr of 0.2-ha plots	120 (24 ha)	53 (10.6 ha)	35 (7 ha)
Nr of stems	3799	1699	1348
Total nr of identified ind.	3531 (93%[a])	1693 (99.5%)	1335 (99%)
Nr of species	186	183	147
Dominant species	GS (8.2%[a])	GS (8.3%)	BW (5.5%)
Nr of families	46	45	42
Dominant family	Annonaceae (15.5%[a])	Annonaceae (20.0%)	Fabaceae (11.9%)
ENS(2)	51.2	50.5	54.1
ENS(100)	79.1	78.5	71.7
Nr of ind. assigned to a RG	3442 (97.5%[a])	1576 (93.1%)	1263 (93.7%)
Nr of sp. assigned to a RG	172 (92.0%[b])	155 (84.7%)	135 (91.8%)
Nr of ind. with WSG	3222 (91.2%[a])	1577 (93.1%)	1217 (90.3%)
Nr of sp. with WSG	170 (91.4%[b])	162 (88.5%)	131 (89.1%)
Nr of P	707 (20.5%[c])	469 (29.8%)	261 (20.7%)
Nr of NPLD	761 (22.1%[c])	306 (19.4%)	261 (20.7%)
Nr of SB	1974 (57.4%[c])	801 (50.8%)	741 (58.7%)

RG = regeneration guild. P = Pioneers. NPLD = Non-Pioneer Light-Demanders. SB = Shade-bearers. ENS(2) or ENS(100) = effective number of species expected for a random sample of 2 or 100 individuals. Ind. = individuals. Sp. = species. WSG = wood-specific gravity (g/cm^3). GS = *Greenwayodendron suaveolens* (Annonaceae). BW = *Blighia welwitschii* (Sapindaceae).

[a]Percentage calculated over the total number of stems in the study area.

[b]Percentage calculated over the total number of species in the study area.

[c]Percentage calculated over the number of individuals assigned to a RG.

WSG was known, as well as the abundance of each RG within each study area is represented in Table 1. A complete list of species, their corresponding RG and WSG, and their abundances in each study area is given in Table S2 in File S1.

Spatial patterns

For each study area, landscape differences between sites were tested using Kruskal-Wallis tests for the following variables calculated at the plot level: (i) mean CAI in two soil layers (see next section of materials and methods), (ii) relative abundance and relative basal area of each RG, (iii) mean WSG and (iv) the abundance of each sufficiently represented species and family ($\geq$ 20 individuals). Tests were performed using function kruskal.test from R stats package [45]. Finer scale spatial patterns were assessed within each of the three areas by measuring the spatial autocorrelation of these variables using Moran's I statistic [46], here defined as:

$$I_{ij} = \frac{(x_i - \bar{x})\,(x_j - \bar{x})}{Var\,(x)} + \frac{1}{(n-1)}$$

where x_i and x_j are the values of variable x for samples i and j, respectively; $\bar{x}$ and $Var(x)$ are, respectively, the mean and variance of variable x; n is the total sample size. The second term is a sample bias correction, ensuring that the average I_{ij} over all existing i, j pairs is equal to zero. Autocorrelograms, representing the mean I_{ij} values for a set of spatial distance intervals against distance, allowed a visual characterization of spatial patterns. Mantel tests [47] between the I_{ij} matrix and the spatial distance matrix were used to assess the significance of spatial structure for each variable.

We analyzed the spatial autocorrelation of charcoal abundance using the CAI index obtained for each soil sample to assess patterns at contrasted spatial scales: within a soil pit (distinguishing adjacent, i.e. in contact, and non-adjacent soil layers), within a plot, within a site and among sites. At the plot level, spatial autocorrelation was also assessed for each RG relative frequency, mean WSG and the abundance of each sufficiently represented species and family. All spatial autocorrelation analyses were performed using software Torocor (Torocor 1.0, http://ebe.ulb.ac.be/ebe/Software.html) (see Appendices S3 and S4).

Testing correlation between light-demanding species and charcoal abundance

^{14}C dating revealed a bimodal distribution in charcoal age: 16 out the 60 (27%) samples were dated between 80 and 860 BP (hereafter named "recent" charcoals, relatively speaking) and were only found in the upper soil layer (up to 25 cm), while 44 of them (73%) were dated more than 1500 BP ("ancient" charcoal) and found in the whole soil profile (Figure 2). Therefore, we estimated the abundance of soil charcoals at the plot level by computing, respectively, (i) the mean CAI value over both the 0–10 and 10–20 cm soil layers (proxy for "recent" fires) and (ii) the mean CAI value over the soil layers deeper than 20 cm (proxy for "ancient" fires).

For each study area taken individually, we then tested the correlation (at the plot level) between the mean CAI above and below 20 cm depth and: (1) the relative frequencies and relative basal areas of each RG, (2) the abundance of each species and family represented by at least 20 individuals (in the study area), and (3) the mean WSG. As classical correlation test is likely to be too liberal when analyzing spatial data, we tested the correlations using (i) a classical parametric test and (ii) a non-parametric torus-translation

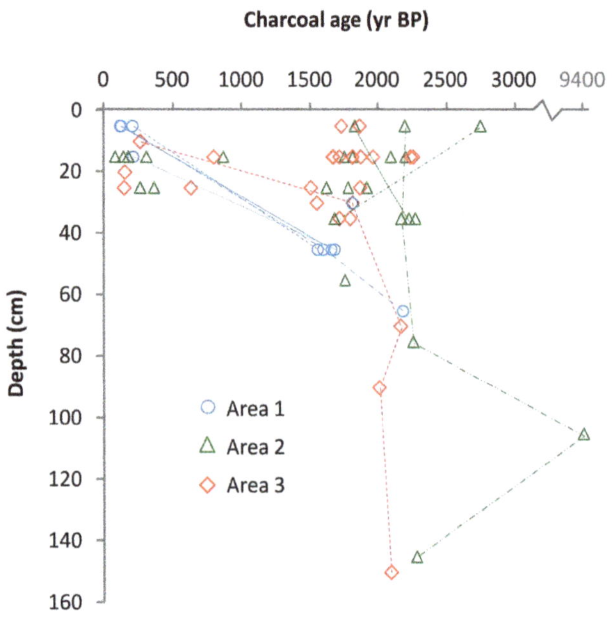

Figure 2. Age-depth relationship for 60 charcoal fragments sampled in the three areas investigated. Charcoals collected in a same soil pit are connected by a dashed line.

procedure taking spatial autocorrelation into account [31]. The latter consisted of generating randomized datasets where the correlation between variables was broken down while the spatial autocorrelation of each variable was kept intact. More specifically, within each site independently, the mean CAI values of the 20 plots were permuted among plots while keeping the original spatial order of plots (except at the extremities). This was like considering that the 20 plots of a site were situated along a ring and that CAI values were shifted by a random number of steps along the circle. 4999 of such permutations, independently repeated for each of the sites of a given area, were conducted to obtain null distributions of Pearson's correlation coefficients (computed using all plots). This allowed defining 95% confidence envelopes under the null hypothesis that there was no correlation between local CAI values and other variables. A test was significant when the observed correlation coefficient (i.e. without permutation) lied outside the 95% confidence envelope. By conserving the local structure (within site) of variables when permuting the objects, the torus-translation procedure tested the correlation at finer scale (i.e. within 3 to 5 km long transects) than the classic correlation test. Torus-translation analyses were performed using software Torocor (see Appendix S4).

Results

Charcoal ages and distribution

Charcoal was found as powder or entire fragments from 1 to 10 mm. It occurred at higher abundance between 20 and 60 cm in Area 1, between 10 and 40 cm in Area 2, and between 10 and 60 cm in Area 3 (see Appendix S3). It was present in 72, 83 and 94% of the plots between 0 and 20 cm deep in Areas 1, 2 and 3, respectively (89, 92 and 94% between 20 and 100 cm).

Radiocarbon dating revealed two distinct periods of ancient fires (Figure 2): 16 charcoal pieces were dated from 80 to 860 BP (hereafter "recent" charcoal), while 43 pieces were dated from 1495 to 2745 BP (hereafter "ancient" charcoal), and one sample was dated 9400 BP. All recent charcoal samples were found in the upper 25 cm soil layer, while ancient charcoal was found at all depths. Beside this trend between recent and ancient charcoal, there was no relationship between age and depth for ancient charcoals, even within a same soil pith (Figure 2). Calibration details for all dated samples are available in Table S3 in File S1.

For each study area, CAI values within soil pits were much more similar between adjacent soil layers than between non-adjacent ones (Figure 3, left). The correlation between mean plot CAI in soil layers above and below 20 cm depth was highly significant in Areas 1 and 2 (r-Pearson $=0.56$ and 0.51, respectively; $P<0.001$ under bilateral torus-translation test) but not in Area 3 (r-Pearson $=-0.09$; $P=0.66$). Among soil pits, there was a regular decrease of Moran's I values with the logarithm of the spatial distance within each study area (Figure 3, right). In particular, Moran's I values were always significantly positive between pits from a same plot, demonstrating a heterogeneity in charcoal abundance among plots within each area, although the relatively weak values found (<0.1) also indicate a high heterogeneity of charcoal abundance at a very local scale (i.e. within plot). At larger spatial scales, considering plot level mean CAI values separately in the 0–20 cm and 20–100 cm soil layers, Mantel tests between Moran's I and spatial distance matrices were highly significant for both soil layers in Area 1 ($P<0.01$; see Appendix S5) while no spatial structure was found in Areas 2 and 3. Consistently, these mean CAI values differed significantly among sites only in Area 1, where charcoal was more abundant in the western part (sites 1 to 3) than in the eastern part (sites 4 to 6; $P<0.001$, see Table 2).

Floristic and functional traits spatial patterns

In Area 1, differences between sites and spatial autocorrelations were significant ($P<0.05$) for each RG relative abundance and relative basal area, and for the mean WSG per plot, while fewer significant differences (Table 2) and spatial structures (Appendix S5) were observed within Areas 2 and 3. More specifically, the relative abundance and basal area of NPLD and P were not spatially structured in Area 2, whereas in Area 3, significant spatial autocorrelation was only observed for NPLD relative abundance and relative basal area. The proportion of species and families displaying significant spatial structure ranged between 35% and 65% within each study area (see Appendix S5). For each area, Kruskal-Wallis tests for difference of species abundances and family abundances between sites are given in Table S4 in File S1.

Correlations between vegetation characteristics and charcoal abundance

In Area 1, the abundance of charcoal in both the 0–20 cm and 20–100 cm soil layers (dominated by "recent" and "old" charcoals, respectively) and the relative abundance and basal area of NPLD decreased from West to East (sites 1 to 6), while the relative abundance and basal area of SB displayed the reverse pattern (Figure S1). Accordingly, Pearson correlations between NPLD relative abundance/relative basal area and CAI in each soil layer (measured at the plot level) were statistically significant using classical correlation tests (Table 3). However, these correlation coefficients became non-significant when using torus-translation tests which tested for within-site correlation and corrected for spatial autocorrelation. In Areas 2 and 3, all tests of correlation between vegetation characteristics and charcoal abundance were non-significant (Table 3). Finally, no significant correlation was observed between CAI and the abundance of any species or family in any study area, regardless of the test considered (see Appendix S6).

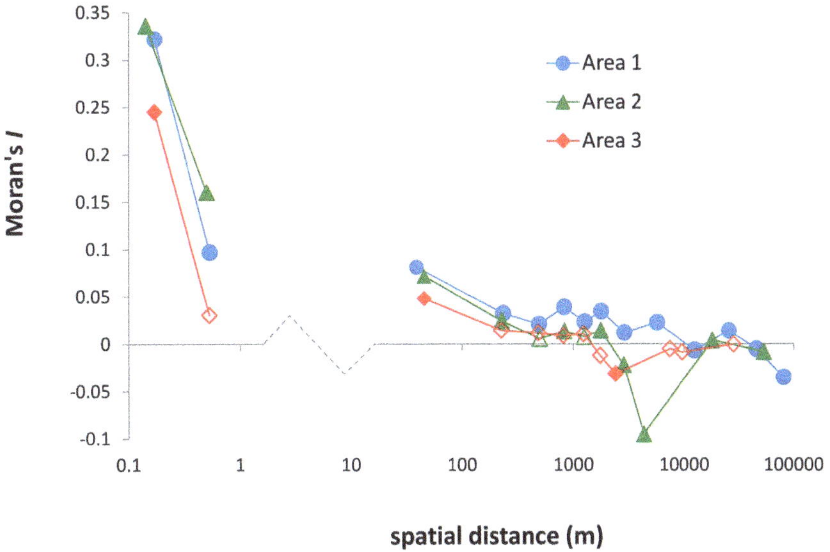

Figure 3. Autocorrelogram of CAI values for each study area: mean Moran's *I* computed for 12 to 14 distance intervals. On the left side (<1 m), the abscissa for the two first symbols represent the vertical distance between soil layers from a same pit, the left and right symbols distinguishing adjacent layers from non-adjacent layers, respectively. On the right side (>10 m), the abscissa corresponds to horizontal distance between soil volumes from different pits located in a same plot (first symbol between 10 m and 100 m), different plots of a same site (between 100 m and 5000 m) or different sites (>5000 m). Full symbols indicate significantly positive or negative Moran's *I* value ($P<0.05$).

Discussion

Origin of soil charcoal: a bimodal age distribution

Charcoal was found in the majority of the plots in each study area and in every soil layers, although at a higher frequency for the layers lying between 10 or 20 cm and 60 cm deep, which is consistent with previous studies [48]. Radiocarbon dating revealed two periods of ancient fires, with 16 samples dated between 80 to 860 BP ("recent" charcoals), and 43 samples between 1500 to 2200 BP (plus one dated 9400 BP; "ancient" charcoals). This bimodal age distribution has already been observed in previous ^{14}C dates distributions from other Central African countries [16,48]. It is also consistent with previous archeological evidence from Gabon and Cameroon pointing out high Human population density during the third millennium BP up to a few centuries afterwards, followed by a gap of Human occupation during the Middle Age [6,19]. This congruence between the abundances of charcoal and archeological artifacts through time supports the hypothesis that charcoal would mainly result from human-mediated fires. Ancient charcoals were found throughout the whole soil profile while recent ones were only found in the superficial soil layers (0–25 cm; Figure 2). For ancient charcoal, the absence of any age-depth relationship might be due to soil profile mixing (pedoturbation) resulting from biotic or abiotic agents. The main biotic agents in tropical soils are termites and ants [49–52] but earthworms [53], rodents, wind-fallen trees or tree uprooting can also play a role [54], while the main abiotic factor is colluvium [55] when located downhill or on a steep slope.

Charcoal abundance displays significant spatial structure

Charcoal was found in all sites, suggesting that Human impacted the whole forest cover in every study areas, especially in Area 3, where charcoal was present in 94% of the plots. Its abundance, however, displayed significant spatial autocorrelation at different scales, reflecting differential intensity of past disturbance across each region. At the smallest spatial scale, charcoal abundance was highly correlated between adjacent layers of a same pit, but the correlation dropped sharply when non-adjacent layers were compared. Hence, charcoal tended to occur as relatively small aggregates in the soil. There was also a regular decay of spatial autocorrelation with the logarithm of the horizontal distance, indicating spatial heterogeneity among plots and sites. In Area 1, a substantial part of the spatial autocorrelation can be explained by an East-West increase in charcoal abundance among sites (Figure S1), which could reflect a landscape gradient (~100 km) in the intensity of past slash-and-burn activities along the transect. In this area, as well as in Area 2, the correlation between charcoal abundance (at the plot level) in the 0–20 cm and the 20–100 cm soil layers was highly significant, according to both classic correlation tests and torus-translation tests (the latter testing within site correlation), indicating that the similarity of charcoal abundance between both layers occurred at the landscape scale (among sites) but also locally (within a site). Unfortunately, it cannot be assessed whether this correlation indicates that slash-and-burn agriculture was concentrated in the same places over the last centuries and c. 2000 years ago because the upper soil layer contains a mixture of both recent and ancient charcoal.

Local distribution of light-demanding species does not seem to be explained by charcoal abundance

In all study areas, most of the identified species could be categorized in a RG and characterized by a WSG. The few uncategorized species may either have a non-well known ecology or very large ecological amplitudes for light conditions. At the plot level, the relative abundance or basal area of each RG, as well as the mean WSG, displayed significant spatial autocorrelation in Area 1, while only two to three of these variables were spatially structured in Areas 2 and 3 (see Appendix S5).

Under our initial hypothesis, slash-and-burn agriculture favored the establishment of pioneer (P) species, which were then replaced by non-pioneer light demanding (NPLD) species within less than a

Table 2. Mean charcoal abundance index (CAI) and functional trait variables per site, and Kruskal-Wallis tests for among sites differences within each study area (using values computed at the plot level).

	CAI[a]		Functional traits						
	0–20	**20–100**	**WSG[b]**	**P-a**	**NPLD-a**	**SB-a**	**P-b**	**NPLD-b**	**SB-b**
Site 1	0.29	0.37	0.62	0.29	0.28	0.43	0.34	0.36	0.31
Site 2	0.29	0.44	0.64	0.18	0.32	0.50	0.24	0.43	0.35
Site 3	0.16	0.27	0.66	0.15	0.20	0.65	0.19	0.35	0.45
Site 4	0.18	.	0.62	0.25	0.27	0.47	0.31	0.37	0.31
Site 5	0.11	0.30	0.65	0.16	0.16	0.68	0.28	0.21	0.51
Site 6	0.14	0.16	0.62	0.24	0.14	0.62	0.41	0.20	0.41
***p*-value[c]**	***	***	***	**	***	***	**	***	**
Site 7	0.47	0.35	0.59	0.32	0.20	0.48	0.47	0.23	0.30
Site 8	0.35	0.28	0.58	0.35	0.22	0.44	0.43	0.26	0.31
Site 9	0.56	0.41	0.64	0.24	0.15	0.61	0.37	0.22	0.41
***p*-value**	0.43	0.39	***	*	0.16	***	0.31	0.77	*
Site 10	0.40	0.38	0.62	0.19	0.27	0.55	0.26	0.42	0.32
Site 11	0.42	0.36	0.65	0.23	0.15	0.62	0.28	0.25	0.47
Site 12	0.38	0.55	0.64	0.22	0.20	0.58	0.27	0.32	0.41
***p*-value**	0.78	0.19	0.39	0.53	**	0.46	0.97	*	0.18

Sites 1 to 6, 7 to 9, and 10 to 12 correspond to Areas 1, 2 and 3, respectively.
[a]Charcoal abundance index (0–20 cm and 20–100 cm).
[b]Wood-specific gravity (g/cm^3).
[c]*P*-value of a Kruskal-Wallis test among sites: *p-value <0.05, **p-value <0.01, *** p-value <0.001. P/NPLD/SB = Pioneers/Non-Pioneer Light-Demanders/Shade-Bearers. a = relative abundance. b = relative basal area.

Table 3. Pearson correlations between CAI in two soil layers (0–20 cm and 20–100 cm) and variables related to species functional traits.

	Area 1	Area 2	Area 3
	r[b]	r	r
P	−0.05	−0.03	0.08
	−0.15	*−0.21*	*0.00*
NPLD	0.20*	−0.26	0.24
	*0.28****	*0.10*	*−0.13*
SB	−0.09	0.19	−0.26
	−0.07	*0.15*	*0.11*
P relative basal area	−0.18	−0.07	−0.04
	−0.20	*−0.12*	*−0.06*
NPLD relative basal area	0.19	−0.08	0.24
	*0.28***	*0.06*	*−0.14*
SB relative basal area	0.00	0.17	−0.21
	−0.06	*0.08*	*0.20*
WSG[a]	0.02	−0.01	0.16
	−0.10	*0.12*	*0.08*

P/NPLD/SB = relative abundance of Pioneers/Non-Pioneer Light-Demanders/Shade-Bearers.
[a]Wood-specific gravity (g/cm^3).
[b]*r*-Pearson correlation between row variable and CAI in the (i) 0–20 cm (upper line) and (ii) the 20–100 cm (bottom line; values in italics) soil layers. "*"indicates significant test with the classic correlation test: *$p \le 0.05$ **$p<0.01$ ***$p<0.001$. All the tests based on toroidal translations were non-significant.

century and, eventually, shade-bearer (SB) species became dominant after a few centuries. Hence, we expected a positive correlation between CAI of the upper layer, reflecting partially the intensity of centuries-old fires, and the relative abundance of NPLD. In Area 1, the relative abundance of NPLD was indeed significantly positively correlated with the abundance of charcoals from the 0–20 cm ($r = 0.2$; $P = 0.03$) soil layer, but also with charcoal from the 20–100 cm ($r = 0.28$; $P = 0.005$) soil layer, when using a classic correlation test, not correcting for spatial autocorrelation. This is due to the parallel decay of charcoal and NPLD abundances among the six sites from West to East, which may have revealed a causal relationship if the intensity of past slash-and-burn agriculture had been stronger in the West. However, as such pattern has not been observed in the other two areas, and as toroidal translation tests also indicated that no correlation between RG relative abundance (and basal area) and soil charcoal abundance occurred within site, the among site correlation in Area 1 may well be spurious.

In Area 2 and Area 3, no significant correlations were detected. In fact, for the specific hypothesis that NPLD trees would have been favored by recent fires, we can note that while there is indeed a positive correlation between CAI in the upper soil layer and the relative abundance or basal area of NDPL in Area 1 and 3 (*r*-Pearson ranging from 0.19 to 0.24), the correlations become negative in Area 2 (*r*-Pearson ranging from −0.08 to −0.26). Hence, variation in floristic composition cannot be clearly explained by the abundance of soil charcoal.

We therefore suggest several explanations for the absence of significant signal. First, assuming a link between tree age and diameter for the entire community, we suspected that most of the trees of our dataset were too young to reflect the last Human disturbances, generating statistical noise. Nevertheless, correlation coefficients between light-demanding species frequencies and CAI did not increase after eliminating the smallest trees of our dataset (dbh <30, <40 <50, <60 and <70 cm, not shown).

Second, we assumed that charcoal from the superficial soil layer may sometimes be removed or covered by allochtonous materials due to colluvium processes, in such a way that it cannot be detected anymore. However, even after removing plots potentially subject to colluvium processes, i.e. located on a steep slope or downhill, correlation tests remained non-significant (not shown). Thus, if pedoturbation affects charcoal abundance, it is rather due to biotic agents (e.g. termites or ants). The latter could have "contaminated" the superficial soil layer (0–20 cm) with "ancient" charcoal, explaining why 59% of the dated charcoals found in the 0–25 cm soil layer were "ancient" (>1500 BP). By mixing with recent charcoals near the soil surface, ancient charcoals inevitably skewed our estimation of the abundance of "young" vs. "old" charcoals" (as recent and ancient material cannot be distinguished in the field) and therefore generated important statistical noise when testing our correlations.

Third, the absence of correlation may be due to an inappropriate study scale. Indeed, under our hypothesis, a correlation between charcoal abundance and the prevalence of light-demanding species is expected provided that there is sufficient spatial heterogeneity in the intensity of slash-and-burn agriculture to imprint a spatial heterogeneity in the abundance of light-demanding species. However, widespread and repeated slash-and-burn agriculture over the landscape may have favored light-demanding species on a large scale without causing substantial spatial heterogeneity on a more local scale.

Finally, it remains possible that the impact of past slash-and-burn agriculture on the floristic composition of forest stands is weak compared to the impact of other factors, like soil properties, climatic gradients (expected to be weak in our study areas), the history of forest establishment (e.g. forest regression ca. 3000 BP), or even stochastic processes structuring the distribution of species.

Conclusions

Investigating the role of past Human disturbances on tropical tree communities had, to our knowledge, never been explored with an adequate statistical design based on soil charcoal. Charcoal remains were found in most of the plots in each study area, revealing frequent and widespread past forest fires in SE and SW Cameroon, presumably prompted by slash-and-burn agricultural activities. Fires occurred during two very distinct periods, one lying from ca. 80 to 850 BP and another from ca. 1500 to 2800 BP. The abundance of charcoal and non-pioneer light-demanding tree species were both spatially structured and positively correlated in only one of the three study areas. However, the reverse pattern was observed in Area 2 and correlations were non-significant at a local scale when correcting for spatial autocorrelation. Testing alternative hypotheses on the drivers of floristic composition should give more insights to confirm or not whether charcoal is an appropriate proxy to study the impacts of past Human disturbances, and whether these disturbances have significantly modified present-day tree species assemblages in Central African moist forests.

Supporting Information

Figure S1 Mean site values of each RG relative abundance and CAI in Area 1. Values were computed at the plot level. Mean CAI was computed over two soil layers: 0–20 and 20–100 cm. Note that data are missing for the 20–100 cm soil layer in site 4. N-P = Non-Pioneer.

File S1 Tables S1-S4. Table S1 (in File S1) Sites Coordinates and number of plots per site. Table S2 (in File S1) Abundances and functional traits of all species inventoried in the three study areas. Table S3 (in File S1) Dates BP of 60 charcoal samples. Table S4 (in File S1) Mean sites values of each variable tested for their difference between sites, in each study Area.

Appendix S1 Contact details of logging companies.

Appendix S2 Datasets used to calibrate the charcoal abundance index.

Appendix S3 Files to be used in Torocor software for testing spatial autocorrelation of charcoal abundance at multiple spatial scales.

Appendix S4 Plot values for spatial autocorrelation and correlation tests on Torocor.

Appendix S5 Results of spatial autocorrelation tests for each variable of Appendix S4.

Appendix S6 Pearson correlations between all variables, tested by a torus-translation procedure.

Acknowledgments

We gratefully acknowledge the following logging companies for their logistic support: Pallisco (M. Rougeron, L. Douaud, P. Lagoute, R. Feteke), WIJMA Cameroun S.A. and SFID (Société Forestière et Industrielle de la Doumé/Doumé Forestry Company). We also thank Adeline Fayolle (Gembloux Agro-Bio Tech) for providing data on species functional traits thanks to the European-funded CoForChange project.

Author Contributions

Analyzed the data: JV TD OH. Contributed reagents/materials/analysis tools: JV JMR KD JFG JLD TD OH. Wrote the paper: JV JMR KD JFG JLD TD OH. Contributed to field inventories: JV JMR ABB KD JFG JLD. Organized radiocarbon dating of charcoal samples: JMR.

References

1. Willis KJ, Gillson L, Brncic TM (2004) How "virgin" is virgin rainforest? Science 304: 402–403.
2. Denham TP, Haberle SG, Lentfer C, Fullagar R, Field J, et al. (2003) Origins of Agriculture at Kuk Swamp in the Highlands of New Guinea. Science 301: 189–193.
3. Kealhofer L (2003) Looking into the Gap: Land Use and the Tropical Forests of Southern Thailand. Asian Perspectives 42(1): 72–95.
4. Ross N (2011) Modern tree species composition reflects ancient Maya "forest gardens" in northwest Belize. Ecological Applications 21: 75–84.
5. Glaser B, Birk JJ (2012) State of the scientific knowledge on properties and genesis of Anthropogenic Dark Earths in Central Amazonia (terra preta de Indio). Geochimica et Cosmochimica Acta 82: 39–51.
6. Oslisly R (1996) The middle Ogooué valley, Gabon: Cultural changes and palaeoclimatic implications of the last four millennia. The growth of farming communities in Africa from the equator southwards, Vol. Azania XXIX-XXX (ed. by J.E.G. Sutton), pp. 324–331. The British Institute in Eastern Africa, London.
7. Oslisly R, White L, Bentaleb I, Favier C, Fontugne M, et al. (2013) Climatic and cultural changes in the west Congo Basin forests over the past 5000 years. Philosophical Transactions of the Royal Scociety, B 368, 20120304.
8. Wotzka H-P (2006) Records of activity: radiocarbon and the structure of Iron Age settlement in Central Africa. In Grundlegungen. Beiträge zur europäischen und afrikanischen Archäologie für Manfred K.H. Eggert (ed. HP . Wotzka), pp. 271–289. Tübingen, Germany.
9. Elenga H, Schwartz D, Vincens A, Bertaux J, De Namur C, et al. (1996) Diagramme pollinique holocène du lac Kitina (Congo): mise en évidence de changements paléobotaniques dans le massif forestier du Mayombe. CR Acad Sc 323: 403–410.
10. Shanahan T, Beck JW, Overpeck JT, McKay NP, Pigati JS, et al. (2012) Late Quaternary sedimentological and climate changes at Lake Bosumtwi Ghana: New constraints from laminae analysis and radiocarbon age modeling. Palaeogeography, Palaeoclimatology, Palaeoecology 361–362: 49–60.
11. Schwart D (1992) Assèchement climatique vers 3 000 B.P. et expansion Bantu en Afrique centrale atlantique: quelques réflexions. Bull. Soc. Géol. France 163(3): 353–361.
12. Maley J (1992) Mise en évidence d'une péjoration climatique entre ca 2500 et 2000 ans BP en Afrique tropicale humide. Bulletin de la Société Géologique de France 163: 363–365.
13. Vincens A, Schwartz D, Bertaux J, Elenga H, de Namur C (1998). Late Holocene Climatic Changes in Western Equatorial Africa Inferred from Pollen from Lake Sinnda, Southern Congo. Quaternary Research 50: 34–45.
14. Ngomanda A, Neumann K, Schweizer A, Maley J (2009) Seasonality change and the third millennium BP rainforest crisis in southern Cameroon (Central Africa). Quaternary Research 71: 307–318.
15. Neumann K, Bostoen K, Höhn A, Khalheber S, Ngomanda A, et al. (2011) First farmers in the Central African rainforest: A view from southern Cameroon. Quaternary International 249: 53–62.
16. Hubau W (2013) Exploring charcoal as a natural archive for palaeofire and vegetation history of the Mayumbe, Democratic Republic of the Congo. PhD thesis, Ghent University.
17. Van Gemerden BS, Olff H, Parren MPE, Bongers F (2003) The pristine rain forest? Remnants of historical Human impacts on current tree species composition and diversity. Journal of Biogeography 30: 1381–1390.
18. Morin-Rivat J, Fayolle A, Gillet J-F, Bourland N, Gourlet-Fleury S, et al. (2014). New evidence of Human activities during the Holocene in the Lowland Forests of the Northern Congo Basin. Radiocarbon 56: 209–220.
19. White LJT, Oates JF (1999) New data on the history of the plateau forest of Okomu, southern Nigeria: an insight into how Human disturbance has shaped the African rain forest. Global Ecology and Biogeography 8: 355–361.

20. Swaine MD, Hall JB (1988) The mosaic theory of forest regeneration and the determination of forest composition in Ghana. Journal of Tropical Ecology 4: 253–269.
21. Clist B (2005) Des premiers villages aux premiers Européens autour de l'estuaire du Gabon: Quatre millénaires d'interactions entre l'homme et son milieu. PhD thesis, Université Libre de Bruxelles, Brussels. 938p.
22. Jones EW (1955) Ecological studies of the rainforest of southern Nigeria IV. The plateau forest of the Okomu Forest Reserve, Part 1. The environment, the vegetation types of the forest, and the horizontal distribution of species. Journal of Ecology 43: 564–594.
23. Jones EW (1956) Ecological studies of the rain forest of southern Nigeria IV. The plateau forest of the Okomu Forest Reserve, Part 2. The reproduction and history of the forest. Journal of Ecology 44: 83–117.
24. Letouzey R (1985) Notice de la carte phytogéographique du Cameroun au 1/500.000. Institut de la Carte Internationale de la Végétation, Toulouse and Institut de la Recherche Agronomique, Yaoundé (Cameroon).
25. Boulvert Y (1986) Carte phytogéographique de la république centrafricaine au 1/1 000 000. Orstom, Paris, France.
26. Fayolle A, Engelbrecht B, Freycon V, Mortier F, Swaine M, et al. (2012) Geological Substrates Shape Tree Species and Trait Distributions in African Moist Forests. PLoS ONE 7(8): e42381. doi:10.1371/journal.pone.0042381.
27. Lebrun J, Gilbert G (1954) Une classification écologique des forêts du Congo. Publication INEAC série scientifique 63, INEAC, Bruxelles.
28. Fay JM (1997) The ecology, social organization, populations, habitat and history of the western lowland gorilla (Gorilla gorilla gorilla Savage and Wyman 1847). PhD thesis, Washington University, Washinghton D.C.
29. Biwolé AB, Morin-Rivat J, Fayolle A, Bitondo D, Dedry L, et al. (2014). New data on the recent history of the littoral forests of southern Cameroon: an insight into the role of historical human disturbances on the current forest composition. Plant Ecology and Evolution, 148 (1). In press.
30. WRB (2006) World reference base for soil resources. 2nd edition. FAO, IUSS, ISRIC (Eds.) World Soil Resources Reports No. 103. FAO, Rome, 145 p.
31. Harms KE, Condit R, Hubbell SP, Foster RB (2001) Habitat association of trees and shrubs in a 50-ha neotropical forest plot. Journal of Ecology 89: 947–959.
32. Talon B, Payette S, Filion L, Delwaide A (2005) Reconstruction of the long-term fire history of an old-growth deciduous forest in Southern Québec, Canada, from charred wood in mineral soils. Quaternary Research 64: 36–43.
33. Talon B (2010) Reconstruction of Holocene high altitude vegetation cover in the French Southern Alps: evidence from soil charcoal. The Holocene 20: 34–44.
34. Chabal L (1997) *Forêts et sociétés en Languedoc (Néolithique final, Antiquité tardive): l'anthracologie, méthode et paléoécologie*. Documents d'archéologie française, 63, éd. de la Maison des sciences de l'Homme, Paris: 189 p., 50 fig., 2 annexes, lex., bib.
35. Bronk Ramsey C (1995) Radiocarbon calibration and analysis of stratigraphy: The OxCal program. Radiocarbon 37(2): 425–430.
36. Bronk Ramsey C (2001) Development of the radiocarbon calibration program OxCal. Radiocarbon 43(2A): 355–363.
37. Bronk Ramsey C, Staff RA, Bryant CL, Brock F, Kitagawa H, et al. (2012) A Complete Terrestrial Radiocarbon Record for 11.2 to 52.8 kyr B.P. Science 338(6105): 370–374.
38. Reimer PJ, Baillie MGL, Bard E, Bayliss A, Beck JW, et al. (2009) IntCal09 and Marine09 radiocarbon age calibration curves, 0–50,000 years cal BP. Radiocarbon 51(4): 1111–1150.
39. Dauby G, Hardy OJ (2011) Sampled-based estimation of diversity sensu stricto by transforming Hurlbert diversities into effective number of species. Ecography 34: 001–012.
40. Hawthorne WD (1995) Ecological profiles of Ghanaian forest trees. Oxford Forestry Institute, Department of Plant Sciences, University of Oxford.
41. Slik JWF (2005) Assessing tropical lowland forest disturbance using plant morphological and ecological attributes. Forest Ecology and Management 205: 241–250.
42. Chave J, Coomes D, Jansen S, Lewis SL, Swenson NG, et al. (2009) Towards a worldwide wood economics spectrum. Ecology Letters 12: 351–366.
43. Zanne AE, Lopez-Gonzalez G, Coomes DA, Ilic J, Jansen S, et al. (2009) Global wood density database. Dryad. Identifier: http://hdl.handle.net/10255/dryad.235.
44. Slik JWF (2006) Estimating species-specific wood density from the genus average in Indonesian trees. Journal of Tropical Ecology 22: 481–482.
45. R Development Core Team. (2009) R: A language and environment for statistical computing. In: R Foundation for Statistical Computing (Ed.), Vienna, Austria.
46. Moran PAP (1948) The interpretation of statistical maps. Journal of the Royal Statistical Society B 10: 243–251.
47. Mantel N (1967) The detection of disease clustering and a generalized regression approach. Cancer Research 27: 209–220.
48. Gillet J-F (2013) Les forêts à Marantaceae au sein de la mosaïque forestière du nord de la république du Congo: origines et modalités de gestion. PhD thesis, Université de Liège - Gembloux Agro-bio Tech.
49. Lee KE, Wood TG (1971) Termites and soils. Academic Press, London, UK. New York, USA.
50. Miklós AAW (1992) Biodynamique d'une couverture pédologique dans la région de Botucatu (Brésil-SP). PhD thesis Université Paris-6, France.
51. Mc Brearty S (1990) Consider the humble termite: Termites as agents of post-depositional disturbance at African archaeological sites. Journal of Archaeological Science 17: 111–143.
52. Araujo AGM (2013) Bioturbation and the upward movement of sediment particles and archaeological materials: comments on Bueno, et al. Journal of Archaeological Science 40: 2124–2127.
53. Lavelle P (1983) The soil fauna of tropical savannas. Elsevier Scientific Publishing Company, Amsterdam, the Netherlands.
54. Johnson DL (1990) Biomantle evolution and the redistribution of earth materials and artifacts. Soil Science 149: 84–102.
55. Gouveia SEM, Pessenda LCR (2000) Datation par le ^{14}C de charbons inclus dans le sol pour l'étude du rôle de la remontée biologique de matière et du colluvionnement dans la formation de latosols de l'état de Sao Paulo, Brésil. C. R. Acad. Sci. Paris, Earth and Planetary Sciences 330: 133–138.

11

Effects of Picoxystrobin and 4-n-Nonylphenol on Soil Microbial Community Structure and Respiration Activity

Marianne Stenrød*, Sonja S. Klemsdal, Hans Ragnar Norli, Ole Martin Eklo

Norwegian Institute for Agricultural and Environmental Research (Bioforsk), Ås, Norway

Abstract

There is widespread use of chemical amendments to meet the demands for increased productivity in agriculture. Potentially toxic compounds, single or in mixtures, are added to the soil medium on a regular basis, while the ecotoxicological risk assessment procedures mainly follow a chemical by chemical approach. Picoxystrobin is a fungicide that has caused concern due to studies showing potentially detrimental effects to soil fauna (earthworms), while negative effects on soil microbial activities (nitrification, respiration) are shown to be transient. Potential mixture situations with nonylphenol, a chemical frequently occurring as a contaminant in sewage sludge used for land application, infer a need to explore whether these chemicals in mixture could alter the potential effects of picoxystrobin on the soil microflora. The main objective of this study was to assess the effects of picoxystrobin and nonylphenol, as single chemicals and mixtures, on soil microbial community structure and respiration activity in an agricultural sandy loam. Effects of the chemicals were assessed through measurements of soil microbial respiration activity and soil bacterial and fungal community structure fingerprints, together with a degradation study of the chemicals, through a 70 d incubation period. Picoxystrobin caused a decrease in the respiration activity, while 4-n-nonylphenol caused an increase in respiration activity concurring with a rapid degradation of the substance. Community structure fingerprints were also affected, but these results could not be directly interpreted in terms of positive or negative effects, and were indicated to be transient. Treatment with the chemicals in mixture caused less evident changes and indicated antagonistic effects between the chemicals in soil. In conclusion, the results imply that the application of the fungicide picoxystrobin and nonylphenol from sewage sludge application to agricultural soil in environmentally relevant concentrations, as single chemicals or in mixture, will not cause irreversible effects on soil microbial respiration and community structure.

Editor: Hauke Smidt, Wageningen University, The Netherlands

Funding: This work was performed as part of the strategic institute programme 'Bioavailability and biological effects of chemicals - Novel tools in risk assessment of mixtures in agricultural and contaminated soil' funded by the Norwegian Research Council, project number 186901/i30. The funders had no role in study design, data collection and analysis, decision to publish, or preparation of the manuscript.

Competing Interests: The authors have declared that no competing interests exist.

* E-mail: marianne.stenrod@bioforsk.no

Introduction

Various practices have been promoted to meet the demands for increased productivity of agricultural areas, including the use of mineral fertilizers and pesticides as well as the use of organic amendments. The potential risk of non-target effects of pesticides in soils is evident, and has been widely studied. Specific attention is now given to the use of organic amendments originating from sewage sludge. Sludge has been shown to contain high levels of many organic chemicals (e.g. nonylphenols, PAHs a.o.) that might exert ecotoxicological effects upon soil addition, e.g., [1,2]. Potentially toxic compounds, single or in mixtures, are added to the soil medium on a regular basis, while the ecotoxicological risk assessment procedures mainly follow a chemical by chemical approach focusing on establishing dose-response relationships for soil fauna. The use of sewage sludge on agricultural fields where pesticides are sprayed regularly as part of conventional farming practices calls for increased attention to the potential combined effects of pesticides and known contaminants in sewage sludge.

Sewage sludge, potentially used for application on agricultural soil, is known to contain considerable amounts of nonylphenol [2–4] – an industrial by-product and degradation product of nonylphenol ethoxilate plasticizers. The occurrence, fate and toxicity of nonylphenol in the environment have been reviewed by Soares and co-workers [4], pointing at knowledge gaps as well as identified challenges with nonylphenol in soil including occurrence and possible accumulation in soil following sludge application, reduced degradation rates in soil due to sorption and reduced bioavailability, and potential toxic effects on soil microorganisms. Nonylphenol is also included on the list of priority substances of the Water Framework Directive [5] and, hence, require further attention to clarify its potential effects in soil. Picoxystrobin is a strobilurin fungicide [6] for spraying in cereals, with a maximum of one spraying per season in Norway. Picoxystrobin inhibits mitochondrial respiration by blocking electron transfer at the Qo centre of cytochrome bc1 [7]. Picoxystrobin is reported to have a high acute toxicity to earthworms (LC_{50} at 6.7 mg kg^{-1}) [8], and field assays of earthworm toxicity indicate acute toxic effects even at recommended doses for use [9], which might be caused by heavy rain shortly after spraying forcing the earthworm to migrate to the surface. Potentially negative effects on soil microbial nitrogen and carbon mineralization activity are shown to be transient (dose: 750 g ha^{-1}, duration: 28 days) [8]. Picoxystrobin require continued attention due to its demonstrated potentially negative effects to earthworms.

The importance of soil microbes and their activity in the functioning of soils, e.g., [10,11] justify their thorough investigation in risk assessments [12–14]. The development of genomic techniques over the last decades has made detailed studies of the soil microbial community possible, beyond the scope of broad-scale measures like substrate induced respiration. DNA extraction from soil followed by different molecular approaches to determine the genetic diversity and quantify the presence of single organisms or groups of related organisms in a soil sample, have been employed successfully in studies of species and functional diversity in agricultural soils [15–17]. According to OECD guidelines for the testing of chemicals carbon [18] and nitrogen [19] transformation tests (with cut-off criteria of 25% effect) are the recommended methods to assess effects concentrations of chemicals on the soil microbial community. In research, soil respiration is commonly used to assess effects of pesticides and other chemicals on soil microbes. These are measures linked to the activity level of the soil microbial community, but are crude measures that do not necessarily reveal all relevant effects. The low percentage of soil microorganisms we are able to culture ex situ stresses the need for employing molecular and genomic methods suitable for terrestrial ecotoxicological studies. T-RFLP-analysis is one much used technique showing good results when looking at effects of different environmental conditions and chemical stressors on soil microbial communities, e.g., [16,17,20]. But there are many important methodological aspects to be considered when interpreting the results, e.g., [21,22], including choice of primers and restriction enzymes, procedure for noise reduction and profile alignment, and statistical analysis using relative abundance or presence/absence data.

The main objective of this study was to assess the effects of picoxystrobin and nonylphenol, as single chemicals and mixtures, on soil microbial community structure and respiration activity in an agricultural sandy loam soil. The choice of compounds is based on their individual occurrence and effects in the environment, and expected potential for co-occurrence in the field, although the latter is scarcely documented. Further, they were chosen to represent an example of combined effects of agricultural and industrial contaminants with their expected independent effect mechanisms due to differences in mode of action. Effects of the chemicals were assessed through measurements of soil microbial respiration activity and soil bacterial and fungal community structure fingerprints, together with a degradation study of the chemicals, through a 70 d incubation period. The different measures showed corresponding results in support of a conclusion that a mixture situation with picoxystrobin and nonylphenol in soil will not increase the potential negative effects of picoxystrobin on the soil microbial community.

Materials and Methods

Soil

Bulk soil sampling in the top 10 cm of the plough-layer of an agricultural field at Norderås, Ås, South East Norway (59°41′14″ N, 10°46′22″ E), was done in middle of August 2008. After sieving (4 mm mesh) the soil was physically and chemically characterized (Table 1; all analyses performed in accordance with recognized laboratory standards at Analycen AS/Eurofins Norway) and stored moist at 4°C for about 3 weeks before use in laboratory experiments. No specific permits were required for the described field soil sampling and subsequent lab studies.

Table 1. Selected physical and chemical characteristics of the studied soil.

Parameter	Value	Unit
pH	6.3	
TOC – Organic C	1.1	%
Kjeldahl N	0.0986	%
C/N	11	
Tot C	2.0	%
Tot N	0.07	%
CEC	11.2	meq/100 g
Coarse (>2 mm)	13.7	%
Sand (2.000-0.060)	68.3	%
Silt (0.060-0.002)	20.7	%
Clay (<0.002)	11.0	%

Analysis performed by Analycen AS/Eurofins Norway.

The Water Holding Capacity (WHC) was estimated on sieved soil gently packed in plastic columns. After wetting with Milli-Q-water overnight, excess water was allowed to drip off during 1 h the next day, and the gravimetric moisture content was measured after drying overnight at 105°C.

Chemicals

Treatment solutions were prepared from analytical grade picoxystrobin (Riedel-de Häen) and 4-n-nonylphenol (Fluka) purchased from Sigma-Aldrich. The chemicals were dissolved in acetone (4 ml portions; Lab-Scan) and added to quartz sand (27 g portions; Sigma-Aldrich). After evaporation of the acetone (about 30 minutes) the quartz sand with chemical was mixed into batches of moist soil (2.7 kg dry weight equivalent (dw eq.), 60% WHC) giving 2 or 10 mg picoxystrobin kg^{-1} soil (dw eq.), or 0.5 or 10 mg 4-n-nonylphenol kg^{-1} soil (dw eq.) as single treatments as well as all mixture combinations (Table 2). A solvent control (acetone added to quartz sand, left to evaporate before addition to soil) and an untreated control (soil with no addition of chemical) were included in the experiment.

The lowest test concentration for picoxystrobin was set from reported levels for transient effects on carbon and nitrogen mineralization (750 g ha^{-1}) [8], considering that picoxystrobin has been shown to sorb to the top cm of soil (50–70% interception, sorption in top 1 cm of soil), although recommended doses in Norway are below these levels. The worst case concentration was set to 10 mg kg^{-1} surpassing reported acute toxicity levels for earthworm (LC_{50} = 6.7 mg kg^{-1}) [8]. The lowest test concentration for nonylphenol was set to the Predicted Initial Environmental Concentration (PIEC) estimated through a risk assessment made by the Norwegian Scientific Committee on Food Safety [2]; application of sludge containing mean expected amounts of nonylphenol (32 mg kg^{-1}) at rates in accordance with Norwegian regulations (40 tonnes ha^{-1}). From this a worst case concentration was set to 10 mg kg^{-1} considering a risk of reduced incorporation depth and possible initial effects.

Experimental Set-up

An incubation experiment was set up with all chemical treatments in five replications and the control treatments in four replications. Subsamples of about 100 g dw eq. of sieved, moist soil

Table 2. Overview of picoxystrobin and 4-n-nonylphenol treatments in the incubation experiment.

Treatment	Picoxystrobin (PI)		4-n-nonylphenol (NP)	
	2 mg kg^{-1} dry soil	10 mg kg^{-1} dry soil	0.5 mg kg^{-1} dry soil	10 mg kg^{-1} dry soil
PI low	x			
PI high		x		
NP low			x	
NP high				x
Mix low	x		x	
Mix NP high	x			x
Mix PI high		x	x	
Mix high		x		x
Control				

(60% WHC) in plastic (polypropylene) containers were placed in air-tight glass jars together with small beakers with NaOH-solution (10 mL, 1 M) to trap CO_2 evolved from the soil. All treatments were set up with five repeats to enable destructive sampling of five and four replicates, of chemical treatments and the control respectively, on each sampling occasion through the 70 d incubation period. A total of 244 samples were incubated at 20°C in the dark. Samples were taken on six time-points; before addition of chemicals (4 samples), and after 1, 7, 14, 28 and 70 days of incubation (48 samples on each occasion). Each sample was homogenized and subsamples taken for chemical residue analysis (25 g), and DNA extraction (30 g). In addition, the general microbial activity level in the soil was monitored by weekly replacement and analysis of the NaOH-traps, giving a total of 11 time-points with measurements.

Total Soil Microbial Activity

The total activity of soil microbial biomass was followed by respirometric measurements [23]. The CO_2 in the NaOH traps was determined by colourimetric continuous flow analysis (AutoAnalyzer3, Bran+Luebbe, Germany). Results are reported as mg CO_2-C kg^{-1} dry soil.

Soil Microbial Community Structure

DNA extraction procedure. Soil samples were homogenized (30 g moist soil +30 mL Milli-Q-water) in a mill (3 min forward +3 min reverse spin; Retsch PM 400) before extraction of DNA according to the Fast DNA Spin Kit for Soil (QBiogene) manual. In short, soil suspension (400 µL) was added to Lysing Matrix E Tubes before homogenization with the FastPrep® instrument (30 sec at speed 5.5; QBiogene, MP Biomedicals) and centrifugation (14000×g, 3 min). DNA was purified with guanidine thiocyanate (5.5 M, Sigma) twice (1 mL and 600 µL) after DNA binding to the Binding Matrix Suspension, before elution (100 µL sterile distilled water) from Spin™Filters. The eluate was further purified through Micro Bio-Spin Chromatography columns (Bio-Rad Laboratories Ltd.) packed with polyvinylpolypyrrolidone (PVPP) (Sigma) (4000×g, 5 min), and stored at −80°C.

PCR procedure. PCR reactions were run with fluorescently labelled PCR primer sets (Applied Biosystems), both forward and reverse primer labelled. After testing of 8 primer sets, chosen from previously published T-RFLP studies, PCR reactions were run for all samples with three selected primer sets and fluorescence labelling (Table 3).

All PCR reactions were run with 1 µL template and 1 µL of forward and reverse primer (10 pmol/µL) in TaqMan® Environmental Master Mix 2.0 (12.5 µL; Applied Biosystems) made up to a total reaction volume of 25 µL with sterile distilled water, on GeneAmp® PCR system 9700 (Applied Biosystems) or PTC-200 (MJ Research) thermal cyclers (40 cycles). An annealing temperature of 55°C was used for all primer combinations. PCR-products were purified with GenElute™PCR clean-up kit (Sigma-Aldrich) according to the instructions from the supplier, before restriction digestion and T-RFLP.

Table 3. Primer sets used in the experiments.

Primer set no.	Primer	Sequence	Reference
1	63F[1]	CAG GCC TAA CAC ATG CAA GTC	[24]
	1087R[2]	CTC GTT GCG GGA CTT ACC CC	[24]
2	EF4[1]	GGA AGG GRT GTA TTT ATT AG	[25]
	Fung5[2]	GTA AAA GTC CTG GTT CCC C	[25]
3	ITS1F[3]	CTT GGT CAT TTA GAG GAA GTA A	[26]
	ITS4[4]	TCC TCC GCT TAT TGA TAT GC	[27]

Fluorescence labeling: [1]6FAM,
[2]VIC,
[3]PET,
[4]NED.

T-RFLP procedure. After initial testing of a range of restriction enzymes (AluI, DdeI, HaeIII, HinfI, MboI, MspI, RsaI, TaqI, HhaI, MvnI, BstUI) on random samples, MspI and HaeIII, and MspI and HinfI were chosen for restriction of PCR products from bacteria and fungi, respectively. Restriction with the different enzymes were run in separate reactions and analysed separately. Analysis was performed on an ABI 3730 DNA Analyzer (Applied Biosystems, Foster City, USA) after mixing of 1 µl sample (template) with 8.75 µl Hi-Di formamid™ (Applied Biosystems) and 0.25 µl GeneScan™ 500 LIZ size standard (Applied Biosystems). The results reported for the T-RFLP analyses are based on relative comparisons of fingerprints, to show relative effects of different chemical treatments and comparison with a soil sample not treated with chemicals.

Degradation Study

Residues of picoxystrobin and 4-n-nonylphenol were extracted from 5 g subsamples of soil after mixing with 1.0 g of dehydrated $MgSO_4$ (purum, Fluka, Sigma-Aldrich GmbH) and 10 mL acetonitrile (Pestiscan, LAB-SCAN POCH SA, Gliwice, Poland) in 50 ml centrifugal tubes. One µg of 4-n-nonylphenol-2,3,5,6-d4-OD (99.4%, Chiron, Trondheim, Norway) was added as internal standard. After a short homogenization (10–15 sec. whirl mix) the samples were extracted by end-over-end shaking (1 h; Reax2, Heidolph). Extraction efficiencies derived from analysis of sterilised (autoclaved) soil at 117±2.0% and 108±0.0% for 4-n-nonylphenol and picoxystrobin, respectively. After centrifugation (1800×g, 5 min), 1.5 ml of the supernatant was transferred to GC-vials for analysis on GC-MS. Calibration standards at 0.001, 0.01, 0.05, 0.2, 1.0 and 5 µg ml^{-1} where prepared by diluting stock solutions of picoxystrobin (98%, Dr. Ehrenstorfer, Augsburg, Germany) and 4-n-nonylphenol (99.2%, Chiron, Trondheim, Norway) with acetonitrile. To balance the matrix in the samples a GC-vial was added 1 ml of a blank soil extract which was evaporated to dryness. One ml of each calibration standard was added to the GC-vial together with 0.1 µg 4-n-nonylphenol-2,3,5,6-d4-OD as internal standard. The measurements were performed on an Agilent 6890 gas chromatograph connected to an Agilent 5973 mass spectrometer using ChemStation Software version D.03.00. The gas chromatograph was equipped with a Gerstel (Mühlheim Ruhr, Germany) programmable temperature vaporizing (PTV) injector with a sintered liner. The separation was performed using a fused silica column (HP-5MSI 30 m width, 0.25 mm internal diameter, 0.25 µm film thickness, J&W Scientific) connected to a 2.5 m methyl deactivated pre column (Varian Inc., Lake Forest CA, USA) of same internal diameter as the analytical column. The temperature program was as follows; 65°C held for 1.5 min, 20°C min^{-1} to 120°C, held for 0 min, 20°C min^{-1} to 300°C, held for 0.5 min, total runtime 13.75 min. The PTV program was as follows: solvent vent temperature 60°C held for 1.40 min with a vent flow at 200 ml min^{-1}. After 1.42 min the split valve was closed and the injector temperature increased by 720°C min^{-1} to 250°C and kept for 2 min. Injection volume 10 µl. The mass spectrometer was operated in selected ion monitoring mode with target/qualifier ions as follows: 4-n-nonylphenol-D4: m/z = 111/224, 4-n-nonylphenol: m/z = 107/220 and picoxystrobin: m/z = 145/335. Transfer line temperature was set at 280°C, ion source temperature at 230°C and quadrupole temperature at 150°C.

Biological degradation of picoxystrobin and 4-n-nonylphenol was verified by spiking selected matrices (quartz sand, unsterile soil, sterilized soil) at 0.05 µg g^{-1} and left in the dark at room temperature (20±2°C) for 24 h before extraction and analysis as described above. The soil was sterilised by autoclaving (121°C, 20 min, x2) (Matachana S1000).

Data Processing and Statistical Analysis

First-order rate constants for organic carbon mineralization were estimated by linear regression (SigmaPlot 11.0, SYSTAT) using mean cumulative values for CO_2-C evolved, from the equation $\ln S = \ln S_0 + kt$, where S is the amount of CO_2-C evolved at time t (mg kg^{-1} dry soil), S_0 is the initial amount of mineralization product, k is the rate constant (d^{-1}), and t is the time (d). Due to the non-normal distribution of our CO_2-measurements, non-parametric analysis was utilized to test for differences between treatments; i.e. Kruskall Wallis test for equality of medians for all treatments, and Mann-Whitney test for equality of medians between selected treatments (Minitab 15, Minitab Inc.).

Handling of T-RFLP data with binning/identification of alleles, normalization (within project; i.e. for all samples independent of plate) and (automatic) alignment of profiles was done with GeneMapper 4.0 (Applied Biosystems) based on a 4 basepair (bp) bin width (assumed to give the more stable number of bins based on a screening of bin widths from 0.5 to 10 bp in 0.5 bp increments) and a threshold of 50 (default). Obvious pull-up peaks were removed from the dataset before further analysis. No further trimming of the data was done (i.e. no small peaks removed) to avoid exclusion of potentially important peaks for the microbial diversity analysis. Size (bp), area and height data were exported to an excel spreadsheet (Excel 2007, Microsoft corp.) for calculation of average size of each identified peak and calculation of relative peak height and area for all peaks within each individual sample. All peak data for each sample was assembled before statistical analysis, keeping the data for each primer set separate. The terminal fragments labelled from the forward primer as well as the fragments labelled from the reverse primer from each of the restriction reactions, were included in a single data row for each sample before analysis. The data were found to be not normally distributed (i.e. skewed due to rarely occurring peaks). Relative peak heights were analysed with non-metric multidimensional scaling (nmMDS) in R-software (v. 2.9.1) [28] using the metaMDS routine in the vegan package [29] with k = 2 or 3 dimensions, the isoMDS routine in the MASS package [30] with k = 2 dimensions, and with principal coordinates analyses using the PCO routine in R. All analyses resulted in similar patterns and the results presented are from metaMDS (k = 3 dimensions) as these gave the best fit to the observed data (STRESS below 0.10) [31,32]. Results are reported as two and three-dimensional plots showing the clustering of treatments at the different time-points of analysis through the incubation period. The samples taken before addition of chemical and onset of the incubation period (original soil sample) were included in the analysis of all time-points as a reference. Further, analysis of similarity was performed with the ANOSIM routine in PAST-software [33] with Bray-Curtis distance measure utilized in all tests. Results are given as ANOSIM R of a value between 0 and 1, where values close to 1 indicate large differences between treatments.

The observed T-RFLP patterns for the untreated and solvent control were not statistically significant different as analysed by nmMDS, and the results presented focus on differences between chemical treatments and the solvent control. This was to avoid overestimating any effects of the studied chemicals. A similar clustering of T-RFLP fingerprints was observed for the results obtained from using primer sets 2 and 3 (i.e. soil fungal community fingerprints; Table 3). Hence, the presented results only include primer set 3.

Soil degradation half-lives (DT50) for picoxystrobin and 4-n-nonylphenol was estimated from results from the degradation study, and calculated according to first-order kinetics from the equation $DT50 = \log_{10}2/b$, where b is the regression coefficient estimated through a linear regression procedure. All measured values were utilized in the estimation of the DT50 values except for two extreme outliers being removed before statistical analysis of the data. Due to the five replicates for each treatment this could be done without compromising the results of the data analyses, as the analyses for the affected treatments were still based on four replicate samples. Due to some differences in the observed degradation pattern for the chemicals at low and high concentrations, the soil half-lives were estimated separately for these. The estimated half-lives are based on data from both the relevant single chemical and mixture treatment, as the observed degradation pattern did not differ significantly between these. As an example

DT50 for 2 mg picoxystrobin kg^{-1} dry soil was estimated using data from both the single chemical treatment and the two mixture treatments with this concentration of picoxystrobin. The estimation was done assuming bi-phasic degradation of both chemicals, with both phases sufficiently described by first-order degradation kinetics and, hence, possible to estimate through linear regression.

A significance level of 5% was used for hypothesis testing.

Results

Total Soil Microbial Activity

No statistically significant differences between the treatments could be detected from observations of cumulative CO_2-C development after 70 d and estimated first-order rate constants for organic C-mineralization (Table 4). However, cumulative respiration curves (Fig. 1) indicated a tendency of reduced respiration activity in treatments with picoxystrobin as a single chemical, and increased respiration activity in treatments with high concentrations of nonylphenol (single chemical and mixture). The untreated control showed a lower respiration rate in the beginning of the period and a higher respiration rate at the end of the period, as compared to the other treatments. This difference did not result in any statistically significant difference in cumulative CO_2-C respiration, but indicated an effect of the solvent (acetone) used in the experiments.

The indications from the latter were supported by analysis of the separate measurements of CO_2 development from the soil samples at 11 time-points through the 70 d incubation period. Kruskall Wallis test for equality of medians between the treatments indicated statistically significant differences ($p<0.001$) when looking at all the individual CO_2 measurements in the analysis separately (i.e. not cumulative values at 70 d). The individual ranking of the treatments (data not shown) gave the solvent control a rank close to the overall rank (i.e. mean), while treatments with picoxystrobin as a single chemical (2 and 10 mg kg^{-1} dry soil) were indicated to cause a reduction in respiration activity, through a rank statistically significant lower than the overall rank. High concentrations of 4-n-nonylphenol (10 mg kg^{-1} dry soil) were indicated to cause an increase in respiration activity, through a rank statistically significant higher than the overall rank. This was also the case for the untreated control.

These indications were further supported by results from Mann-Whitney tests for equality of treatment medians, comparing the treatments two by two (test criteria: $p<0.05$). High concentration of 4-n-nonylphenol (10 mg kg^{-1} dry soil) in mixture with low concentration of picoxystrobin (2 mg kg^{-1} dry soil) gave statistically significant higher CO_2 development than treatments with picoxystrobin as a single chemical (2 and 10 mg kg^{-1} dry soil), low concentration of 4-n-nonylphenol (0.5 mg kg^{-1} dry soil), low concentration mixture (2 and 0.5 mg kg^{-1} dry soil of picoxystrobin and 4-n-nonylphenol, respectively) as well as the solvent control. Further, treatments with picoxystrobin as a single chemical (2 and 10 mg kg^{-1} dry soil) gave statistically significant reduction in CO_2 development compared to all mixture treatments, high concentration of 4-n-nonylphenol (10 mg kg^{-1} dry soil) as a single chemical, and the untreated control.

Soil Microbial Community Structure

The fungicide picoxystrobin was found to have a statistically significant effect on the soil microbial community structure, as found from nmMDS and ANOSIM of relative peak height data from T-RFLP analyses (Table S1). Looking at the effects of single chemical treatments in bacteria (Fig. S1) the largest segregation between treatments was observed after 7 d, with an overall ANOSIM R of 0.65 ($p=0.0001$). Here there was a statistically significant segregation between the low and high concentration treatment ($R=0.42$; $p=0.018$), indicating a concentration effect, and also a significant effect compared to the solvent control at low levels of picoxystrobin in the soil ($R=0.44$, $p=0.04$). Large variability was found in the data for the effects of picoxystrobin on the soil fungal community structure, with the largest spread in the data as shown by nmMDS after 7 d (Fig. 2a, Fig. S2) ($R=0.47$, $p=0.0001$). However, an effect due to different treatment concentrations of picoxystrobin was evident until 28 d ($R=0.38$, $p=0.017$) (Fig. S2). There was still a statistically significant difference between the single chemical treatments and the solvent control at 70 d for both bacteria (Fig. S1) and fungi (Fig. 2b), with an overall ANOSIM R of 0.65 and 0.4 ($p=0.0001$), respectively.

Less evident effects were observed for the single chemical treatments with 4-n-nonylphenol. Analysis of the bacterial community structure indicated an increasing change in the T-RFLP fingerprints with time, with statistically significant differences due to chemical treatment in nmMDS-analysis at 28 d (overall ANOSIM $R=0.56$, $p=0.0001$) and 70 d (overall ANOSIM $R=0.67$, $p=0.0001$) (Fig. 3, Fig. S3). A concentration effect was also evident at 28 d ($R=0.42$, $p=0.035$). The T-RFLP fingerprints of the fungal community upon treatment with 4-n-nonylphenol showed a similar pattern (Fig. S4), however, with larger variation between replicate samples giving lower ANOSIM R-values (range: 0.25–0.52).

The effects of the mixture treatments (Fig. S5 and S6) could only be separated from the solvent control after 28 and 70 d for the T-RFLP fingerprints of the soil bacterial community ($R=0.73$, $p=0.007$ and $R=0.44$, $p=0.04$, respectively) and after 70 d for the soil fungal community ($R=0.72$, $p=0.007$). When comparing the mixture treatments with the single treatments, the more prominent alteration in T-RFLP fingerprints originated from picoxystrobin as a single chemical (10 mg kg^{-1} dry soil), with the largest separation between treatment effects observed after 28 d both for the soil bacterial (Fig. 4a) and soil fungal (Fig. 4b) community structure (overall ANOSIM $R=0.61$ for both analyses). At this time-point also the single chemical treatment with 4-n-nonylphenol (10 mg kg^{-1} dry soil) could be singled out, with statistically significant changes in T-RFLP fingerprint as compared to the solvent control and other chemical treatments.

The duration of the observed changes in T-RFLP fingerprints of the soil microbial communities under the different chemical treatments was examined through nmMDS analysis of the grouping of the time-points for T-RFLP analyses. In general, these results indicated that the observed changes were transient. There was a more rapid re-establishment of the original T-RFLP fingerprints in the soil bacterial (i.e. closer resemblance between time-points 0 and 70) as compared to the soil fungal community (Fig. 5).

Degradation of Chemicals

Soil chemical residue was analysed day 0, 1, 7, 14, 28 and 70 of the incubation period. A lag phase was observed before the degradation of picoxystrobin commenced (Fig. 6a), while 4-n-nonylphenol was rapidly degraded in the soil (Fig. 6b). Calculations of soil half-lives (Table 5) were based on data from both single chemicals and mixtures as the degradation patterns were similar for these. Due to very rapid initial degradation of 4-n-nonylphenol in the low concentration treatments (0.5 mg kg^{-1} dry soil), the soil half-lives were here estimated for the 0–14 days and 14–70 days separately. For the other treatments the estimates were made for 0–28 days and 28–70 days. Soil half-life for picoxystrobin for the 28–70 day period was not calculated due to a measured

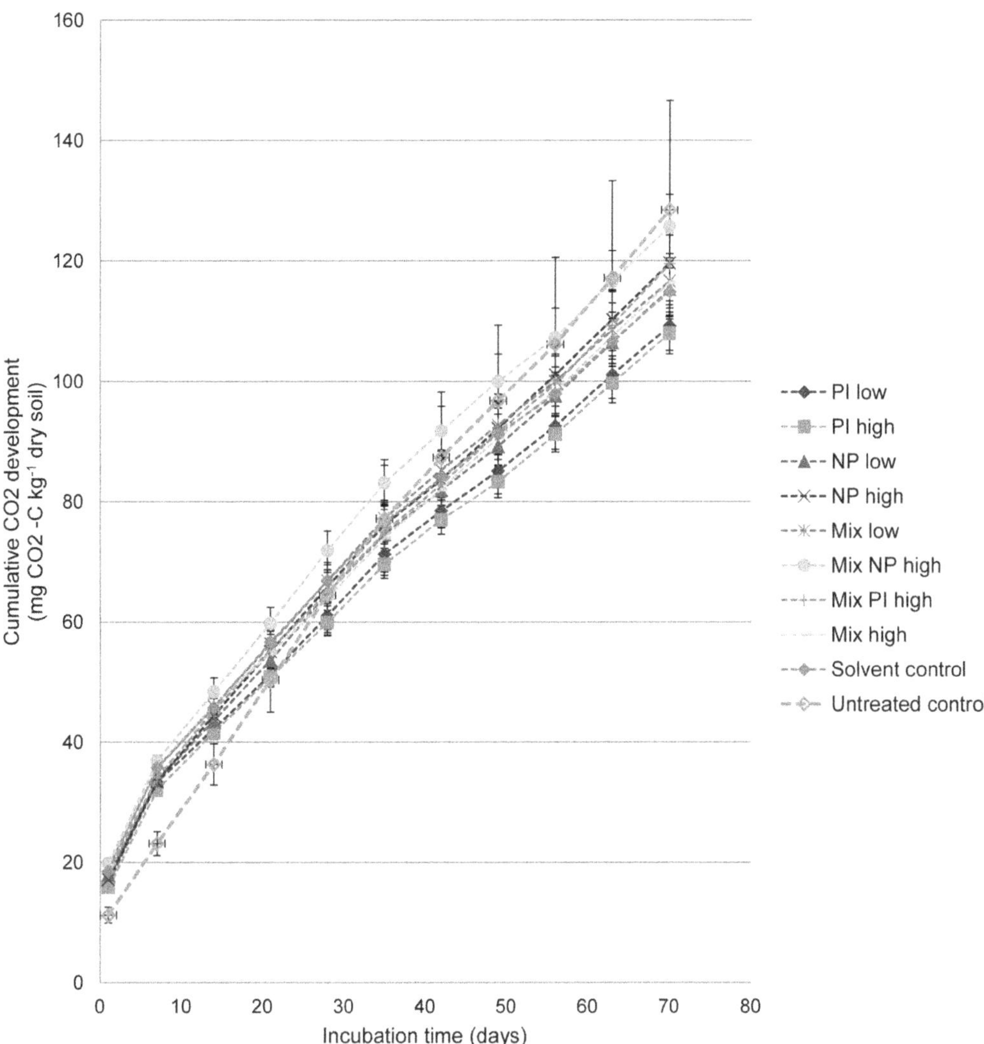

Figure 1. Effects of picoxystrobin and 4-n-nonylphenol on soil microbial respiration activity. Cumulative CO_2 development in a sandy loam soil during a 70 d incubation period after treatment with picoxystrobin and 4-n-nonylphenol as single chemicals and in mixture. Mean values ± SD. Legends: Treatments with picoxystrobin (PI) or 4-n-nonylphenol (NP) or both chemicals (Mix), in low (2 mg kg^{-1} PI, 0.5 mg kg^{-1} NP) or high (10 mg kg^{-1}) concentrations. Cf. Table 2 for full details.

increase in the soil residual concentrations at the end of the incubation period, hence, not allowing a reliable estimate of the long term degradation rate.

The rapid initial degradation of 4-n-nonylphenol was confirmed and shown to be microbially mediated (Fig. 7).

Discussion

Significance and Duration of Observed Effects

Total soil microbial activity. Our findings of a negative effect of picoxystrobin on individual measurements of CO_2 development, but not cumulative values, were in accordance with previous reports. Approval guidelines for plant protection products require documentation of effects on carbon and nitrogen mineralization, and only transient effects on nitrogen and carbon mineralization have been reported for picoxystrobin [8]. This fungicide inhibits mitochondrial respiration by blocking electron transfer at the Qo centre of cytochrome bc1 [7], and, hence, can be expected to affect the soil fungal community. We did not, however, expect an adverse effect in bacteria. The observed lag-phase before on-set of degradation of picoxystrobin (Fig. 6a) could correspond with a negative initial effect on respiration activity and microbial degradation activity. This may, however, not be concluded from the results as bacterial growth on anthropogenic substrates in soil is generally preceded by a lag-phase to allow adaptation to a new substrate. The observed slowing of the degradation rates after 28 days (Table 5) indicates limitations to microbial activity (e.g. oxygen limitation, occurrence of inhibiting factors), reduced bioavailability of the compound (e.g. time-dependent sorption) and/or a negative effect of picoxystrobin on the soil microbial community.

Our results indicated stimulating effects of high levels of 4-n-nonylphenol on soil respiration that could, in part, be explained by an increase in soil microbial activity level. As an endocrine-disrupting agent, we did not expect adverse effects of 4-n-nonylphenol on the soil bacterial community. Our findings are in accordance with others reporting low risk of adverse effects of nonylphenol in environmentally relevant concentrations on soil fungi, but with a stated need for more in-depth studies [34]. This study shows stimulation of specific fungal strains during long-term

Table 4. Organic C mineralization in soil.

Treatment	Organic C mineralization		
	Cumulative CO_2-C	Rate constant	
	(mg CO_2-C/kg dry soil 70d)	(k_{totC}, 10^{-4} day^{-1})	r^2
PI low	109.2±4.1[a]	1.7±0.17[b]	0.82
PI high	108.0±3.5	1.6±.0.17	0.83
NP low	115.3±4.6	1.7±0.18	0.83
NP high	119.6±5.1	1.8±0.17	0.85
Mix low	116.6±4.5	1.8±0.19	0.81
Mix NP high	125.7±5.3	1.9±0.20	0.83
Mix PI high	119.5±6.8	1.8±0.18	0.84
Mix high	116.4±7.9	1.8±0.19	0.81
Solvent control	114.9±3.9	1.8±0.19	0.80
Untreated control	128.5±18.2	1.7±0.19	0.96
Total		1.8±0.05	0.83

[a]SD (standard deviation of measurements),
[b]SE (standard error of estimate).

exposure, as well as appreciable sorption of nonylphenol in soil. Others have reported toxic effects of nonylphenol on soil microbes [4]. However, the general picture is that of rapid and complete mineralization in a wide range of soils [35].

Soil microbial community structure and degradation of chemicals. The observed changes in the T-RFLP patterns of the soil bacterial and fungal communities during the lag-phase before onset of picoxystrobin degradation (Fig. 6a), are in agreement with the general assumption of adaptation of the microbial community to a new substrate. The strobilurin mode of action is, however, typically very fast acting [6], and it may not be ruled out that the observed rapid effects on T-RFLP patterns are, in part, caused by negative effects of the fungicide. We observed a bi-phasic degradation, with initial degradation rate (DT50) in correspondence with previous reports [8]. However, in our studies the levels of picoxystrobin stabilized after 28 days, possibly due to sorption and reduced bioavailability [8,36]. The estimated degradation half-lives (DT50) for picoxystrobin assume that the analysed fraction is the total residual fraction available for degradation. The test for initial rapid degradation (Fig. 7, left) did not indicate a rapid sorption in moist soil, but specific sorption studies were not performed. Picoxystrobin is, however, classified as only slightly mobile [36], and non-bioavailable residues are expected to be above 20% after 100 days [8]. This would explain the lack of a concentration effect beyond 28 days. Alternatively this could be due to a lasting effect on the soil microbial community, as indicated by the lasting significant differences in T-RFLP-fingerprints in picoxystrobin treated soils compared to the solvent control. Similar percentage levels of degradation of picoxystrobin were observed for the two concentrations tested, meaning that actual concentration levels remaining in the treatments with high concentrations of picoxystrobin were much higher after 28 days than for the treatments with low concentrations. There was, however, no observable concentration effect at 70 days between the T-RFLP-profiles.

This observed levelling out of the residual soil concentration of picoxystrobin after 28 days together with the lack of a concentration effect at 70 days, indicated that the observed transience of the effects on the soil microbial community was coupled to decreasing bioavailability of the chemical. Time-dependent sorption might be the reason for reduced degradation and bioavailability of the pesticide after 70 days [37]. Observed major metabolites of picoxystrobin from degradation under aerobic conditions in soil are reported to have mean half-lives in the range of 14–29 days [8]. The chemical analyses performed in our study did not include metabolites, and we may, hence, not rule out any effects of these in our short-term laboratory experiment.

Although 4-n-nonylphenol was found to stimulate soil respiration, the effects on the microbial community structure were less evident. The observed changes in the T-RFLP fingerprints for the soil bacterial community developed more slowly after 4-n-nonylphenol treatment than for picoxystrobin. This indicated a change induced by microbial growth on the chemical, in accordance with the rapid disappearance of 4-n-nonylphenol from the soil through microbial degradation (Fig. 7, right). Residual amounts approached zero both when measured in percentage and in actual amounts. Nonylphenol originates from anthropogenic activity and accumulates and persists in environmental compartments characterized by high organic content, such as sewage sludge [3]. We observed rapid degradation in agricultural soil of comparatively low organic carbon content (quite on the average for large parts of the Norwegian agricultural area). This rapid degradation in an aerobic soil environment is in accordance with previous studies [35,38]. There are, however,

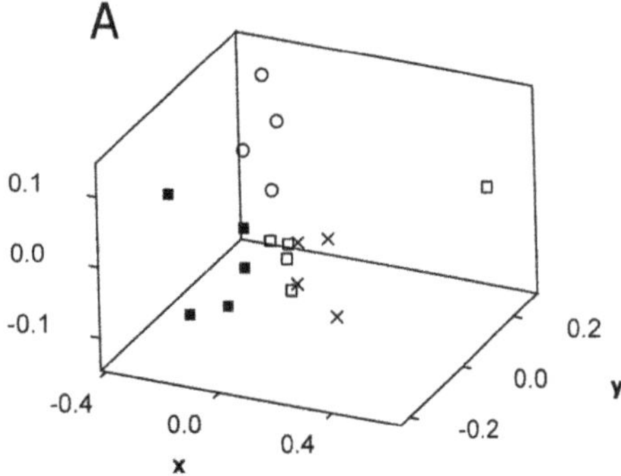

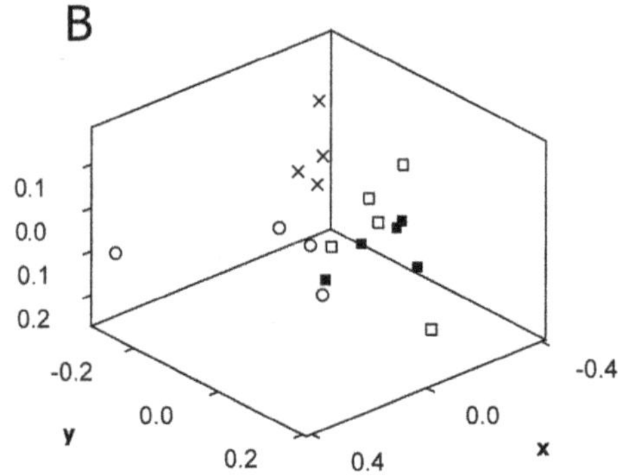

Figure 2. Effects of picoxystrobin on soil fungal community structure. Effects of picoxystrobin (PI) on soil fungal community structure, as shown from nmMDS of data from T-RFLP analyses after 7 (a) and 70 (b) days of incubation. Legends: Treatment PI low (□; 2 mg kg^{-1} dry soil) and PI high (■; 10 mg kg^{-1} dry soil), solvent control (×; no chemical added) and original soil sample (○).

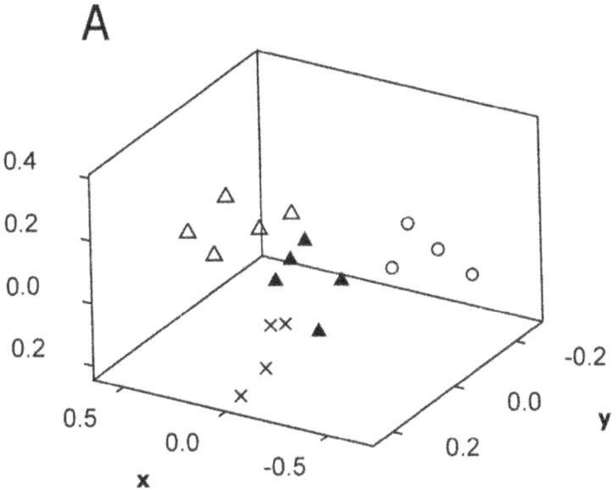

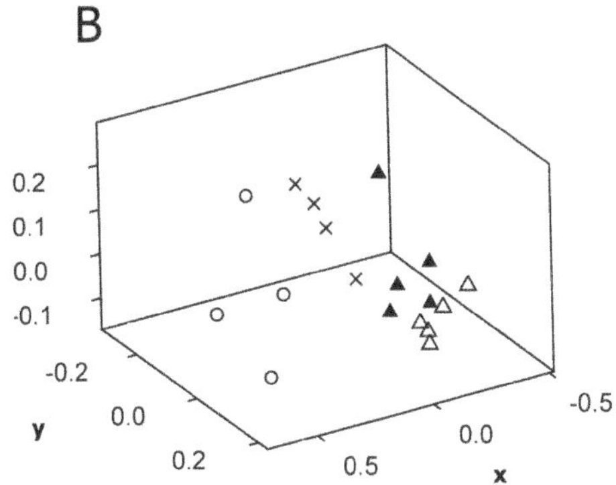

Figure 3. Effects of 4-n-nonylphenol on soil bacterial community structure. Effects of 4-n-nonylphenol (NP) on soil bacterial community structure, as shown from nmMDS of data from T-RFLP analyses after 28 (a) and 70 (b) days of incubation. Legends: Treatment NP low (Δ; 0.5 mg kg^{-1} dry soil) and NP high (▲; 10 mg kg^{-1} dry soil), solvent control (×; no chemical added) and original soil sample (○).

reports of field studies with fast initial degradation, while some residues might persist [4]. The occurrence of metabolites of 4-n-nonylphenol was not measured in this study, but the observed stimulation of respiration activity by 4-n-nonylphenol addition as compared to the solvent control, was not large enough to account for the observed disappearance of the chemical through the incubation period. This indicated that the chemical was not completely mineralized. Results from a mineralization study with a range of soils show the conversion of around 40% of applied nonylphenol to CO_2 during a 40 day incubation period [39].

Our results indicated shorter duration of the effects of the chemicals in the soil bacterial as compared to the soil fungal community. This was expected due to shorter generation times enabling a faster re-establishment of the community structure. Overall, our results were in accordance with this, showing smaller differences in T-RFLP fingerprints between start and end of the experiment for the former (Fig. 5). The more statistically significant effects of the treatments were however shown for bacteria, while the changes in the soil fungal community were less consistent and varied a lot between replicate samples.

Implications of a Mixture Situation on the Effects of Picoxystrobin on the Soil Microbial Community

Analyses of the effects of mixture treatments on the soil microbial community structure indicated that the two chemicals might have antagonistic effects. Established concepts to estimate biological effects of chemical mixtures rely on data available for single chemicals [40,41]. Single chemicals vary in how they excerpt their effect and the effects of a mixture cannot be found directly from the different constituents' independent effects. One can observe a wide variety of effects due to synergism, antagonism or other forms of interactions, and the common concepts of concentration addition and independent action often come short. Despite the expected and indicated differences in modes of action of the two chemicals, we observed less evident changes in T-RFLP patterns in soils treated with the chemicals in mixture, as compared to the single chemical treatments. Our results showed that statistically significant different T-RFLP patterns, for both the soil bacterial and fungal communities, resulted from treatments with picoxystrobin, 4-n-nonylphenol, and a mixture of these. However, the changes in T-RFLP patterns evolved more slowly and were less pronounced for the mixture than for the single chemical treatments, as compared to the solvent control. Further, the negative effects of low levels of picoxystrobin on respiration were apparently remediated when mixing with high concentration

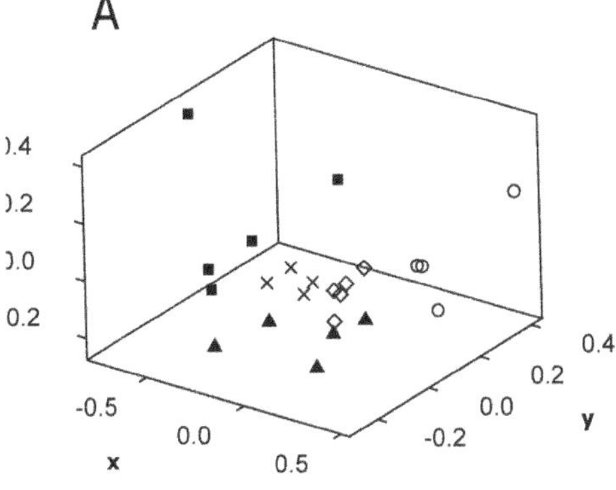

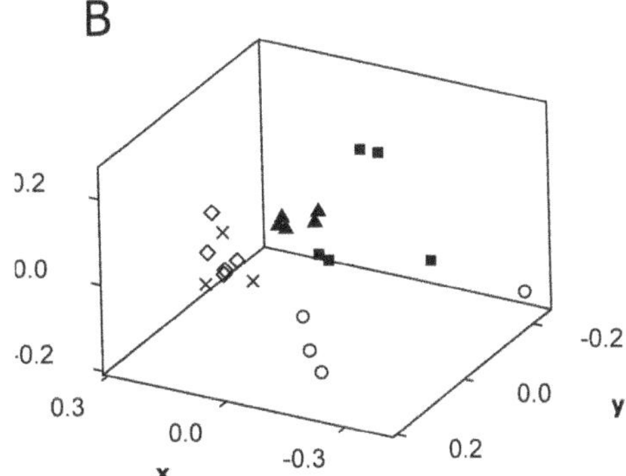

Figure 4. Effects of picoxystrobin and 4-n-nonylphenol on soil microbial community structure. Effects of picoxystrobin (PI) and 4-n-nonylphenol (NP) as single chemicals and in mixture on soil bacterial (a) and fungal (b) community structure, as shown from nmMDS of data from T-RFLP analyses after 28 days of incubation. Legends: Treatment PI high (■; 10 mg kg^{-1} dry soil), NP high (▲; 10 mg kg^{-1} dry soil), Mix high (◇; 10 mg picoxystrobin and 10 mg 4-n-nonylphenol kg^{-1} dry soil), solvent control (×; no chemical added) and original soil sample (○).

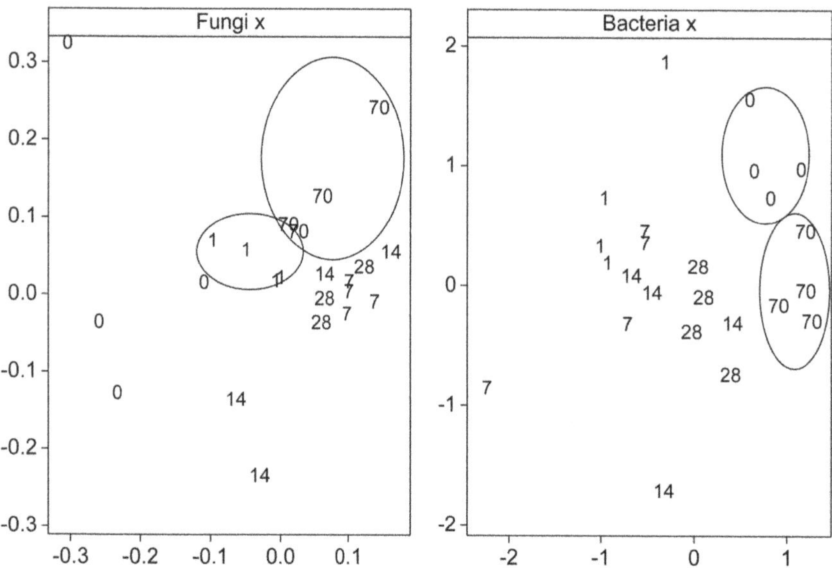

Figure 5. Duration of effects of chemicals on soil microbial community structure. Representative pattern (exemplified by the solvent control treatment) for observed clustering of T-RFLP fingerprints of soil microbial communities at day 0, 1, 7, 14, 28, and 70 after addition of chemicals, shown for fungi (left panel) and bacteria (right panel). Small distance between start (0) and end (70) of the incubation period indicates only minor differences in T-RFLP fingerprints, and imply transient effects of the chemicals.

of 4-n-nonylphenol (Mix NP high), while high levels of picoxystrobin nulled the positive effects of high levels of 4-n-nonylphenol (Mix high). In a practical mixture situation in the field, the expected environmental concentrations of nonylphenol can from this not be expected to increase the potentially adverse effects of picoxystrobin.

To assure the validity in extrapolation of the observed effects to a risk assessment situation in the field, the results need to be verified in field/semi-field/mesocosm trials. These lab studies do not take into account the additional stress the soil microbial community experience in nature from climatic conditions (drought, frost), predators, competition, or food shortage, nor the effects of alteration in the soil microbial community structure on other parts of the soil fauna. The studied mixture is relevant in an agricultural context and the results should be possible to extrapolate to a certain degree to other nonylphenols and strobilurin fungicides. However, care should be taken to consider the variety of physico-chemical and toxicological properties of the different strobilurins [6] as well as the differences in estrogenicity, sorption properties and degradation rates of different isomers of linear and branched nonylphenols [42]. Further, due to the general importance of soil properties for the sorption of chemicals, extrapolation of results from this sandy loam soil with low organic carbon content should be done with care.

Table 5. Estimated degradation half-lives (DT50) for picoxystrobin and 4-n-nonylphenol in a sandy loam soil.

Start concentrations	DT50 initial	r^2	DT50 2nd phase	r^2
2 mg picoxystrobin kg^{-1} dry soil[a]	20.0±2.5	0.87	300±170	0.74
10 mg picoxystrobin kg^{-1} dry soil[a]	26.0±4.0	0.74		
0.5 mg 4-n-nonylphenol kg^{-1} dry soil[b]	3.0±1.0	0.85	42±2	0.58
10 mg 4-n-nonylphenol kg^{-1} dry soil[a]	6.5±0.5	0.92	52±21	0.89

[a]Initial phase 0–28 days, 2nd phase 28–70 days,
[b]Initial phase 0–14 days, 2nd phase 14–70 days.
Calculations are based on results from both single chemical and mixture experiments. (The test statistic of $p<0.05$ is valid for all estimated DT50-values.).

Methodological Considerations

Our results showed that both soil microbial respiration activity and community structure was affected by the chemicals. They also illustrate the need for the use of several measures to be able to assess effects of chemicals in soil with a minimum degree of certainty. We found statistically significant negative effects of picoxystrobin on activity levels as shown by CO_2 development, but no statistically significant differences on cumulative CO_2-values (Table 4) due to too large variability between replicate samples. This was despite observable trends in the data (Fig. 1). Further, the T-RFLP results indicated a need for frequent sampling shortly after addition of chemicals, and possibly the resolution of the sampling was too low to capture the effects on the soil fungal community structure as the largest spread in these data as shown by nmMDS was found after 7 days (Fig. 2). The community fingerprints arising from T-RFLP analyses depend on the PCR-primers and restriction enzymes utilized, and cannot easily be interpreted in terms of ecological relevance of the observed effects. This was ameliorated through correspondence with effects on soil respiration activity. Our studies showed that the T-RFLP technique could be used as a valuable tool in elucidating how rapid and to what extent effects (positive or negative) of the chemicals in the soil microbial community can be expected.

Our results for the fungicide picoxystrobin show that the mere measurement of changes in soil chemical concentrations for this moderately sorbing chemical could not be used as an indication of when to expect effects, while exposure assessment in ecological risk assessment is often restricted to external exposure like concentrations in water, soil or sediment. However it has been recognized that in natural ecosystems, in particular in soils and sediments, the amount of chemicals truly available for uptake into organisms is frequently only a fraction of the total amount present, due to a

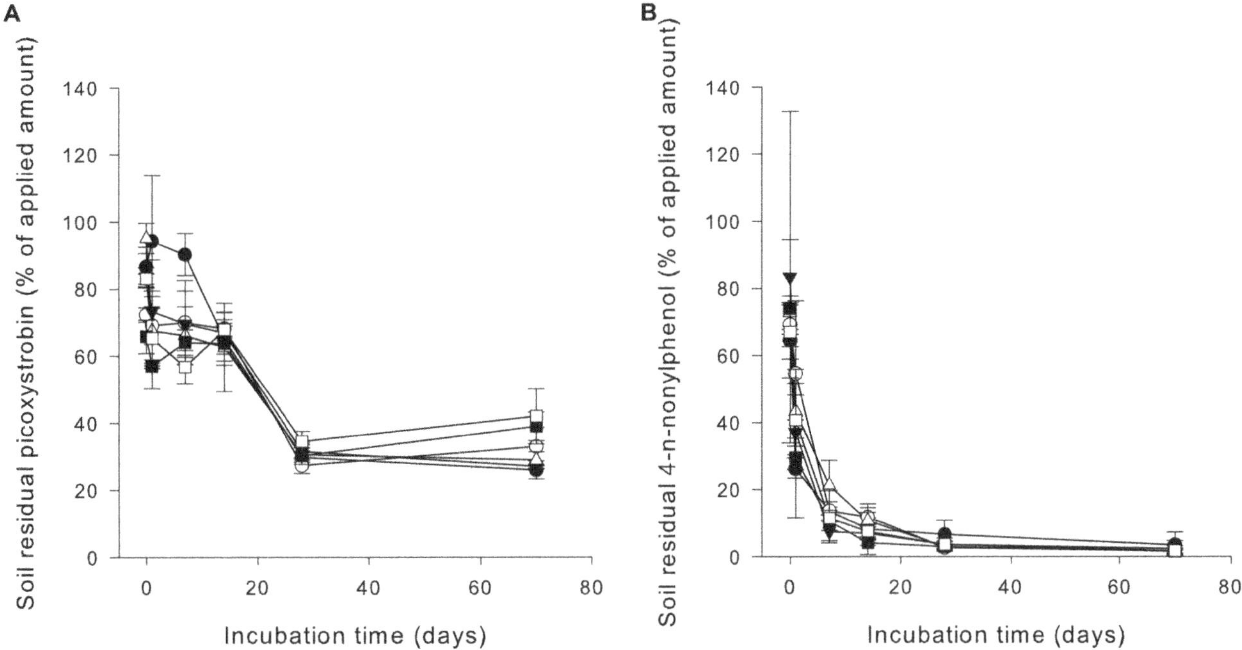

Figure 6. Degradation of picoxystrobin and 4-n-nonylphenol in soil. Degradation curve for picoxystrobin (a) and 4-n-nonylphenol (b) in soil, showing similar pattern and similar percentage degradation during the incubation period for all treatments with picoxystrobin and 4-n-nonylpyhenol, respectively, regardless of initial concentration or mixture situation. Mean values ± SD. Legends: Treatment PI low (a) and NP low (b) (●; PI = 2 mg and NP = 0.5 mg kg^{-1} dry soil), PI high (a) and NP high (b) (○; PI and NP = 10 mg kg^{-1} dry soil), Mix low (▼; PI = 2 mg and NP = 0.5 mg kg^{-1} dry soil), Mix NP high (△;PI = 2 mg and NP = 10 mg kg^{-1} dry soil), Mix PI high (■;PI = 10 mg and NP = 0.5 mg kg^{-1} dry soil), and Mix high (□; PI and NP = 10 mg kg^{-1} dry soil).

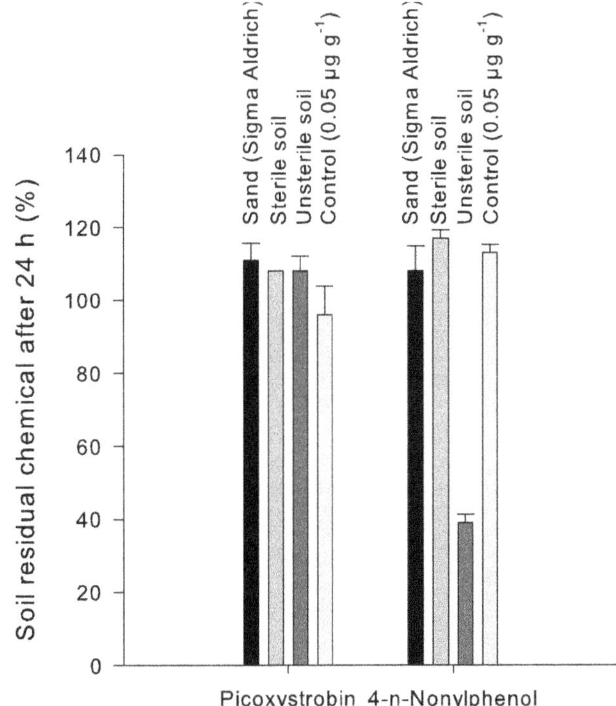

Figure 7. Microbial degradation of picoxystrobin and 4-n-nonylphenol in soil. Laboratory test results for degradation of picoxystrobin and 4-n-nonylphenol in selected media during 24 h, showing rapid degradation of 4-n-nonylphenol in unsterile soil.

complex environmental behaviour [43]. Measurements of residual chemical concentrations in our test system was, however, shown to be a valuable tool for explaining the observed changes in soil microbial community structure and respiration activity and elucidating the concept of a chemicals bioavailability, and should be included in effects assessments.

Conclusions

In summary, we have studied the degradation and effects of picoxystrobin and 4-n-nonylphenol on the soil microbial community structure and respiration activity in a sandy loam soil. The fungicide picoxystrobin was shown to decrease the soil microbial activity, while 4-n-nonylphenol caused an increase. These effects were accompanied by statistically significant changes in the T-RFLP fingerprints for the soil microbial community that were still detectable, but small, after 70 days. A mixture situation relevant for assessing the environmental risk of fungicide application in years with sewage sludge amendment of agricultural soil was tested. These results implied that a mixture of picoxystrobin and 4-n-nonylphenol will not have more adverse effects on soil respiration or the soil microbial community structure than the single chemicals. They indicated that the chemicals affected different parts of the soil microbial community, and resulted in a low net effect of the chemicals in mixture. The presented results do, however, not allow us to propose an explanation for the mechanisms causing this response.

In conclusion, our results imply that the application of picoxystrobin and nonylphenol to agricultural soil in environmentally relevant and worst-case concentrations will not cause irreversible effects on soil microbial respiration and community structure. Further, a mixture situation with the fungicide picoxystrobin and nonylphenol from sewage sludge application, will not increase any potentially adverse effects of picoxystrobin on the soil microbial community. In a wider perspective, these results illustrate that there is a need and possibility to refine today's risk assessment procedures to encompass the study of both soil bacterial and fungal communities, both broad-scale measures and genomics, and interpret the results in relation to residual soil chemical concentrations.

Supporting Information

Figure S1 Effects of picoxystrobin on soil bacterial community structure. Effects of picoxystrobin (PI) on soil bacterial community structure, as shown from nmMDS of data from T-RFLP analyses. Legends: Treatment PI low (□; 2 mg kg^{-1} dry soil) and PI high (■; 10 mg kg^{-1} dry soil), solvent control (X; no chemical added) and original sample (○).

Figure S2 Effects of picoxystrobin on soil fungal community structure. Effects of picoxystrobin (PI) on soil fungal community structure, as shown from nmMDS of data from T-RFLP analyses. Legends: Treatment PI low (□; 2 mg kg^{-1} dry soil) and PI high (■; 10 mg kg^{-1} dry soil), solvent control (X; no chemical added) and original sample (○).

Figure S3 Effects of 4-n-nonylphenol on soil bacterial community structure. Effects of 4-n-nonylphenol (NP) on soil bacterial community structure, as shown from nmMDS of data from T-RFLP analyses. Legends: Treatment NP low (Δ; 0.5 mg kg^{-1} dry soil) and NP high (▲; 10 mg kg^{-1} dry soil), solvent control (X; no chemical added) and original sample (○).

Figure S4 Effects of 4-n-nonylphenol on soil fungal community structure. Effects of 4-n-nonylphenol (NP) on soil fungal community structure, as shown from nmMDS of data from T-RFLP analyses. Legends: Treatment NP low (Δ; 0.5 mg kg^{-1} dry soil) and NP high (▲; 10 mg kg^{-1} dry soil), solvent control (X; no chemical added) and original sample (○).

Figure S5 Effects of picoxystrobin and 4-n-nonylpyhenol on soil bacterial community structure. Effects of picoxystrobin (PI) and 4-n-nonylphenol (NP) on soil bacterial community structure, as shown from nmMDS of data from T-RFLP analyses. Legends: Treatment PI high (■; 10 mg kg^{-1} dry soil), NP high (▲; 10 mg kg^{-1} dry soil), Mix high (◇; 10 mg picoxystrobin and 4-n-nonylphenol kg^{-1} dry soil), solvent control (X; no chemical added) and original sample (○).

Figure S6 Effects of picoxystrobin and 4-n-nonylpyhenol on soil fungal community structure. Effects of picoxystrobin (PI) and 4-n-nonylphenol (NP) on soil fungal community structure, as shown from nmMDS of data from T-RFLP analyses. Legends: Treatment PI high (■; 10 mg kg^{-1} dry soil), NP high (▲; 10 mg kg^{-1} dry soil), Mix high (◇; 10 mg picoxystrobin and 4-n-nonylphenol kg^{-1} dry soil), solvent control (X; no chemical added) and original sample (○).

Table S1 Summary of results from nmMDS (STRESS; k = 3 dimensions, Bray-Curtis distance measure) and ANOSIM (overall R) of T-RFLP fingerprints for the soil microbial community after different chemical treatments.

Acknowledgments

Monica Skogen, Hege Særvold Steen, Grete Lund and Elameen Abdelhameed, Norwegian Institute for Agricultural and Environmental Research, are acknowledged for technical assistance in the lab. Many thanks to Tim Daniell, James Hutton Institute, for advice on the experimental set-up, and analyses of T-RFLP-data. We thank the anonymous reviewers for a thorough review and helpful comments.

Author Contributions

Conceived and designed the experiments: MS SSK OME. Performed the experiments: MS HRN. Analyzed the data: MS HRN. Contributed reagents/materials/analysis tools: MS SSK HRN OME. Wrote the paper: MS SSK HRN OME.

References

1. JRC (2001). Organic contaminants in sewage sludge for agricultural use. Study coordinated by European Commission and Joint Research Centre and the Institute for Environment and Sustainable Soil and Waste Unit. Report prepared by UMEG Center for Environmental Measurements, Environmental Inventories and Product Safety, 18 October 2001. 73 p. Available: http://ec.europa.eu/environment/waste/sludge/pdf/organics_in_sludge.pdf. Accessed 21 May 2013.
2. Norwegian Scientific Committee for Food Safety (2009). Risk assessment of contaminants in sewage sludge applied on Norwegian soils – Opinion from the Panel on Contaminants in the Norwegian Scientific Committee for Food Safety. Oslo: VKM. 208 p. Available: http://www.vkm.no/dav/2ae7f1b4e3.pdf. Accessed 21 May 2013.
3. Hansen AB, Lassen P (2008). Screening of phenolic substances in the Nordic environments. TemaNord 2008:530. Copenhagen: Nordic Council of Ministers. 145 p.
4. Soares A, Guieysse B, Jefferson B, Cartmell E, Lester JN (2008). Nonylphenol in the environment: A critical review on occurrence, fate, toxicity and treatment in wastewaters. Environ Int 34, 1033–1049.
5. EC (2000). Directive 2000/60/EC of the European Parliament and of the Council of 23 October 2000 establishing a framework for Community action in the field of water policy.
6. Bartlett DW, Clough JM, Godwin JR, Hall AA, Hamer M, Parr-Dobrzanski B (2002). The strobilurin fungicides. Pest Manag Sci 58: 649–62.
7. BCPC (2011). ePesticide Manual V5.2 2011–2012.
8. EC (2003). Review report for the active substance picoxystrobin. SANCO/10196/2003-Final. Available: http://ec.europa.eu/food/plant/protection/evaluation/newactive/list1_picoxystrobin_en.pdf. Accessed 21 May 2013.
9. Norwegian Food Safety Autority (2005). Vurdering av plantevernmidlet Acanto Prima – pikoksystrobin+cyprodinil – vedrørende søknad om godkjenning. (In Norwegian). Available: http://www.mattilsynet.no/mattilsynet/multimedia/archive/00014/Plantevernmidler__Ra_14615a.pdf. Accessed 21 May 2013.
10. Coleman DC (2008). From peds to paradoxes: Linkages between soil biota and their influences on ecological processes. Soil Biol Biochem 40: 271–289.
11. Wu L, Wang H, Zhang Z, Lin R, Zhang Z, et al. (2011). Comparative metaproteomic analysis on consequtively *Rehmannia glutinosa*-monocultured rhizosphere soil. PloS ONE 6(5): e20611. Doi:10.1371/journal.pone.0020611.
12. Winding A, Hund-Rinke K, Rutgers M (2005). The use of microorganisms in ecological soil classification and assessment concepts. Ecotox Environ Safe 62: 230–248.
13. EFSA (2010). EFSA Panel on Plant Protection Products and their Residues (PPR); Scientific Opinion on the development of specific protection goal options for environmental risk assessment of pesticides, in particular in relation to the revision of the Guidance Documents on Aquatic and Terrestrial Ecotoxicology (SANCO/3268/2001 and SANCO/10329/2002). EFSA Journal 8: 1821. 55 p. Available: http://www.efsa.europa.eu/en/efsajournal/doc/1821.pdf. Accessed 21 May 2013.
14. Imfeld G, Vuilleumier S (2012). Measuring the effects of pesticides on bacterial communities in soil: A critical review. Eur J Soil Biol 49: 22–30.
15. Øvreås L, Torsvik V (1998). Microbial diversity and community structure in two different agricultural soil communities. Microbial Ecol 36: 303–315.
16. Bending GD, Rodrígues-Cruz MS, Lincoln SD (2007). Fungicide impacts on microbial communities in soils with contrasting management histories. Chemosphere 69: 82–88.
17. Griffiths RI, Thomson BC, James P, Bell T, Bailey M, et al. (2011). The bacterial biogeography of British soils. Environ Microbiol 13: 1642–1654.
18. OECD (2000). Test No. 217: Soil Microorganisms: Carbon Transformation Test, OECD Guidelines for the Testing of Chemicals, Section 2, OECD Publishing. doi: 10.1787/9789264070240-en.

19. OECD (2000). Test No. 216: Soil Microorganisms: Nitrogen Transformation Test, OECD Guidelines for the Testing of Chemicals, Section 2, OECD Publishing. doi: 10.1787/9789264070226-en.
20. MacDonald CA, Clark IM, Zhao F-J, Hirsch PR, Singh BK, et al. (2011). Long term impacts of zinc and copper enriched sewage sludge additions on bacterial, archeal and fungal communities in arable and grassland soils. Soil Biol Biochem 43: 932–941.
21. Schütte UME, Abdo Z, Bent SJ, Shyu C, Williams CJ, et al. (2008). Advances in the use of terminal restriction fragment length polymorphism (T-RFLP) analysis of 16S rRNA genes to characterize microbial communities. Appl Microbiol Biot 80: 365–380.
22. Aiken JT (2011). Terminal restriction fragment length polymorphism for soil microbial community profiling. Soil Sci Soc Am J 75: 102–111.
23. Chaussod R, Nicolardot B, Catroux G (1986). Mesure en routine de la biomasse microbienne des sols par la méthode de fumigation au chloroforme. Science du Sol - Bulletin AFES 2: 201–211.
24. Singh BK, Nunan N, Ridgeway KP, McNicol J, Peter W, et al. (2008). Relationship between assemblages of mycorrhizal fungi and bacteria on grass roots. Environ Microbiol 10: 534–541.
25. Anderson IC, Cairney JWG (2004). Diversity and ecology of soil fungal communities: increased understanding through the application of molecular techniques. Environ Microbiol 6: 769–779.
26. Gardes M, Bruns TD (1993). ITS primers with enhanced specificity for basidiomycetes: application to the identification of mycorrhizae and rusts. Mol Ecol 2: 113–118.
27. White TJ, Bruns T, Lee S, Taylor J (1990). Amplification and direct sequencing of fungal ribosomal RNA genes for phylogenetics. In: Innis MA, Gelfand DH, Sninsky JJ, White TJ, editors. PCR Protocols: a guide to methods and applications. New York: Academic Press. 315–322.
28. R Core Team (2012). R: A language and environment for statistical computing. R Foundation for Statistical Computing, Vienna, Austria. ISBN 3–900051–07–0, URL http://www.R-project.org/. Accessed 28 February 2013.
29. Oksanen J, Blanchet FG, Kindt R, Legendre P, Minchin PR, et al. (2013). vegan: Community Ecology Package. R package version 2.0–6. Available: http://CRAN.R-project.org/package = vegan. Accessed 28 February 2013.
30. Venables WN, Ripley BD (2002). Modern Applied Statistics with S. Fourth Edition. New York: Springer. ISBN 0-387-95457-0.
31. Kruskal JB (1964a), Multidimensional Scaling by Optimizing Goodness of Fit to a Nonmetric Hypothesis. Psychometrika 29: 1–27.
32. Kruskal JB (1964b). Nonmetric multidimensional scaling: a numerical method. Psychometrika 29: 115–129.
33. Hammer Ø, Harper DAT, Ryan PD (2001). PAST: Paleontological statistics software package for education and data analysis. Palaeontol Electron 4(1): art. 4. 9 p. Available: http://palaeo-electronica.org/2001_1/past/issue1_01.htm. Accessed 28 Februart 2013.
34. Kollmann A, Brault A, Touton I, Dubroca J, Chaplain V, et al. (2003). Effect of nonylphenol surfactants on fungi following the application of sewage sludge on agricultural soils. J Environ Qual 32: 1269–1276.
35. Corvini PFX, Chäffer A, Schlosser D (2006). Microbial degradation of nonylphenol and other alkylphenols – our evolving view. Appl Microbiol Biot 72: 223–243.
36. PPDB (2009). The Pesticide Properties Database (PPDB) developed by the Agriculture & Environment Research Unit (AERU), University of Hertfordshire, funded by UK national sources and the EU-funded FOOTPRINT project (FP6-SSP-022704). http://sitem.herts.ac.uk/aeru/footprint/en/index.htm. Accessed 21 May 2013.
37. Defra (2010). Development of guidance on the implementation of aged soil sorption studies into regulatory exposure assessments. Research report for DEFRA project PS 2235. The Food and Environment Research Agency and Alterra.
38. Hesselsøe M, Jensen D, Skals K, Olesen T, Moldrup P, et al. (2001). Degradation of 4-Nonylphenol in homogeneous and nonhomogeneous mixtures of soil and sewage sludge. Environ Sci Technol 35: 3695–3700.
39. Topp E, Starratt A (2000). Rapid mineralization of the endocrine-disrupting chemical 4-nonylphenol in soil. Environ Toxicol Chem 19: 313–318.
40. Loewe S (1927). Die Mischarznei. Versuch einer allgemeinen Pharmakologie der Arzneikombinationen. Klin Wochenschr 6: 1077–1085.
41. Bliss CI (1939). The toxicity of poisons applied jointly. Ann Appl Biol 26: 585–615.
42. Shan J, Jiang B, Yu B, Li C, Sun Y, et al. (2011). Isomer-Specific Degradation of Branched and Linear 4-Nonylphenol Isomers in an Oxic Soil. Environ Sci Technol 45: 8283–8289.
43. Alexander M (1995). How toxic are toxic chemicals in soil? Environ Sci Technol 29: 2713–2717.

12

Field Evidence of Cadmium Phytoavailability Decreased Effectively by Rape Straw and/or Red Mud with Zinc Sulphate in a Cd-Contaminated Calcareous Soil

Bo Li[1,2], Junxing Yang[1,3], Dongpu Wei[1], Shibao Chen[1], Jumei Li[1], Yibing Ma[1]*

1 National Soil Fertility and Fertilizer Effects Long-term Monitoring Network, Institute of Agricultural Resources and Regional Planning, Chinese Academy of Agricultural Sciences, Beijing, P. R. China, **2** Institute of Plant Nutrition and Environmental Resources, Liaoning Academy of Agricultural Sciences, Shenyang, P. R. China, **3** Centre for Environmental Remediation, Institute of Geographic Sciences and Natural Resources Research, Chinese Academy of Sciences, Beijing, P. R. China

Abstract

To reduce Cd phytoavailability in calcareous soils, the effects of soil amendments of red mud, rape straw, and corn straw in combination with zinc fertilization on Cd extractability and phytoavailability to spinach, tomato, Chinese cabbage and radish were investigated in a calcareous soil with added Cd at 1.5 mg kg^{-1}. The results showed that water soluble and exchangeable Cd in soils was significantly decreased by the amendments themselves from 26% to 70%, which resulted in marked decrease by approximately from 34% to 77% in Cd concentration in vegetables. The amendments plus Zn fertilization further decreased the Cd concentration in vegetables. Also cruciferous rape straw was more effective than gramineous corn straw. In all treatments, rape straw plus red mud combined with Zn fertilization was most effective in decreasing Cd phytoavailability in soils, and it is potential to be an efficient and cost-effective measure to ensure food safety for vegetable production in mildly Cd-contaminated calcareous soils.

Editor: Wenju Liang, Chinese Academy of Sciences, China

Funding: The authors thank the Special Fund for Public Industry in China (Agriculture, 200903015) and the Natural Science Foundation of China (41201312 and 41401361) for financial supports. The funders had no role in study design, data collection and analysis, decision to publish, or preparation of the manuscript.

Competing Interests: The authors have declared that no competing interests exist.

* Email: ybma@caas.ac.cn

Introduction

Recently, increasing cadmium (Cd) accumulation in vegetables is a growing concern globally because of increased fertilizer- and biosolids-borne Cd in soils [1–3]. As a consequence, international trade organizations have sought to limit the concentration of Cd in some crops sold in international markets. The National Food Hygiene Standard of China (NFHSC, GB 15201-94) proposed maximum levels of 0.05 mg kg^{-1} of Cd for spinach and other vegetables. However, there are some areas where Cd concentrations in vegetables are over the limit [4,5]. Therefore, reduction of Cd uptake by vegetables and translocation to edible parts is one of the important strategies for proper use of mildly Cd-contaminated soils and safeguarding the safety of farm produce [1].

Nowadays, "in situ" remediation techniques of mildly Cd-contaminated soils are regarded as possible effective approaches to address the issues of excessive vegetable Cd concentrations. During the last decade, the possibility of Cd immobilization in soils through the addition of different amendments or sorbent, has been extensively investigated in order to reduce the risk of groundwater contamination, plant uptake, and exposure to living organisms [6–9]. Among these amendments or sorbents, red mud (RM), a by-product of aluminium (Al) manufacturing, can be very effective in increasing Cd sorption and decreasing soluble Cd concentrations in Cd-contaminated and acidic soils under pot trials [10–13] and field studies [14,15], and lead to a reduction in Cd uptake by plants. Lombi et al. [16] indicated that the specific sorption of Cd by Fe and Al oxides in RM was the main mechanisms of fixation. It is therefore important to select cost-effective and feasible amendments to immobilize Cd by specific sorption. It has been documented that thiol (-SH) can reduce Cd bioavailabilty by mechanisms of chelation [17,18]. The cruciferous rape (Brassica napus L.) exhibited higher concentration of thiol (-SH) in straw than other crops [19,20]. However, the effect of the incorporation of rape straw (RS) into Cd-contaminated soil has never been investigated under field condition. In addition, use of zinc (Zn) fertilizers in soils, such as $ZnSO_4$, has been reported to decrease the accumulation of Cd in crops [21–24]. Abdel-Sabour et al. [21] reported that the Cd/Zn ratio in plant tops was significantly affected by both Cd and Zn concentrations in soil. Yang et al. [23] also found that the application of foliar Zn or seed Zn fertilizer could significantly decrease the Cd concentration in cucumber shoots by about 12–36% in Cd-contaminated soils. Köleli et al. [24] also expressed that Cd toxicity in the shoots of bread and durum wheat was alleviated by Zn treatment.

Therefore, we hypothesized that the Cd immobilizing amendments of RS plus RM in combination with Zn fertilizer might be more effective in reducing Cd accumulation in vegetables grown in Cd-contaminated and calcareous (high pH) soils under field conditions. The present study was conducted to investigate the efficiency of these amendments with Zn fertilization on Cd accumulation in the edible parts of four common vegetables (spinach, tomato, Chinese cabbage and radish) grown in mildly Cd-contaminated and calcareous soil under field condition. Furthermore, the effect of Cd immobilizing amendments (RS, RM, RS+RM) with Zn fertilizer on the Cd fractions associated with different soil components was also studied. The results will be helpful to find practical and cost-effective measures to reduce Cd accumulation in crops.

Materials and Methods

Soil characteristics and amendments used

The field experiment was conducted at long-term experiment station of the Chinese Academy of Agricultural Sciences, Dezhou (DZ) city, Shandong Province, China (37°20′N, 116°29′E). The soil used in the field experiment had a pH (1:5 soil/water suspension) of 8.9 and contained 1.2% organic matter, 6.17% $CaCO_3$, 0.08% total N and 0.1% total P as measured by the standard methods given in Jackson [25]. The soil contained 64% sand, 18% clay and 18% silt. Total Zn and Cd concentrations of the soil were 54 mg Zn kg^{-1} and 0.11 mg Cd kg^{-1}, measured as described by Jackson [25], while DTPA-extractable [26] concentrations of Zn and Cd were 0.11 mg Zn kg^{-1} and <0.005 mg Cd kg^{-1}, respectively.

Red mud, RS, and corn straw (CS) were used as immobilizing amendments in the field experiment. Red mud (pH = 11.1) was from Zibo City, Shandong Province, China. The mineralogical composition of RM sample (XRD analysis) is a mixture of SiO_2 (20%), Fe_2O_3 (28%), Al_2O_3 (21%), CaO (6.2%), MgO (1.3%), TiO_2 (3.3%), K_2O (0.26%) and Na_2O (11%). The specific surface area, determined by the BET/N_2-adsorption method (Sorptomatic CarloErba), was 12.2 m^2 g^{-1} for RM. Zinc and Cd concentrations in RM were 94 mg kg^{-1} and <0.01 mg Cd kg^{-1}, respectively. The RM sample was dried overnight at 105°C, finely ground and sieved to <1 mm. The rape straw sample was obtained from rape (Allium cepa L. cv. Zheshuang No. 6) grown at long-term experiment station of the Chinese Academy of Agricultural Sciences, Jiaxing city, Zhejiang Province, China (30°15′N, 120°20′E), which was oven-dried at 70°C to constant weight and then finely ground in a Retsch-grinder (Type: 1 mm, made in Germany) using a 1 mm mesh screen to ensure uniform plant tissue disruption and distribution in soil during the field experiment. The rape straw with pH of 6.41 and electrical conductivity (EC) of 398 µS cm^{-1} (straw:solution ratio 1:10), contained 19 mg Cu kg^{-1}, 23 mg Zn kg^{-1}, 0.86 mg Pb kg^{-1} and 0.67 mg Cd kg^{-1} dry weight. The corn straw sample was obtained from corn (Zea mays L. cv. Jingdan No. 28) grown at long-term experiment station of the Chinese Academy of Agricultural Sciences, Changping, Beijing, China (40°13′N 116°15′E). The sample was also oven-dried at 70°C to constant weight and then finely ground using a 1 mm mesh. The corn straw with pH of 6.10 and EC of 187 µS cm^{-1}, contained 36 mg Cu kg^{-1}, 59 mg Zn kg^{-1}, 0.76 mg Pb kg^{-1} and 0.84 mg Cd kg^{-1} dry weight.

Field experiment and plant analysis

The field experiment was a randomized complete block split-spot design with 3 replications for the control and amendment treatments (main treatments) and 2 replications for the control and amendment treatments (sub-treatments) in combination with Zn fertilization (12 g $ZnSO_4$ per plot, based on the mass of top 20 cm soil). The size of each plot was 4 m^2 (2 m×2 m). Before amendment addition, soils in the plots were added with 1.5 mg Cd kg^{-1} in the form of $CdSO_4$ on 1 February, 2009. The concentration of Cd added to soil was chosen based on preliminary experiments and represented mildly Cd contamination. To decrease the variability, the salts of $CdSO_4$ were mixed with topsoil samples (0–20 cm) separately in a container, after which the spiked soils were returned to the plots and equilibrating for 2 months. Fertilizers were then applied to all plots according to local farming practices. The equivalent nitrogen (0.2 g N kg^{-1} soil as urea), phosphorus (0.06 g P_2O_5 kg^{-1} soil as superphosphate), potassium (0.06 g K_2O kg^{-1} soil as potassium sulfate) were applied as basal fertilizers to each plot before the spinach and tomato seeding. After spinach and tomato harvest, the same equivalent phosphorus, potassium and nitrogen were applied as basal fertilizers to each plot before Chinese cabbage and radish seeding. All nutrients were mixed homogenously with soil before sowing.

The main treatments were (1) control, (2) 0.5% RM (W/W), (3) 0.1% RS (W/W), (4) 0.5% RM+0.1% RS, (5) 0.1% CS (W/W), (6) 0.5% RM+0.1% CS. The sub-treatments were applied with Zn fertilization (12 g $ZnSO_4$ per plot) before the Cd-contaminated soils in the plots had been added with the amendments mentioned above except CS treatment. The amendments were applied to the surface of each plot before being ploughed into the soil to a depth of 20 cm.

Four commonly cultivated vegetable varieties (Spinach (Spinacia oleracea L. cv. Huabo No. 1), Tomato (Lycopersicum esculentum Mill. cv. Lufen No. 3), Chinese cabbage (Brassica campestris L. cv. Degao No. 16) and Radish (Raphanus sativus Linn. cv. Qianxi No. 2) in DZ were selected in this experiment and were sown directly into the soil according to different growth periods. Spinach was sown on 3 April and harvested on 10 May, 2009. Tomato was sown on 5 May, and harvested on 15 September, 2009. Radish and Chinese cabbage were sown simultaneously on 16 October, 2009 after harvest of spinach and tomato. Sufficient seed was sown to guarantee healthy germination, then seedlings were thinned after germination). Only the edible portions were sampled after all vegetables were grown to maturity as the study was focused on the food safety. At each plot, 10 subsamples of the edible parts of vegetables were collected and combined for chemical analysis. The fresh vegetable samples were put in clean plastic bags and transported to the laboratory for sample treatment. The samples were washed with 0.2% HCl solution followed by tap water and de-ionized water, then oven-dried (not peeled) at 70°C for 6 h to constant weight and dry weights (DW) were recorded. The plant samples were ground using a Retsch-grinder (Type: 0.5 mm, made in Germany), then weighted 0.5 g to 200 mL digestion tubes with 10 ml of concentrated nitric acid (HNO_3) and digested for 9 h at 110°C after standing overnight [27]. Cadmium concentrations were determined using inductively coupled plasma mass spectrometry (ICP-MS). Blank and bush leaf material (BGW-07603) (China Standard Materials Research Center, Beijing, PR China) were used for quality control. The Cd recovery rates were 90±10%.

Soil analysis by sequential extraction procedure

After harvest, 10 subsamples of soils (0–20 cm) were evenly collected from each plot, bulked together, air-dried, and ground to pass a 0.26-mm sieve. Soil pH was measured using de-ionized water (1:5 soil/water suspension) with an ORION combined electrode. The fractions of Cd bound to the soil were determined by a sequential extraction procedure according to Basta and Gradwohl [28], in order to study the effects of the different amendments on Cd fractions. To extract the water soluble fraction (WS-Cd), each sample collected from plots (2 g) was treated with 25 mL of de-ionized water (pH 6.5) and shaken for 2 h at room temperature. It was then treated with 25 mL of 0.1 N $Ca(NO_3)_2$ solution to extract the exchangeable fraction (Exch-Cd), and with 25 mL 0.02 M EDTA solution to extract the complexed fraction

Table 1. The concentration of Cd (C_{Cd}, mg Cd kg^{-1} in dry weight) and yield (g $plant^{-1}$ in dry weight) for edible part of spinach, tomato, Chinese cabbage and radish in Cd-contaminated soils (added Cd at 1.5 mg kg^{-1}) with different amendments with (+Zn) and without Zn fertilization.

Treatment	Spinach		Tomato		Chinese cabbage		Radish	
	C_{Cd} (mg kg^{-1})	Yield (g $plant^{-1}$)	C_{Cd} (mg kg^{-1})	Yield (g $plant^{-1}$)	C_{Cd} (mg kg^{-1})	Yield (g $plant^{-1}$)	C_{Cd} (mg kg^{-1})	Yield (g $plant^{-1}$)
CK	0.75±0.05 a	0.96±0.03 c	0.35±0.03 a	4.31±0.03 d	0.49±0.03 a	161±1 h	0.56±0.02 a	94±1 d
CK+Zn	0.53±0.03 b	1.11±0.02 b	0.31±0.03 b	4.58±0.06 bc	0.35±0.02 b	164±2 h	0.45±0.02 b	102±1 c
RM	0.40±0.04 c	1.14±0.03 b	0.18±0.02 d	4.61±0.03 bc	0.19±0.02 c	175±1 f	0.19±0.01 d	115±1 b
RM+Zn	0.29±0.03 d	1.21±0.04 ab	0.16±0.02 de	4.69±0.06 b	0.16±0.01 d	191±2 c	0.16±0.01 e	117±1 b
RS	0.41±0.05 c	0.91±0.03 c	0.14±0.02 e	4.51±0.04 c	0.16±0.01 d	179±2 e	0.17±0.01 de	103±1 c
RS+Zn	0.36±0.03 c	0.93±0.04 c	0.13±0.01 e	4.77±0.07 ab	0.12±0.01 e	185±2 d	0.16±0.02 e	108±3 c
RM+RS	0.18±0.02 e	1.25±0.02 a	0.12±0.02 e	4.83±0.05 a	0.12±0.01 e	192±3 c	0.13±0.01 f	117±2 b
RM+RS+Zn	0.12±0.02 f	1.31±0.03 a	0.09±0.02 f	4.89±0.07 a	0.09±0.01 f	201±3 a	0.10±0.01 g	123±1 a
CS	0.47±0.05 bc	0.99±0.03 c	0.23±0.02 c	4.48±0.04 c	0.20±0.02 c	171±2 g	0.22±0.01 c	95±1 d
RM+CS	0.37±0.04 c	1.25±0.04 a	0.16±0.01 de	4.69±0.05 b	0.18±0.01 cd	189±1 c	0.18±0.01 de	113±1 b
RM+CS+Zn	0.26±0.04 d	1.27±0.03 a	0.13±0.02 e	4.62±0.09 bc	0.15±0.01 d	196±1 b	0.15±0.02 e	120±1 a

The application rates in soils were at 0.5% (W/W) for red mud (RM), 0.1% (W/W) for rape straw (RS) and corn straw (CS), and 3 g $ZnSO_4$ per square meter, respectively.
Note: Within each column in the same vegetable, mean values ± standard errors with the same letter do not differ significantly at 5% level ($P<0.05$) according to the Fisher's least significant test.

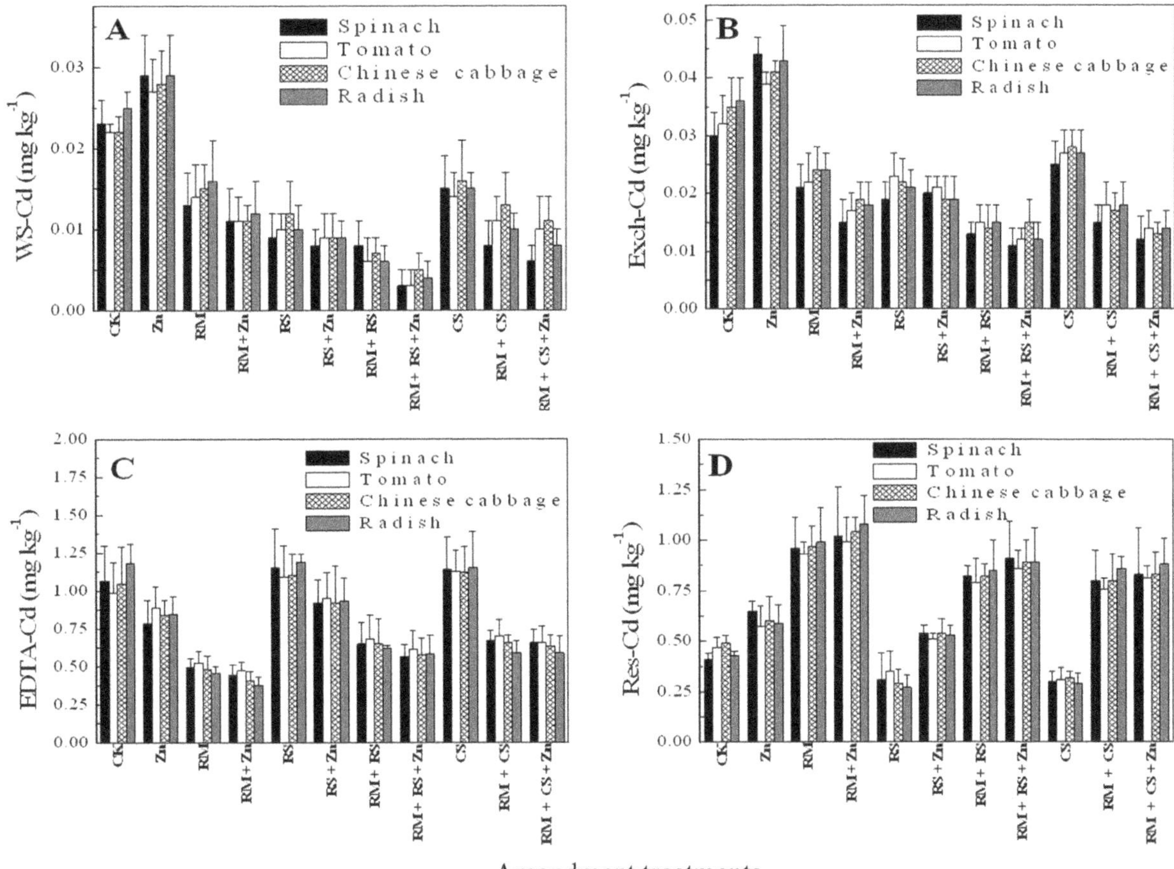

Figure 1. Concentrations (mg Cd kg^{-1} soil) of water soluble (WS-Cd, A), exchangeable (Exch-Cd, B), EDTA extractable (EDTA-Cd, C) and residual Cd (Res-Cd, D) in the soils with and without different amendments after spinach, tomato, Chinese cabbage and radish cultivation.

(EDTA-Cd). After each step of the extraction process the samples were centrifuged at 10,000 rpm for 0.5 h and filtered to separate the liquid and solid phases. After the third extraction, the residual form of Cd (Res-Cd) was determined by drying the solid phase overnight at 105°C and digesting it with HNO_3 and HCl (ratio 1:3) in a Microwave Milestone MLS 1200. The Cd concentrations in each extract or digest were determined using inductively coupled plasma mass spectrometry (ICP-MS).

Data analysis

All results were presented as arithmetic means with standard errors and analyzed by SPSS 11.0 statistical package. Statistical comparisons of means of plant data were analyzed with one way ANOVA followed by the Fisher's least significant test. Correlation coefficient analyses were conducted using program of Origin 7.0.

Results

Effects on Cd in vegetables

The concentrations of Cd and yield for edible parts of spinach, tomato, Chinese cabbage and radish with both the unamended and amended soil combined with Zn fertilization are presented in Table 1. Compared with the vegetable grown in unamended soil, the concentrations of Cd in the edible parts of the four vegetables were reduced with amendment treatments, and the reduction (% of control) was significantly different ($P<0.05$) among the different treatments. The reduction of Cd in vegetables ranged from 37% to 76% for spinach, and from 34% to 63% for tomato, and from 59% to 76% for Chinese cabbage, and from 61% to 77% for radish, with the lowest for CS treatment and the highest for RS+ RM treatment. Although the yield of edible parts of spinach, tomato, Chinese cabbage and radish treated with amendments were increased by 3–36%, 4–13%, 6–25% and 1–31%, respectively (Table 1), the total Cd uptake in edible parts of the four vegetables was still significantly decreased with amendment treatments (data can be calculated using the Cd concentration multiplying by yield of vegetables in Table 1) as there was "dilution effect" of Cd in edible parts of plant. Combined with amendment treatments, Zn application further decreased the Cd concentration in vegetables up to 74–84% of those in unamended treatment.

Changes in Cd fractions in soils

The concentrations of Cd in different fractions of soils with different treatments after plant harvest were shown in Figure 1. The proportion of Cd in the control (total Cd 1.5 mg kg^{-1}) soil were 1.46% (0.023 mg kg^{-1}) in the WS-Cd, 2.11% in the Exch-Cd, 67.9% in the EDTA-Cd, and 28.5% in the Res-Cd. The concentrations of WS-Cd and Exch-Cd were significantly lower ($P<0.05$) in the amended soil than in the unamended soil except Zn treatment (Fig. 1). In general, addition of amendments combined with Zn fertilization to the soil can significantly decrease

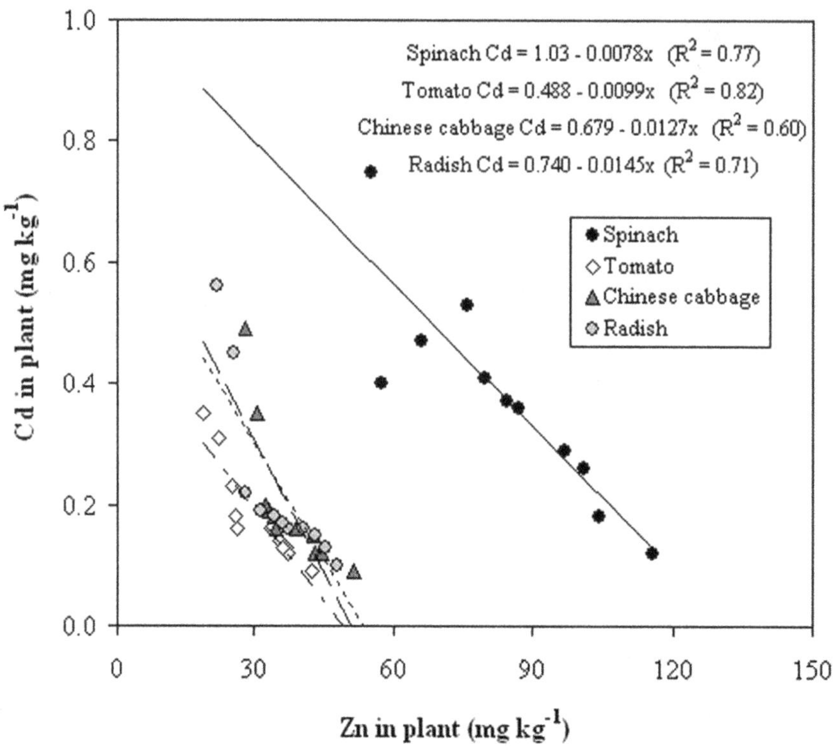

Figure 2. Concentration of Cd in edible parts of spinach, tomato, Chinese cabbage and radish as a function of the Zn concentration in these plants in Cd-contaminated soils (1.5 mg added Cd kg^{-1} soil) with different amendments with and without Zn fertilization.

WS-Cd from 34% (CS) to 84% (RS+RM+Zn) among the four plants. The addition of the RM, RM+CS and RM+RS combined with Zn fertilization also remarkably increased the Res-Cd. As for RS and CS treatments, the EDTA-Cd increased about 5%. Among the fractions, WS-Cd and Exch-Cd were more pronouncedly affected by the treatment of amendments than the other fractions, which suggested that WS-Cd and Exch-Cd were transformed to the non-extractable form in the amended soil.

Discussion

Accumulative evidence from pot trials [27,29] and field samples [4,14,15,30] clearly showed that vegetables grown on Cd-contaminated soils results in elevated Cd levels in edible parts of the vegetables, exceeding the maximum allowable limit (0.05 mg kg^{-1} fresh weight) of NFHSC. Our results showed that Cd concentrations of the edible of parts of the spinach, tomato, Chinese cabbage and radish grown in the unamended soil (1.5 mg kg^{-1} Cd exposure) was 0.75 mg kg^{-1} (0.06 mg kg^{-1} fresh weight), 0.35 mg kg^{-1} (0.06 mg kg^{-1} fresh weight), 0.49 mg kg^{-1} (0.053 mg kg^{-1} fresh weight) and 0.56 mg kg^{-1} (0.08 mg kg^{-1} fresh weight), respectively, being 1.06–1.60 fold as high as the NFHSC value (Table 1). When different amendments were added to soil, the Cd concentrations in edible parts of the four vegetables were almost decreased to the values of $<$0.05 mg kg^{-1} fresh weight. The present results also indicated that distinctive differences in Cd accumulation when comparing one vegetable to another, following the order: spinach (leafy vegetables, Chenopodiaceae) $>$ radish (root vegetables, Cruciferae) $>$ Chinese cabbage (kale vegetables, Cruciferae) $>$ tomato (fruit vegetables, Solanaceae), which is similar with the result in soil with Cd added less than 2 mg kg^{-1} from the study of Yang et al. [31]. The differences in Cd accumulation are probably because the soil-to-plant transfer factor of Cd (TF = M(edible part)/M(soil), M is the Cd concentration in edible part or soil) for the leafy vegetables were higher than those for the non-leafy vegetables [27,30]. When grown in the unamended soil (1.5 mg kg^{-1} Cd exposure), the TF of edible parts of spinach, radish, Chinese cabbage and tomato was 0.50, 0.37, 0.33 and 0.23, respectively. When applied with the amendments and/or Zn fertilization, the decline in TF of the four vegetables treated with the amendments was significantly different, ranging from 0.08 (RS+RM+Zn) to 0.35 (CK+Zn) for spinach, from 0.07 (RS+RM+Zn) to 0.30 (CK+Zn) for radish, from 0.06 (RS+RM+Zn) to 0.23 (CK+Zn) for Chinese cabbage, and from 0.08 (RS+RM+Zn) to 0.21 (CK+Zn) for tomato. Although all amendments played an important role of decreasing Cd transfer from soil to plant, however, in comparison with other three vegetables, the transfer ability of spinach for Cd were always stronger than others (Table 1). Not only for Cd, spinach could also accumulate other heavy metals (e.g., Ni, Pb and Cu) intensely [31–33], for instance, nickel level in spinach shoot was found to be 1.5–4.9 fold as high as those in other six plant species (including tomato and cabbage) in high-pH soil [33]. These results might be related to the characteristic of leafy vegetables, easy uptake/translocation of heavy metals from soil to shoots by passive uptake - transpiration based on bigger surface area of plant leaves and stomatal aperture [34]. Generally, accumulation for heavy metals of different parts of plant is in the order of root $>$ shoot $>$ fruit, which might be the reason that radish (root vegetables) could accumulate more Cd than Chinese cabbage (kale vegetables) and tomato (fruit vegetables).

Previous studies showed that addition of red mud [10–12,14,15,35] and plant materials [36,37] to Cd-contaminated and acidic soils could effectively reduce Cd bioavailability in soils. However, little information was available about the addition of red mud and/or plant materials to mildly Cd-contaminated and calcareous soils (high pH) under field conditions. Results from the present study showed that Cd concentration and uptake in edible parts of the four vegetables treated with RM, RS and CS under mildly Cd-contaminated and calcareous soils was significantly ($P < 0.05$) reduced. The reason might be because the RM, RS and CS application markedly reduced Cd mobility in the Cd-contaminated soil. As shown in Fig. 1, the WS-Cd and Exch-Cd in the amended soils averagely decreased from 26% (CS) to 70% (RS+RM) except Zn treatment, while the Res-Cd averagely increased from 35% (RM+CS) to 108% (RM) among RM, RM+CS and RM+RS treatment, and the EDTA-Cd averagely increased at 5% for RS and CS treatments, suggesting that the addition of RM could lead to Cd transformation from WS-Cd and Exch-Cd to Res-Cd, while the addition of CS and RS could lead to Cd transformation from WS-Cd and Exch-Cd to EDTA-Cd. Furthermore, among the amendments RS+RM was the most effective, with the greatest reductions in WS-Cd (70%) and Exch-Cd (57%). Yang et al. [38] also showed that RS and nano-treated RM were the two best amendments in decreasing Exch-Cd in alkaline soil and total Cd in cucumber plants. The reason might be ascribed mainly to different mechanisms of bindings of Cd between these amendments. In the present study, th e application of RM (0.5%, W/W), RS (0.1%, W/W) and CS (0.1%, W/W) to the calcareous DZ soil (pH 8.9) had no obvious increase on soil pH (data not shown). The transformation of WS-Cd and Exch-Cd to Res-Cd was probably due to the specific sorption of Cd by Fe and Al oxides in RM, while the transformation of WS-Cd and Exch-Cd to EDTA-Cd was probably due to the complexation with RS or CS. Luo et al. [39] further investigated the sorption mechanism of cadmium on red mud as same as used in the study using batch sorption experiments, sequential extraction analysis and X-ray absorption near edge structure (XANES) spectroscopy and supplied evidence of the formation of inner-sphere complexes of Cd similar to XCdOH (X represents surface groups on red mud) on the red mud surfaces although outer-sphere complexes of Cd were the primary species.

Some studies revealed that plant materials could be sorbent materials for Cd due to the tendency of Cd to form stable complexes with organic ligands [36,40]. Wu et al. [41] found that the Cd concentrations in grains of rice by rotation with rape were decreased approximately by 46–80% of those for rice cultivation only, the decreasing of WS-Cd plus Exch-Cd and increasing of Org-Cd was might be related to the abundance of organic material secreted from rape roots in soil. Harada et al. [17] also found that Cd stress could result in a 3-fold increase in total thiols mainly contributing to synthesis of cysteine, glutathione and phytochelatins in *Arabidopsis*. In the present study, an obvious decrease of WS-Cd in soil was displayed with the addition of RS (Fig. 1), which might be ascribed to the high affinity for Cd induced by sulfur compounds (thiol) in rape straw. However, for CS, a relative lower affinity for Cd could be resulted from (semi)cellulose as main components in straw [42].

Results from the present study indicated that Cd concentrations in the edible parts of vegetables were significantly lower for amendment treatments with Zn fertilization than those with amendments only (Table 1). Furthermore, the results (Fig. 2) also showed that there were significant negative correlations between concentrations of Cd and Zn in edible parts of vegetables with $R^2 \geq 0.60$ ($P < 0.01$, $n = 11$), and supplied the evidence of Zn antagonistic effect on Cd uptake by plants in the calcareous soils with amendments. Oliver et al. [22] also found Zn fertilization markedly reduced Cd concentration in wheat grain in areas where it was marginal to severe Zn deficiency in South Australia. These results suggested that Zn fertilization could be a practical measure combined with soil amendments to decrease Cd concentration in plants in soils where the Zn availability is low.

Recently, there is evidence that RM combined with other amendment, such as gravel sludge (GS), a waste product of the gravel industry, had higher long-term efficiency in immobilizing Cd than only RM or GS treatment under field condition [15]. Similarly, our results provided clearly evidence for a synergistic interaction between RM and RS leading to highly significant ($P < 0.01$) reductions in the Cd accumulation of edible parts of the four vegetables. Considering all of the above-mentioned facts, the RS+RM is suggested to act as an efficient, economic and practical measure for mildly Cd-contaminated and calcareous soil, best in combination with Zn fertilization if soils with low Zn availability.

Conclusions

This study clearly demonstrated that water soluble and exchangeable Cd in soils was significantly decreased by red mud, rape and corn straw, which resulted in significant decrease by about 34% to 77% in Cd concentration in vegetables in Cd-contaminated and calcareous soils. Combined with the amendments, Zn fertilization further decreased Cd concentration in the edible part of vegetables up to 74% to 84%. The effect of rape and corn straw could be ascribed to formation of stable complexes with organic ligands while red mud to specific sorption of Cd. Also cruciferous rape straw was more effective than gramineous corn straw. In all treatments, rape straw plus red mud combined with Zn fertilization was most effective in decreasing Cd phytoavailability in soils, and it is potential to be an efficient and cost-effective measure to ensure food safety for vegetable production in mildly Cd-contaminated and calcareous soils.

Acknowledgments

We thank Yubao Zuo for field management of the experiment.

Author Contributions

Conceived and designed the experiments: YM BL. Performed the experiments: BL JY DW. Analyzed the data: BL JY SC JL. Contributed reagents/materials/analysis tools: DW SC. Contributed to the writing of the manuscript: BL YM JY.

References

1. McLaughlin MJ, Parker DR, Clarke JM (1999) Metals and micronutrients-food safe issues. Field Crop Res 60: 143–163.
2. Satarug S, Baker JR, Urbenjapol S, Haswell-Elkin M, Reilly PEB, et al. (2003) A global perspective on cadmium pollution and toxicity in non-occupationally exposed population. Toxicol Lett 137: 65–83.
3. Hooda PS (2010) Trace Elements in Soils. New York: Wiley-Blackwell.
4. Demirezen D, Aksoy A (2006) Heavy metal levels in vegetables in Turkey are within safe limits for Cu, Zn, Ni and exceeded for Cd and Pb. J Food Quality 29: 252–265.
5. Li J, Xie ZM, Xu JM, Sun YF (2006) Risk assessment for safety of soils and vegetables around a lead/zinc mine. Environ Geochem Hlth 28: 37–44.
6. Castaldi P, Santona L, Melis P (2005) Heavy metals immobilization by chemical amendments in a polluted soil and influence on white lupin growth. Chemosphere 60: 365–371.
7. Castaldi P, Melis P, Silvetti M, Deiana P, Garau G (2009) Influence of pea and wheat growth on Pb, Cd, and Zn mobility and soil biological status in a polluted amended soil. Geoderma 151: 241–248.
8. Tandy S, Healey JR, Nason MA, Williamson JC, Jones DL (2009) Remediation of metal polluted mine soil with compost: co-composting versus incorporation. Environ Pollut 157: 690–697.
9. Liu YJ, Naidu R, Ming H (2011) Red mud as an amendment for pollutants in solid and liquid phases. Geoderma 163: 1–12.

10. Lombi E, Zhao FJ, Wieshammer G, Zhang G, McGrath SP (2002) In situ fixation of metals in soils using bauxite residue: biological effects. Environ Pollut 118: 445–452.
11. Lombi E, Zhao FJ, Zhang GY, Sun B, Fitz W, et al. (2002) In situ fixation of metals in soils using bauxite residue: chemical assessment. Environ Pollut 118: 135–443.
12. Friesl W, Lombi E, Horak O, Wenzel W (2003) Immobilization of heavy metals in soils using inorganic amendments in a greenhouse study. J Plant Nutr Soil Sc 166: 191–196.
13. Lee SH, Kim EY, Park H, Yun J, Kim JG (2011) In situ stabilization of arsenic and metal-contaminated agricultural soil using industrial by products. Geoderma 161: 1–7.
14. Friesl W, Friedl J, Platzer K, Horak O, Gerzabek MH (2006) Remediation of contaminated agricultural soils near a former Pb/Zn smelter in Austria: Batch, pot and field experiments. Environ Pollut 144: 40–50.
15. Friesl W, Platzer K, Horak O, Gerzabek MH (2009) Immobilising of Cd, Pb, and Zn contaminated arable soils close to a former Pb/Zn smelter: a field study in Austria over 5 years. Environ Geochem Hlth 31: 581–594.
16. Lombi E, Hamon RE, McGrath SP, McLaughlin MJ (2003) Lability of Cd, Cu, and Zn in polluted soils treated with lime, beringite, and red mud and identification of a non-labile colloidal fraction of metals using isotopic techniques. Environ Sci Technol 37: 979–984.
17. Harada E, Yamaguchi Y, Koizumi N, Hiroshi S (2002) Cadmium stress induces production of thiol compounds and transcripts for enzymes involved in sulfur assimilation pathways in Arabidopsis. J Plant Physiol 159: 445–448.
18. Herbette S, Taconnat L, Hugouvieux V, Piette L, Magniette MLM (2006) Genome-wide transcriptome profiling of the early cadmium response of Arabidopsis roots and shoots. Biochimie 88: 1751–1765.
19. Jones MG, Hughes J, Tregova A, Milne J, Tomsett AB, et al. (2004) Biosynthesis of the flavour precursors of onion and garlic. J Exp Bot 55: 1903–1918.
20. Wang LQ. Remediation of Cd-contaminated soils by in situ immobilization techniques. Dissertation for the Doctoral Degree. Beijing: Capital Normal University, 2009 (in Chinese).
21. Abdel-Sabour MF, Mortvedt JJ, Kelsoe JJ (1988) Cadmium-zinc interactions in plants and extractable cadmium and zinc fractions in soil. Soil Sci 145: 424–431.
22. Oliver DP, Hannam R, Tiller KG, Wilhelm NS, Merry RH, et al. (1994) The effects of zinc fertilization on cadmium concentration in wheat grain. J Environ Qual 23: 705–711.
23. Yang JX, Wang LQ, Wei DP, Chen SB, Ma YB (2011) Foliar spraying and seed soaking of zinc fertilizers decreased cadmium accumulation in cucumbers grown in Cd-contaminated soils. Soil Sediment Contam 20: 400–410.
24. Köleli N, Eker S, Cakmak I (2004) Effect of zinc fertilization on cadmium toxicity in durum and bread wheat grown in zinc-deficient soil. Environ Pollut 131: 453–459.
25. Jackson ML (1958) Soil Chemical Analysis (2nd ed.). Baton Rouge: CRC Press.
26. Lindsay WL, Norvell WA (1978) Development of a DTPA soil test for zinc, iron, manganese and copper. Soil Sci Soc Am J 42: 421–428.
27. Alexander PD, Alloway BJ, Dourado AM (2006) Genotypic variations in the accumulation of Cd, Cu, Pb and Zn exhibited by six common grown vegetables. Environ Pollut 144: 736–745.
28. Basta NT, Cradwohl M (2000) Estimation of Cd, Pb, and Zn bioavailability in smelter-contaminated soils by a sequential extraction procedure. J Soil Contam 9: 149–164.
29. Kuboi T, Noguchi A, Yazaki J (1986) Family-dependent cadmium accumulation in higher plants. Plant Soil 92: 405–415.
30. Wang G, Su MY, Chen YH, Lin FF, Luo D, et al. (2006) Transfer characteristics of cadmium and lead from soil to the edible parts of six vegetable species in southeastern China. Environ Pollut 144: 127–135.
31. Yang JX, Guo HT, Ma YB, Wang LQ, Wei DP, et al. (2010) Genotypic variations in the accumulation of Cd exhibited by different vegetables. J Environ Sci 22: 1246–1252.
32. Li B (2010) The phytotoxicity of added copper and nickel to soils and predictive models. Ph D dissertation. Chinese Academy of Agricultural Sciences, Beijing, China.
33. Giordani C, Cecchi S, Zanchi C (2005) Phytoremediation of soil polluted by nickel using agricultural crops. Environ Manage 36: 675–681.
34. Marchiol L, Sacco P, Assolari S, Zerbi G (2004) Reclamation of polluted soil: phytoremediation potential of crop-related *Brassica* species. Water Air Soil Pollut 158: 345–356.
35. Garau G, Silvetti M, Deiana S, Deiana P, Castaldi P (2011) Long-term influence of red mud on As mobility and soil physico-chemical and microbial parameters in a polluted sub-acidic soil. J Hazard Mater 185: 1241–1248.
36. Cui YS, Du X, Weng LP, Zhu YG (2008) Effect of rice straw on the speciation of cadmium (Cd) and copper (Cu) in soils. Geoderma 146: 370–377.
37. Tlustoš P, Vostál J, Száková J, Balík J (1995) Direct and subsequent efficiency of selected measures on the Cd and Zn content in the biomass of spinach. Rostlinná Výroba 41: 31–37 (in Czech).
38. Yang JX, Wang LQ, Li JM, Wei DP, Chen SB, et al. (2014) Effects of rape straw and red mud on extractability and bioavailability of cadmium in a calcareous soil. Front Environ Sci Eng (in press).
39. Luo L, Ma CY, Ma YB, Zhang SZ, Lv JT, et al. (2011) New insights into the sorption mechanism of cadmium on red mud. Environ Pollut 159: 1108–1113.
40. Almas A, Singh BR, Salbu B (1999) Mobility of cadmium-109 and zinc-65 in soil influenced by equilibration time, temperature, and organic matter. J Environ Qual 28: 1742–1750.
41. Wu FL, Lin DY, Su DC (2011) The effect of planting oilseed rape and compost application on heavy metal forms in soil and Cd and Pb uptake in rice. Agr Sci China 10: 267–274.
42. Chen SB, Sun C, Wei W, Lin L, Wang M (2012) Difference in cell wall components of roots and its effect on the transfer factor of Zn by plant species. China Environ Sci 32: 1309–1313 (in Chinese).

13

Mechanisms Controlling Arsenic Uptake in Rice Grown in Mining Impacted Regions in South China

Junhui Li[1], Fei Dong[1,2], Ying Lu[1]*, Qiuyan Yan[3], Hojae Shim[4]*

1 College of Natural Resources and Environment, South China Agricultural University, Guangzhou, China, 2 Agricultural Bureau of Xiangfen County, Shanxi Province, Xiangfen, China, 3 Institute of Wheat Research, Shanxi Academy of Agricultural Sciences, Linfen, China, 4 Department of Civil and Environmental Engineering, Faculty of Science and Technology, University of Macau, Macau SAR, China

Abstract

Foods produced on soils impacted by Pb-Zn mining activities are a potential health risk due to plant uptake of the arsenic (As) associated with such mining. A field survey was undertaken in two Pb-Zn mining-impacted paddy fields in Guangdong Province, China to assess As accumulation and translocation, as well as other factors influencing As in twelve commonly grown rice cultivars. The results showed that grain As concentrations in all the surveyed rice failed national food standards, irrespective of As speciation. Among the 12 rice cultivars, "SY-89" and "DY-162" had the least As in rice grain. No significant difference for As concentration in grain was observed between the rice grown in the two areas that differed significantly for soil As levels, suggesting that the amount of As contamination in the soil is not necessarily the overriding factor controlling the As content in the rice grain. The iron and manganese plaque on the root surface curtailed As accumulation in rice roots. Based on our results, the accumulation of As within rice plants was strongly associated with such soil properties such as silicon, phosphorus, organic matter, pH, and clay content. Understanding the factors and mechanisms controlling As uptake is important to develop mitigation measures that can reduce the amount of As accumulated in rice grains produced on contaminated soils.

Editor: Ivan Baxter, United States Department of Agriculture, Agricultural Research Service, United States of America

Funding: The authors have no support or funding to report.

Competing Interests: The authors have declared that no competing interests exist.

* Email: luying@scau.edu.cn (YL); hjshim@umac.mo (HS)

These authors contributed equally to this work.

Introduction

Arsenic (As) is a carcinogenic metalloid ubiquitous in the environment, and is obtained from natural and anthropogenic sources [1,2]. Anthropogenic activities such as metal mining and smelting, the use of As-containing pesticides, herbicides, wood preservatives, feed additives, and irrigation with As-rich groundwater, have resulted in elevated As levels in soil [1,3]. The transfer of As in soil-plant systems represents one of the principal pathways for human exposure to As [4]. A recent cohort study [5] showed that daily consumption of 500 g cooked rice containing As content above 200 μg/kg can give rise to genotoxic effects in humans.

Rice is the staff of life for 3 billion people, predominantly in Asia [6], contributing over 70% of the energy and 50% of the protein provided by their daily food intake [7]. China is the world's top rice producer, producing 36.9% of the world's rice yield on 22.8% of world rice cropping area [8], and a top rice consumer with more than 60% of the Chinese population relying on rice as a dietary staple [2]. Unfortunately, among grain crops, rice is particularly efficient in As accumulation as it is generally cultivated in flooded paddy fields where As is more soluble and available to plant uptake [9,10].

Some studies have revealed that As concentrations in rice grains were associated with the As concentrations in irrigated groundwater and/or soil [11–13], although rice grain can accumulate relatively large amounts of As even from soils not contaminated by As [14]. Others have shown that As in rice does not directly depend on total As concentration in soil and groundwater but may be due to various other factors controlling As solubility, bioavailability and uptake in the soil-rhizosphere-plant system [15–17]. Redox chemistry is one of the predominant factors controlling As speciation and solubility in soil [15]. Arsenic in paddy soil is taken up by plant roots via macro-nutrient transporters; arsenate via the phosphate transporters, and arsenite via silicon transporters [3,18]. Iron (Fe), through forming iron plaque on rice root surfaces, has strong influence on As-uptake by rice roots [19]. The soil physiochemical properties, e.g., redox condition, pH, organic matter, soil texture, Fe and Mn oxides, and sulfur, also affect the solubility and bioavailability of As [15,20]. In addition, the As concentration in various rice tissues varies between rice genotypes [21]. Understanding the genetics associated with grain As concentration is crucial for developing mitigation measures to counter the problem of food-chain contamination by As.

Arsenic is a natural component of Pb, Zn, Cu, and Au ores. Therefore, As is commonly found in soils in mine impacted regions at elevated levels, posing a risk to human and ecosystem health [1]. Paddy rice is one of the most important grain crops in South China [22]. The present study is centered on the Lechang and Renhua Pb-Zn mining regions located in the north of the Guangdong Province, South China. In this study, As levels for soil, root surface, root, straw, and grain were obtained so that grain As levels could be reviewed in regard to both soil and straw As levels. The objectives of this study were: to characterize concentrations of As in rice grains grown in the mining regions with elevated and non-elevated soil As levels, to explore the transfer of As from (rhizosphere) soil through the plant to grain, to identify rice cultivars with low As accumulation in grains, and to understand how the iron and manganese plaque on root surfaces, other macro- and micro-nutrients within the paddy soil and other edaphic properties influence As uptake, assimilation and redistribution, in order to develop potential strategies for reducing As accumulation in rice grains.

Materials and Methods

Ethics statement

No specific permits were required for the described field studies. No specific permissions were required for these locations. We confirm that the location is not privately-owned or protected in any way. We confirm that the field studies did not involve endangered or protected species.

Study area

The present study was conducted in two mining regions in the northern part of Guangdong Province, China (Fig. 1). This research area has a humid subtropical climate with a long-term average annual temperature of 19.6°C and an average annual precipitation of 1,522 mm [23]. The Fankou Pb-Zn mine is an extremely large mine located in Renhua County (Fig. 1). It is geologically situated in the northern part of the central Guangdong Hercynian trough of the South China parageosyncline. The mine was put into production in September 1968, and currently produces 4,500 t of ore per day. Major ore minerals in the Fankou ore mine are pyrite, sphalerite, and galena [24,25]. The mine is classified as a submarine hydrothermal spring effusion type lead/zinc mine, which is relevant to reformed sedimentary rock. [25,26]. Lechang Pb-Zn mine is located in Lechang County (Fig. 1). The major ore minerals are sphalerite, galena, pyrite, and chalcopyrite. As a conventional underground operation, this ore mine was opened in 1959 and is still in operation with a cover area of 1.5 km^2 and produces of 250,000 tons of waste rocks and 30,000 tons of tailings per year occupying respective 8,300 and 60,000 m^2 [23]. The ore of both mines is finely disseminated and complicated and the flotation technology to treat the core is so complex that it is difficult to remove contaminants and make use of the wastewater [27]. The surrounding paddy fields were seriously affected by the continuing year-round irrigation usage of untreated mining wastewater lifted from mines and filtrated from tailings [23]. The rice cropping system in the study area is double-season rice.

Soil and plant samples collection

A total of 28 soil and 28 rice plant (including root, straw, and grain) samples were collected at maturity from 28 paddy fields within or adjacent to the two Pb-Zn mining areas, i.e., eight from Lechang and twenty from Renhua (Fig. 1). The fields were chosen primarily to reflect different rice cultivars being commonly grown by local farmers in these regions. The fields were not irrigated and drained 5 days prior to harvest. Soil samples, 0–15 cm depth, were collected from the base of the rice stem using a soil auger at harvest. At the same time, individual plants of twelve commonly grown cultivars of rice, i.e., Shanyou (SY)-82, SY-86, SY-89, SY-122, SY-162, SY-428, Tianyou (TY)-10, Meixiangzhan (MXZ)-2, Mabei-Youzhan (MBYZ), Diyou (DY)-162, Jinyou (JY)-118, and Fengyou (FY)-998, were collected from the sites where soil samples were taken (Table S1). Composite soil and plant samples were derived by mixing sub-samples from 5 random sites within 25 m^2 per paddy field [22]. The six hybrid SY cultivars were genetically related in that they were all bred using Zhenshan 97A as the female parent. An entire single plant was dug up from 5 sites per each of the 28 fields. However, with the majority of the cultivars being grown only in Renhua or Lechang (three were grown in both), genotypic effects and regional effects were not distinct, but confounded. We therefore analyzed relationships between genotypic and environmental data across the two regions.

Sample preparation

After harvesting, collected plants were washed thoroughly in tap water, followed by deionized water, before extracting the iron plaque from fresh root surfaces using dithionite-citrate- bicarbonate (DCB) as described by Liu et al. [28]. After DCB extraction, plants were separated into their respective tissue components (ear, straw, and root) with stainless steel scissors, weighed to determine fresh weight, oven-dried at 80°C for 72 h, then weighed again (dry weight). Dry spikelets were dehusked by hand and divided into grain and husk. The oven-dried root, straw, and grain samples were powdered using a model MM200 ball mill (Retsch, Germany). Soil samples were air-dried, crushed to pass through a 2 mm nylon sieve (10-mesh), and homogenized. The sub-samples were then ground with an agate grinder to pass through a 150 μm nylon sieve (100-mesh).

Sample analyses

For soil digestion, 0.2 g soil samples were weighed into quartz glass tubes and 5 mL of 12.0 mol/L hydrochloric acid plus 5 mL of 15.2 mol/L nitric acid added, then left to stand overnight at room temperature before being digested on the block digester at 100°C for 1 h, then at 120°C for 1 h, and finally at 140°C for 4 h [22]. For digestion of plant tissues (root, straw, and grain), 0.2 g of grain samples, and 0.1 g of rice root and straw samples, were weighed into 50-mL polypropylene digest tubes and 2 mL of nitric acid added and left to stand overnight. Then 2 mL of hydrogen peroxide was added, and the samples digested using a microwave oven at 50% power (approximately 600 watts). The temperature was raised to 55°C held for 10 min, then to 75°C held for 10 min, and finally to 95°C for 30 min, and then allowed to cool to room temperature [22]. Total As concentrations in solution were determined by hydride generation atomic fluorescence spectrometry (AFS-8130, Beijing). Quality assurance and quality control of metal analyses were carried out by using duplicates (10% of the samples), reagent blanks and standard reference materials (National Environmental Monitoring Centre of China). The recoveries of As in standard reference materials were within ±10% of recommended values, and the relative standard deviation of duplicate measurements was less than 10%.

Selected soil properties, including soil pH, organic matter (OM), available phosphorus (AP), available silicon (ASi), cation exchange capacity (CEC), electrical conductivity (EC), and iron and manganese oxides, were analyzed according to the standard methods recommended by the Soil Science Society of China [29]. Free Fe and Mn, i.e., the bioavailable pool, were extracted by sodium sulfite-sodium citrate-sodium bicarbonate (DCB, mixture of 0.03 $mol \cdot L^{-1}$ $Na_3C_6H_5O_7 \cdot 2H_2O$, 0.125 $mol \cdot L^{-1}$ $NaHCO_3$ and 0.5 g $Na_2S_2O_4$) solution and determined by flame atomic absorption spectrophotometer (FAAS) (Hitachi Z-5300).

Iron plaque on fresh root surface was digested by DCB. For root digestion, 3 g mixed root sample per paddy was weighed into quartz glass tubes (100 mL) and steeped in 30 mL of DCB solution for 1 hr at 25°C before being transferred into quartz glass tubes

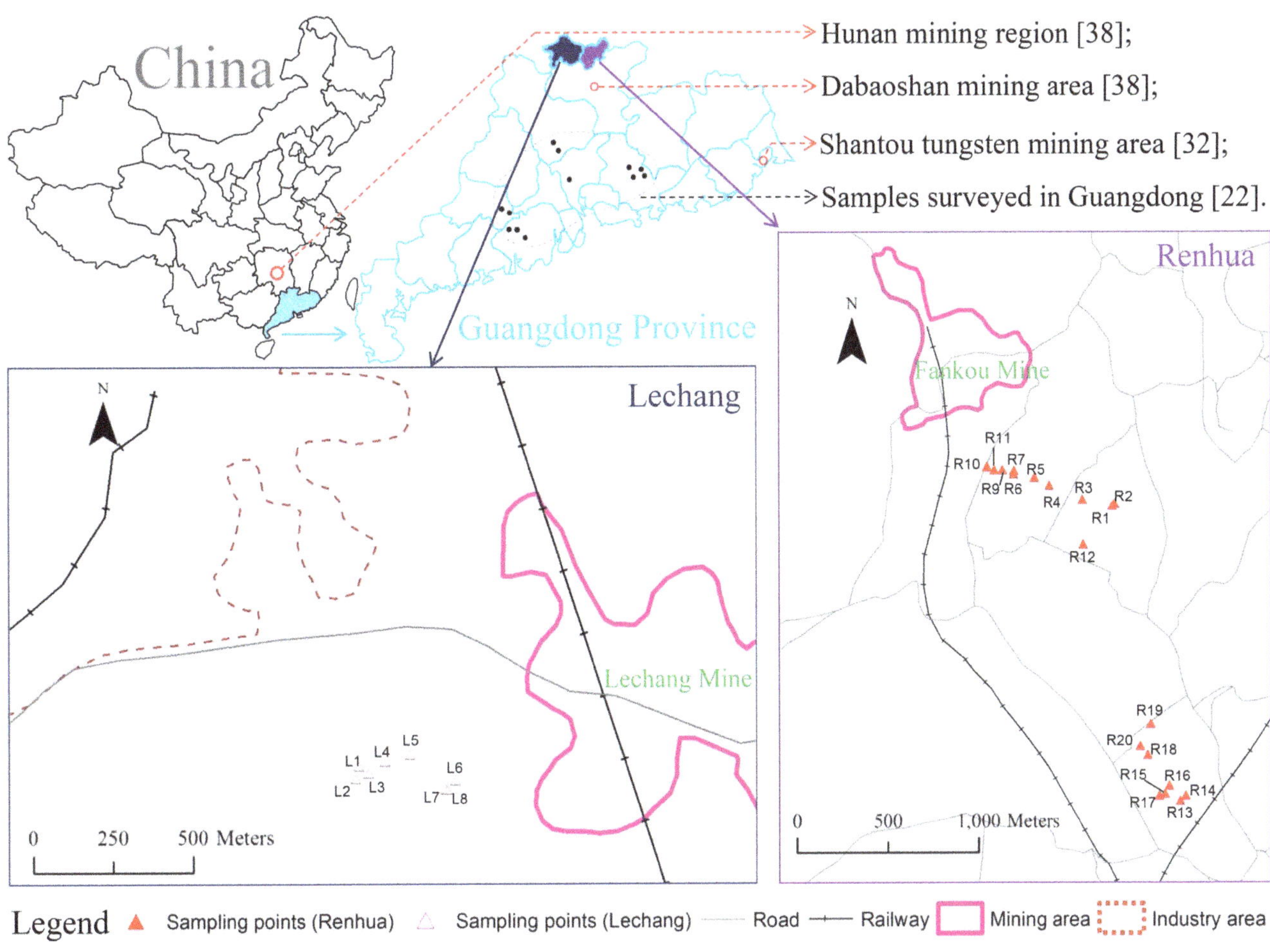

Figure 1. Location map of the study area and distribution of sampling sites.

(100 mL). The root surface As, Fe, and Mn digested by DCB were measured by AFS and FAAS respectively.

Calculation of As transfer factors

Transfer factors (TFs) were determined using the expression C_p/C_s, where C_p is the concentration of As in straw or grain and C_s is the concentration of As in corresponding soil or straw.

Straw/soil transfer factors (S^{traw}/S_{oil} TFs) = Straw As concentrations/Soil As concentrations.

Grain/soil transfer factors (G^{rain}/S_{oil} TFs) = Grain As concentrations/Soil As concentrations.

Grain/straw transfer factors (G^{rain}/S_{traw} TFs) = Grain As concentrations/Straw As concentrations.

Statistical analyses

Descriptive analyses were conducted with SPSS V13.0 for Windows. Principal component analysis (PCA), based on the correlation matrix, was carried out with XLStat-Pro 7.5.2 software, used as a Microsoft Excel plug-in. A probability level of $p<0.05$ was considered as significant difference.

Results and Discussion

Arsenic concentrations in soils

Soil characteristics are presented in Table 1. Arsenic concentrations in the surface paddy soil ranged from 3.7 to 120 $mg{\cdot}kg^{-1}$ with an average of 30.4 $mg{\cdot}kg^{-1}$ in Pb-Zn across the two mining areas in our survey (Table 1). Although uncommon, cultivated paddy soils in Hunan Province, China have been found to contain up to 1,226.5 $mg{\cdot}kg^{-1}$ [30]. The As concentration of soils in this survey would be in the lower part of the range, compared with other mining impacted arable land, e.g., the Hengyang Pb-Zn mine area in Hunan Province (with an average of 253 $mg{\cdot}kg^{-1}$) [31], the Chenzhou Pb-Zn mine area in Hunan Province (405.7 $mg{\cdot}kg^{-1}$) [30], the Shantou abandoned tungsten mine region in Guangdong Province, China (129 $mg{\cdot}kg^{-1}$) [32], and the Rodalquilar Au-(Cu-Pb-Zn) mining district in Almería province, Southeastern Spain (180 $mg{\cdot}kg^{-1}$) [33].

Mean As concentrations in soils collected from Renhua were 10.2 $mg{\cdot}kg^{-1}$ (Table 1), similar to previously reported background concentration of 10.4 $mg{\cdot}kg^{-1}$ in Guangdong surface soils [34] and the median surface soil As concentrations in the paddy fields in Guangdong [22]. In contrast, soil As concentrations collected from Lechang were elevated, averaging 80.8 $mg{\cdot}kg^{-1}$, 8-fold higher compared to that of Renhua (Table 1). All eight soil samples collected from Lechang had As concentrations higher than the maximum allowable concentration (35 $mg{\cdot}kg^{-1}$ for soil with pH ≤5.5, 30 for soil with pH 5.5–6.5, 25 for soil with pH 6.5–7.5, and 20 for soil with pH ≥7.5) for agricultural soil in accordance with the Chinese Environmental Quality Standard for Soils [35]. Soil As concentrations exceeding 40 $mg{\cdot}kg^{-1}$ may be harmful to exposed organisms [36], and soil As exceeding

Table 1. Descriptive statistics of soil properties of Renhua and Lechang.

Characteristics	Probability	Renhua (n = 20)		Lechang (n = 8)		Total (n = 28)
		Mean±SD	Range	Mean±SD	Range	Mean±SD
pH	0.0096	4.8±0.4	*4.4*–6.0	6.7±0.7	5.9–**7.6**	5.4±1.0
As-soil (mg·kg^{-1})	<0.0001	10.2±4.2	*3.7*–20.0	80.8±25.9	53.7–**120.3**	30.4±35.2
Fe_2O_3 (mg·kg^{-1})	0.93	9.3±4.3	*3.6*–21.3	26.0±4.7	21.3–**35.6**	14.1±8.8
Mn_2O_3 (mg·kg^{-1})	<0.0001	17.8±14.0	*7.9*–62.1	214.8±100.4	94.8–**361.3**	74.1±104.7
AP (mg·kg^{-1})	0.012	26.5±14.9	*9.2*–**57.5**	17.1±5.5	10.4–25.0	23.8±13.5
ASi (mg·kg^{-1})	0.013	53.3±13.8	*29.8*–88.1	100.3±26.7	73.1–**150.0**	66.7±28.1
CEC (cmol·kg^{-1})	0.91	6.6±1.1	*4.9*–8.4	16.4±1.2	14.8–**18.3**	9.4±4.6
OM (mg·kg^{-1})	0.61	30.7±5.6	*18.3*–43.7	47.0±6.6	37.4–**57.3**	35.4±9.5
EC (μs·cm^{-1})	0.0054	106.9±47.1	*49.6*–210	460.1±110.2	307–**631**	207.8±176.4
Sand (%)	0.39	31.3±4.6	*17.4*–36.4	26.4±5.9	18.8–**37.4**	29.9±5.4
Silt (%)	0.040	50.3±4.7	*42.7*–**63.6**	49.5±1.9	46.4–52.0	50.1±4.1
Clay (%)	0.010	18.3±2.3	*14.7*–23.9	24.1±4.7	16.2–**30.6**	20.0±4.1

Results are presented as arithemic mean ± SD; probability indicates the differences between Renhua and Lechang; n represents Number of samples; the *italic* number represents the minimum value of the characteristic in all the 28 surveyed samples; the **bold** number represents the maximum value of the characteristic in all the 28 surveyed samples.

100 mg·kg^{-1} poses a severe risk to the pregnant women and their offspring [32,37]. All of the eight samples from Lechang were over 40 mg·kg^{-1} As, two of which were over 100 mg·kg^{-1} As in this survey, indicating severe As contamination in soils around the Lechang Pb-Zn mine.

Arsenic accumulation and translocation in tissues of rice plant

The overall mean total concentration of grain As was 0.26 mg·kg^{-1} (Table 2), which is comparable to field-collected unpolished rice from non-mining-impacted fields in Guangdong (0.29 mg·kg^{-1}, n = 12, rice cultivar Peizha-Taifeng) [22] and Hunan mining region (0.30 mg·kg^{-1}, n = 22) [38] (Fig. 1). In contrast, the average grain As in this survey was much higher than the field-collected unpolished rice from Dabaoshan mining areas in Guangdong (0.19 mg·kg^{-1}, n = 95) [38] yet around half the grain As concentration reported in field-collected brown rice from Shantou tungsten mining area in Guangdong (0.56 mg·kg^{-1}, n = 33) [32] (Fig. 1). All of the presently surveyed rice grains possessed As concentrations >0.17 mg·kg^{-1}, exceeding Chinese maximum contaminant levels (MCLs) of 0.15 mg·kg^{-1}, irrespective of As speciation [39], indicating that rice from this region would be a significant source of dietary As for the population. Chinese standards for As in rice are probably the strictest in the world, which have been designed to protect a nation with high rice intakes [38]. Compared to the global 'normal' range of 0.08–0.20 mg·kg^{-1} for As concentration in rice grain [12], 25 out of these 28 samples exceeded the 'normal' range. The mean As concentrations for these rice grain samples were much higher compared to that in rice from U.S. and Europe (both 0.198 mg·kg^{-1}) [12].

It has been demonstrated that different rice cultivars showed significant differences for concentrations of As in straw, husk and grain [12,19]. While we did not detect significant differences for rice grain As among the Lechang varieties, we did find difference between rice cultivars harvested from Renhua (Table 3). Although the genetic differences for As-root and As-straw were not significant, the cultivars SY-89 and DY-162 showed the lowest As concentrations for all tissues when grown in Renhua (Table 3), while the cultivar SY-122 showed the highest or second highest concentrations for all three tissues in both Renhua and Lechang. What makes this especially interesting is that, though the soils within each site were not significantly different for As concentration, the SY-122 with the highest tissue As was grown in a field with relatively lower soil As per site. Despite the fact that Lechang was higher than Renhua for As in soils, higher As concentrations of soil, DCB extracts, and straw, the As concentration in grains from the two sites was comparable (0.28 and 0.25 mg·kg^{-1}) (Table 2). When data were analyzed among just the varieties grown in both Renhua and Lechang, As concentrations in grain were again comparable, i.e., SY-428 0.27 (Renhua)-0.25 (Lechang) mg·kg^{-1}, MBYZ 0.24–0.29, and SY-122 0.39-0.32, in spite of their having more As in soil, DCB extracts, and straw in samples from Lechang compared to Renhua (Table 3). Although the mean soil As concentration in Renhua was lower compared to the national soil background, the rice grain As concentrations exceeded the Chinese MCLs, i.e., rice grain can accumulate relatively large amounts of As even from soils having very low level of As. Williams et al. [41] also reported that there were elevated grain As concentrations even with background soil levels. It is clear that the amount of As added by contamination to soil is not necessarily the overriding factor controlling the As concentration in rice grain. Other researchers [11,12] reported that the high As levels in rice were associated with As-contaminated irrigation water. As uptake by rice mainly depends on As availability rather than total As in soil [14].

Similar to the total As in soils, the DCB-extracted As concentrations from the root surfaces were highly variable from one paddy field to another, and there was an approximate 3-fold difference in mean DCB-extracted As concentration between Lechang and Renhua (Table 2). Interestingly, SY-122 as one of the three cultivars grown in both locations had the highest recorded mean root As concentrations in both locations (Table 3), yet in Renhua it had the lowest plaque concentration of As (Table 3). The root As concentrations for SY-122 grown in Lechang (52.0 mg·kg^{-1}) and Renhua (48.6 mg·kg^{-1}) were simi-

Table 2. Descriptive statistics of rice plant accumulation and transfer factors.

Characteristics	Probability	Renhua (n = 20)		Lechang (n = 8)		Total (n = 28)
		Mean±SD	**Range**	**Mean±SD**	**Range**	**Mean±SD**
TF (soil-straw)	0.0047	0.36±0.18	0.13–**0.90**	0.079±0.041	*0.018*–0.13	0.28±0.20
TF (straw-grain)	0.62	0.084±0.30	*0.032*–**0.17**	0.057±0.023	0.036–0.11	0.076±0.030
TF (soil-grain)	0.0030	0.028±0.013	0.012–**0.057**	0.0037±0.0010	*0.0020*–0.0047	0.021±0.016
As-root ($mg·kg^{-1}$)	0.49	25.6±12.5	*11.1*–**59.4**	35.0±9.4	24.8–52.0	28.3±12.3
As-straw ($mg·kg^{-1}$)	0.029	3.3±1.3	*1.6*–6.9	5.8±2.8	2.1–**10.4**	4.0±2.1
As-grain ($mg·kg^{-1}$)	0.37	0.25±0.051	0.18–**0.39**	0.28±0.067	*0.17*–0.38	0.26±0.06
As-DCB ($mg·kg^{-1}$)	0.013	25.4±14.4	*6.4*–59.4	70.7±28.9	27.7–**101.8**	38.3±28.2
Mn-DCB ($mg·kg^{-1}$)	<0.0001	25.6±22.5	*7.2*–93.4	189.1±109.4	75.6–**341.0**	72.3±95.5
Fe-DCB ($g·kg^{-1}$)	0.044	29.7±11.4	*4.3*–**48.6**	36.3±5.0	26.5–44.6	31.6±10.4

Results are presented as arithemic mean ± SD; probability indicates the differences between Renhua and Lechang; n represents Number of samples; the *italic* number represents the minimum value of the characteristic in all the 28 surveyed samples; the **bold** number represents the maximum value of the characteristic in all the 28 surveyed samples.

lar, which is in stark contrast to the 10-fold differences for the corresponding DCB-extracted As and soil As levels observed in this survey (Table 3). In this regard, although the As concentrations in soil and root surface for Lechang were significantly higher, there was not a significant difference for concentration of As inside the root, suggesting that As in rice does not directly depend on the total As concentration in the soil and root surface but may be due to other factors and uptake mechanisms.

Regardless of rice cultivars and locations, the As concentrations in soil and DCB extracts were much higher than As concentrations in rice plants, excepting rice root, whereas no obvious trend was observed between the As concentrations in soil, DCB extracts, and roots (Tables 1–3). Rice roots contained considerably higher concentrations of As compared to any other parts of the plant, regardless of soil As concentration and rice cultivars (Tables 1–3). The levels of root As were found to be on average 7.7 times higher than their corresponding straw, a trend which was maintained throughout several orders of magnitude in grain As. Previous researches [31,41,42] also observed that much more As accumulated in rice root than other parts. In the current study, the levels of As in straw were found to be, on average, 15.4 times higher than their corresponding grain samples.

Both straw As concentration and mean As straw/soil transfer factors (S^{traw}/S_{oil} TFs) based on total As concentrations were highly variable between and/or within Renhua and Lechang locations (Table 2). Mean S^{traw}/S_{oil} TFs for Renhua and Lechang were significantly different, with the values of 0.36 and 0.079, respectively (Table 2), and in both location, the highest and the lowest mean S^{traw}/S_{oil} TFs were seen for SY-122 and SY-428, respectively (Table 3).

The range of grain/soil transfer factors (G^{rain}/S_{oil} TFs) were 0.012–0.057 for Renhua, and 0.0020–0.0047 for Lechang. There was an over 7-fold difference in mean G^{rain}/S_{oil} TFs between Renhua and Lechang respectively, probably related to the significant difference of soil As between the two locations (Table 1). Similar to S^{traw}/S_{oil} TF, the G^{rain}/S_{oil} TF for SY-122 was significantly higher compared to SY-428, regardless of location (Table 3). Mean As grain/straw transfer factors (G^{rain}/S_{traw} TFs) ranged from 0.038 to 0.11, averaging 0.076, which was a little higher compared to the As G^{rain}/S_{traw} TFs in rice surveyed in Guangdong [22]. Differences in As G^{rain}/S_{traw} TFs were not apparent between locations (Table 1) and cultivars respectively (Tables 3).

Factors affecting As transfer

As discussed above, As uptake by rice plants appears more affected by As availability than total As in the soil. The bioavailability of As to plants is governed by key edaphic physiochemical properties (e.g., pH, Eh, organic matter, texture, Fe/Mn-oxides/hydroxides, and phosphorus, silicon, and sulfur concentrations); environmental conditions and modification of the soil in the rhizosphere; these factors interact to influence As speciation in the soil [16,41]. Rice is normally cultivated in flooded paddy soil, an environment that leads to a mobilization and, hence, a much enhanced bioavailability of As to rice plants. Rice is also a strong accumulator of the macro-nutrient silicon, an element that plays an important role in the defense against a range of biotic and abiotic stresses [10]. The principal component analysis (PCA) (Fig. 2) was performed with the concentrations of As, Fe and Mn in DCB extracts from root surfaces, the concentration of total As in soil, the concentrations of Fe and Mn oxides, AP and ASi, selected soil properties, and the concentrations of As in rice tissues in order to analyze the relationships among these indices and identify the factors affecting As transfer. The first 2 principal components accounted for 67.6% of the variability observed among all the cultivars and across all harvest sites. The results from PCA is in agreement with predictions that plant As is determined more strongly by external soil properties affecting As availability than by differences in internal plant processes.

The Fe and Mn oxide phases are common in various soils and are very efficient in sorbing As [15]. Manganese plaque and Fe and Mn oxides positively correlated with As in rice tissues respectively (Fig. 2), indicating Mn plaque and Fe and Mn oxides in soil may inhibit As transfer from soil to rice plant. There were 1.2 times, 12.2 times, and 27.4 times difference in iron plaque, manganese plaque, and Mn_2O_3 level respectively between Lechang and Renhua (Table 1), which might account for fact that the higher As in the Lechang soils did not result in significantly greater As in plant tissues compared with the rices grown in Renhua region.

The mechanism for arsenate uptake, the dominant inorganic As species under aerobic conditions, is through phosphate transport-

Table 3. Cultivar means for each parameter as observed in Renhua and Lechang.

Location	Cultivar	S^{traw}/S_{oil}	G^{rain}/S_{traw}	G^{rain}/S_{oil}	As-root	As-straw	As-grain	As-DCB	Mn-DCB	Fe-DCB	As-soil
		TF	TF	TF	$mg \cdot kg^{-1}$	$mg \cdot kg^{-1}$	$mg \cdot kg^{-1}$	$mg \cdot kg^{-1}$	$mg \cdot kg^{-1}$	$g \cdot kg^{-1}$	$mg \cdot kg^{-1}$
Renhua	SY-428	0.19±0.060b	0.11±0.053a	0.022±0.017b	22.6±11.2a	2.9±1.0a	0.27±0.052bc	25.4±7.8b	10.5±3.2a	28.0±4.4ab	16.0±6.8a
	MXZ	0.34±0.035ab	0.083±0.019a	0.028±0.0058ab	26.0±7.5a	2.8±0.4a	0.230.036cd	19.1±8.9b	35.3±30.3a	32.6±8.8ab	8.2±0.6b
	SY-162	0.37±0.10ab	0.086±0.015a	0.031±0.0047ab	22.2±5.2a	3.1±0.7a	0.26±0.016bc	24.9±16.3b	39.5±46.7a	29.8±19.9ab	8.7±0.8b
	MBYZ	0.57±0.46a	0.080±0.047a	0.034±0.010ab	38.1±30.1a	3.6±2.2a	0.24±0.0029bcd	14.5±5.6b	24.9±15.7a	32.9±13.9ab	7.2±2.0b
	TY-10	0.37±0.11ab	0.094±0.047a	0.037±0.027ab	19.5±5.3a	2.7±1.5a	0.22±0.012bcd	31.2±5.7b	31.9±11.4a	30.8±0.4ab	8.3±6.5ab
	SY-86	0.35±0.21ab	0.058±0.037a	0.016±0.00052b	30.0±14.7a	4.8±3.0a	0.22±0.00093bcd	57.5±2.7a	17.0±4.7a	38.0±6.1a	13.7±0.4ab
	SY-122	0.70a	0.081a	0.057a	48.6a	4.8a	0.39a	6.4b	16.5a	4.3b	6.9ab
	SY-89	0.21ab	0.084a	0.018b	11.1a	2.2a	0.19d	26.5b	21.5a	41.9a	10.8ab
	DY-162	0.23ab	0.079a	0.018ab	14.6a	2.3a	0.18d	23.5b	16.1a	23.1ab	9.9ab
	JY-118	0.40ab	0.064a	0.025ab	23.7a	4.8a	0.31ab	17.4b	18.9a	17.3ab	12.1ab
Lechang	SY-428	0.039±0.018c	0.077±0.026a	0.003±0.00062b	29.8±6.8b	3.6±2.2a	0.25±0.090a	68.2±33.1a	228.0±116.0a	37.4±1.3a	96.7±36.0a
	SY-82	0.12±0.010a	0.039±0.0043a	0.005±0.00011a	28.7±2.3b	8.6±2.6a	0.33±0.064a	64.8±52.4a	213.2±177.3a	35.6±12.8a	72.5±15.5a
	MBYZ	0.052bc	0.061a	0.003ab	43.0ab	4.7a	0.29a	67.0a	111.1a	35.7a	89.0a
	SY-122	0.12ab	0.038a	0.005a	52.0a	8.3a	0.32a	65.6a	75.6a	35.2a	68.3a
	FY-998	0.10ab	0.044a	0.005a	38.6ab	5.6a	0.25a	98.4a	215.4a	36.4a	53.7a

Results are presented as arithemic mean ± SD; means within a row for a certain genotype grown in Lechang or Renhua followed by different letters are significantly different at the 0.05 level; the comparisons are based on estimated marginal means.

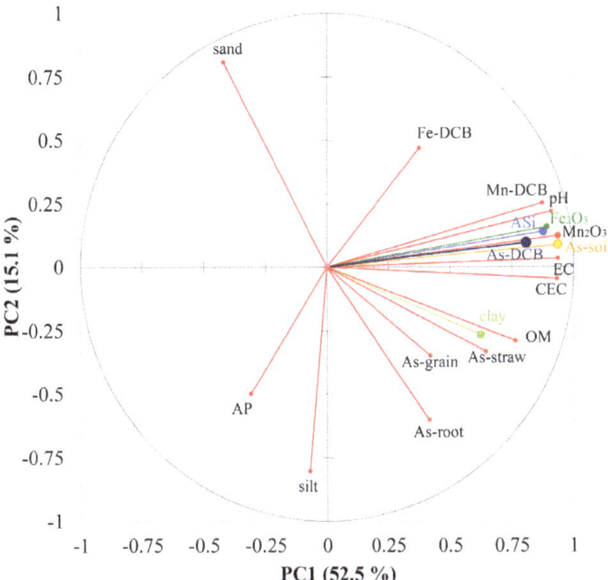

Figure 2. Plot of the first two principal components from Principal Component Analysis (PCA). The PCA was performed with As, Fe and Mn in DCB extracts, total As in soil, Fe_2O_3 and Mn_2O_3, available phosphorus (AP) and Si (ASi), selected soil properties, As in rice tissues.

ers, as arsenate is an analogue of phosphate [43]. Addition of phosphate to the soil may decrease arsenate uptake and consequently may reduce As toxicity, depending on soil conditions etc. [22]. In the current study, however, no remarkable relationships were observed between the AP in soil and rice As (Fig. 2), which is incompatible with the arsenate uptake mechanism discussed above, probably due to arsenite being the predominant form of As in flooded paddy soils [18], which doesn't compete with phosphate for transport as arsenate. Arsenite shares the same transport system responsible for silicon uptake, both influx and efflux transporters mediate transport of arsenite [28], and arsenite associated with iron plaque may be much more easily desorbed than arsenate [15,16]. Therefore, the application of silica fertilizer to soil can decrease the transfer of As from the soil and irrigation water to rice. The significantly higher available silicon in Lechang ($p = 0.013$), as shown in Table 1, may also be responsible for the fact that the higher soil As in Lechang did not result in significantly greater As in rice grain compared with the rices grown in Renhua region.

The PCA effects of root As, straw As, and grain As were close to each other, indicating the significant positive relationships between these rice tissues, while the As in rice tissues showed negative relationship with DCB-extracted As and soil As, respectively (Fig. 2). This further verifies that a rise in the soil As may not increase the accumulation of As in rice tissues.

Soil texture is another important factor affecting As bioavailability [15]. In general, soils with a clayey texture have less availability of As compared with sandy soils [17]. As observed in Table 1, the significant higher soil clay content in Lechang ($p = 0.010$) may decrease the availability of its soil As, so that even though these soils contained more As, they consequently inhibited its uptake by rice plants.

The solubility and bioavailability of As can be affected by soil pH because it controls the As speciation and leachability [15]. The soil pH differed significantly between Lechang and Rehua ($p = 0.0096$). The soils in Lechang ranged from neutral (pH 5.9) to slightly alkaline (pH 7.6) and in Renhua from strongly acidic (pH 4.4) to neutral (pH 6.0), respectively (Table 1). The soils collected from the Lechang mine region were expected to be more acidic and similar to those in Renhua. The industries in the northeast part of the paddy field (Fig. 1), including chemical plant, cement plant, textile mill, metal processing factory, plastic products factory, and bulb factory, might be affecting the soil pH. Arsenite solubility increases as the pH decreases within the range commonly found in soil (pH 3–9), while the pattern is reversed in the case of arsenate. Arsenite predominates in flooded paddy soils. In this regard, for the current study, a decline in soil pH can increase the mobilization of As in soils, which explain why we observed nearly equal concentrations of As in rice grain regardless of the soil As levels. The significant higher soil pH in Lechang may decrease the availability of its soil As.

By understanding the factors controlling bioavailability of As to rice plants and mechanisms of As uptake in plants, one could develop proper strategies for limiting As accumulation in rice grains. Possibilities include altering farm practices, e.g., growing rice aerobically in raised beds instead of in the traditional flooded paddy fields, which offers an opportunity to reduce the mobilization of soil arsenite and curtail As transfer from soil to grain. This approach would require, however, a fundamental change in farming practices in Asia [6,14], and aerobically produced rice is generally lower yielding [16,44] and more susceptible to rice blast disease and heat stress [45]. In addition, silica and phosphate fertilizations can be applied in soil to decrease As accumulation in rice, dependent on soil conditions [16]. Another tack would be the selection of rice cultivars with low accumulation of As in grains [40]. To be a success on the farm, any new cultivars will have to have decent yields. A hypothetical cancer risk pales in comparison with an empty stomach [6].

Conclusions

The results indicated both environmental and genetic effects caused diversity for grain As concentration among different rice cultivars grown in two locations in China with mining-contaminated soils. All the grain samples in this study exceeded national food standards for grain As. The fact that grain As levels were not significantly different in the rices from the two areas differing significant for soil As levels, suggests that As uptake by rice is determined more by As availability rather than by total As in the soil. The As behaviour in the soil-rice system was found associated with various factors, i.e., iron and manganese plaque, iron and manganese oxides in the soil, soil available silicon and phosphorus, soil pH, soil organic matter, and soil texture. Understanding the mechanisms controlling As uptake would improve our understanding of how soil As sometimes but not always increases rice grain As, and to develop genetic and physico-chemical strategies for reducing As accumulation in rice grains.

Author Contributions

Conceived and designed the experiments: FD YL JL QY. Performed the experiments: FD QY YL. Analyzed the data: JL YL HS FD QY. Contributed reagents/materials/analysis tools: YL HS. Wrote the paper: JL YL HS FD.

References

1. Smith E, Naidu R, Alston A (1998) Arsenic in the soil environment: a review. Adv Agron 64: 149–195.
2. Duan G, Liu W, Chen X, Hu Y, Zhu Y (2013) Association of arsenic with nutrient elements in rice plants. Metallomics 5: 784–792.
3. Zhao FJ, Ma JF, Meharg AA, McGrath SP (2009) Arsenic uptake and metabolism in plants. New phytol 181: 777–794.
4. Dave R, Singh P, Tripathi P, Shri M, Dixit G, et al. (2013) Arsenite tolerance is related to proportional thiolic metabolite synthesis in rice (*Oryza sativa* L.). Arch Environ Contam Toxicol 64: 235–242.
5. Banerjee M, Banerjee N, Bhattacharjee P, Mondal D, Lythgoe PR, et al. (2013) High arsenic in rice is associated with elevated genotoxic effects in humans. Sci Rep 3.
6. Stone R (2008) Arsenic and paddy rice: a neglected cancer risk? Science 321: 184–185.
7. Kennedy D (2002) The importance of rice. Science 296: 13–13.
8. Fan JB, Zhang YL, Turner D, Duan YH, Wang DS, et al. (2010) Root physiological and morphological characteristics of two rice cultivars with different nitrogen-use efficiency. Pedosphere 20: 446–455.
9. Xu XY, McGrath SP, Meharg AA, Zhao FJ (2008) Growing rice aerobically markedly decreases arsenic accumulation. Environ Sci Technol 42: 5574–5579.
10. Zhao FJ, Zhu YG, Meharg AA (2013) Methylated arsenic species in rice: geographical variation, origin, and uptake mechanisms. Environ Sci Technol 47: 3957–3966.
11. Williams PN, Islam MR, Adomako EE, Raab A, Hossain SA, et al. (2006) Increase in rice grain arsenic for regions of Bangladesh irrigating paddies with elevated arsenic in groundwaters. Environ Sci Technol 40: 4903–4908.
12. Zavala YJ, Duxbury JM (2008) Arsenic in rice: I. Estimating normal levels of total arsenic in rice grain. Environ Sci Technol 42: 3856–3860.
13. Meharg AA, Rahman MM (2003) Arsenic contamination of Bangladesh paddy field soils: implications for rice contribution to arsenic consumption. Environ Sci Technol 37: 229–234.
14. Ahmed ZU, Panaullah GM, Gauch H, McCouch SR, Tyagi W, et al. (2011) Genotype and environment effects on rice (*Oryza sativa* L.) grain arsenic concentration in Bangladesh. Plant Soil 338: 367–382.
15. Sahoo PK, Kim K (2013) A review of the arsenic concentration in paddy rice from the perspective of geoscience. Geosci J 17: 107–122.
16. Zhao FJ, McGrath SP, Meharg AA (2010) Arsenic as a food chain contaminant: mechanisms of plant uptake and metabolism and mitigation strategies. Annu Rev Plant Biol 61: 535–559.
17. Heikens A, Panaullah GM, Meharg AA (2007) Arsenic behaviour from groundwater and soil to crops: impacts on agriculture and food safety. Rev Environ Contam Toxicol 189: 43–87.
18. Ma JF, Yamaji N, Mitani N, Xu X-Y, Su Y-H, et al. (2008) Transporters of arsenite in rice and their role in arsenic accumulation in rice grain. P Natl Acad Sci 105: 9931–9935.
19. Liu WJ, Zhu YG, Hu Y, Williams PN, Gault AG, et al. (2006) Arsenic sequestration in iron plaque, its accumulation and speciation in mature rice plants (*Oryza sativa* L.). Environ Sci Technol 40: 5730–5736.
20. Xu J, Tian YZ, Zhang Y, Guo CS, Shi GL, et al. (2013) Source apportionment of perfluorinated compounds (PFCs) in sediments: Using three multivariate factor analysis receptor models. J Hazard Mat 260: 483–488.
21. Moreno-Jiménez E, Esteban E, Peñalosa J (2012) The fate of arsenic in soil-plant systems. In: Whitacre DM, editor: Springer New York. 1–37.
22. Lu Y, Dong F, Deacon C, Chen HJ, Raab A, et al. (2010) Arsenic accumulation and phosphorus status in two rice (*Oryza sativa* L.) cultivars surveyed from fields in South China. Environ Pollut 158: 1536–1541.
23. Yang QW, Lan CY, Shu WS (2008) Copper and Zinc in a paddy field and their potential ecological impacts affected by wastewater from a lead/zinc mine, P. R. China. Environm Monit Assess 147: 65–73.
24. Lu HZ (1983) Fluid inclusion study of Fankou Pb-Zn deposit, Fankou, Guangdong, China. Geochemistry 2: 45–57.
25. Wei X, Cao J, Holub RF, Hopke PK, Zhao S (2013) TEM study of geogas-transported nanoparticles from the Fankou lead–zinc deposit, Guangdong Province, South China. J Geochem Explor 128: 124–135.
26. Deng J, Yang L, Chen X, Wang Q, Liu Y (2005) Fluid system and ore-forming dynamics of the Yuebei Basin, China. In: Mao J, Bierlein F, editors. Mineral Deposit Research: Meeting the Global Challenge: Springer Berlin Heidelberg. 107–109.
27. Hu Y, Sun W, Liu R, Dai J (2012) Water recycling technology in Fankou lead-zinc mine of China. In: Drelich J, editor. Water in Mineral Processing. Englewood, Colorado, USA: Society for Mining Metallurgy & Exploration 371–388.
28. Liu WJ, Zhu YG, Smith FA, Smith SE (2004) Do iron plaque and genotypes affect arsenate uptake and translocation by rice seedlings (*Oryza sativa* L.) grown in solution culture? J Exp Bot 55: 1707–1713.
29. Lu RK (2000) Analytical methods for soils and agricultural chemistry. Beijing, China: China agricultural science and technology Press (In Chinese).
30. Liu HY, Probst A, Liao BH (2005) Metal contamination of soils and crops affected by the Chenzhou lead/zinc mine spill (Hunan, China). Sci Total Environ 339: 153–166.
31. Williams PN, Lei M, Sun G, Huang Q, Lu Y, et al. (2009) Occurrence and partitioning of cadmium, arsenic and lead in mine impacted paddy rice: Hunan, China. Environ Sci Technol 43: 637–642.
32. Liu CP, Luo CL, Gao Y, Li FB, Lin LW, et al. (2010) Arsenic contamination and potential health risk implications at an abandoned tungsten mine, southern China. Environ Pollut 158: 820–826.
33. Oyarzun R, Cubas P, Higueras P, Lillo J, Llanos W (2009) Environmental assessment of the arsenic-rich, Rodalquilar gold-(copper-lead-zinc) mining district, SE Spain: data from soils and vegetation. Environm Geol 58: 761–777.
34. Zhang HH, Yuan HX, Hu YG, Wu ZF, Zhu LA, et al. (2006) Spatial distribution and vertical variation of arsenic in Guangdong soil profiles, China. Environ Pollut 144: 492–499.
35. Chinese Environmental Protection Agency (2008) Environmental quality standards for soils. GB 15618–2008.
36. Dudka S, Miller WP (1999) Permissible concentrations of arsenic and lead in soils based on risk assessment. Water Air Soil Pollut 113: 127–132.
37. DeSesso JM, Jacobson CF, Scialli AR, Farr CH, Holson JF (1998) An assessment of the developmental toxicity of inorganic arsenic. Reprod Toxicol 12: 385–433.
38. Zhu YG, Sun GX, Lei M, Teng M, Liu YX, et al. (2008) High percentage inorganic arsenic content of mining impacted and nonimpacted Chinese rice. Environ Sci Technol 42: 5008–5013.
39. Chinese Food Standards Agency (2005) Maximum levels of contaminants in food. GB 2762–2005.
40. Norton GJ, Pinson SR, Alexander J, McKay S, Hansen H, et al. (2012) Variation in grain arsenic assessed in a diverse panel of rice (*Oryza sativa*) grown in multiple sites. New phytol 193: 650–664.
41. Williams PN, Villada A, Deacon C, Raab A, Figuerola J, et al. (2007) Greatly enhanced arsenic shoot assimilation in rice leads to elevated grain levels compared to wheat and barley. Environ Sci Technol 41: 6854–6859.
42. Liao XY, Chen TB, Xie H, Liu YR (2005) Soil As contamination and its risk assessment in areas near the industrial districts of Chenzhou City, Southern China. Environ Int 31: 791–798.
43. Norton GJ, Adomako EE, Deacon CM, Carey A-M, Price AH, et al. (2013) Effect of organic matter amendment, arsenic amendment and water management regime on rice grain arsenic species. Environ Pollut 177: 38–47.
44. Peng S, Bouman B, Visperas RM, Castañeda A, Nie L, et al. (2006) Comparison between aerobic and flooded rice in the tropics: Agronomic performance in an eight-season experiment. Field Crops Res 96: 252–259.
45. Farooq M, Siddique KHM, Rehman H, Aziz T, Lee D-J, et al. (2011) Rice direct seeding: Experiences, challenges and opportunities. Soil Till Res 111: 87–98.

14

Interact to Survive: *Phyllobacterium brassicacearum* Improves Arabidopsis Tolerance to Severe Water Deficit and Growth Recovery

Justine Bresson[1,2], François Vasseur[1,3], Myriam Dauzat[1], Marc Labadie[1,2], Fabrice Varoquaux[2], Bruno Touraine[2], Denis Vile[1]*

1 Laboratoire d'Ecophysiologie des Plantes sous Stress Environnementaux (LEPSE), UMR759, Institut National de la Recherche Agronomique-SupAgro, Montpellier, France, **2** Laboratoire des Symbioses Tropicales et Méditerranéennes (LSTM), UMR113, Université Montpellier 2-IRD-CIRAD-INRA-SupAgro, Montpellier, France, **3** Max Planck Institute for Developmental Biology, Tübingen, Germany

Abstract

Mutualistic bacteria can alter plant phenotypes and confer new abilities to plants. Some plant growth-promoting rhizobacteria (PGPR) are known to improve both plant growth and tolerance to multiple stresses, including drought, but reports on their effects on plant survival under severe water deficits are scarce. We investigated the effect of *Phyllobacterium brassicacearum* STM196 strain, a PGPR isolated from the rhizosphere of oilseed rape, on survival, growth and physiological responses of *Arabidopsis thaliana* to severe water deficits combining destructive and non-destructive high-throughput phenotyping. Soil inoculation with STM196 greatly increased the survival rate of *A. thaliana* under several scenarios of severe water deficit. Photosystem II efficiency, assessed at the whole-plant level by high-throughput fluorescence imaging (F_v/F_m), was related to the probability of survival and revealed that STM196 delayed plant mortality. Inoculated surviving plants tolerated more damages to the photosynthetic tissues through a delayed dehydration and a better tolerance to low water status. Importantly, STM196 allowed a better recovery of plant growth after rewatering and stressed plants reached a similar biomass at flowering than non-stressed plants. Our results highlight the importance of plant-bacteria interactions in plant responses to severe drought and provide a new avenue of investigations to improve drought tolerance in agriculture.

Editor: Jin-Song Zhang, Institute of Genetics and Developmental Biology, Chinese Academy of Sciences, China

Funding: JB was funded by French Ministry of Higher Education and Research. The funders had no role in study design, data collection and analysis, decision to publish, or preparation of the manuscript.

Competing Interests: The authors have declared that no competing interests exist.

* Email: denis.vile@supagro.inra.fr

Introduction

Drought is a global concern and episodes of severe drought will most probably be more frequent with dramatic consequences on agriculture [1]. Severe water stress greatly reduces plant biomass production and can lead to plant mortality [2]. Over the last decade, it has been shown that plants can largely benefit from their interactions with soil microorganisms; especially with plant growth promoting rhizobacteria (PGPR) that colonize the rhizosphere of many plants species [3]. The stimulation of growth by PGPR is often associated with lower plant susceptibility to various biotic and abiotic stresses [4,5] and there is a growing interest in the use of these rhizobacteria in agriculture [6,7].

Survival to drought events is found in plants that are able to maintain key cellular functions under severe water stress and recover similar pre-stress values when conditions become favorable again [8]. The capacity to tolerate low leaf water status, or dehydration tolerance, is widely variable among species [9]. The most spectacular adaptation to severe drought is illustrated by resurrection plants [10]. These plants display rapid physiological responses and metabolic adjustments [11], and tolerate nearly complete tissue dehydration. During mild drought or water stress of limited duration, plants that maintain a good water status can complete their life cycle, although often with reduced performance. However, when stress becomes more drastic or is prolonged the leaf water potential drops and leaf damages occur [12]. Then, dramatic reduction of biomass production and even plant mortality appear [2]. To prevent tissue damages, and survive at low leaf water content, many processes and signaling pathways are involved [13]. Osmotic adjustments and accumulation of specific protective osmolytes such as proline [14], glycine betaine [15] or trehalose [16] allow stabilizing cellular structures. One of the most rapid responses to prevent hydraulic failure is stomatal closure. However under severe water stress, stomatal closure can diminish photosynthetic uptake and induce carbon starvation [17] that can lead to total or partial leaf senescence. Drought-induced senescence of older leaves can contribute to water saving, while allowing the reallocation of nutrient to the younger leaves [18]. However, leaf senescence alters photosynthetic functioning and chlorophyll (Chl) properties [19]. Chl-fluorescence is a powerful, rapid and minimally invasive indicator of plant health [20]. In particular, dark-adapted measurements of the ratio of variable to maximal fluorescence (F_v/F_m) give the potential quantum yield (or efficiency) of the photosystem II (PSII) photochemistry, which varies with plant water status [21,22]. Decrease in F_v/F_m is due to an increase in leaf damages that may to some extent be reversible [22]. After a period of water stress, it has been shown that plants have the capacity to recover progressively, but sometimes incompletely, their photosynthetic [23,24] and growth [25] potential. During stress, plant growth rate is reduced, even stopped, but leaf cells retain their ability to expand when conditions become favorable again [25].

Rhizobacteria can help plants to cope with negative effects of water deficit. Under water stresses of moderate intensity, some PGPR can improve resistance to water deficit through i) modifications in phytohormones content and/or signaling, notably ethylene, auxin, cytokinin, and abscisic acid (*e.g.*, [26,27,28,29]), ii) enhanced cells detoxification by increasing antioxydase activities such as catalase [30] or superoxide dismutase [31], iii) changes in plant functional traits such as photosynthetic capacity through changes in chlorophyll content [31] and in photosynthetic PSII efficiency [32,33], or iv) the formation of a biofilm which enhances soil aggregation and improves water stability in the soil [34]. Even though a rich literature exists on plant responses to rhizobacteria under water stress (for reviews see [4,35]), studies of PGPR effects on plant survival are surprisingly limited.

The aim of this study was to investigate the effects of the free-living PGPR, *Phyllobacterium brassicacearum* strain STM196, on survival, growth and physiological responses of *A. thaliana* during the time-course of severe drought progression. The STM196 strain belongs to the *Phyllobacteriaceae* family in the *Rhizobiales*, order of *α-Proteobacteria* [36]. This strain was the most efficient PGPR isolated from the rhizoplan of field-grown *Brassica napus* roots [37,38]. We have recently shown that STM196 improves *A. thaliana* resistance to moderate water deficit through a reproductive delay and changes in transpiration rate correlated to modifications of leaf ABA content [29]. Moreover, previous *in vitro* studies showed that STM196 modifies root architecture and hormonal signaling [39,40,41,42]. Here, our main experimental goals were (*i*) to determine whether plant-PGPR interaction mitigate the negative consequences of severe drought on plant survival, (*ii*) to assess how biotic interactions with PGPR influence physiological mechanisms of plants (*iii*) to evaluate the benefits of inoculation on growth and productivity of plants after stress. *A. thaliana* plants were subjected to five scenarios of severe soil water deficit, with progressive soil drying and rewatering treatments. The use of the plant phenotyping platform PHENOPSIS allowed fine-tuning of soil water content and daily acquisition of images of plants [43]. The dynamics of physiological changes in plants were investigated independently in surviving and perishing plants under severe drought by estimating survival with non-invasive chlorophyll fluorescence measurements at high throughput levels. This approach is broadly applicable to investigate survival of plants under various stresses affecting chlorophyll properties and leaf functioning.

Materials and Methods

Bacteria material, bacterial inoculum and soil inoculation

The *Phyllobacterium brassicacearum* STM196 strain was grown for three days in Petri dishes on a sterile (20 min at 120°C) 1.5% agar (w/v; Sigma-Aldrich) medium (E′) containing 2.87 mM K_2HPO_4, 0.81 mM $MgSO_4$, 1.71 mM NaCl, 7.91 mM KNO_3, 0.34 mM $CaCl_2$, 30 µM $FeCl_3$, 1% mannitol (w/v) and 0.3% yeast extract (w/v; Sigma-Aldrich), adjusted to pH 6.8. Next, the bacteria were grown aerobically in liquid E′ medium on a rotary shaker (145 rpm) at 25°C for 24 h to reach the exponential phase of growth. Culture of bacteria cells was pelleted by centrifugation (3200 g, 15 min, 20°C) and resuspended in deionized water. To obtain 3.10^7 colony forming units (cfu) per gram of soil, the volume was adjusted based upon a correspondence with the absorbance measured at 595 nm (WPA UV 1101, Biotech Photometer, Cambridge, UK). This inoculum was directly put into the non-sterilized soil substrate (see Table S1 in File S1 for soil chemical properties), which was then manually homogenized.

Plant material, growth conditions and irrigation treatments

All experiments were realized with *A. thaliana* (L.) Heynh accession Col-0. Five seeds were sown at the soil surface in 260 mL culture pots filled with a damped mixture (1:1, v:v) of loamy soil and organic compost (Neuhaus N2; see Table S1 in File S1 for soil chemical properties) inoculated with STM196 or not. Non-inoculated soil was previously damped with deionized water to avoid difference in initial soil humidity with inoculated soil. Soil water content was controlled during pot filling by determining soil fresh weight (FW_{soil}) and soil dry weight (DW_{soil}, after 5 d at 60°C) every ten pots. Initial soil relative water content was determined as $RWC_{soil} = (FW_{soil} - DW_{soil}) \times 100 \times DW_{soil}^{-1}$. The pots were kept in the dark for two days in the PHENOPSIS growth chamber [43] and were damped with sprayed deionized water three times a day until germination. Then, plants were cultivated under 12 h day length (180 µmol m^{-2} s^{-1} photosynthetic photon flux density, PPFD, at plant height). During germination phase (7 d), air temperature was set to 20°C day and night, and air relative humidity was adjusted in order to maintain constant water vapor pressure deficit (VPD) at 0.6 kPa. Then, plants were grown at 20/17°C day/night and 0.8 kPa of VPD. Seedlings with similar sizes and developmental stages were selected and thinned to one to four plants per pot just before the beginning of water stress (see Figure 1A, C and Table 1, for watering scenarios and details on replicate numbers). Soil water content was daily adjusted with a modified one-tenth-strength Hoagland solution [44]. Soil water content was maintained at 0.35 g H_2O g^{-1} dry soil in the well-watered treatment (35%, WW) and it was decreased progressively to the desired RWC_{soil} by stopping irrigation in the water deficit treatments (WD; Table 1). Continuous moderate water deficit ($20\%_c$) was maintained at 0.20 g H_2O g^{-1} dry soil during the whole plant life cycle. In the case of severe punctual stresses, when the soil reached the desired RWC_{soil} level depending on the experiment (*i.e.*, 0.10, 0.07 or 0.06 g H_2O g^{-1} dry soil), irrigation was resumed after 1 day (for $10\%_p$, $7\%_p$ and $6\%_p$ stresses) or after 10 days (for $10\%_{p\text{-}10d}$) to progressively reach the WW soil condition (avoiding no more than 10 ml of the modified Hoagland solution per day to avoid soil leaching). Soil water content was then maintained at WW until final harvests at first flower open (stage 6.00; [45]).

Soil water potential was determined during soil drying (from 0.35 to 0.06 g H_2O g^{-1} dry soil, Table 1 and Figure S1 in File S1; WP4-T dewpoint meter, Decagon Devices, Pullman, WA 99163, USA).

Plant survival

Plants that failed to develop after rewatering and deteriorated until the complete senescence of the rosette were considered as dead. Survival percentage was scored in three consecutive experiments that were carried out following the same experimental procedure (Table 1). In experiments 1 and 2, water stresses ($20\%_c$, $10\%_{p\text{-}10d}$, $10\%_p$ and $7\%_p$) were started at the emergence of the

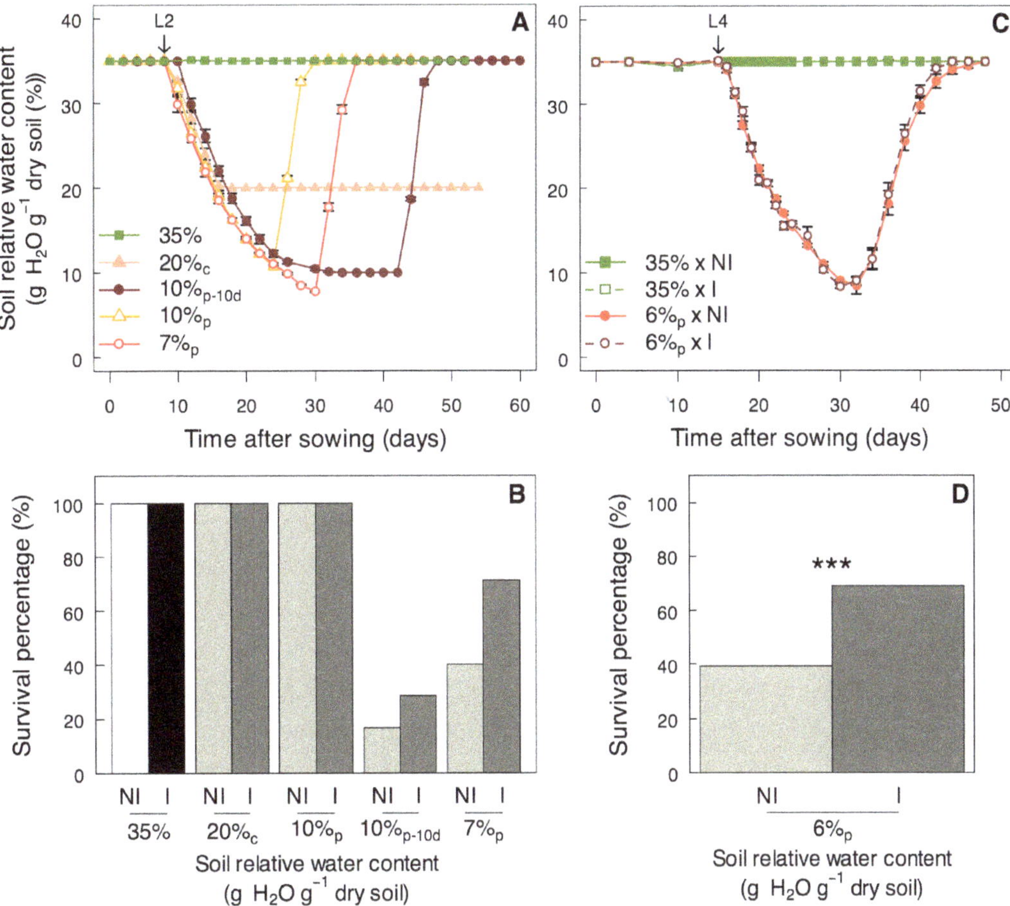

Figure 1. *Phyllobacterium brassicacearum* STM196 enhances *A. thaliana* survival under severe water deficits. A) Average soil relative water content and **B)** survival percentage of non-inoculated (NI) and inoculated plants (I) in five watering scenarios including constant well-watered conditions (35% g H_2O^{-1} dry soil; 35%), water withdrawing from the two firsts leaves (L2) followed by constant moderate water deficit (20% g H_2O^{-1} dry soil; 20%$_c$), punctual severe water deficits with rewatering after 10 days at 10% g H_2O^{-1} dry soil (10% g H_2O^{-1} dry soil; 10%$_{p\text{-}10d}$) or after1 day (10%$_p$), and after 1 day at 7% g H_2O^{-1} dry soil (7%$_p$). **C)** Average soil relative water content and **D)** survival percentage of NI (closed symbols) and I (open symbols) plants in two watering scenarios including constant well-watered conditions (35% g H_2O^{-1} dry soil; 35%), and water withdrawing from the four-leaves stage (L4) followed by rewatering after 1 day at 6% g H_2O^{-1} dry soil (6%$_p$). Asterisks indicate significant differences following Chi^2 test between NI and I plants (***: $P<0.001$).

first two leaves (L2; stage 1.02; [45]). In experiment 3, the number of replicates was increased in order to maximize the statistical power, water stress was started at four-leaf stage (L4; stage 1.04; [45]) to allow precise measurements on early developmental stages and RWC_{soil} was decreased to reach 0.06 g H_2O g^{-1} dry soil (6%$_p$ stress) before rewatering to reach well-watered soil condition (RWC_{soil} = 35%).

Measurements of whole-plant traits

Detailed plant phenotyping of growth and physiological status was performed in experiment 3 (6%$_p$ stress) throughout the whole plant cycle.

Measurement of photosynthetic efficiency. Measurement of photosynthetic efficiency was daily performed from early developmental stages to the emergence of the flowering stem (*i.e.*, bolting stage) under WW and water deficit. The maximum quantum yield of PSII was estimated by the ratio of variable to maximal chlorophyll fluorescence (F_v/F_m) on dark-adapted plants, after 8–12 h of dark (IMAGING-PAM; Maxi-version; W-IMAG-K6 camera implemented in PHENOPSIS; Imaging Win software; Walz; Effeltrich, Germany). F_v/F_m is given by $(F_m\text{-}F_0)/F_m$ [46], where F_0 is the basal fluorescence in the dark adapted state and F_m is the maximal fluorescence obtained after saturating light pulse (Si 9, width 800 ms). For unstressed plants, the value of F_v/F_m around 0.83 measured for most plant species and values lower than this indicate that plants are stressed [46]. Whole-rosette F_v/F_m values were extracted by image analyses using ImageJ (ImageJ 1.47V, Rasband, Bethesda, Maryland, USA).

Determination of plant water status. To determinate plant relative water content (RWC_{leaf}), plants were harvested at different levels of RWC_{soil} during establishment of water stress (*i.e.*, at 0.35, 0.20, 0.10 and 0.06 g H_2O g^{-1} dry soil) and after rewatering (*i.e.*, at 0.20r, 0.10r and 0.35r g H_2O g^{-1} dry soil and at first flower open). Rosettes were cut and immediately weighted, after the removal of inflorescence stems for plants harvested at stage 6.00 [45], to determine aboveground vegetative fresh mass (FM). The rosettes were wrapped in moist paper and placed into Petri dishes at 4°C in darkness overnight to achieve complete rehydration. Water-saturated fresh mass (SM) was then determined. The rosettes were oven-dried at 65°C for 48 h, and rosette dry mass (DM) was determined. From these measurements, relative water content (RWC_{leaf} = (FM – DM)×100×(SM –

Table 1. Description of watering scenarios and design of experiments.

			Water stress characteristics				Experiment		
Treatment	**Watering scenario**	**Soil water potential (Mpa)**	**Stage of irrigation withdrawing**	**Days to reach desired RWC_{soil}**	**Days before rewatering**	**Number of plants**	**1**	**2**	**3**
$35\%_c$; WW	RWC_{soil} maintained at 35% by daily irrigation during the whole plant life cycle	0.07	No stress			13–16	x	x	x
$20\%_c$	Irrigation stopped at stage L2 to decrease RWC_{soil} at 20% and RWC_{soil} thereafter maintained constant during the whole plant life cycle	−0.28	L2	6	No rewatering	9–20		x	
$10\%_{c\text{-}10d}$	Irrigation stopped at stage L2 to decrease RWC_{soil} at 10%. RWC_{soil} then maintained constant during 10 days before rewatering to reach well-watered level ($RWC_{soil}=35\%$)	−3.19	L2	15	10	7–12	x		
$10\%_p$	Irrigation stopped at stage L2 to decrease RWC_{soil} at 10%. RWC_{soil} then maintained constant during 1 day before rewatering to reach well-watered level	−3.19	L2	14	1	7–12	x		
$7\%_p$	Irrigation stopped at stage L2 to decrease RWC_{soil} at 7%. RWC_{soil} then maintained constant during 1 day before rewatering to reach well-watered level	−9.52	L2	20	1	7–10	x	x	
$6\%_p$	Irrigation stopped at stage L4 to decrease RWC_{soil} at 6%. RWC_{soil} then maintained constant during 1 day before rewatering to reach well-watered level	–	L4	30	1	48–52			x

WW: well-watered condition (control); RWC_{soil}: soil relative water content (%; g H_2O g^{-1} dry soil); L2: emergence of the first two leaves (stage 1.02; [45]) and L4: emergence of the four-leaf stage (stage 1.04; [45]). All treatments were performed in soil inoculated or not with *Phyllobacterium brassicacearum* STM196. See Figure 1A, C for a graphical representation of watering scenarios.

$DM)^{-1}$) was calculated at the rosette level. Water content (WC_{leaf}) was calculated as FM × DM^{-1} ratio.

Rosette expansion during time course. Projected area of the rosettes (RA_{proj}) was determined every days from semi-automated analysis (ImageJ 1.43C [47]) of zenithal images of the plants (Sony SSC-DC393P camera). A sigmoid curve was fitted for each plant following $RA_{proj} = a/[1+\exp-[(d-a/2)/b]]$ where a is the maximum area, and d is the number of days after sowing. The maximum rate of leaf expansion (R_{max}, $mm^2\ d^{-1}$) was calculated from the first derivative of this logistic model at d_0 as $R_{max} = a/(4b)$.

Flowering time was determined as the number of days from germination until visualization of the first flower open.

Measurements of leaf morphology at flowering. Surviving individuals were harvested at first flower open. Rosettes were cut and immediately weighted after the removal of inflorescence stems to determine aboveground vegetative FM. SM was then determined as describe above. Total leaf number was determined, and the leaf blades were separated from their petiole in order of leaf emergence and scanned for measurements of individual leaf area (ImageJ 1.43C). Leaf blades, petioles and reproductive structures were then separately oven-dried at 65°C for 48 h, and their dry mass was determined. Rosette DM was calculated as the sum of blades and petioles dry masses and RWC_{leaf} was calculated at the rosette level. All phenotypic data were stored in the PHENOPSIS database [47].

Quantification of bacteria in the soil

To analyze bacterial growth under water stress in soil, a natural mutant of STM196 strain was selected in a selection medium E′ containing 100 µg ml^{-1} of rifampin and then, was transformed using pCH60 vector. The vector pCH60 encodes for tetracycline resistance and contains the *gfp* gene that is constitutively expressed [48]. Bacterial concentration was estimated during soil drying at 0.35, 0.20, 0.10 and 0.06 g $H_2O\ g^{-1}$ dry soil and after rewatering at 0.35r g $H_2O\ g^{-1}$ dry soil. Quantification of bacteria was performed in soil without plant. The concentration of colony-forming units (cfu/mg) was estimated using the most probable number method (MPN; [49]). 100 mg of inoculated soil were put in 1 ml of physiological water (8.5 g l^{-1} de NaCl) on a rotary shaker (145 rpm) at 25°C for 2:30. The solubilized soil samples were serially diluted until 10^{-7}, and 100 µl were spread in Petri dishes on a sterile (20 min at 120°C) 1.5% agar (w/v; Sigma-Aldrich) medium (E′) with addition of 50 µl of rifampin and tetracycline. Bacteria were then counted after 6 days at 25°C.

Statistical analyses and determination of the lethal F_v/F_m threshold

All analyses were performed using R 2.15 [50]. Comparisons of mean trait values between treatments were performed with Kruskal–Wallis non-parametric tests. Plant survival was analyzed by Chi^2 tests. To estimate the survival of harvested plants during water stress (only for 6_p% stress), a 90% lethal threshold was determined just before rewatering (*i.e.*, at 0.06 g $H_2O\ g^{-1}$ dry soil) from plants with known survival, in a dose-response analysis of survival as a function of F_v/F_m values. The relationships between survival probability and whole-rosette F_v/F_m values were modeled using a binomial logistic regression. The effect of inoculation was tested by Chi^2 tests on deviance ratio. The 90%-mortality threshold (*i.e.*, 10% survival probability) of F_v/F_m value was inferred from the regression. Plants with F_v/F_m values above this threshold were considered as able to survive the stress imposed whereas plants with F_v/F_m values below this threshold were considered as perishing plants. Estimated mortality ratios (*i.e.*, proportion of perishing plants) were compared by Chi^2 tests.

Results

Phyllobacterium brassicacearum STM196 strain increases *A. thaliana* survival under multiple scenarios of severe water deficit

Arabidopsis thaliana Col-0 was grown under five scenarios of soil water availability to determine a level of stress that induced plant mortality and then analyze the effects of STM196 strain on plant survival. Soil relative water content was maintained at 0.35 g $H_2O\ g^{-1}$ dry soil in the well-watered (WW) treatment until flowering and it was decreased progressively to the desired RWC_{soil} by stopping irrigation in the water deficit treatments followed by rewatering or not (see Table 1 and Figure 1A, C for a description of the watering treatments). Under WW conditions, all plants survived and reached the reproductive stage (Figure 1A, B). All plants also survived a continuous moderate WD ($20\%_c$; Figure 1A, B), *i.e.* irrigation withdrawn from two first leaves emerged (L2) and RWC_{soil} then maintained at 20% g $H_2O\ g^{-1}$ dry soil until flowering. Decreasing RWC_{soil} punctually to 10% g $H_2O\ g^{-1}$ dry soil ($10\%_p$) did not affect plant survival, but when this RWC_{soil} level was prolonged for 10 days ($10\%_{p\text{-}10d}$) more than 80% of the non-inoculated plants died (Figure 1A, B). Decreasing RWC_{soil} punctually to 7% g $H_2O\ g^{-1}$ dry soil ($7\%_p$) resulted in 40% of non-inoculated plants that survived and reproduced after stepwise rewatering to WW conditions (Figure 1A, B).

To perform accurate measurements of plant development and physiology during soil drying, the beginning of water stress was delayed to four leaves emerged (L4), and RWC_{soil} was punctually decreased to 6% ($6\%_p$; Figure 1C). Under this scenario, plant survival rate of non-inoculated plants was 40%, *i.e.* similar to the rate observed under punctual $7\%_p$ stress (Figure 1B, D). In all watering scenarios causing plant mortality ($10\%_{p\text{-}10d}$, $7\%_p$ and $6\%_p$), soil inoculation by STM196 strain resulted in a great increase in plant survival rate (Figure 1B, D). For instance, 70% of inoculated plants survived against only 40% of non-inoculated plants under $6\%_p$ stress ($P<0.001$). This stress level was reached 1.7 days earlier in inoculated plants than in non-inoculated plants (the mean ± SE number of days to reach 6% RWC_{soil} was 16.8±1.9 ($n = 50$) and 18.5±2.2 ($n = 48$) for inoculated and non-inoculated plants, respectively; $P<0.001$). To decipher the effects of STM196 (only under $6\%_p$ stress) at similar RWC_{soil} levels, the traits of stressed plants were analyzed and presented independently of time but as a function of soil humidity during soil drying and after rewatering. The growth of STM196 in the soil was also analyzed during the WD treatment, without plant. Bacterial growth was not affected by WD and the concentration of bacteria remained constant during the experiment (Figure S2 in File S1).

STM196 strain delays and reduces plant mortality under severe water deficit

Non-destructive measurements of Chl-fluorescence were used as a sensitive indicator of photosynthetic performance (efficiency of PSII) from early developmental stages to the emergence of flowering stem. Under WW conditions, whole-rosette mean F_v/F_m was 0.80 during the entire life cycle and was not affected by soil inoculation with STM196 ($P = 0.57$; see Figure S3 in File S1). As expected, F_v/F_m decreased significantly under severe WD (6_p% stress). Mean F_v/F_m just before rewatering (*i.e.*, $RWC_{soil} = 6\%$ g $H_2O\ g^{-1}$ dry soil) was equal to 0.7 for surviving plants whereas it was equal to 0.3 for the plants that failed to develop and perished after rewatering, for both non-inoculated and inoculated plants

(Figure 2A, C). A 90% lethal threshold was then determined with F_v/F_m values of these latter plants in order to estimate the mortality of harvested plants with unknown survival (Figure 2A, B). There was no difference between the logistic regressions of survival on F_v/F_m performed on non-inoculated and inoculated plants ($P = 0.518$; Figure 2B) and the average fit was therefore used. The 90%-mortality threshold was inferred at $F_v/F_m = 0.398$. In further analyses, plants with F_v/F_m values above this threshold were considered as able to survive the stress imposed and plants with F_v/F_m values below this threshold were considered as perishing. The distinction between surviving and perishing plants was crucial to avoid errors of interpretation of the results due to a higher number of inoculated surviving plants, and could help to differentiate the behavior of plants according to their ability to survive to WD. This threshold showed that estimated mortality rate (*i.e.*, the proportion of perishing plants) tended to increase at 20% g H_2O g^{-1} dry soil in non-inoculated plants and never before 6% g H_2O g^{-1} dry soil in inoculated plants (Figure 2D). From $RWC_{soil} = 6\%$ g H_2O g^{-1} dry soil and after rewatering, the estimated mortality rate of inoculated plants was significantly lower than that of non-inoculated plants (Figure 2D; note that at the end of the experiment most senescing plants were no more detectable because decomposition started, which explains the biased decrease of mortality rate observed).

Delayed dehydration of tissues confers a higher tolerance to photosynthetic damages in STM196-inoculated plants

Whole-rosette Chl-fluorescence was then analyzed independently in surviving and perishing plants inoculated or not with STM196. At the whole-rosette level, the decrease in mean F_v/F_m was not progressive in plants exposed to stress but was dramatically affected beyond 10% RWC_{soil} in both surviving and perishing plants, with a higher magnitude for the latter (Figure 3A, B). At the maximum of stress severity (*i.e.*, 6% RWC_{soil}), lowering of whole-rosette mean F_v/F_m was more pronounced in surviving inoculated plants than in non-inoculated plants ($P<0.05$), and F_v/F_m of inoculated plants was closer to the mortality threshold (see grey points and dashed line in Figure 3A). Upon rewatering, whole-rosette mean F_v/F_m of both non-inoculated and inoculated surviving plants recovered progressively F_v/F_m values to reach initial mean F_v/F_m (0.8), similar to non-stressed plants (Figure 3A and Figure S3 in File S1). Both inoculated and non-inoculated perishing plants reached an equivalent mean F_v/F_m (0.49) at 6% RWC_{soil} (Figure 3B). This result suggests that inoculation by STM196 induced a slight decrease in photosynthetic performance but surviving inoculated plants had higher tolerance to photosynthetic damages under WD.

Severe WD in the soil unequivocally led to reduced water content in plant tissues (Figure 4A). RWC_{leaf} was progressively affected by soil drying and 6% RWC_{soil} resulted in a great decrease causing a RWC_{leaf} as low as 25% in non-inoculated plants compared to 82% in plants grown under WW conditions (Figure 4A). At 10% RWC_{soil}, surviving inoculated plants displayed higher RWC_{leaf} ($P<0.001$) than non-inoculated plants, which suggested that soil inoculation by STM196 slowed the loss of water in the leaves. At 6% RWC_{soil} the effect of inoculation was opposite and the RWC_{leaf} of surviving inoculated plants was lower than that of non-inoculated plants ($P<0.05$). This result showed that soil inoculation by STM196 allowed plants to withstand higher leaf dehydration than non-inoculated plants. RWC_{leaf} and Chl-fluorescence were closely related (Figure 4B). The relationship between F_v/F_m and RWC_{leaf} showed that inoculated plants displayed a lower decline of F_v/F_m for lower values of RWC_{leaf} (*e.g.*, around 20%). Fitting a logistic regression to the relationship between F_v/F_m and WC_{leaf} also showed that the decrease of F_v/F_m in response to WD was delayed in inoculated plants compared to non-inoculated plants and appeared for lower values of WC_{leaf} (see Figure S4 in File S1). Moreover, inoculated plants displayed higher survival probability (estimated from whole-rosette F_v/F_m values; Figure 1B) at very low RWC_{leaf} (*e.g.*, around 20%), and the decline of survival as a function of RWC_{leaf} was delayed in inoculated plants compared to non-inoculated plants (Figure 5). Together these results showed that STM 196 induced a higher plant survival during stress through a good maintenance of photosynthetic efficiency at worst leaf dehydration levels.

STM196 improves growth recovery of surviving plants, and increases biomass production

Establishment of WD ($6\%_p$ stress) resulted in reduced leaf growth, and total leaf area declined until rewatering compared to plants under WW conditions (Figure 6A). Upon rewatering, leaf growth of stressed surviving plants resumed and the plants reached the reproductive stage. At flowering, WD induced a decrease by 50% of total leaf area in non-inoculated plants (insert in Fig 6C).

The plant growth promotion effect of STM196 was not effective under WW conditions but strongly occurred under WD (Figure 6). Under WD, soil inoculation by STM196 induced a significant increase in the maximum rate of leaf expansion after rewatering (R_{max}; insert in Figure 6A; $P<0.01$) that led to a larger total leaf area at flowering (insert in Figure 6C; $P<0.01$). This was associated with a significant 45% increase of shoot dry biomass in inoculated plants under WD (Figure S5A in File S1). The increase in total leaf area of inoculated plants under WD was associated with larger individual leaves than non-inoculated plants (Figure 6C). At flowering, inoculated plants displayed also a higher number of leaves (Figure 6C). Flowering time was delayed by 15 d under WD but it was not affected by inoculation (Figure S5B in File S1). At flowering, all surviving plants recovered a complete rehydration of tissues (Figure S5C in File S1). Taken together, all traits demonstrated a better tolerance of inoculated plants to severe WD and an improvement of biomass accumulation upon rewatering.

Discussion

Severe water stress induces dehydration of plant tissues and can cause irreversible cellular damages leading to death [17]. Nonetheless, plants are able to some extent to withstand periods in a dried status and restart their metabolic functions after rehydration (*e.g.*, [25,51,52]). Several genes in Arabidopsis have been shown to be implicated in plant survival to water deficit and transgenic modifications could improve plant survival [53]. In addition, some soil bacteria such as PGPR strains can improve tolerance to water deficit, but reports on their effects on plant survival are scarce [54,55], specifically in response to severe water stress.

We recently showed that the PGPR *Phyllobacterium brassicacearum* strain STM196, previously isolated from the rhizosphere of oilseed rape *Brassica napus* [37,38], improved Arabidopsis resistance to moderate water deficit through delayed developmental transitions and modifications of plant physiology, notably by a decrease of leaf transpiration through an increase of leaf abscisic acid (ABA) content [29]. Here, we show that inoculation by STM196 strain consistently induces a significant increase in survival rate under multiple scenarios of severe water deficit. We highlight that STM196 delayed and reduced mortality rate during water stress establishment through a better tolerance to leaf

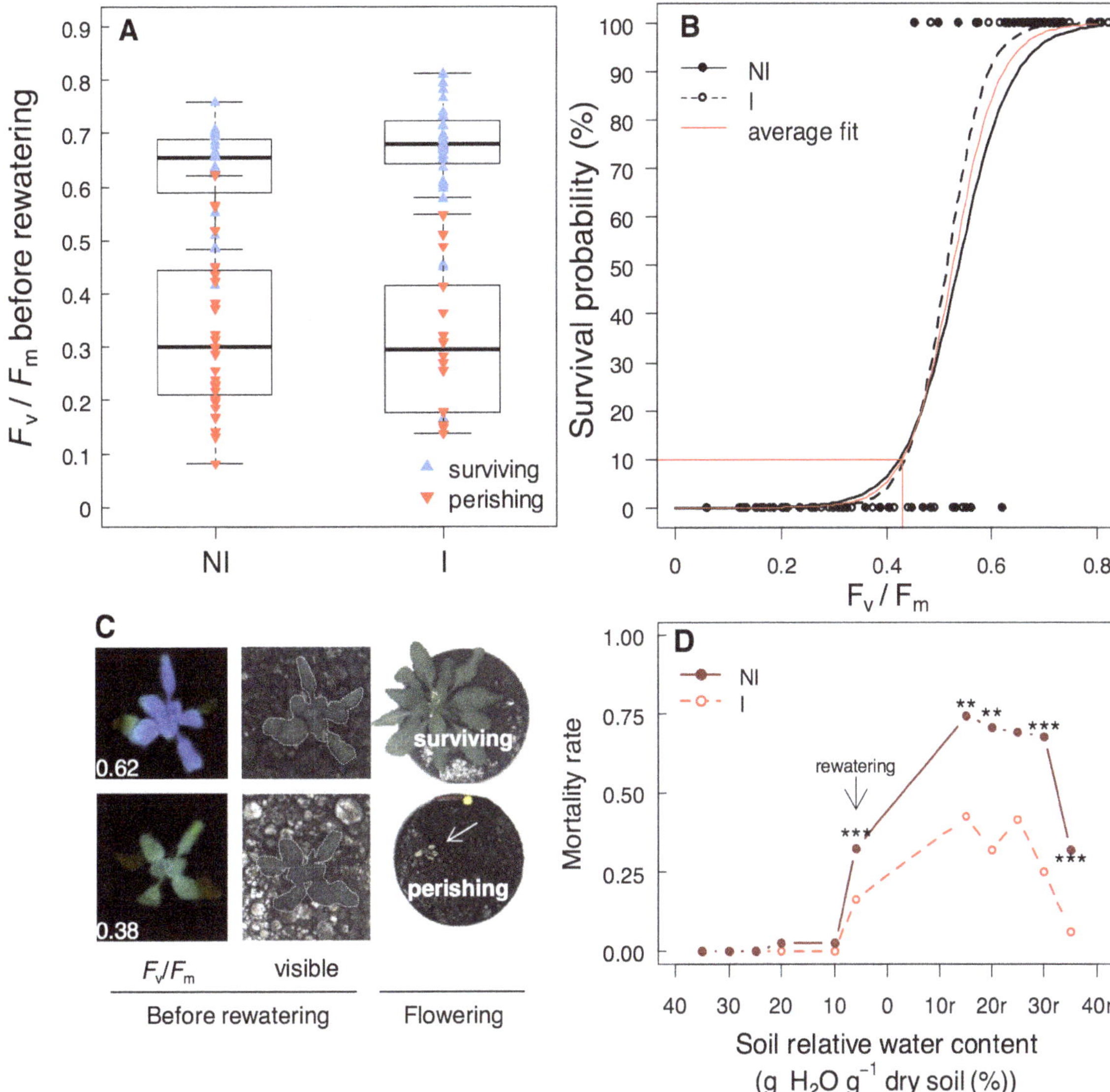

Figure 2. ***P. brassicacearum*** **STM196 does not affect** ***A. thaliana*** **mortality threshold but delays and reduces mortality rate during soil drying. A)** Whole-rosette F_v/F_m just before rewatering (*i.e.*, 0.06 g H_2O^{-1} dry soil) of non-inoculated (NI) and inoculated (I), and surviving (▲; n = 19–36) and perishing (▼; n = 16–29) plants as observed at the end of the experiment conducted with water withdrawing followed by rewatering at 6% g H_2O^{-1} dry soil (6%$_p$ stress in Figure 1). **B)** Relationships between F_v/F_m and survival probability (same data as in A); the 90%-mortality threshold (F_v/F_m = 0.398) is shown. **C)** F_v/F_m false-colour images (left) and visible images (middle) of vegetative rosettes before rewatering and of surviving flowering and perishing plant at the end of the experiment (right). **D)** Mortality rate of stressed NI (closed symbols) and I (open symbols) plants during soil drying and rewatering, as estimated from the 90%-mortality threshold. Asterisks indicate significant differences following Chi^2 test between NI (n = 28–242) and I (n = 16–187) plants (**: P<0.01; ***: P<0.001).

dehydration and leaf photosynthetic damages. Contrary to common findings where rhizobacteria enhance physiological plant status (*e.g.*, leaf water content or photosynthetic performance), here we show that STM196-inoculated plants can survive under stress with more leaf damages. Importantly, STM196 not only increased plant survival but also increased growth recovery in surviving plants and led to a higher biomass production at flowering.

Inoculation by STM196 allows a better tolerance to leaf damages and conservation of leaf water content during stress, and a better growth recovery after rewatering

Although some studies detailed the mechanisms underlying the improvement of plant resistance to water stress by PGPR inoculation, a very few studies have showed that PGPR could improve plant survival under drastic conditions. It has been shown that some rhizobacteria, genetically modified to overproduce trehalose in their cells, can improve survival of plants under severe

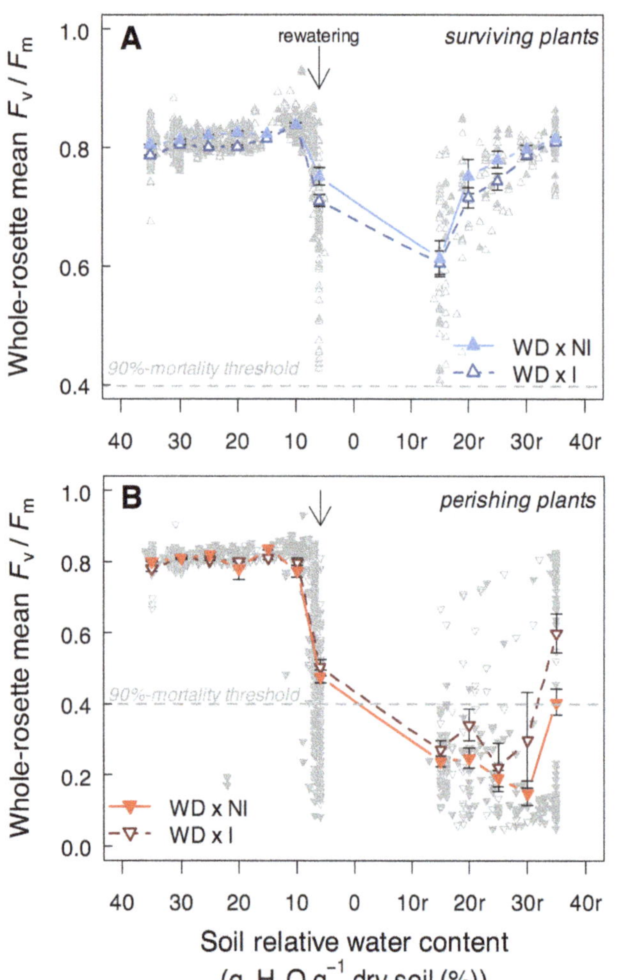

Figure 3. ***P. brassicacearum*** **STM196 improves** ***A. thaliana*** **tolerance to higher levels of leaf photosynthetic damages under severe water deficit (WD; 6%$_p$).** Whole-rosette F_v/F_m of **A)** surviving and **B)** perishing non-inoculated (NI; $n_{surviving}=7–147$; $n_{perishing}=6–137$) and inoculated (I; $n_{surviving}=10–152$; $n_{perishing}=6–78$) plants during soil drying and during rewatering. Dashed lines show the 90%-mortality threshold and arrows indicate the beginning of soil rewatering. Grey points represent individuals for each condition (NI; closed symbols and I; open symbols).

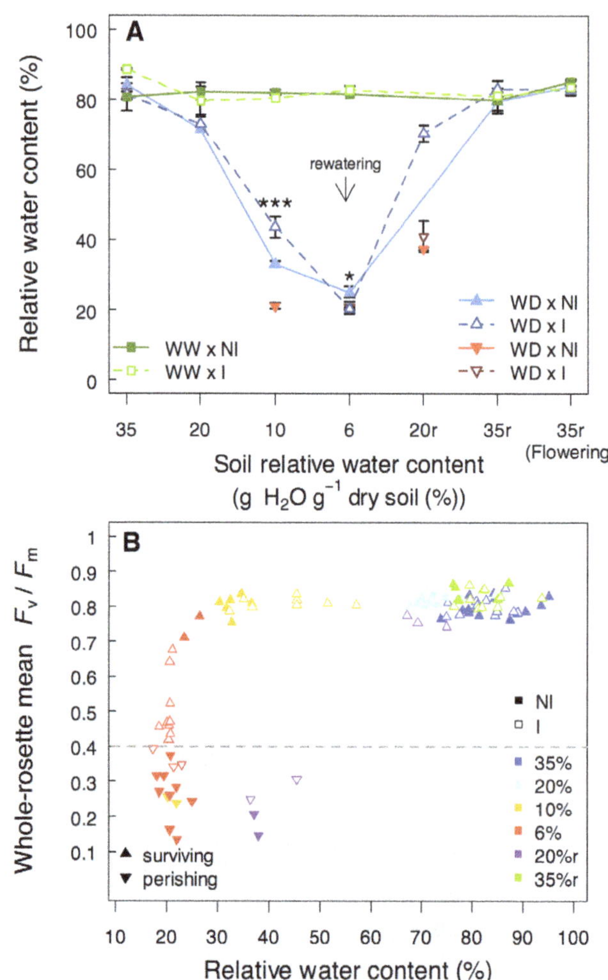

Figure 4. ***P. brassicacearum*** **STM196 induces a delayed dehydration of tissues and increases tolerance to severe water deficit. A)** Leaf relative water content and **B)** relationship between whole-rosette F_v/F_m and leaf relative water content of non-inoculated (NI; closed symbols) and STM196-inoculated (I; open symbols) plants under well watered (WW) and water deficit (WD; 6%$_p$) during soil drying (35%, 20%, 10% and 6%) and after rewatering (20%r, 35%r and 35%r at flowering). Arrow in A indicates the beginning of soil rewatering. Dashed line in B represents the 90%-mortality threshold. Surviving plants with mean F_v/F_m values above the threshold, are represented by triangles (▲; n = 3–10 and n = 3–19 for NI and I plants, respectively) and perishing plants, with mean F_v/F_m below the threshold, are represented by upside-down triangles (▼; n = 3–9 and n = 3 for NI and I plants, respectively). Asterisks indicate significant differences following Kruskal-Wallis tests between NI and I plants (*: $P<0.05$ and ***: $P<0.001$).

water-limiting conditions, notably by increasing leaf water content or by inducing the accumulation of trehalose content in the plant [54,55]. Here, we used the automated phenotyping platform PHENOPSIS, that allows the precise control of soil watering [43], to analyze the effects of STM196, a natural PGPR, on the physiology and growth of *A. thaliana* under multiple scenarios of severe water deficit throughout the whole plant cycle. The scenarios of water deficit used in this study induced a large decrease in plant survival from 60 to 83%, which is comparable to a previous report using a similar procedure (water stress/rewatering from stage 1.04) and similar intensities of soil drying [53]. Plants inoculated by STM196 strain consistently presented a higher survival rate in comparison with non-inoculated plants.

It is well established that severe water stress strongly affects plant growth, water status and causes decline of photosynthetic capacity [17], specifically through stomatal closure and leaf senescence. Dedicated measurements require a precise knowledge of the dynamics of stress establishment and are often highly time-consuming. For this reason, non-destructive measurements based on chlorophyll fluorescence imaging have been extensively used to decipher the effects of different stresses on plant physiology (*e.g.*, [56,57,58]) but have rarely been used at high throughput (but see [21]). In this paper, we used chlorophyll fluorescence measurements at high throughput in order to unravel the effects of rhizobacteria on the dynamic plant responses to severe water deficit. Amongst the different photosynthetic parameters existing, dark-adapted F_v/F_m, reflects the maximal efficiency of PSII and is therefore one of the most used parameters for measuring leaf physiological status [20]. Most often the mean F_v/F_m of a photosynthetic organ or a whole-plant is used to characterize the

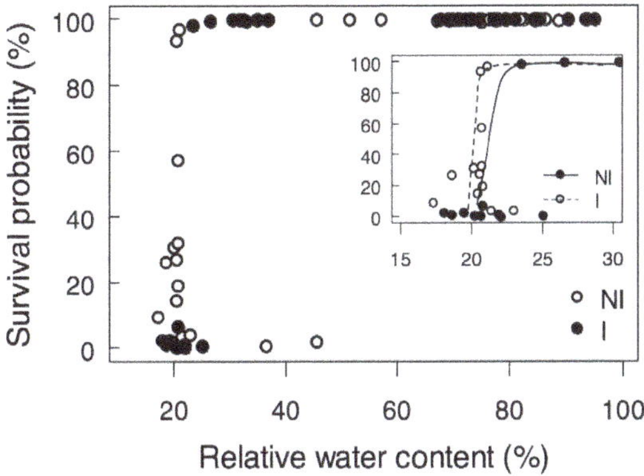

Figure 5. *P. brassicacearum* STM196 increases plant survival to severe leaf dehydration. Relationship between plant survival (estimated from whole-rosette F_v/F_m values; Figure 1B) and leaf relative water content of non-inoculated (NI; closed circle; n = 36) and STM196-inoculated (I; open circle; n = 44) under severe water deficit (6%$_p$). Insert represents fitting of logistic regression at very low leaf relative water content (solid and dashed lines for NI and I plants, respectively).

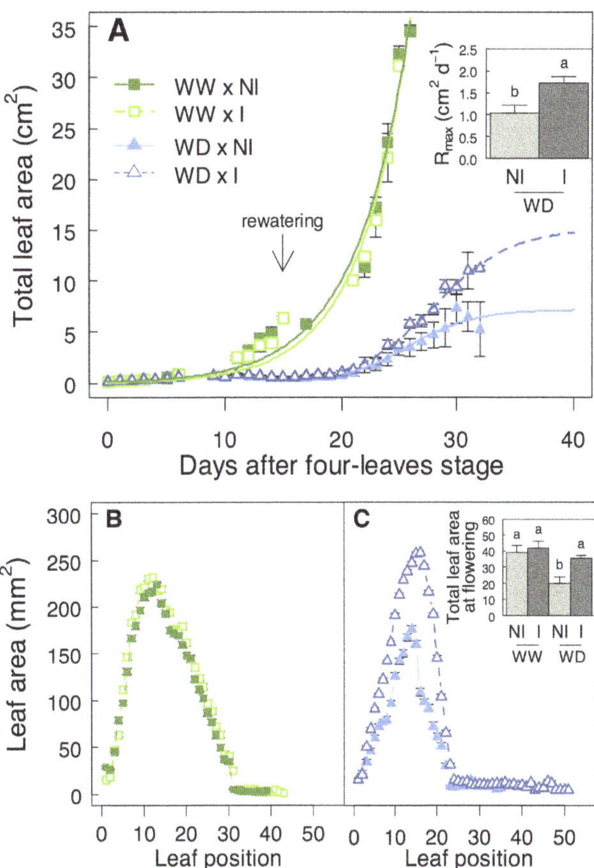

Figure 6. *P. brassicacearum* STM196 increases growth rate of surviving plants after rewatering. A) Total projected leaf area of non-inoculated (NI; closed symbols) and STM196-inoculated (I; open symbols) plants under well-watered condition (WW) and severe water deficit (WD; 6%$_p$ in Figure 1) as a function of days after four-leaves stage. Arrow in A indicate the beginning of soil rewatering. Insert in **A** represents the maximum rate of leaf expansion (R_{max}) after rewatering of surviving stressed plants. Area of individual leaves of I and NI plants under **B)** WW condition and **C)** WD. Insert in **C** shows total leaf area of surviving plants at flowering. Data are means (±SE) of 11-27 plants. Different letters indicate significant differences between means following Kruskal-Wallis tests ($P<0.05$).

response to a stressor (*e.g.*, [22]). Here, we first showed that the whole-rosette mean F_v/F_m was related to the probability of survival to severe water deficit. The determination of a mortality threshold allowed the estimation of survival of harvested plants and thus, the discrimination between surviving and perishing plants. The mortality threshold also allowed following the variation of plant mortality during time course. Moreover, this method was necessary to decipher the effect of an exogenous treatment that induced differences in sample size. Then, we showed that improvement of plant survival by STM196-inoculation was not related to changes in mortality threshold as determined by whole-rosette F_v/F_m values but was associated to differences in tolerance to WD of surviving plants. During stress, plants are able to some extent to endure leaf photosynthetic damages. Surviving inoculated plants tolerate lower values of whole-rosette mean F_v/F_m just before rewatering. The "Point of no return", the limit point that once passed a plant dies, seemed to appear for lower values of F_v/F_m in inoculated plants. The large decline in mean F_v/F_m during prolonged water deficit is consistently associated with exacerbated leaf senescence [59,60]. STM196-inoculated plants could survive with a higher proportion of leaf senescence and thus, presented a higher tolerance to leaf photosynthetic damages. Therefore, inoculated plants displayed a delayed and reduced mortality rate during water stress establishment. Leaf senescence is a common way to saving resources [18]. It allows reallocation of nutrients to reproductive organs and reduces water consumption by older and less productive leaves [61]. Leaf senescence is therefore an adaptive trait that may allow plant survival under stressful conditions [61,62]. It has been reported that some microorganisms are able to affect photosynthetic efficiency, especially by an increase of whole-rosette F_v/F_m. For instance, inoculation by the PGPR *Pseudomonas fluorescens* Aur6 strain in *P. halepensis* increased mean F_v/F_m value and lead to the improvement of tree growth under well-watered conditions [33]. The increase in chlorophyll content could participate to the PGPR-triggered improvement of plant photosynthetic performance [31]. Under water stress, a positive correlation between tolerance to water deficit and maintenance of PSII efficiency has been observed in rice inoculated by an arbuscular mycorrhizal fungus [63]. By contrast, it has been recently shown that inoculation by the PGPR *Bukholderia phytofirmans* PsJn strain induces a higher number of senescent leaves in *A. thaliana* at flowering under well watered conditions [64]. Here in accordance with this finding, we found that plants inoculated by PGPR could survive with more critical physiological status.

The improvement of tolerance to leaf damages by STM196-inoculation could be related to a delayed dehydration of tissues and an improved tolerance to low water status. PSII efficiency and leaf relative water content were tightly related, as previously reported by Woo et al., [22]. Traits related to leaf water status are often measured in response to rhizobacteria and drought. In response to PGPR-inoculation, it is widely accepted that rhizobacteria increase leaf water content that leads to increase plant resistance under water deprivation (*e.g.*, [65,66,67]). Here, inoculation by STM196 led to delayed leaf dehydration and then, at the maximum of stress severity, inoculated plants displayed a higher tolerance to low water status. Contrary to common findings, we show that STM196-inoculated plants were more likely to survive at very low water status compared to non-inoculated plants. Moreover, during water-stress establishment, STM196-inoculated plants displayed a lower decline of F_v/F_m for a given leaf water content, and non-inoculated plants began to die at lower soil humidity compared to non-inoculated plants. Delayed leaf dehydration induced by STM196-inoculation could explain the delayed mortality. Dehydration delay and dehydration-tolerance are important in survival strategy [52]. These involve traits that increase access to water and decrease water losses and could result from osmolytes accumulation [68,69], changes in stomatal conductance [70] and a large and deep root system [71]. It has been reported that inoculation by *Bacillus spp.* could alleviate negative effects of drought by affecting osmo-regulation through increasing osmoprotectors such as proline, sugars and free amino acids [72]. In the case of STM196 strain, our previous studies under moderate water deficit have shown that inoculation improves *A. thaliana*'s strategy of water saving by a developmental slowdown, a two-fold increase in root biomass and a significant decrease of transpiration rate related to an increase of ABA concentration in the leaf [29]. ABA plays a crucial role in plant responses to water stress and is involved in water loss regulation by control of stomatal closure. Modifications in leaf ABA content by STM196-inoculation could participate to delay and improve tolerance to dehydration and may be a cause of a better survival of plants under severe water stress. Moreover, it has been showed that changes in ABA content could also play a crucial role in the carbon remobilization from senescing leaves of drought-stressed plants [61]. Some other bacteria have also the capacity to modulate ABA metabolism in plants. For instance, recent work showed that inoculation by *Bacillus licheniformis* induces delayed water losses in grapevine that was correlated to an increase of ABA in leaf tissues [73]. Inoculation by STM196 may allow plants to be more efficient to cope with water scarcity in soils.

After rewatering, plant processes such as photosynthesis [23], transpiration [74], plant water status and growth [25] progressively recover their potential. Leaf growth rate followed the variation of soil water availability, and thus its decrease occurred progressively during water stress establishment. Upon rewatering, surviving plants resumed their growth and developed new leaves. We showed that inoculation by STM196 induced a better growth rate after rewatering and led to a large increase in biomass at flowering. Inoculated plants reached a similar biomass at flowering than non-stressed plants. This is due to an increase in both the number and size of leaves. This result was in accordance with our previous findings under moderate water deficit [29], where inoculation by STM196 allowed a 2-fold increase in plant biomass related to an increase in number and size of individual leaves. However, contrary to the findings under moderate water deficit, improvement of plant biomass by STM196 was not related to a delayed flowering time after rewatering.

STM196 may therefore allow a better conservation of leaf water content during stress establishment and help maintaining physiological integrity in a dried state, and then a better growth recovery when soil conditions become suitable for plant growth. The underlying physiological and molecular processes that could be involved in cells viability and growth potential remain to be elucidated.

Conclusion

Overall our findings indicate that inoculation by *Phyllobacterium brassicaceraum* STM196 strain reinforced the survival strategy of *A. thaliana* under conditions of severe water stress. STM196 induced a better tolerance to leaf damages through delayed leaf dehydration during water stress establishment that could allow a better conservation of cell integrity and thus, growth recovery when soil conditions became favorable again. Remarkably, STM196 allowed a production of plant biomass similar to non-stressed plants. Improvement of plant tolerance to water stress is a real challenge for crop breeding, especially under global climate change. The use of plant-bacteria interactions to enhance plant tolerance to abiotic stresses in the field offers valuable and promising prospects in addition or in complement to the classical strategies of genetic selection.

Supporting Information

File S1 Supporting information. Table S1, Soil chemical properties of the compost (Neuhaus N2), soil and two mixtures of both. Mixture 1 was sampled before experimentation and mixture 2 was sampled after experimentations. nd: not determined. Soil analysis was performed by ALFA Agricultural Service and Research Building, Soil Testing Laboratory of Auburn University. Figure S1, Soil water potential during soil drying. Soil water potential was determined using a potentiometer (WP4-T dewpoint meter, Decagon Devices, Pullman, WA 99163, USA) during soil drying (from 0.35 to 0.06 g H_2O g^{-1} dry soil). Figure S2, Growth of *P. brassicacearum* STM196 strain is not affected by soil water deficit. Growth of STM196 strain was represented by cfu/mg of soil under well-watered condition (WW) and water deficit (WD). Data are means (±SE) of 3 replicates. Figure S3, Whole-rosette mean F_v/F_m is not affected by inoculation under well watered condition (WW). Mean F_v/F_m of non-inoculated plants (NI; closed squares) and inoculated plants (I; open squares) during time courses. Data are means (±SE) of 3–32 plants. Grey points represent individuals for each condition (NI; closed symbols and I; open symbols). Figure S4, *P. brassicacearum* STM196 induces a delayed decrease of F_v/F_m in response to WD. Relationship between whole-rosette F_v/F_m and leaf relative water content of non-inoculated (NI; closed symbols) and STM196-inoculated (I; open symbols) plants under well watered (WW) and water deficit (WD; $6\%_p$) during soil drying (35%, 20%, 10% and 6%) and after rewatering (20%r, 35%r and 35%r at flowering). The dashed line represents the 90%-mortality threshold. Surviving plants with mean F_v/F_m values above the threshold, are represented by triangles (n = 3–10 and n = 3–19 for NI and I plants, respectively) and perishing plants, with mean F_v/F_m below the threshold, are represented by upside-down triangles (n = 3–9 and n = 3 for NI and I plants, respectively). Figure S5, Effect of *P. brassicacearum* STM196 strain and water deficit on growth, physiology and development of *A. thaliana* at flowering. A) Dry mass of rosette leaves, B) days to flowering and C) leaf relative water content of non-inoculated (NI) and inoculated (I) plants under well watered (WW) and severe water deficit (WD; $6\%_p$). Data are means (±SE) of 11–27 plants. Different letters indicate significant differences following Kruskal-Wallis test ($P<0.05$).

Acknowledgments

The authors are grateful to Gaëlle Rolland, Alexis Bediee, Crispulo Balsera and Garance Koch for help during the experiments. We thank Antoine LeQuéré and Isabelle Babic for providing us with *gfp*-tagged and rifampin-tetracycline-resistant STM196 strain. We also thank two anonymous reviewers for helpful comments on this paper.

Author Contributions

Conceived and designed the experiments: JB DV F. Vasseur BT. Performed the experiments: JB DV MD ML. Analyzed the data: JB DV F. Vasseur. Contributed reagents/materials/analysis tools: DV F. Varoquaux BT. Contributed to the writing of the manuscript: JB F. Vasseur F. Varoquaux BT DV.

References

1. Grayson M (2013) Agriculture and drought. Nature 501: S1–S1.
2. McDowell N, Pockman WT, Allen CD, Breshears DD, Cobb N, et al. (2008) Mechanisms of plant survival and mortality during drought: why do some plants survive while others succumb to drought? New Phytologist 178: 719–739.
3. Lugtenberg B, Kamilova F (2009) Plant-growth-promoting rhizobacteria. Annu Rev Microbiol 63: 541–556.
4. Yang J, Kloepper JW, Ryu CM (2009) Rhizosphere bacteria help plants tolerate abiotic stress. trends in Plant Science 14: 1–4.
5. Selosse MA, Baudoin E, Vandenkoornhuyse P (2004) Symbiotic microorganisms, a key for ecological success and protection of plants. Comptes rendus biologies 327: 639–648.
6. Lucy M, Reed E, Glick BR (2004) Applications of free living plant growth-promoting rhizobacteria. Antonie Van Leeuwenhoek 86: 1–25.
7. Babalola OO (2010) Beneficial bacteria of agricultural importance. Biotechnol Lett 32: 1559–1570.
8. Lawlor DW (2012) Genetic engineering to improve plant performance under drought: physiological evaluation of achievements, limitations, and possibilities. J Exp Bot 64: 83–108.
9. Thomas AK, Bettina MJE, Amy B, Melvin TT, Bouchra EIO, et al. (2009) Tolerance to low leaf water status of tropical tree seedlings is related to drought performance and distribution. Functional Ecology 23: 93–102.
10. Moore JP, Le NT, Brandt WF, Driouich A, Farrant JM (2009) Towards a systems-based understanding of plant desiccation tolerance. Trends in Plant Science 14: 110–117.
11. Kranner I, Beckett RP, Wornik S, Zorn M, Pfeifhofer HW (2002) Revival of a resurrection plant correlates with its antioxidant status. Plant Journal 31: 13–24.
12. Verslues PE, Agarwal M, Katiyar-Agarwal S, Zhu J, Zhu JK (2006) Methods and concepts in quantifying resistance to drought, salt and freezing, abiotic stresses that affect plant water status. Plant J 45: 523–539.
13. Farooq M, Wahid A, Kobayashi N, Fujita D, Basra SMA (2009) Plant drought stress: Effects, mechanisms and management sustainable agriculture. In: Lichtfouse E, Navarrete M, Debaeke P, Véronique S, Alberola C, editors: Springer Netherlands. pp. 153–188.
14. Gruszka Vendruscolo EC, Schuster I, Pileggi M, Scapim CA, Correa Molinari HB, et al. (2007) Stress-induced synthesis of proline confers tolerance to water deficit in transgenic wheat. Journal of Plant Physiology 164: 1367–1376.
15. Sakamoto A, Murata N (2002) The role of glycine betaine in the protection of plants from stress: clues from transgenic plants. Plant Cell Environ 25: 163–171.
16. Elbein AD, Pan YT, Pastuszak I, Carroll D (2003) New insights on trehalose: a multifunctional molecule. Glycobiology 13: 17R–27R.
17. McDowell NG (2011) Mechanisms linking drought, hydraulics, carbon metabolism, and vegetation mortality. Plant Physiol 155: 1051–1059.
18. Chaves MM (1991) Effects of water deficits on carbon assimilation Journal of Experimental Botany 42: 1–16.
19. Lim PO, Kim HJ, Nam HG (2007) Leaf senescence. Annu Rev Plant Biol 58: 115–136.
20. Murchie EH, Lawson T (2013) Chlorophyll fluorescence analysis: a guide to good practice and understanding some new applications. J Exp Bot: ert208.
21. Jansen M, Gilmer F, Biskup B, Nagel KA, Rascher U, et al. (2009) Simultaneous phenotyping of leaf growth and chlorophyll fluorescence via GROWSCREEN FLUORO allows detection of stress tolerance in *Arabidopsis thaliana* and other rosette plants. Functional Plant Biology 36: 902–914.
22. Woo NS, Badger MR, Pogson BJ (2008) A rapid, non-invasive procedure for quantitative assessment of drought survival using chlorophyll fluorescence. Plant Methods 4: 27.
23. Galmes J, Abadia A, Cifre J, Medrano H, Flexas J (2007) Photoprotection processes under water stress and recovery in Mediterranean plants with different growth forms and leaf habits. Physiologia Plantarum 130: 495–510.
24. Xu Z, Zhou G, Shimizu H (2009) Are plant growth and photosynthesis limited by pre-drought following rewatering in grass? Journal of Experimental Botany 60: 3737–3749.
25. Lechner L, Pereyra-Irujo GA, Granier C, Aguirrezabal LA (2008) Rewatering plants after a long water-deficit treatment reveals that leaf epidermal cells retain their ability to expand after the leaf has apparently reached its final size. Ann Bot 101: 1007–1015.
26. Figueiredo MVB, Burity HA, Martínez CR, Chanway CP (2008) Alleviation of drought stress in the common bean (*Phaseolus vulgaris* L.) by co-inoculation with *Paenibacillus polymyxa* and *Rhizobium tropici*. Applied Soil Ecology 40: 182–188.
27. Belimov AA, Dodd IC, Hontzeas N, Theobald JC, Safronova VI, et al. (2009) Rhizosphere bacteria containing 1-aminocyclopropane-1-carboxylate deaminase increase yield of plants grown in drying soil via both local and systemic hormone signalling. New Phytol 181: 413–423.
28. Liu F, Xing S, Ma H, Du Z, Ma B (2013) Cytokinin-producing, plant growth-promoting rhizobacteria that confer resistance to drought stress in *Platycladus orientalis* container seedlings. Applied microbiology and biotechnology: 1–10.
29. Bresson J, Varoquaux F, Bontpart T, Touraine B, Vile D (2013) The PGPR strain *Phyllobacterium brassicacearum* STM196 induces a reproductive delay and physiological changes that result in improved drought tolerance in Arabidopsis. New Phytologist 200: 558–569.
30. Kohler J, Hernandez JA, Caravaca F, Roldan A (2008) Plant-growth-promoting rhizobacteria and arbuscular mycorrhizal fungi modify alleviation biochemical mechanisms in water-stressed plants. Functional Plant Biology 35: 141–151.
31. Wang CJ, Yang W, Wang C, Gu C, Niu DD, et al. (2012) Induction of drought tolerance in cucumber plants by a consortium of three plant growth-promoting rhizobacterium strains. PLoS One 7.
32. Heidari M, Golpayegani A (2012) Effects of water stress and inoculation with plant growth promoting rhizobacteria (PGPR) on antioxidant status and photosynthetic pigments in basil (*Ocimum basilicum L.*). Journal of the Saudi Society of Agricultural Sciences 11: 57–61.
33. Rincon A, Valladares F, Gimeno TE, Pueyo JJ (2008) Water stress responses of two Mediterranean tree species influenced by native soil microorganisms and inoculation with a plant growth promoting rhizobacterium. Tree Physiol 28: 1693–1701.
34. Timmusk S, Timmusk K, Behers L (2013) Rhizobacterial plant drought stress tolerance enhancement: towards sustainable water resource management and food security. Journal of Food Security 1: 6–10.
35. Dimkpa C, Weinand T, Asch F (2009) Plant-rhizobacteria interactions alleviate abiotic stress conditions. Plant Cell Environ 32: 1682–1694.
36. Mantelin S, Saux MF, Zakhia F, Bena G, Bonneau S, et al. (2006) Emended description of the genus *Phyllobacterium* and description of four novel species associated with plant roots: *Phyllobacterium bourgognense* sp. nov., *Phyllobacterium ifriqiyense* sp. nov., *Phyllobacterium leguminum* sp. nov. and *Phyllobacterium brassicacearum* sp. nov. International journal of systematic and evolutionary microbiology 56: 827–839.
37. Bertrand H, Nalin R, Bally R, Cleyet-Marel JC (2001) Isolation and identification of the most efficient plant growth-promoting bacteria associated with canola (*Brassica napus*). Biology and Fertility of Soils 33: 152–156.
38. Larcher M, Muller B, Mantelin S, Rapior S, Cleyet-Marel JC (2003) Early modifications of *Brassica napus* root system architecture induced by a plant growth-promoting *Phyllobacterium* strain. New Phytologist 160: 119–125.
39. Mantelin S, Desbrosses G, Larcher M, Tranbarger TJ, Cleyet-Marel JC, et al. (2006) Nitrate-dependent control of root architecture and N nutrition are altered by a plant growth-promoting *Phyllobacterium sp.* Planta 223: 591–603.
40. Contesto C, Milesi S, Mantelin S, Zancarini A, Desbrosses G, et al. (2010) The auxin-signaling pathway is required for the lateral root response of Arabidopsis to the rhizobacterium *Phyllobacterium brassicacearum*. Planta 232: 1455–1470.
41. Galland M, Gamet L, Varoquaux F, Touraine B, Desbrosses G (2012) The ethylene pathway contributes to root hair elongation induced by the beneficial bacteria *Phyllobacterium brassicacearum* STM196. Plant Science 190: 74–81.
42. Kechid M, Desbrosses G, Rokhsi W, Varoquaux F, Djekoun A, et al. (2013) The *NRT2.5* and *NRT2.6* genes are involved in growth promotion of Arabidopsis by the plant growth-promoting rhizobacterium (PGPR) strain *Phyllobacterium brassicacearum* STM196. New Phytologist 198: 514–524.
43. Granier C, Aguirrezabal L, Chenu K, Cookson SJ, Dauzat M, et al. (2006) PHENOPSIS, an automated platform for reproducible phenotyping of plant responses to soil water deficit in *Arabidopsis thaliana* permitted the identification of an accession with low sensitivity to soil water deficit. New Phytol 169: 623–635.
44. Hoagland DR, Arnon DI (1950) The water-culture method for growing plants without soil. California Agricultural Experiment Station Circular 347: 1–32.
45. Boyes DC, Zayed AM, Ascenzi R, McCaskill AJ, Hoffman NE, et al. (2001) Growth stage-based phenotypic analysis of Arabidopsis: A model for high throughput functional genomics in plants. Plant Cell 13: 1499–1510.
46. Maxwell K, Johnson GN (2000) Chlorophyll fluorescence - a practical guide. Journal of Experimental Botany 51: 659–668.
47. Fabre J, Dauzat M, Negre V, Wuyts N, Tireau A, et al. (2011) PHENOPSIS DB: an Information System for Arabidopsis thaliana phenotypic data in an environmental context. BMC Plant Biology 11.
48. Cheng HP, Walker GC (1998) Succinoglycan is required for initiation and elongation of infection threads during nodulation of alfalfa by *Rhizobium meliloti*. Journal of bacteriology 180: 5183–5191.
49. Halvorson H, Ziegler N (1933) Application of statistics to problems in bacteriology: I. A means of determining bacterial population by the dilution method. Journal of Bacteriology 25: 101.
50. R Development Core Team (2009) R: a language and environment for statistical computing. Vienna, Austria: R Foundation for Statistical Computing. 409 p.
51. Rivero RM, Kojima M, Gepstein A, Sakakibara H, Mittler R, et al. (2007) Delayed leaf senescence induces extreme drought tolerance in a flowering plant. Proceedings of the National Academy of Sciences of the United States of America 104: 19631–19636.

52. Tyree MT, Engelbrecht BMJ, Vargas G, Kursar TA (2003) Desiccation tolerance of five tropical seedlings in Panama. Relationship to a field assessment of drought performance. Plant Physiology 132: 1439–1447.
53. Skirycz A, Vandenbroucke K, Clauw P, Maleux K, De Meyer B, et al. (2011) Survival and growth of Arabidopsis plants given limited water are not equal. Nature Biotechnology 29: 212–214.
54. Suarez R, Wong A, Ramirez M, Barraza A, Orozco MD, et al. (2008) Improvement of drought tolerance and grain yield in common bean by overexpressing trehalose-6-phosphate synthase in rhizobia. Molecular Plant-Microbe Interactions 21: 958–966.
55. Rodriguez-Salazar J, Suarez R, Caballero-Mellado J, Iturriaga G (2009) Trehalose accumulation in *Azospirillum brasilense* improves drought tolerance and biomass in maize plants. Fems Microbiology Letters 296: 52–59.
56. Ehlert B, Hincha DK (2008) Chlorophyll fluorescence imaging accurately quantifies freezing damage and cold acclimation responses in Arabidopsis leaves. Plant Methods 4: 12.
57. Gray GR, Hope BJ, Qin XQ, Taylor BG, Whitehead CL (2003) The characterization of photoinhibition and recovery during cold acclimation in *Arabidopsis thaliana* using chlorophyll fluorescence imaging. Physiologia Plantarum 119: 365–375.
58. Sperdouli I, Moustakas M (2012) Spatio-temporal heterogeneity in *Arabidopsis thaliana* leaves under drought stress. Plant Biology 14: 118–128.
59. Wingler A, Purdy S, MacLean JA, Pourtau N (2006) The role of sugars in integrating environmental signals during the regulation of leaf senescence. J Exp Bot 57: 391–399.
60. Diaz C, Purdy S, Christ A, Morot-Gaudry JF, Wingler A, et al. (2005) Characterization of markers to determine the extent and variability of leaf senescence in Arabidopsis. A metabolic profiling approach. Plant Physiol 138: 898–908.
61. Munné-Bosch S, Alegre L (2004) Die and let live: leaf senescence contributes to plant survival under drought stress. Functional Plant Biology 31: 203–216.
62. Ono K, Nishi Y, Watanabe A, Terashima I (2001) Possible mechanisms of adaptive leaf senescence. Plant Biology 3: 234–243.
63. Ruiz-Sanchez M, Aroca R, Munoz Y, Polon R, Ruiz-Lozano JM (2010) The arbuscular mycorrhizal symbiosis enhances the photosynthetic efficiency and the antioxidative response of rice plants subjected to drought stress. Journal of Plant Physiology 167: 862–869.
64. Poupin MJ, Timmermann T, Vega A, Zuniga A, Gonzalez B (2013) Effects of the plant growth-promoting bacterium *Burkholderia phytofirmans* PsJN throughout the life cycle of *Arabidopsis thaliana*. PLoS One 8: e69435.
65. Arzanesh MH, Alikhani HA, Khavazi K, Rahimian HA, Miransari M (2011) Wheat (*Triticum aestivum L.*) growth enhancement by *Azospirillum* sp under drought stress. World Journal of Microbiology & Biotechnology 27: 197–205.
66. Creus CM, Sueldo RJ, Barassi CA (2004) Water relations and yield in *Azospirillum*-inoculated wheat exposed to drought in the field. Canadian Journal of Botany 82: 273–281.
67. Marulanda A, Barea JM, Azcon R (2009) Stimulation of plant growth and drought tolerance by native microorganisms (AM fungi and bacteria) from dry environments: mechanisms related to bacterial effectiveness. Journal of Plant Growth Regulation 28: 115–124.
68. Morgan PW (1990) Effects of abiotic stresses on plant hormone systems. In: Wiley-Liss I, editor. in: Stress Responses in plants: adaptation and acclimation mechanisms. pp. pp. 113–146.
69. Taiz LZE (2006) Plant Physiology; Ed. t, editor. Massachusetts: Sinauer Associates Inc. Publishers.
70. Farquhar GD, Sharkey TD (1982) Stomatal Conductance and Photosynthesis. Annual Review of Plant Physiology 33: 317–345.
71. Padilla FM, Pugnaire FI (2007) Rooting depth and soil moisture control Mediterranean woody seedling survival during drought. Functional Ecology 21: 489–495.
72. Vardharajula S, Ali SZ, Grover M, Reddy G, Bandi V (2011) Drought-tolerant plant growth promoting *Bacillus spp.*: effect on growth, osmolytes, and antioxidant status of maize under drought stress. Journal of Plant Interactions 6: 1–14.
73. Salomon MV, Bottini R, de Souza Filho GA, Cohen AC, Moreno D, et al. (2013) Bacteria isolated from roots and rhizosphere of Vitis vinifera retard water losses, induce abscisic acid accumulation and synthesis of defense-related terpenes in in vitro cultured grapevine. Physiologia Plantarum.
74. Bengtson C, Falk SO, Larsson S (1977) The after-effect of water stress on transpiration rate and changes in abscisic acid content of young wheat plants. Physiologia Plantarum 41: 149–154.

15

New Insights into How Increases in Fertility Improve the Growth of Rice at the Seedling Stage in Red Soil Regions of Subtropical China

Yilin Li[1], Weiming Shi[1], Xingxiang Wang[1,2*]

1 State Key Laboratory of Soil and Sustainable Agriculture, Institute of Soil Science, Chinese Academy of Sciences, Nanjing, China, 2 Jiangxi Key Laboratory of Ecological Research of Red Soil, Yingtan, Jiangxi, China

Abstract

The differences in rhizosphere nitrification activities between high- and low- fertility soils appear to be related to differences in dissolved oxygen concentrations in the soil, implying a relationship to differences in the radial oxygen loss (ROL) of rice roots in these soils. A miniaturised Clark-type oxygen microelectrode system was used to determine rice root ROL and the rhizosphere oxygen profile, and rhizosphere nitrification activity was studied using a short-term nitrification activity assay. Rice planting significantly altered the oxygen cycling in the water-soil system due to rice root ROL. Although the oxygen content in control high-fertility soil (without rice plants) was lower than that in control low-fertility soil, high rice root ROL significantly improved the rhizosphere oxygen concentration in the high-fertility soil. High soil fertility improved the rice root growth and root porosity as well as rice root ROL, resulting in enhanced rhizosphere nitrification. High fertility also increased the content of nitrification-induced nitrate in the rhizosphere, resulting in enhanced ammonium uptake and assimilation in the rice. Although high ammonium pools in the high-fertility soil increased rhizosphere nitrification, rice root ROL might also contribute to rhizosphere nitrification improvement. This study provides new insights into the reasons that an increase in soil fertility may enhance the growth of rice. Our results suggest that an amendment of the fertiliser used in nutrient- and nitrification-poor paddy soils in the red soil regions of China may significantly promote rice growth and rice N nutrition.

Editor: Wagner L. Araujo, Universidade Federal de Vicosa, Brazil

Funding: This research was financially supported by the National Natural Science Foundation of China (No. 31471948), the Special Fund for Agro-scientific Research in the Public Interest (No. 201203050-3) and the GanPo 555 Talents Program of Jiangxi Province, China. The funders had no role in study design, data collection and analysis, decision to publish, or preparation of the manuscript.

Competing Interests: The authors have declared that no competing interests exist.

* Email: xxwang@issas.ac.cn

Introduction

Nitrification, an important component of the nitrogen cycle, involves two aerobic microbial processes. In nitrification, ammonia (NH_3) is first converted to nitrite (NO_2^-); then, NO_2^- is further converted to nitrate (NO_3^-) [1,2]. Nitrification occurs in a variety of environments, such as sediments and soils. However, in waterlogged environments, such as paddy soils and flooding waters, nitrification is limited by the low availability of oxygen, which results in low NO_3^- availability. An increased number of hydroponic experiments have demonstrated that rice absorbs and assimilates NO_3^- in preference to ammonium (NH_4^+) [3,4]. Rice growth, yield, net N acquisition and N use efficiency (NUE) have been shown to be superior using mixed NH_4^+ and NO_3^- nutrition compared with pure NH_4^+ cultivation [3,5,6]. In paddy fields, NO_3^- concentrations often fluctuate and are difficult to control compared with hydroponic culturing, primarily because of denitrification in the bulk soil and nitrification in the rhizosphere. In our previous soil culture study [7], different rice cultivars (Indica and Japonica) showed a significant difference in the nitrification activity in their rhizospheres, and rhizosphere nitrification was closely related to rice growth and N nutrition as well as rice N accumulation and NUE.

The oxidation of NH_3 is a microbial process and is thought to be the rate-limiting step for nitrification [8]. Recently, according to metagenomic [9,10] and cultivation [11] methods, another type of NH_3 oxidiser, the ammonia-oxidising archaea (AOA) group, has been found to be abundant in the natural environment [12] and has been hypothesised to be a dominant group among the ammonia-oxidising prokaryotes in the soil [13]. Because ammonia-oxidising bacteria (AOB) and AOA are both strictly aerobic chemolithoautotrophic microorganisms, nitrification can only occur in the presence of oxygen and is restricted in the oxic upper few millimetres of flooded soils [14–16].

However, rice plants form numerous adventitious roots containing aerenchyma. These roots promote radial oxygen loss (ROL) from the roots to the rhizosphere, supporting aerobic microbial nitrification [17,18]. Furthermore, ROL alleviates the phytotoxicity of reduced compounds (e.g., Fe^{2+}, Mn^{2+}, H_2S) in rice [19]. ROL is of biological significance for submerged plants and has been investigated extensively using various methods [20–23]. A Clark-type oxygen microelectrode has been used to measure the rice ROL and the detailed oxygen vertical profiles in the field, indicating that rice plants at 21 d had 20% air saturation at the root surface. Diffusion of O_2 away from the root was able to create oxic conditions up to a distance of 0.4 mm away from the root [23]. The diameter of this oxygen microelectrode was only 25 µm, which should have caused minimal damage to the soil and rice plants.

Two paddy soils (representing low and high fertilities) in red soil regions with low nitrification activity were examined by oxygen microelectrode techniques to study the relationship between rice ROL and rhizosphere nitrification and the contribution of root ROL to rice N nutrition. Red soils, which can be classified as acrisols and ferralsols according to the Food and Agriculture Organization (FAO) nomenclature [24], are widespread in tropical and subtropical China, covering 11.8% of the nation's land area. The red soil region contains abundant water resources and heat resources and is one of the principal rice production areas in China [25]. Due to intensive leaching and weathering, long-term inappropriate utilisation and management, and the inherent fragility of hill ecosystems, the red soils are characterised by low pH and fertility, as well as low nitrification activity [26,27]. When natural factors, such as parent materials, topography, and climate, are the same in the paddy soil-forming process, time and human factors, such as long-term fertilisation, are the primary influencing factors in improving soil fertility. Two paddy soils collected from paddy fields with different fertility levels were defined as low fertility (i.e., reclamation of approximately 15 years, long-term inorganic and organic fertilisation in the rice season) and high fertility (i.e., reclamation great than 50 years, long-term inorganic and organic fertilisation in the rice season and manure application in the fallow winter). Although increasing soil fertility may improve rice root growth, little is known about whether it could change the rice root aerenchyma development and ROL. Our previous study showed that oxygen concentrations in paddy soil (without rice plants) with high fertility were much lower than those measured in low-fertility paddy soil collected from red soil regions (unpublished data). In this study, we hypothesised that rice plants grown in high-fertility paddy soil would improve the root aerenchyma development and ROL, which would lead to increased oxygen in the rhizosphere soil, and, thus, enhance rhizosphere nitrification activity.

Materials and Methods

Paddy soils and rice plant cultivars

Paddy soils derived from Quaternary red clay (hydragric anthrosols according to the FAO nomenclature [24]) were collected from the Chinese Academy of Sciences Red Soil Ecological Experimental Station, located in Yingtan (28°15′20″ N, 116°55′30″ E), Jiangxi Province, which is one of the major rice production areas in China. The cultivation practice in the area is two crops per year. The Yingtan region has a typical subtropical monsoon climate with an annual precipitation of 1785 mm and a mean annual temperature of 17.8°C. Two paddy soils (0–20 cm) collected from paddy fields with different fertility levels were defined as low fertility and high fertility, and their properties are described in Table 1. The soil samples were air-dried, ground, sieved through 0.71-mm mesh, and stored for further incubation experiments. The plant residues were carefully removed by hand. One of the most popular Indica rice cultivars in Jiangxi Province, Hesheng 10, was used in the experiment.

Soil incubation and sampling

A rhizobox, as described by Li *et al.* [28], was used to collect soil samples at various distances from the rice root surface. Two pieces of nylon nets (with a 30-µm mesh size) were used to divide the rhizobox into three compartments to facilitate the isolation of the rice root from the rhizosphere soil. Three rice plants were planted in the middle compartment of each rhizobox. The mesh of the nylon net was fine enough to allow the water and nutrient elements to pass through. A 600-g sample of the paddy soil (air dried and sieved) was mixed thoroughly with urea (120 mg N kg^{-1}) and KH_2PO_4 (93 mg kg^{-1}) to fill each rhizobox (the fertiliser applications were performed according to the conventional amounts used by the local farmers to support rice growth adequately). A rhizobox without rice plants was used as the control (CK). One centimetre of surface water was maintained after rice germination, using deionised water every morning and evening throughout the incubation. The rice plants were incubated at 25°C with a 16-h photoperiod during the experiment. Three replicate rhizoboxes for each treatment were used, and soil samples were collected at 40, 50 and 60 d after rice germination; an additional six replicate rhizoboxes were used for in situ measurements of soil oxygen concentration and rice root ROL 50 d after germination.

A previous experiment in this laboratory showed that between 40 and 60 days after rice germination, the rice roots were fully developed and filled the middle compartment of the rhizobox so that the nylon nets would not influence the rice root growth. Therefore, the soil adhering to the rice roots in the inner rhizobox was defined as root surface soil, and the soil in the outer rhizobox within a 4-mm distance from the nylon nets was defined as rhizosphere soil. To prevent the water layer from interfering with the soil mineral N and nitrification activity determination, no watering occurred on the evening of the day prior to sampling to maintain a thin (1–2 mm) water layer. The sampling was conducted at 10:00 am, and each treatment (three replicates) was first frozen at −20°C for 2 h to harden the paddy soil for subsequent slicing. First, the rhizobox was carefully split along the two nylon nets into three compartments: middle, right, and left parts. Next, a piece of organic glass board (2 mm thick, 8 cm×8 cm) was inserted into the inner side of the right or left part of the rhizobox. Then, a soil sample of 2-mm thickness was pushed aside, and the soil was sliced to obtain a soil sample located 2 mm away from the root surface. Next, another piece of organic glass board (2 mm thick, 8 cm×8 cm) was inserted to obtain a layer located 4 mm away from the root surface. Using the same procedure, soil samples were obtained at distances of 6, 8, 10, 20, 30 and 40 mm from the root surface. The soil samples collected in the right and left compartments at the same distance were mixed together for the assessments. The root surface soil was separated from the rice roots using sterilised tweezers. All soil samples were collected at the same time as the plant sampling to assay for mineral N and nitrification activity.

Mineral N assays

Fresh soil samples were extracted with 2 mol L^{-1} of KCl (soil:solution ratio: 1:10), and the extracts were measured colourimetrically for NH_4^+ and NO_3^- [26] using a continuous flow autoanalyser (model Autoanalyzer 3, Bran + Luebbe, Hamburg, Germany).

Short-term nitrification activity assays

Short-term nitrification estimates are typically used for the nitrification activity assay. The principle underlying this method is based on the determination of NO_2^- after the incubation of soil

Table 1. Properties of the paddy soils used in the experiment.

Property	Low fertility	High fertility
pH (water:soil, 2.5:1)	4.75	5.02
Organic matter (g kg^{-1})	17.5	39.1
Total N (g kg^{-1})	0.85	1.74
Total P (g kg^{-1})	0.32	0.67
NH_4^+-N (mg kg^{-1})	0.72	2.27
NO_3^--N (mg kg^{-1})	0.17	0.21
Clay (%)	36.3	20.9
Average yield of double-cropping rice per year (kg hm^{-1})	7500	10000
Reclamation years	15	>50

samples with $NaClO_3$ in the absence of NH_4^+ for 24 h at 25°C [29,30]. Briefly, each soil sample (5 g) was shaken with 2.5 mL of $NaClO_3$ (75 mmol L^{-1}, which prevents oxidation of NO_2^-) at 170 rpm on a rotary shaker (model HZ-9611K, Hualida Laboratorial Equipment Co., Ltd., Taicang, China) for 30 min, and then incubated aerobically in an incubator for 24 h at 25°C. After incubation, NO_2^- was extracted from the soil samples into a total volume of 15 mL using two solutions (5 mL of deionised H_2O followed by 10 mL of 2 mol L^{-1} KCl) by shaking at 170-rpm for 30 min on a rotary shaker, as described previously. The contents were mixed thoroughly and immediately filtered. Five millilitres of the clear filtrate was pipetted into glass test tubes, followed by 3 mL of the buffer (0.19 mol L^{-1} NH_4Cl, pH 8.5) and 2 mL of the reagent (1% sulphanilamide, 0.05% N-(1-naphthyl) ethylene-diamine dihydrochloride and 10% phosphoric acid mixed solution), for NO_2^- determination. The contents were again vigorously shaken and allowed to stand for 15 min at room temperature. For a control measurement, soil samples were extracted as previously described after incubation with $NaClO_3$ at −20°C. The amount of NO_2^- produced was measured based on the absorbance at 520 nm in a spectrophotometer. Pre-experiments using the paddy soil without planted rice (waterlogged for 40 d) to observe the kinetics of ammonia oxidation in the first 24 hours were performed and revealed that nitrite was produced linearly over time ($p<0.05$).

In situ measurements of soil oxygen concentrations

An additional three replicate rhizoboxes were used for the in situ measurement of soil oxygen concentrations 50 d after germination. The incubation conditions were the same as described above except that a 1-cm water layer was maintained in the rhizoboxes for the in situ measurement.

The oxygen microelectrode was a miniaturised Clark-type oxygen electrode with a guard cathode (OXY25, ø = 20–30 µm, Unisense, Aarhus, Denmark [31]). The soil oxygen profile measurements were performed in the right and left compartments of each rhizobox and in the centre of the range at a distance of 2 mm from the nylon nets for all treatments. All measurements were performed with a 50-µm depth interval using a microelectrode with a diameter of approximately 25 µm, and the periods for "wait before measure" and "measure" were both set as 3 s.

In vitro measurements of rice root oxygen loss in an agar microcosm

Three rhizosphere boxes were split into three compartments each, and the rice plants with soil were placed into a plastic bucket filled with deionised water. The soil adhering to the rice roots was removed by repeated immersion in the deionised water. After the soil was removed, the rice roots were separated according to the individual plants and were then individually placed into beakers with International Rice Research Institute (IRRI) rice nutrient solution (pH 5.5) [32] for 2 h. The roots of each individual rice plant were then inserted into a container filled with IRRI rice nutrient solution (pH 5.5) containing 0.9% agar [23] before the agar was allowed to solidify at 38°C. The container was combined with one small culture dish (6 cm in diameter, 2.5 cm in depth) and one large culture dish (15 cm in diameter, 2.5 cm in depth). The rice root (to be measured) was inserted into the small culture dish, whereas the other roots were evenly distributed in the large culture dish. Next, paraffin oil was poured onto the surface of the solidified agar in the small culture dish to prevent air penetration. To obtain comparable results, the sample root was placed horizontally in the agar. Because the solidified agar was a clear medium, it allowed observation of the rice roots and provided fixation support for the roots. To measure the oxygen levels of the outer and inner rice roots, the microelectrode had to be sufficiently thin to longitudinally penetrate the root. Therefore, a much slimmer microelectrode (OXY10, ø = 8–10 µm, Unisense, Aarhus, Denmark [31]) was used for the measurement. Newly formed roots with lengths of approximately 45 mm were selected and measured at 25 mm from the root tip. The measurements were conducted after the rice roots had been fixed in the agar for approximately 2 h; these measurements were repeated for different rice plants with different newly formed roots with the same length as described above. All measurements were conducted with a 50-µm depth interval, and the periods for "wait before measure" and "measure" were both set to 3 s.

Because oxygen concentration is very sensitive to temperature, soil and agar temperatures were monitored and recorded during the oxygen measurements. All of the oxygen measurements, including soil oxygen concentration and root ROL, were performed in a specialised laboratory without electrical noise interference at a temperature of 20°C.

Plant dry matter and N accumulations assays

The rice tissues were separated into shoots and roots and heated at 105°C for 30 min to terminate enzyme activity. The samples were then oven-dried at 70°C for 48 h to a constant weight. The dried plant material was ground and digested for total N determination using the Kjeldahl method. A 5-mL aliquot from the 100-mL digested solution was then analysed for N using a continuous flow auto-analyser (model Autoanalyser 3, Bran + Luebbe, Hamburg, Germany).

Plant NRA and GSA assays

Fresh leaves and roots of both treatments were immediately frozen with liquid N and then stored in a refrigerator at −70°C for the later measurement of glutamine synthetase activity (GSA) and NO_3^- reductase activity (NRA).

The level of GSA was assayed using an in vitro method. Frozen plant material (0.5 g) was ground in a precooled mortar with sand and a pestle after adding liquid N and then homogenised in an extraction buffer (pH 7.2, 10 mL g^{-1} fresh weight) containing 0.5 mmol L^{-1} of EDTA and 50 mmol L^{-1} of K_2SO_4. The homogenates were centrifuged at 20,000 *g* for 20 min. Next, 1.2 mL of the clear filtrate was pipetted into a centrifuge tube, to which was then added 0.6 mL of imidazole-HCl (pH 7.0, 0.25 mol L^{-1}), 0.4 mL of sodium glutamate (pH 7.0, 0.3 mol L^{-1}), 0.4 mL of ATP-Na (pH 7.0, 15 mmol L^{-1}), 0.2 mL of $MgSO_4$ (0.5 mol L^{-1}) and 0.2 mL of hydroxylamine (1 mol L^{-1}). After the mixture was incubated at 25°C for 20 min, the reaction was terminated by adding 0.8 mL of acidic $FeCl_3$ [24% (W/V) trichloroacetic acid and 10% (W/V) $FeCl_3$ in 18% HCl]. The amount of γ-glutamyl hydroxamate was measured with a spectrophotometer at 540 nm. One unit of GS activity was defined as the amount of enzyme that catalysed the formation of 1 μmol γ-glutamyl hydroxamate per min at 25°C [33].

The level of NRA was assayed by another in vitro method. Frozen plant material (0.5 g) was ground in a precooled mortar with sand and a pestle after adding liquid N and then homogenised in an extraction buffer (10 mL g^{-1} fresh weight) containing 5 mmol L^{-1} of EDTA, 5 mmol L^{-1} of cysteine and 25 mmol L^{-1} of potassium phosphate buffer (pH 8.7). The homogenates were centrifuged at 20,000 *g* for 20 min. Next, 0.4 mL of the clear filtrate was pipetted into a centrifuge tube, and the filtrate was added to 1.0 mL of KNO_3 (0.1 mol L^{-1}) and 0.6 mL of NADH (2 mg mL^{-1}). After the mixture was incubated at 25°C for 30 min, the excess NADH was oxidised by the addition of 0.5 mL of sulphanilamide (1%). The mixture was centrifuged at 20,000 *g* for 15 min. The amount of NO_2^- produced was measured after the addition of N-(1-naphtyl) ethylene-diamine dihydrochloride (0.02%) using an absorbance of 540 nm in a spectrophotometer [34].

Plant POR assay

For this assay, 0.4–0.6 g of fresh adventitious roots was sampled according to published methods [13,35,36]. POR was determined as , where POR is the root porosity (%), P_{gr} is the mass of the pycnometer with water and ground roots (g), P_r is the mass of the pycnometer with water and roots (g), r is the mass of roots (g), and P is the mass of the pycnometer with water (g).

Data analysis

All statistical analyses were performed using SPSS version 13.0. A one-way ANOVA with a homogeneity of variance test was used, followed by a least significant difference (LSD) test to check for quantitative differences between treatments. The statistically significance level was set at $p<0.05$.

Results

Distribution of soil nitrification activity

The nitrification activities of the root surface and rhizosphere soils increased over time, whereas those in the bulk soil did not change significantly (Fig. 1A, B and C). The maximal nitrification activities were found 2 mm away from the root surface in the rhizosphere soil, and their values were 2.76 (0.33) μg kg^{-1} h^{-1} for the low-fertility treatment and 4.32 (0.92) μg kg^{-1} h^{-1} for the high-fertility treatment for the last sampling period (Fig. 1C). Minimal nitrification activities appeared in the root surface and bulk soils (20–40 mm away from the root surface). Throughout the sampling period, the nitrification activities in the paddy soil with high fertility were always significantly higher than those in the soil with low fertility, except for the root surface soil 40 d after germination (Fig. 1A). Compared with the control, rice planting generally increased the rhizosphere soil nitrification activity significantly, but no increased effect of rice planting on nitrification occurring in the root surface or bulk soil was observed in either treatment. For example, the nitrification activity 2 mm away from the root surface was 1.8, 2.8 and 5.2 times higher than that in the control in the low-fertility soils at 40, 50 and 60 d after germination, respectively, and the corresponding activity in the high-fertility soil was 1.7, 2.9 and 3.9 times higher than the control, respectively (Fig. 1A, B, C and D).

Distribution of soil nitrate concentration

The NO_3^- concentration did not change over time. Although there was a strong absorption of NO_3^- by the rice roots, the horizontal distribution of soil NO_3^- at different distances from the root surface showed no concentration gradient (Fig. 2A, B and C). This result might have been due to the strong mobility of NO_3^- in the soil. The NO_3^- concentrations in the high-fertility soil were always higher than those in the low-fertility soil. This difference which might have contributed to the higher nitrification activity of the high-fertility soil described above (Fig. 1A, B, and C). The NO_3^- concentration in the soil planted with rice did not differ significantly from that in the control (Fig. 2A, B, C and D).

Distribution of the soil ammonium concentration

The NH_4^+ concentration in the soil decreased over time (Fig. 3), suggesting that a large quantity of NH_4^+ was absorbed by the rice roots. This conclusion is partially supported by the distinct depletion of NH_4^+ in the zone near the root surface (Fig. 3A, B, and C), where the NH_4^+ concentration increased with increasing distance from the rice roots. In the period 40–60 d after germination, the NH_4^+ concentrations of the high-fertility treatment in the bulk soil were always higher than those of the low-fertility treatment, except for the measurement taken 40 d after germination (Fig. 3A). This difference increased over time (Fig. 3A, B, and C). As the NH_4^+ concentrations in the control increased slightly with time, the NH_4^+ concentrations in the control for the high-fertility treatment were significantly higher than those in the control for the low-fertility treatment (Fig. 3D). Due to the NH_4^+ uptake by rice roots, the content of NH_4^+ in the treatments with rice was significantly lower than that in the control (without rice), especially on the last sampling day. For example, the NH_4^+ concentrations 40 mm away from the rice root (bulk soil) were 19.9% and 26.9% of those in the control at 60 d after germination in the low- and high-fertility soils, respectively, and those in the root surface soil were only 9.75% and 8.86% of the control.

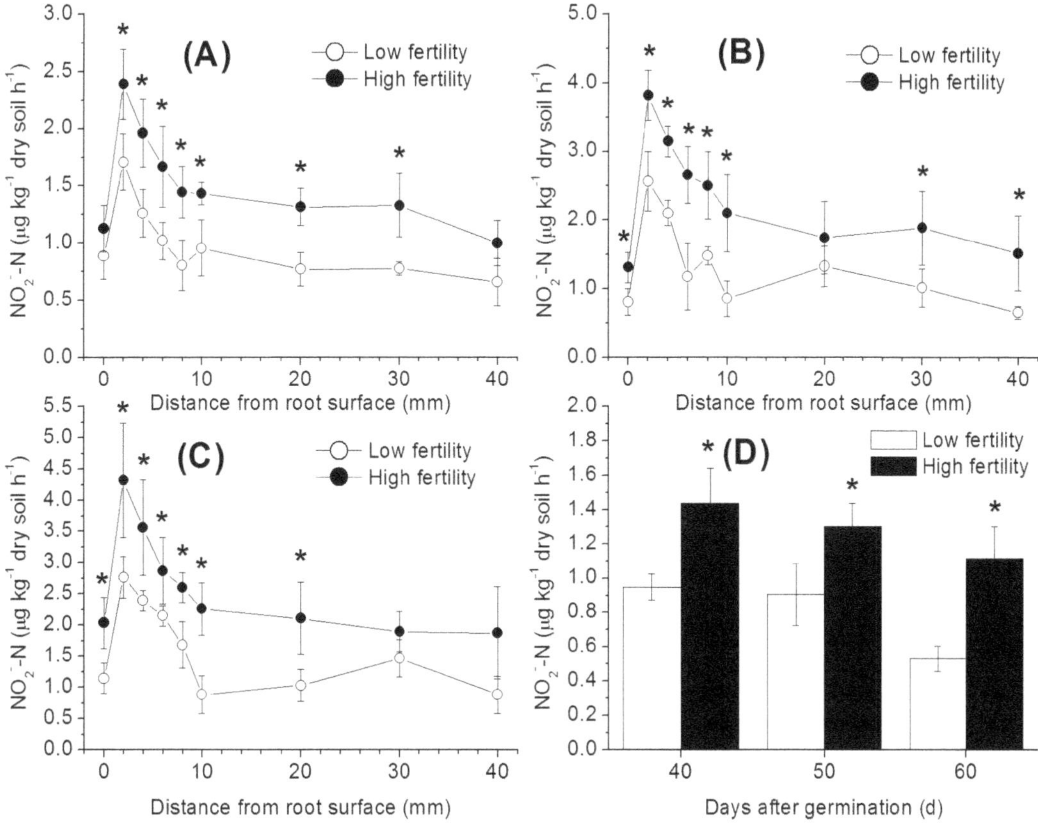

Figure 1. Nitrification activities measured at various distances from the rice root surface in paddy soils with different fertility levels (A) 40 d after germination, (B) 50 d after germination, (C) 60 d after germination, relative to the (D) control (without rice plants). The bars indicate ±SD. Dots and bars with * indicate a significant difference ($p<0.05$) between the two fertility levels.

Rice plant nitrogen assimilation, growth, and nitrogen accumulation

The NRA and GSA in the rice leaves and roots did not change over time, and the levels of NRA and GSA in the rice leaves were significantly higher than those in the roots (Table 2). In addition, high soil fertility ameliorated the NRA and GSA levels significantly; the leaf and root NRAs increased by 48.2% and 49.7% in the high-fertility soil compared to those in the low-fertility soil, whereas the leaf and root GSA levels in the high-fertility soil were, respectively, 37.6% and 103.1%, greater than those in the low-fertility soil (Table 2).

The rice biomass and N accumulation measurements differed significantly between the two fertility levels. The high-fertility soil increased the rice plant biomass by 90.4%, 65.2%, and 45.6% compared with the low-fertility soil when measured 40, 50, and 60 d after germination, respectively. The corresponding N accumulations increased by 47.9%, 48.6%, and 29.9% at the measurements taken 40, 50, and 60 d after germination, respectively (Table 2). High soil fertility markedly improved the growth and N uptake of the rice.

Rice root morphological structure

We observed large changes in the morphological structure of the rice roots and found that the major characteristics of the developmental level of the root system such as root biomass, adventitious root number and root diameter were much greater in the high-fertility soil than in the low-fertility soil (Table 3). Furthermore, the characteristics of the rice root, such as POR, which indicated the developmental degree of aerenchyma, were significantly higher in the high-fertility soil than in the low-fertility soil (Table 3). In summary, the high-fertility soil markedly improved the growth of the rice root and the root porosity.

Distribution of the soil oxygen concentration

The oxygen dissolved in the water layer remained stable. The average values were approximately 231 and 243 µmol L^{-1} in the low-fertility and high-fertility treatments (n = 6; Fig. 4A), respectively, and 218 and 209 µmol L^{-1} in their respective controls (n = 6; Fig. 4B). Rice planting significantly increased the content of dissolved oxygen in the water layer. The oxygen concentration decreased with soil depth so rapidly that it became zero at a depth of approximately 2.6–4.2 mm below the soil surface (Fig. 4A, B). Interestingly, the oxygen concentration and oxygen penetration depth in the rhizosphere soil with high fertility were significantly higher than the corresponding values in the rhizosphere soil with low fertility, whereas a reverse trend existed in the controls of these two fertility levels, although certain values at certain depths showed no significant difference due to the presence of abnormal peaks. For example, at a depth of 2 mm from the soil surface, the average dissolved oxygen concentrations in the low- and high-fertility rhizosphere soils were 33.1 and 68.9 µmol L^{-1}, respectively (n = 6; Fig. 4A), and their corresponding controls were 24.5 and 13.7 µmol L^{-1}, respectively (n = 6; Fig. 4B). Among the six repetitions, the abnormal peaks appeared more often in the rice planting treatments than in the control. This finding may be a result of the rice root ROL.

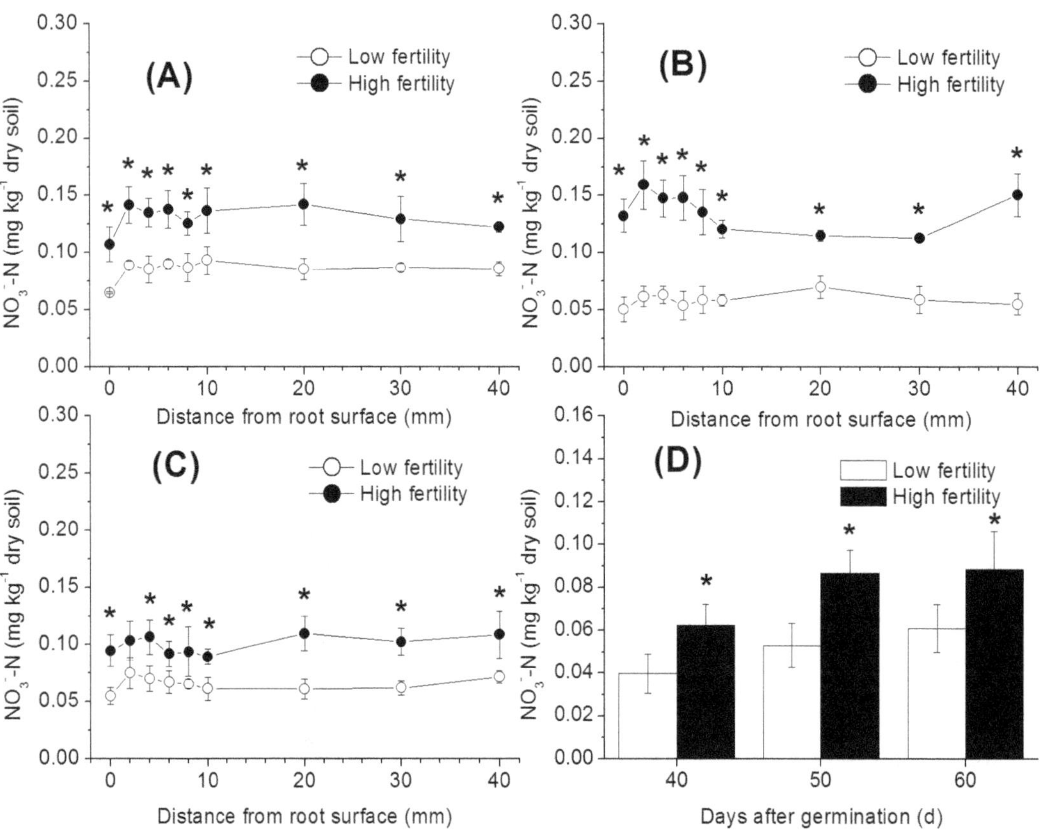

Figure 2. NO_3^--N measured at various distances from the rice root surface in paddy soils with different fertility levels (A) 40 d after germination, (B) 50 d after germination, and (C) 60 d after germination, relative to the (D) control (without rice plants). The bars indicate ±SD. Dots and bars with * indicate a significant difference ($p<0.05$) between the two fertility levels.

Root radial oxygen loss

The oxygen in the agar was detected at an average distance of 400 μm from the root surface in the low-fertility treatment (Fig. 5A), in comparison with an average distance of 600 μm from the root surface in the high-fertility treatment (Fig. 5B). This finding indicated that the root ROL level of rice in the high-fertility soil was significantly higher than that in the low-fertility soil. The oxygen concentration decreased rapidly with distance near the rice root surface. When the microelectrode was in contact with the root surface, the average oxygen concentrations were 50.2 and 97.7 μmol L^{-1} in the low- and high-fertility treatments, respectively. The oxygen concentration remained unchanged inside the rice root, and the oxygen concentrations in the roots of the low-fertility treatment were lower than those in the high-fertility treatment (Fig. 5A, B). For example, the average oxygen concentrations in the low-and high-fertility treatments were 55.4 and 103.3 μmol L^{-1}, respectively.

Discussion

The penetration depths of oxygen in the soil surface and in the rhizosphere generally depend on both rice root ROL and the oxygen consumption caused by soil organisms and chemical oxidation [37]. In fact, rice planting could significantly alter (and, most likely, improve) the cycling of oxygen in the water and soil system due to rice root ROL, and our results confirmed this hypothesis (see Fig. 4A, B). For example, the dissolved oxygen concentration in the water layer was approximately 231 and 243 μmol L^{-1} in the low- and high-fertility treatments (Fig. 4A), respectively, compared with 218 and 209 μmol L^{-1} in their corresponding controls (Fig. 4B). In planted microcosms (planting with 110-d-old rice seedlings), oxygen is still detectable to a depth of at least 40 mm [38], which appears to be the maximum depth reported to date [23]. Interestingly, our results showed that the oxygen concentration in the rhizosphere soil with high fertility was significantly higher than that in the rhizosphere soil with low fertility (Fig. 4A), whereas the soil oxygen concentration in the high-fertility soil control was significantly lower than that in the low-fertility soil control (Fig. 4B). Because the control paddy soil without rice plants had no exogenous replenishment of oxygen as occurred for the rice root ROL, except for the dissolved oxygen provided by the flooding water, the high organic matter content in the high-fertility soil (see Materials and Methods) led to more severe oxygen depletion. This result may have been due to the higher oxygen consumption required by the degradation of organic matter and reduced compounds in the high-fertility soil compared with the low-fertility soil [37]. Although the oxygen content in the control high-fertility soil was lower than that in the control low-fertility soil (Fig. 4B), the soil oxygen concentration in the high-fertility paddy soil with rice plants was higher than that in the low-fertility soil (Fig. 4A), potentially due to the high rice root ROL in the high-fertility soil (Fig. 5B).

The difference in dissolved oxygen concentrations in rhizosphere soils between the high- and low-fertility soils appeared to be related to the differences in rice root ROL in these soils. When rice plants were grown in the high-fertility soil, the rice root growth

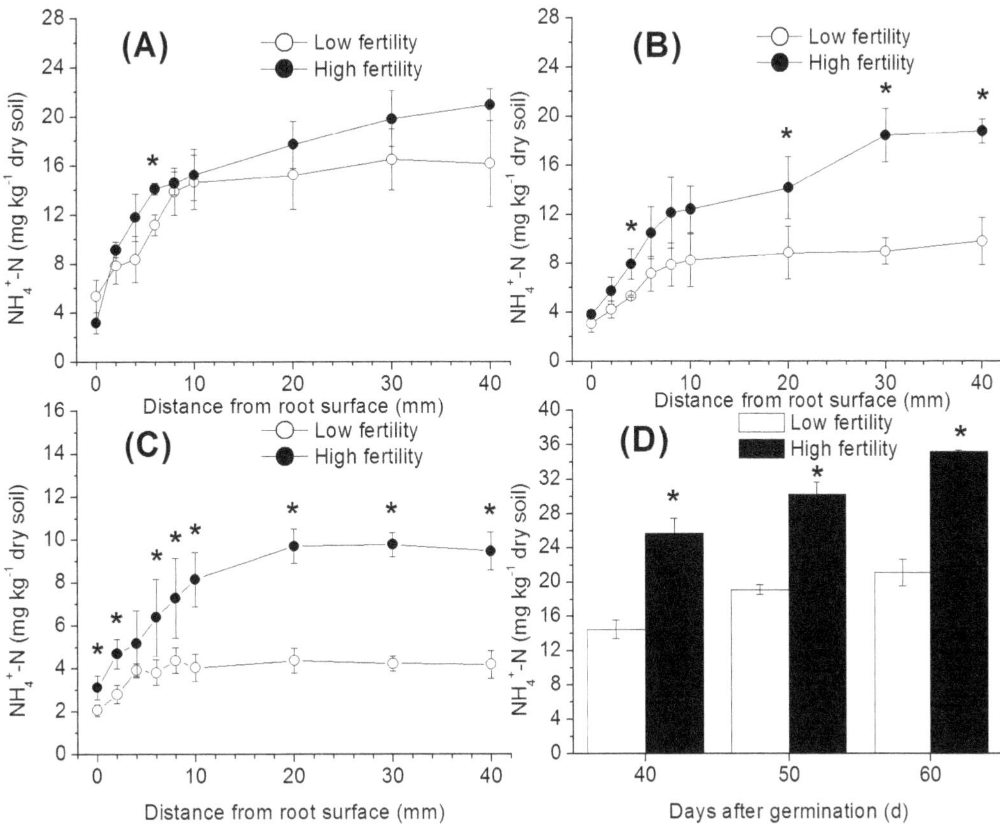

Figure 3. NH_4^+-N measured at various distances from the rice root surface in paddy soils with different fertility levels (A) 40 d after germination, (B) 50 d after germination, and (C) 60 d after germination, relative to the (D) control (without rice plants). The bars indicate ±SD. Dots and bars with * indicate a significant difference ($p<0.05$) between the two fertility levels.

was greatly improved; the root biomass, adventitious root number, and root diameter of the rice grown in the high-fertility soil were much greater than the corresponding values for the rice grown in the low-fertility soil (Table 3). A similar observation has previously been reported for wheat [39]. Not only did the rice root biomass, adventitious root number and root diameter change, but root porosity (see Table 3) was also improved by high soil fertility. This phenomenon does not appear to have been previously reported, as its positive feedback mechanism is unclear. ROL from rice roots estimated by Kludze *et al.* [19], showed a significant and positive correlation with POR. Oxygen efflux from root to soil represented 30%–40% of oxygen transportation from shoot to root [17]. Well-developed root systems and aerenchyma formation resulted in higher rice root ROL in the high-fertility soil (Fig. 5) [7] and consequently increased the oxygen content in the rhizosphere soil (Fig. 4A). Mugnai et al. [40] investigated maize roots and found that the root apex transition zone (intercalated between the apical division zone and the elongation zone [41]) plays central roles in both sensing and adapting to root hypoxia and emits the greatest amount of nitric oxide. This discovery motivated us to further investigate whether the root apex transition zone of rice plays a key role in altering the ROL and, correspondingly, the rhizosphere soil oxygen distribution.

An agar microcosm was used to further quantify the rice root ROL. The results indicated that the root ROL of rice planted in the high-fertility soil was greater than that in the low-fertility soil. The average oxygen concentrations at the root surface were 50.2 and 97.7 µmol L^{-1}, and they decreased rapidly to nearly zero at distances of approximately 400 µm and 600 µm from the root surfaces in the low- and high-fertility treatments, respectively (Fig. 5A, B). The average oxygen concentrations in the interior of the rice roots in the low- and high-fertility treatments were approximately 55.4 and 103.3 µmol L^{-1}, respectively (Fig. 5A, B). Similar results have been obtained by Revsbech *et al.* [23], who found that oxygen leakage created an oxic zone extending approximately 150 µm away from the fine lateral root (for 21-d-old rice); the inner oxygen concentration of the root (42-d-old rice) was approximately 75 µmol L^{-1}. The rice root ROL can also promote the development of aerobic niches in the rhizosphere to restrict the accumulation of phytotoxic compounds (e.g., Fe^{2+}, Mn^{2+}, H_2S) as well as denitrification [42] and maintain aerobic microbial processes, such as nitrification [14,19].

Due to long-term fertiliser changes, soil fertility increased significantly and promoted soil nitrification, as widely reported in previous studies of paddy soil [8,16] and upland soil [2,43]. A low clay content contributed to a high soil porosity, which promoted gas transport by diffusion and advection, thereby supporting soil nitrification. The high ambient NH_4^+ pool in the high-fertility soil also improved its nitrification activity (Fig. 1 and 5). The maximal nitrification activities were found in the rhizosphere soil located 2 mm away from the root surface, and decreased with distance from the root surface (Fig. 1). In our previous study, the nitrification activity gradient also revealed nitrification activity differences in different rice cultivars. It also showed that the maximum nitrification activity detected in the rhizosphere soil, i.e., the maximum nitrification activities of the Indica- and

Table 2. Rice plant N assimilation, growth, and nitrogen accumulation[a].

Fertility level	Sampling date (d)	NRA (μg g^{-1} FW h^{-1})		GSA (mmol g^{-1} FW min^{-1})		Biomass (g DW plant^{-1})	N accumulation (mg N plant^{-1})
		Leaf	Root	Leaf	Root		
Low fertility	40	7.21 ± 0.93^{b}	4.06 ± 0.33^{c}	1.85 ± 0.14^{c}	0.08 ± 0.02^{b}	0.49 ± 0.10^{d}	16.8 ± 1.77^{d}
	50	6.84 ± 0.76^{b}	4.61 ± 0.82^{bc}	2.09 ± 0.05^{bc}	0.09 ± 0.01^{b}	0.85 ± 0.09^{c}	23.4 ± 0.93^{c}
	60	7.38 ± 0.30^{b}	4.14 ± 1.08^{c}	2.35 ± 0.15^{bc}	0.11 ± 0.04^{b}	1.32 ± 0.15^{b}	34.8 ± 3.23^{b}
High fertility	40	10.6 ± 1.38^{a}	5.85 ± 0.69^{ab}	2.56 ± 0.45^{b}	0.18 ± 0.02^{a}	0.93 ± 0.14^{c}	24.8 ± 3.59^{c}
	50	10.7 ± 2.41^{a}	6.86 ± 0.49^{a}	2.58 ± 0.25^{b}	0.18 ± 0.01^{a}	1.41 ± 0.23^{b}	34.8 ± 3.88^{b}
	60	10.5 ± 0.49^{a}	6.45 ± 1.22^{a}	3.51 ± 0.41^{a}	0.21 ± 0.06^{a}	1.92 ± 0.15^{a}	45.1 ± 3.24^{a}

[a]Values represent means ± SDs with 3 replicates (with each replicate composed of rice plant tissue sampled from different rhizoboxes of each treatment). Different letters in the same column indicate a significant difference at the $p<0.05$ level.

Japonica- planted soils occurred 6 and 2 mm away from the rice roots, respectively [28]. The use of a rhizobox is convenient for distinguishing between the root surface and rhizosphere soil, but this design produces a dense root-zone and thus leads to a generally more oxidised environment, which may enhance nitrification. In contrast, more competition for NH_4^+ in this design may lower nitrification.

Based on our previous experience, a water layer of 1 cm that is maintained with applications of deionised water every morning and evening throughout the incubation would make it difficult to slice the soil. Additionally, mineral N, especially NO_3^-, would be dissolved in the overlying water because the oxygen concentration was much higher in the water layer than in the soil profile. If we continued to maintain a 1-cm water layer, the concentration of NO_3^- in the soil would be very low and might even remain undetected because the paddy soil that we used is derived from Quaternary red clay and has relatively low pH values (usually 4–5.5). The nitrification activity in this paddy soil is much weaker than that in neutral paddy soil [26]. The NO_3^- concentration in the paddy soil is correspondingly low. For this reason, we chose not to water on the evening of the day prior to the sampling date. The water layer was actually 1–2 mm deep when we measured the mineral N and nitrification activity. The advantage of this approach is that the soil is easy to slice, moreover, the NO_3^- concentration can be measured more effectively. However, the disadvantage of this approach is that it imperceptibly stimulates nitrification and results in overestimates of the soil NO_3^- concentration. In this experiment, the NO_3^- concentration in the soil planted with rice did not differ significantly from that in the control. Because the nitrification activity in the rice planting soil was much higher than that in the control (Fig. 1D), the majority of the NO_3^- resulting from increased nitrification might have been absorbed primarily by the rice roots. NO_3^- can be absorbed rapidly by rice roots or can diffuse rapidly to other parts of the soil. Therefore, the NO_3^- distribution in flooded paddy soil is relatively even. However, advection caused by evaporation from the leaves and following water transport to the roots increases transport processes in the soil to a greater extent than that resulting from diffusion. This pattern, was also confirmed by our previous experiment [28]. The low pH caused by acidic root exudates [44] and NH_4^+ (nitrification substrate) exhaustion in the root surface soil (Fig. 3A, B, and C) strongly inhibited root surface nitrification in both of the studied soil types (Fig. 1A, B, and C) [14]. Due to the high nitrification activity in the high-fertility soil, the NO_3^- concentrations in the high-fertility soil were higher than those in the low-fertility soil (Fig. 2). This difference, might have caused more NO_3^- uptake and assimilation by the rice plants grown in the high-fertility soil (the NRA results in Table 2 may support this hypothesis).

The NO_3^- that is converted via nitrification in the root surface and rhizosphere soils is very important for N nutrition of rice [3,4]. Extensive hydroponic experiments have demonstrated that rice growth, yield, net N acquisition, N translocation to the shoot, and NUE are greater with a mixed NH_4^+ and NO_3^- supply than with NH_4^+ alone [3,5,6]. In the present study, the NO_3^- concentration in the rice-planted soil was not significantly different from that measured in the control (Fig. 2), although the nitrification activity in the rice-planted soil was much higher than that in the control (Fig. 1), thus indicating that greater amounts of NO_3^- absorbed by the rice root might have produced the results described above. NO_3^- reductase is a key enzyme involved in NO_3^- assimilation in crops, and its activity is strongly dependent on the external NO_3^- concentration [28]. In this experiment, the rice leaf and root NRA in the high-fertility soil increased by 48.2% and 49.7%,

Table 3. Rice root growth and morphology[a].

Fertility level	Sampling date (d)	Root biomass (g DW plant^{-1})	Adventitious root number (plant^{-1})	Root diameter (mm)	POR (%)
Low fertility	40	0.11±0.02^{d}	22±3^{d}	0.64±0.05^{d}	14.1±1.46^{e}
	50	0.20±0.03^{c}	42±3^{c}	0.78±0.06^{c}	16.2±1.13^{d}
	60	0.24±0.01^{c}	49±3^{c}	0.90±0.02^{b}	16.9±1.38^{cd}
High fertility	40	0.21±0.04^{c}	28±3^{d}	0.75±0.04^{c}	19.6±1.44^{bc}
	50	0.35±0.02^{b}	61±13^{b}	0.90±0.03^{b}	22.0±2.92^{ab}
	60	0.42±0.06^{a}	82±7^{a}	1.06±0.09^{a}	23.1±2.36^{a}

[a]Values represent means (means ± SDs), and different letters in the same column indicate a significant difference at the $p<0.05$ level. Data were pooled for n = 3, except for adventitious root diameter (n = 12), which was measured at the middle of each root selected for the average length of the whole root.

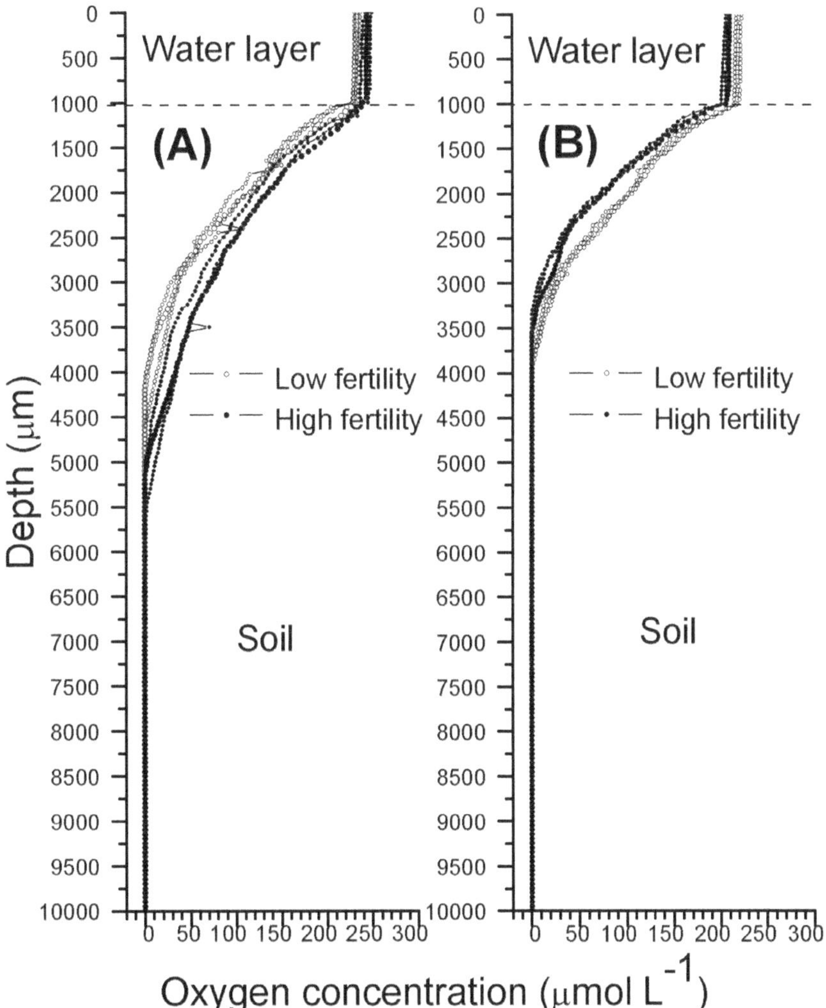

Figure 4. A representative graph showing oxygen profiles of (A) the rhizosphere soil (2 mm away from the root surface) and (B) the control (without rice plants) 50 d after germination in paddy soils with different fertility levels at 20°C. The oxygen microelectrode used was 20–30 μm in diameter (OXY25, Unisense, Aarhus, Denmark [31]). Three rhizoboxes for each treatment were used in the in situ measurements, and the oxygen concentration profile was determined in the right and left compartments of each rhizobox. All measurements were conducted in the centre of the range marked by a 2-mm distance from the nylon nets in the low- and high-fertility soils with and without the rice plant treatments. Hence, the measurements were repeated six times in each treatment. Three results (measured in the right compartment) are shown, and each curve shows one individual soil oxygen profile.

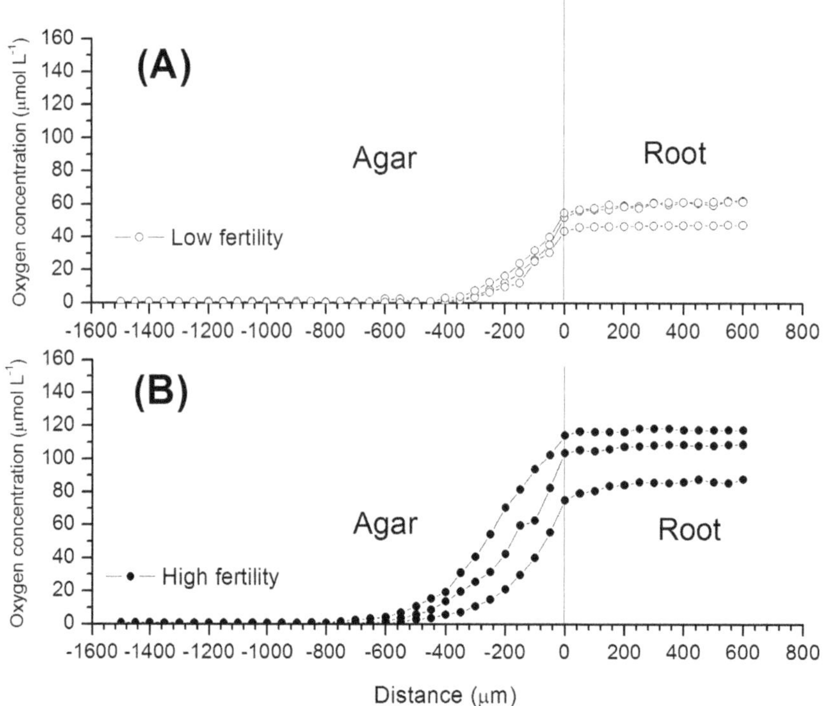

Figure 5. A representative graph showing oxygen profiles of the rhizosphere and within the rice root, 50 d after germination, growing in the (A) low- and (B) high-fertility soils at 20°C. The rice plant roots were separated from the soil and placed in agar with a nutrient solution. Newly formed roots with uniform 45-mm lengths were selected, and both were measured at a position 25 mm away from the root tips. The average diameters of the roots growing in the low- and high-fertility conditions were 0.74 and 0.93 mm at the measured points, respectively. The measurements were repeated in different rice plants and with different newly formed roots, with lengths ranging from 30 to 55 mm. An OXY10 oxygen microelectrode (Unisense, Aarhus, Denmark [31]) 8–10 µm in diameter was used to penetrate the rice root longitudinally. Three rhizoboxes were used for the individual rice root ROL measurements. Each curve shown in the figure indicates an individual rice root.

respectively, compared with those in the low-fertility soil (Table 2), which also confirmed that more NO_3^- was absorbed and assimilated by the rice planted in the high-fertility soil compared with that planted in the low-fertility soil. Greater amounts of NO_3^- absorbed and assimilated by rice plants will also accelerate NH_4^+ uptake and assimilation, perhaps through the NO_3^--specific induction of additional pathways for NH_4^+ assimilation [3], as also confirmed by the GSA results (Table 2). Hence, the rice N accumulation and rice growth of plants grown in the high-fertility soil were improved compared with those of plants grown in the low-fertility soil (Table 2).

Conclusions

The present study investigated reasons that increasing soil fertility would enhance the growth of rice. The study adopted a new perspective focusing on rice root ROL and its corresponding soil oxygen profile. The fertile soil had high porosity (because of the low clay content) and a high NH_4^+ content, which supported soil nitrification. A high rhizosphere oxygen pressure distribution caused by the rice root ROL, which was attributed to the improved rice root growth and root porosity observed in the high-fertility soil, also contributed to the promotion of rhizosphere nitrification. The mechanism accelerating the development of rice root aerenchyma in the high-fertility soil requires further exploration. Investigation of the mechanisms of N loss and the quantification of N loss were not performed in this study but will be addressed by our laboratory in future research. Long-term tillage and fertilisation (especially organic fertiliser) would significantly increase soil fertility, but a large amount of organic matter would decrease the soil oxygen content. Fortunately, rice planting altered the water layer and soil oxygen status via root ROL, and rice in the fertile soil further improved the rhizosphere soil oxidisation environment. Our results suggest that increased fertilisation could lead to increased soil aeration in red soil regions in China, which might promote rice growth and rice N nutrition.

Acknowledgments

We thank Dr. O. Pedersen at the Freshwater Biological Laboratory, Institute of Biology, University of Copenhagen, Denmark, for his patience and experienced instruction in oxygen microelectrode measurement and careful English revisions of this manuscript.

Author Contributions

Conceived and designed the experiments: YLL WMS XXW. Performed the experiments: YLL. Analyzed the data: YLL. Contributed reagents/materials/analysis tools: YLL XXW. Wrote the paper: YLL WMS XXW.

References

1. Abbasi MK, adams WA (1998) Loss of nitrogen in compacted grassland soil by simultaneous nitrification and denitrification. Plant and Soil 200: 265–277.
2. He J Z, Shen JP, Zhang LM, Zhu YG, Zheng YM, et al. (2007) Quantitative analyses of the abundance and composition of ammonia-oxidizing bacteria and

ammonia-oxidizing archaea of a Chinese upland red soil under long-term fertilization practices. Environmental Microbiology 9: 2364–2374.
3. Kronzucker HJ, Siddiqi MY, Glass ADM, Kirk GJD (1999) Nitrate-ammonium synergism in ice: a subcellular flux analysis. Plant Physiology 119: 1041–1046.
4. Kronzucker HJ, Glass ADM, Siddiqi MY, Kirk GJD (2000) Comparative kinetic analysis of ammonium and nitrate acquisition by tropical lowland rice: implications for rice cultivartion and yield potential. New Phytologist 145: 471–476.
5. Raman DR, Spanswick RM, Walker LP (1995) The kinetics of nitrate uptake from flowing nutrient solutions by rice: Influence of pretreatment and light. Bioresourse Technology 53: 125–132.
6. Duan YH, Zhang YL, Ye LT, Fan XR, Xu GH, et al. (2007) Responses of rice cultivars with different nitrogen use efficiency to partial replaced nitrate nutrition. Annals of Botany 99: 1153–1160.
7. Li YL, Fan XR, Shen QR (2008) The relationship between rhizosphere nitrification and nitrogen-use efficiency in rice plants. Plant, Cell and Environment 31: 73–85.
8. Chen XP, Zhu YG, Xia Y, Shen JP, He JZ (2008) Ammonia-oxidizing archaea: important players in paddy rhizosphere soil? Environmental Microbiology 10: 1978–1987.
9. Venter JC, Remington K, Heidelberg JF, Halpern AL, Rusch D, et al. (2004) Environmental genome shotgun sequencing of the Sargasso Sea. Science 304: 66–74.
10. Schleper C, Jurgens G, Jonuscheit M (2005) Genomic studies of uncultivated archaea. Nature Reviews Microbiology 3: 479–488.
11. Könneke M, Bernhard AE, de la Torre JR, Walker CB, Waterbury JB, et al. (2005) Isolation of an autotrophic ammonia-oxidizing marine archaeon. Nature 437: 543–546.
12. Francis CA, Roberts KJ, Beman JM, Santoro AE, Oakley BB (2005) Ubiquity and diversity of ammonia-oxidizing archaea in water columns and sediments of the ocean. Proceedings of the National Academy of Sciences of the United States of America 102: 14683–14688.
13. Leininger S, Urich T, Schloter M, Schwark L, Qi J, et al. (2006) Archaea predominate among ammonia-oxidizing prokaryotes in soils. Nature 442: 806–809.
14. Arth I, Frenzel P (2000) Nitrification and denitrification in the rhizosphere of rice: the detection of processes by a new multi-channel electrode. Biology and Fertility of Soils 31: 427–435.
15. Briones AM, Okabe S, Umemiya Y, Ramsing NB, Reichardt W, et al. (2002) Influence of different cultivars on populations of ammonia-oxidizing bacteria in the root environment of rice. Applied and Environmental Microbiology 68: 3067–3075.
16. Wang YN, Ke XB, Wu LQ, Lu YH (2009) Community composition of ammonia-oxidizing bacteria and archaea in rice field soil as affected by nitrogen fertilization. Systematic and Applied Microbiology 32: 27–36.
17. Armstrong W (1979) Aeration in higher plants. Advances in Botanical Research 7: 225–332.
18. Armstrong W, Justin SHFW, Beckett PM, Lythe S (1991) Root adaptation to soil waterlogging. Aquatic Botany 39: 57–73.
19. Kludze HK, DeLaune RD, Patrick WH (1993) Aerenchyma formation and methane and oxygen exchange in rice. Soil Science Society of America Journal 57: 386–391.
20. Armstrong W (1964) Oxygen diffusion from the roots of some British Bog plants. Nature 204: 801–802.
21. Armstrong W (1971) Radial oxygen losses from intact rice roots as affected by distance from the apex, respiration and waterlogging. Physiologia Plantarum 25: 192–197.
22. Kludze HK, DeLaune RD, Patrick WH (1994) A colorimetric method for assaying dissolved oxygen loss from container-grown rice roots. Agronomy Journal 86: 483–487.
23. Revsbech NP, Pedersen O, Reichardt W, Briones A (1999) Microsensor analysis of oxygen and pH in the rice rhizosphere under field and laboratory conditions. Biology and Fertility of soils 29: 379–385.
24. FAO (1998) FAO, ISRIC, ISSS World Reference Base for Soil Resources. World Soil Resources Reports 84. Rome, Italy.
25. Zhong WH, Cai ZC (2007) Long-term effects of inorganic fertilizers on microbial biomass and community functional diversity in a paddy soil derived from quaternary red clay. Applied Soil Ecology 36: 84–91.
26. Yuan F, Ran W, Shen QR, Wang DZ (2005) Characterization of nitrifying bacteria communities of soils from different ecological regions of China by molecular and conventional methods. Biology and Fertility of Soils 41: 22–27.
27. Zhang MK, Xu JM (2005) Restoration of surface soil fertility of an eroded red soil in southern China. Soil and Tillage Research 80: 13–21.
28. Li YL, Zhang YL, Hu J, Shen QR (2007) Contribution of Nitrification Happened in Rhizospheric Soil Growing with Different Rice Cultivars to N Nutrition. Biology and Fertility of Soils 43: 417–425.
29. Berg P, Rosswall T (1985) Ammonium oxidizer numbers, potential and actual oxidation rates in two Swedish arable soils. Biology and Fertility of Soils 1: 131–140.
30. Alef K, Nannipieri P (1995) Methods in Applied Soil Microbiology and Biochemistry. In: Berg P, Rosswall T, Editors.Assay of nitrification (short-term estimations).New York: Harcourt Brace Academic Press. pp.241–242.
31. Revsbech NP (1989) An oxygen microsensor with a guard cathode. Limnology and Oceanography 34: 474–478.
32. Fan XR, Jia LJ, Li YL, Smith SJ, Miller AJ, et al. (2007) Comparing nitrate storage and remobilization in two rice cultivars that differ in their nitrogen use efficiency. Journal of Experimental Botany 58: 1729–1740.
33. Zhang CF, Peng SB, Peng XX, Chavez A Q, Bennett J (1997) Response of glutamine synthetase isoforms to nitrogen sources in rice (*Oryza sativa* L.) roots. Plant Science 125: 163–170.
34. Ogawa T, Fukuoka H, Yano H, Ohkawa Y (1999) Relationships between nitrite reductase activity and genotype-dependent callus growth in rice cell cultures. Plant Cell Reports 18: 576–581.
35. Jensen CR, Luxmoore RJ, Van Gundy SD, Stolzy LH (1969) Root air-space measurements by a pycnometer method. Agronomy Journal 61: 474–475.
36. Kim JD, Jugsujinda A, Carbonell-Barrachina AA, DeLaune RD, Patrick WH (1999) Physiological functions and methane and oxygen exchange in Korean rice cultivars grown under controlled soil redox potential. Botanical Bulletin of Academia Sinica 40: 185–191.
37. Ratering S, Schnell S (2001) Nitrate-dependent iron (II) oxidation in paddy soil. Environmental. Microbiology 3: 100–109.
38. Frenzel P, Rothfuss F, Conrad R (1992) Oxygen profiles and methane turnover in a flooded rice microcosm. Biology and Fertility of Soils 14: 84–89.
39. Wang YF, Yu ZW, Li SX, Yu SL (2003) Effects of nitrogen rates and soil fertil ity levels on root nitrogen uptake and assimilation and grain protein content of winter wheat. Plant Nutrition and Fertilizer Science 9: 39–44.
40. Mugnai S, Azzarello E, Baluška F, Mancuso S (2012) Local root apex hypoxia induces NO-mediated hypoxic acclimation of the entire root. Plant and Cell Physiology 53: 912–920.
41. Baluška F, Mancuso S, Volkmann D, Barlow PW (2010) Root apex transition zone: a signaling-response nexus in the root. Trends in Plant Science 15: 402–408.
42. Sorrell BK (1999) Effect of external oxygen demand on radial oxygen loss by *Juncus* roots in titanium citrate solutions. Plant, Cell and Environment 22: 1587–1593.
43. Chu HY, Fujii T, Morimoto S, Lin XG, Yagi K, et al. (2007) Community structure of ammonia-oxidizing bacteria under long-term application of mineral fertilizer and organic manure in a sandy loam soil. Applied and Environmental Microbiology 73: 485–491.
44. Colmer TD, Bloom AJ (1998) A comparison of NH_4^+ and NO_3^- net fluxes along roots of rice and maize. Plant, Cell and Environment 21: 240–246.

16

Integrated Analysis of Climate, Soil, Topography and Vegetative Growth in Iberian Viticultural Regions

Helder Fraga[1]*, Aureliano C. Malheiro[1], José Moutinho-Pereira[1], Rita M. Cardoso[2], Pedro M. M. Soares[2,3], Javier J. Cancela[4], Joaquim G. Pinto[5,6], João A. Santos[1]

1 Centre for the Research and Technology of Agro-Environmental and Biological Sciences, Universidade de Trás-os-Montes e Alto Douro, UTAD, Vila Real, Portugal, **2** Instituto Dom Luiz, CGUL, Universidade de Lisboa, Lisbon, Portugal, **3** ADEC, Instituto Superior de Engenharia de Lisboa, Lisbon, Portugal, **4** GI-1716, Proyectos y Planificación. Dpto. Ingeniería Agroforestal, Universidad de Santiago de Compostela, Escuela Politécnica Superior, Lugo, Spain, **5** Department of Meteorology, University of Reading, Reading, United Kingdom, **6** Institute for Geophysics and Meteorology, University of Cologne, Cologne, Germany

Abstract

The Iberian viticultural regions are convened according to the Denomination of Origin (DO) and present different climates, soils, topography and management practices. All these elements influence the vegetative growth of different varieties throughout the peninsula, and are tied to grape quality and wine type. In the current study, an integrated analysis of climate, soil, topography and vegetative growth was performed for the Iberian DO regions, using state-of-the-art datasets. For climatic assessment, a categorized index, accounting for phenological/thermal development, water availability and grape ripening conditions was computed. Soil textural classes were established to distinguish soil types. Elevation and aspect (orientation) were also taken into account, as the leading topographic elements. A spectral vegetation index was used to assess grapevine vegetative growth and an integrated analysis of all variables was performed. The results showed that the integrated climate-soil-topography influence on vine performance is evident. Most Iberian vineyards are grown in temperate dry climates with loamy soils, presenting low vegetative growth. Vineyards in temperate humid conditions tend to show higher vegetative growth. Conversely, in cooler/warmer climates, lower vigour vineyards prevail and other factors, such as soil type and precipitation acquire more important roles in driving vigour. Vines in prevailing loamy soils are grown over a wide climatic diversity, suggesting that precipitation is the primary factor influencing vigour. The present assessment of *terroir* characteristics allows direct comparison among wine regions and may have great value to viticulturists, particularly under a changing climate.

Editor: Inés Álvarez, University of Vigo, Spain

Funding: This study was supported by national (FCT – Fundação para a Ciência e a Tecnologia) and European Union Funds (FEDER/COMPETE - Operational Competitiveness Programme), under the projects PTDC/AGR-ALI/110877/2009, FCOMP-01-0124-FEDER-022692, PEst-OE/AGR/UI4033/2014, by PRODER – Inovação (GreenVitis project IF0018), by ON2 ENOEXEL (NORTE-07-0124-FEDER-000032), and by EUPORIAS - 7th Framework Programme for Research, Grant Agreement 308291. The funders had no role in study design, data collection and analysis, decision to publish, or preparation of the manuscript.

Competing Interests: The authors have declared that no competing interests exist.

* Email: hfraga@utad.pt

Introduction

The most renowned viticultural regions in the Iberian Peninsula (Portugal and Spain) have a long standing tradition in winemaking and are considered world-class grapevine (*Vitis vinifera* L.) producing regions. Spain, which currently has the largest vineyard area in the world (over 1×10^6 ha), is the 3rd wine producer worldwide, while Portugal ranks in the 11th place, with internationally acclaimed wines, such as the Port wine [1]. In these countries, viticultural regions are convened according to Denominations of Origin (DO), or Qualified Denomination of Origin (DOCa), which are imposed by governmental institutions and controlled by strict regulations. Although the spatial distribution and limits of each DO are subjected to different rules in Portugal and Spain, the significance of the DO is nearly the same. In particular, the law enforcements are similar to both Portugal and Spain, even if they still differ in certain aspects of national legislation. The Portuguese Douro/Porto DO (henceforth Douro DO) was the first viticultural region worldwide to implement these regulations in 1756 [2]. Such regulations aim at obtaining a superior wine quality, while establishing the specific wine characteristics of each region [3]. The concept of *terroir*, which includes specific soil, topography, climate, landscape characteristics and biodiversity features of each winemaking region [4], is entrenched within the classification of a DO. Each DO is expected to be a recognized trademark, whilst other vineyards/products not included in the DO are not allowed to bear this denomination.

Being part of the *terroir*, soil is one of the most important factors for viticulture [2]. It supports the root system, which accumulates carbohydrates, absorbs water and other nutrients, being crucial for grapevine growth, physiology and yield attributes [5,6]. Soil structure and chemistry can influence grapevine composition and consequently wine quality [7]. Compact and shallow soils can obstruct root access to oxygen, water and nutrients, limiting root growth and development [8]. In grapevines, nutrient and water uptake occur mostly within 0.5–1.0 m soil profile [2,9]. Therefore, deep soils with good drainage (either natural or manmade) are usually preferred for vineyard installation [10]. Additionally, soil water retention properties are also important, as they can affect grapevine performance [11,12]. A high soil water storage capacity is indeed important in regions where grapevines are subjected to excessive heat and water stress, as is the case of the Mediterranean regions [13].

Climate, also an important component of the *terroir*, is widely acknowledged as one of the most important factors for grapevine development and growth [8,14–17]. During the growing season (April–October in the Northern Hemisphere), climatic conditions exert a significant influence on vine physiological processes. One of the most well-known climatic limitations of grapevine is the 10°C base temperature, needed for the onset of its yearly cycle [5]. Throughout its different stages of development, sunlight, heat and water demands vary. In fact, it has been shown that the timings and duration of the grapevine phenological stages are deeply tied to the prevailing atmospheric conditions [18,19], which also contribute to variability in grapevine yield [20,21], wine production [14,22,23] and quality [24,25]. All these climatic factors limit the geographic distribution of grapevine [15,26,27], being also key factors in determining the suitability of a given region for specific varieties and wine types [27–29].

The topographic elements represent yet another key factor that influences viticultural and oenological characteristics of a given region. Amongst the most important topographic elements for viticulture are elevation, slope degree and aspect/exposure [11,30]. Elevation can have a significant impact on vineyard temperatures (i.e. vertical temperature gradient), thus exerting a strong influence in site and varietal selection [2]. The slope degree of the terrain impacts on canopy microclimate (e.g. through solar exposure), soil erosion, water drainage [31] and viticultural management. The aspect refers to the compass direction to which the terrain faces (e.g. northern/southern exposure), influencing the surface net incoming solar radiation flux [31,32], thus being determinant for site selection. These elements further enhance the singularity of viticultural regions, since they influence cultural and management practices [2]. As an example, growers tend to select row orientation according to the geographical aspect of the terrain (e.g. northern/southern exposure) in order to optimize solar radiation intercepted by the canopy. Another example is the implementation of walled terraces to overcome slope degree in steep mountainous areas, allowing mechanization and decreasing soil erosion [2].

Vine vigour is related to the growth dynamics of grapevines. It is used as an indicator of grapevine performance, affecting yield, wine production and grape quality [33]. For instance, relationships between vegetative growth and remote sensing derived metrics are broadly recognised [34]. Spectral vegetation indices have shown a good agreement with grapevine vigour, phenology, grape production and wine attributes [34–37], proving a suitable metric of grapevine spatial variability and performance.

The Iberian Peninsula presents a wide range of all these site-related elements influencing grapevine performance. From a climatic perspective it delivers a relatively large set of mesoclimates, spanning from dryer regions, in the inner south, to more humid regions, in the north and northwest [38,39]. Topography and soils are also quite distinct throughout the peninsula, ranging from extended flatland areas to steep mountainous regions [40], each with very different soil characteristic [2], which may influencing crop selection and settlements in each region. All these elements are reflected in the different varieties grown throughout the peninsula [41]. Iberia presents a large number of autochthonous grapevine varieties, according to their adaptation to the different climates, soils and topographic conditions [42], with red varieties usually prevailing in the south and white ones in the north [43]. Given the heterogeneous conditions in which grapevines are grown in Iberia, understanding the complex relationships between all these factors represents a serious concern for grapevine growers and winemakers.

The present study aims to evaluate the conditions of the viticultural regions in Iberia, regarding the main features of the *terroir*. This is first integrated analysis of this kind over the entire Iberian Peninsula. Therefore, the objectives of this work are three-fold: 1) to assess the prevailing conditions in terms of climate, soil and topography in the Iberian viticultural regions; 2) to develop an integrated analysis of the previous three elements and their impact on vegetative growth; and 3) to establish a zoning of homogeneous climate-soil-topography-vegetative growth areas.

Materials and Methods

Viticultural regions and vineyard area

To assess the spatial characteristics of each DO region, boundaries of each DO or DOCa (Rioja and Priorat) were defined using data available in the Portuguese *'Instituto do Vinho e da Vinha'* (IVV; http://www.ivv.min-agricultura.pt) and Spanish *'Ministerio de Agricultura, Alimentación y Medio Ambiente'* (MAGRAMA; http://sig.magrama.es). The viticultural regions in the islands of Madeira, Azores (Portugal) and Canarias (Spain) were not assessed due to limitations in the soil and climatic datasets. Other viticultural regulated regions, such as, quality wine with specific geographical indication, estate wine, qualified estate wine and country wines, Indicação de Proveniencia Regulamentada (in Spain) and Vinho Regional (in Portugal), are out of the scope of the current study, since DO regions are usually considered of higher importance. Note that the DO regions of Málaga and Sierras de Málaga (in Spain) are two different DO regions that geographically coincide and were therefore treated jointly (henceforth DO Málaga & Sierras de Málaga). As a result, the spatial boundaries of 81 DO regions (82 effectively), 25 in Portugal and 56 (57) in Spain, were identified within Iberian Peninsula (Fig. 1a).

In order to analyse the spatial distribution of the vineyards in Iberia, the Corine Land Cover Map, version 13–2012, was used [44,45]. This dataset, last updated in 2012, provides an inventory of the land cover over Europe and has previously proven to have high accuracy in representing the land cover over mainland Portugal and Spain [46,47]. The vineyard fraction of the land cover over Iberia was extracted from this dataset for subsequent analysis (Fig. 1b).

Topography

For the topographic analysis, elevation and geographical aspect were selected as the leading topographic elements to be studied, due to their importance in Iberian viticulture. Although slope degree (degree of inclination of the land) is also considered a key landscape element, it was not included in the current analysis, as in Iberian vineyards certain practices, such as walled terraces, are often implemented to flatten steep-slopes. As detailed information of such practices was not available, this cannot be taken into account. For the elevation assessment, the GTOPO30 digital elevation model (DEM) was used at the 30 arc-seconds spatial resolution (https://lta.cr.usgs.gov/GTOPO30). Elevation ranges are isolated inside each region (according to Fig. 1a, b). Aspect was calculated using the same GTOPO30 DEM and geographical information systems.

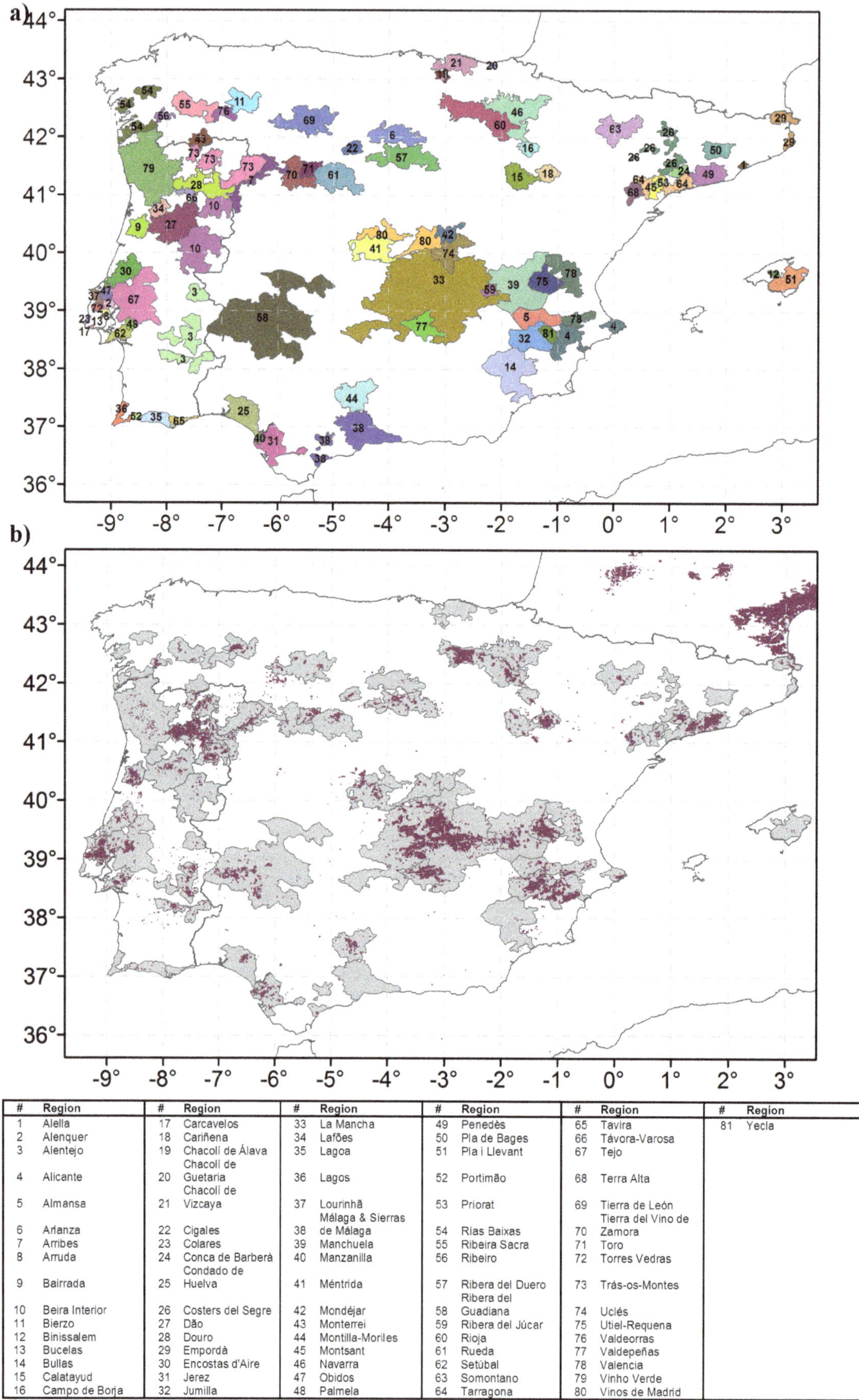

#	Region	#	Region	#	Region	#	Region	#	Region	#	Region
1	Alella	17	Carcavelos	33	La Mancha	49	Penedès	65	Tavira	81	Yecla
2	Alenquer	18	Cariñena	34	Lafões	50	Pla de Bages	66	Távora-Varosa		
3	Alentejo	19	Chacolí de Álava	35	Lagoa	51	Pla i Llevant	67	Tejo		
4	Alicante	20	Chacolí de Guetaria	36	Lagos	52	Portimão	68	Terra Alta		
5	Almansa	21	Chacolí de Vizcaya	37	Lourinhã	53	Priorat	69	Tierra de León		
6	Arlanza	22	Cigales	38	Málaga & Sierras de Málaga	54	Rías Baixas	70	Tierra del Vino de Zamora		
7	Arribes	23	Colares	39	Manchuela	55	Ribeira Sacra	71	Toro		
8	Arruda	24	Conca de Barberà	40	Manzanilla	56	Ribeiro	72	Torres Vedras		
9	Bairrada	25	Condado de Huelva	41	Méntrida	57	Ribera del Duero	73	Trás-os-Montes		
10	Beira Interior	26	Costers del Segre	42	Mondéjar	58	Ribera del Guadiana	74	Uclés		
11	Bierzo	27	Dão	43	Monterrei	59	Ribera del Júcar	75	Utiel-Requena		
12	Binissalem	28	Douro	44	Montilla-Moriles	60	Rioja	76	Valdeorras		
13	Bucelas	29	Empordà	45	Montsant	61	Rueda	77	Valdepeñas		
14	Bullas	30	Encostas d'Aire	46	Navarra	62	Setúbal	78	Valencia		
15	Calatayud	31	Jerez	47	Obidos	63	Somontano	79	Vinho Verde		
16	Campo de Borja	32	Jumilla	48	Palmela	64	Tarragona	80	Vinos de Madrid		

Figure 1. Vitcultural regions in Iberia. a) Location of the viticultural regions in Iberia, along with their denomination. b) Spatial distribution of the vineyard land cover over Iberia (dark-red), assessed using the Corine Land Cover, version 13- 2012, along with the viticultural regions (light-grey).

Climate

In order to analyse the climatic conditions of each region, a categorized bioclimatic index for viticultural zoning (CatI) was used [29]. CatI establishes climatic categories by combining three bioclimatic indices (Table 1): Huglin Index [48], Dryness Index [49] and Cool Night Index [50]. The Huglin Index expresses the thermal potential of a given region and relates the heat accumulation to the physiological development of grapevines. The Dryness Index assesses water availability for grapevines, by estimating potential water balance over the growing season. The Cool Night Index accounts for minimum temperatures at the end of the vegetative cycle (September in the Northern Hemisphere), as lower nocturnal temperatures during this stage tend to be favourable for wine quality [50]. Thus, CatI allows determining the optimum climatic suitability in terms of phenological development, heat and water availability conditions during the growth season, as well as ripening conditions (Table S1).

For CatI calculation (and combining indices), data from a regional climate model (RCM) Weather Research and Forecast model (WRF) [51] version 3.1.1 was used. The WRF model is a non-hydrostatic model and has been widely used for dynamical downscaling regional climate. For the present climate, a simulation with a horizontal grid resolution of 9-km (nested in a 27-km grid) was used, with both grids centred in the Iberian Peninsula. The RCM simulation started at 00 00 UTC 1 January 1989 and ended at 18 00 UTC 31 January 2013, with initial, lateral and lower boundary conditions derived from ERA-Interim. From the model output, precipitation and temperature over Iberia were considered for this study. A more detailed description of the model set-up can be found in Soares et al. [52] and Cardoso et al. [53], where the simulation results were validated for inland maximum and minimum temperatures and precipitation, showing a good agreement with observations. Patterns of the simulated CatI were then compared to previously established patterns using observational data [29], showing a good agreement. Additionally, solar radiation (surface net downward shortwave flux), from the Modern Era Retrospective-analysis for Research and Applications (MERRA; http://gmao.gsfc.nasa.gov/merra/) at a $0.6^{\circ}\times0.6^{\circ}$ longitude/latitude spatial resolution, was also obtained. Mean values over the growing season were calculated for 1989–2012.

Soils

For soil analysis, the predominant soil texture (SoilT) was assessed according to the United States Department of Agriculture soil textural classification [54]. The texture of a soil refers to its relative content of clay, sand and silt particles (Table 2). Each soil texture class presents its own properties in terms agricultural applicability. Clay soils have fine particles and retain large amounts of water, but are poorly drained and usually difficult to manage [55,56]. Conversely, sandy soils are coarse and usually excessively drained, with low water retention capacity [55,56]. With relatively even proportions between particles, the loamy soils are typically well drained and provide sufficient nutrient retention and are thus usually preferable for agricultural use [55,56]. Soil texture is a fundamental soil property used as a qualitative classification tool to determine other soils properties [54], such as soil plasticity, drainage and available water content [56]. Soil texture classes (SoilT) were obtained from the Harmonized World Soil Database (HWSD) [57].

Table 1. Categorized Index (CatI), along with the corresponding classes of the combined indices: Huglin, Dryness and Cool Night indices, according to [29].

CatI	Huglin Index (°C) [48]	Dryness Index (mm) [49]	Cool Night Index (°C) [50]	Description
0	<900	<−100		Unsuitably cold or excessively dry
1	900 - 1500	−100 - 50	<14	Cool, dry with cool nights
2	900 - 1500	−100 - 50	> 14	Cool, dry with warm nights
3	900 - 1500	> 50	<14	Cool, humid with cool nights
4	900 - 1500	> 50	> 14	Cool, humid with warm nights
5	1500 - 2100	−100 - 50	<14	Temperate, dry with cool nights
6	1500 - 2100	−100 - 50	> 14	Temperate, dry with warm nights
7	1500 - 2100	> 50	<14	Temperate, humid with cool nights
8	1500 - 2100	> 50	> 14	Temperate, humid with warm nights
9	2100 - 2700	−100 - 50	<14	Warm, dry with cool nights
10	2100 - 2700	−100 - 50	> 14	Warm, dry with warm nights
11	2100 - 2700	> 50	<14	Warm, humid with cool nights
12	2100 - 2700	> 50	> 14	Warm, humid with warm nights
13	> 2700	−100 - 50	<14	Very warm, dry with cool nights
14	> 2700	−100 - 50	> 14	Very warm, dry with warm nights
15	> 2700	> 50	<14	Very warm, humid with cool nights
16	> 2700	> 50	> 14	Very warm, humid with warm nights

Table 2. Soil texture categories, along with the respective percentages of Clay, Silt and Sand, according to USDA soil textural classification [54].

Soil Category	Clay (%)	Silt (%)	Sand (%)	Texture
1	60–100	0–40	0–45	heavy clay
2	40–60	40–60	0–20	silty clay
3	40–60	40–60	0–45	clay
4	27–40	40–73	0–20	silty clay loam
5	27–40	15–52	20–45	clay loam
6	0–12	88–100	0–20	silt
7	0–27	74–88	20–50	silty loam
8	35–55	0–20	45–65	sandy clay
9	7–27	28–50	23–52	loam
10	20–35	0–28	45–80	sandy clay loam
11	0–20	0–50	50–70	sandy loam
12	0–15	0–30	70–86	loamy sand
13	0–10	0–14	86–100	sand

Vegetative growth

The Enhanced Vegetation Index (EVI) was used for the analysis of the grapevine vigour. Spectral vegetation indices are based on visible and near-infrared radiation fluxes, captured by sensors on-board of polar orbiting satellites, and are a measure of the concentration of green leaf vegetation in space [58]. The EVI algorithm accounts for canopy background (e.g. soil and bare earth) and atmospheric effects (e.g. clouds), while also being barely affected by manmade structures [58,59].

In this study, the EVI from the Moderate Resolution Imaging Spectroradiometer (MODIS - MOD13A2 Collection 5) was extracted from the National Aeronautic and Space Administration (NASA) Land Processes Distributed Active Archive Center (LP DAAC; https://lpdaac.usgs.gov/). The EVI is described through the following equation (Eq. 1):

$$EVI = G \times \frac{NIR - RED}{NIR + C1 \times RED - C2 \times BLUE + L}, \qquad \text{(Eq.1)}$$

where NIR is the near-infrared band (841–876 nm), RED is the red band (620–670 nm), BLUE is the blue band (459–479 nm), L is the canopy background coefficient (L = 1), C1 and C2 are aerosol resistance and influence coefficients of the blue and red bands, respectively (C1 = 6 and C2 = 7.5), and G is a gain factor (G = 2.5) [60,61]. Four MODIS tiles (h17v04, h17v05, h18v04 and h18v05), covering all of the Iberian mainland were obtained at a 1-km spatial resolution for the 2012 growing season. This single year was selected to match the latest available land cover update (described in section 2.1), which ensures land cover changes do not interfere with the analysis of grapevine vegetative growth. The mean growing season EVI was then calculated (April-October mean) and the spatial average for the vineyard areas over Iberia was 0.23. For categorization of each DO according to the vegetative growth, two EVI classes (EVIc-1 and 2) were defined: EVIc-1 (EVI $\leq$ 0.23), for low vegetative growth areas, and EVIc-2 (EVI$>$ 0.23), for high vegetative growth.

Results

Mesoscale patterns

Overall, topography over Iberia displays large differences in both elevation and aspect (Fig 2a, b). Effective solar radiation depicts a strong north/south contrast, with higher solar radiation values in the south (Fig. 2c). Noticeable is the relatively low solar radiation in centre western Iberia (near La Mancha DO, #33), when compared to the surrounding areas. Grapevines in these areas are generally less affected by excessive solar radiation.

CatI (Fig. 3a) depicts a clear distinction between the northern regions, generally cooler, and the southern regions, with higher thermal accumulation. In fact, most of northern Iberia is situated in CatI-3 (cool, humid, with cool nights), making this the predominant category regarding all of Iberia. Also in the north, CatI-0 climates are present, suggesting lack of viticultural suitability owing to insufficient thermal accumulation. The lower category, CatI-1 (Cool, dry, with cool nights), is observed in an isolated area near the centre of the peninsula. Conversely, the warmer climatic region CatI-14 is located in the centre-south. CatI-10/9 (warm, dry, with warm/cool nights), represent the 2nd/3rd dominant climatic categories, scattered across the south and northeast. CatI-5 (Temperate, dry with cool nights) occurs over a large area in centre/northern Iberia, being the 4th dominant category (taking into account all of the mesoscale pattern in Iberia). Other secondary categories appear in transitional areas, between cool and warm climates. For example, CatI-11/12, with warm and humid climates, appear isolated in opposite sides of the peninsula (east/west), depicting the singularity of these regions.

Soil textural classes present a very homogeneous pattern (Fig. 3b and Table 2). Loamy soils (SoilT-9), which are commonly considered highly suitable for agriculture, are prevalent in most of the Iberian Peninsula. However, in central Iberia, sandy-loam soils (SoilT-11) are more frequent, whereas in southwestern Iberia loamy-sand soils (SoilT-12) are also common. Clay loam soils (SoilT-5) are isolated in a small region in southwestern Iberia. Clay soils (SoilT-3) dominate most of southern and eastern Iberia. In northwest Iberia, some intrusions of silt-loam soil (SoilT-7) are also depicted.

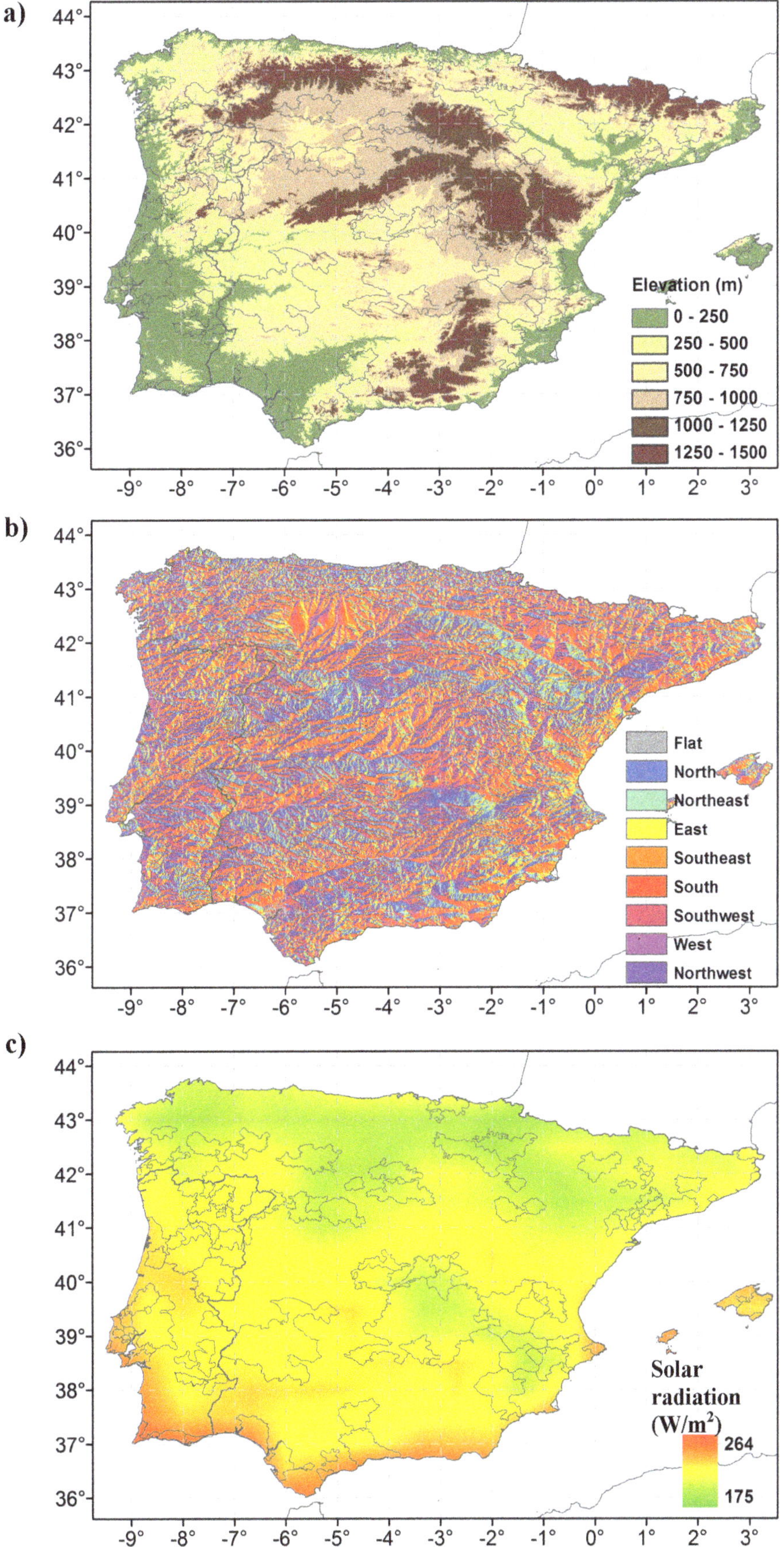
a)
Elevation (m)
0 - 250
250 - 500
500 - 750
750 - 1000
1000 - 1250
1250 - 1500
b)
Flat
North
Northeast
East
Southeast
South
Southwest
West
Northwest
c)
Solar radiation (W/m²)
264
175
44°
43°
42°
41°
40°
39°
38°
37°
36°
-9° -8° -7° -6° -5° -4° -3° -2° -1° 0° 1° 2° 3°

Figure 2. The topography in Iberia. a) Elevation (m) in the Iberian Peninsula, calculated using the GTOPO30 dataset. b) As in (a) but for the aspect. c) Solar radiation over Iberia, mean growing season values in 1989–2012 calculated using MERRA data at a 0.6º spatial resolution.

Regarding the vegetative development, represented by EVI and EVIc (Fig. 3c), the north/south and west/east contrasts are clear. Northern and western areas generally present a much higher EVI than southern and eastern areas. This was particularly clear in the EVIc, where the coastal areas in northern and western Iberia are classified as EVIc-2 (high vigour), while the rest of Iberia was generally keyed to EVIc-1 (low vigour).

Spearman ranked correlation coefficients between the previous mesoscale patterns (Table S2) reveal weak to moderate correlations. The highest positive correlation was found between CatI and surface net solar radiation flux (0.61), as both indirectly reflect the latitudinal effect over temperature and incoming solar radiation. The strongest negative correlation (−0.66) found was between CatI and elevation, undoubtedly reflecting the temperature lapse rate already embedded into this climatic index. A negative moderate correlation (−0.35) found was between EVI and CatI, while a weak positive correlation was found between EVI and SoilT (0.12).

Integrated analysis

An integrated analysis of the selected *terroir* elements was performed taking only the grapevine growing areas into account (Fig. 4). Fig. 4a depicts the grapevine vegetative growth in Iberia (EVIc) as a function of their climates (CatI) and soil characteristics (SoilT). Nearly all vineyards situated in SoilT-3 (clay) tend to show lower vegetative growth (EVIc-1), whereas higher vigour (EVIc-2) prevails in SoilT-11/12 (Warm, humid with cool/warm nights). Much more diverse vigour conditions are keyed to SoilT-9 (loam), which is the most frequent soil type in Iberia (Fig. 4a). In this soil type, vineyards show either high vigour (EVIc-2), in regions with temperate climates (CatI-6, 7 and 8), or low vigour (EVIc-1), in regions with cool (CatI-1 to 4) or warm (CatI-9 to 16) climates (Fig. 4a). However, there are some exceptions: in SoilT-9 and CatI-5 (temperate climate) vineyards depict low vigour, while for SoilT-9 (loam) under cool/warm climates high vigour only occurs when combined with humid conditions. Thus, the influence of dryness/humidity (a CatI component) on SoilT-9 vineyards is highlighted for cool/warm conditions, which was not verified in other soil types.

Concerning the climatic influences on vegetative growth, in cool climatic regions (CatI-1 to 4) low vigour (EVIc-1) prevails, with the exception of the more humid regions (CatI-3) with loam/sandy-loam soils (SoilT-9/11). On the other hand, CatI-6 and 8 regions (Temperate, warm nights and dry or humid respectively) predominantly show high vigour (EVIc-2) (Fig. 4a), regardless of SoilT. The same applies to CatI-7 vineyards, with the exception of SoilT-3 areas. For warmer regions (CatI-9, 10, 12 and 14), low vigour is also dominant, apart from sandy loam or loamy sand soils (SoilT-11/12), which mostly have high vigour vineyards.

The same integrated analysis was also performed taking into account grapevine vegetative growth as a function of elevation and aspect (Fig. 4b). The largest vineyard concentrations are located at 600–800 m elevation ranges. Low elevation vineyards (most of them along coastal strips) present much higher vegetative growths, which can be partly explained by the more humid coastal local climates, or by the existence of deeper soils in these regions. Regarding the geographical aspect, growers tend to prefer south-facing vineyards, but no strong relationship can be established between aspect and vegetative growth.

DO regional analysis

With respect to vineyard locations (Fig. 1b), they are mostly confined by the DO boundaries, with the largest visible concentrations of vineyards in La Mancha and Rioja (Spain, #33 and #60) and Douro (Portugal, #28). Overall, 81 DO are depicted in Fig. 1a. These regions show large differences in terms of shape and area (Fig.1a; Table S3). The largest DO is La Mancha (Spain) and the smallest is Carcavelos (Portugal, #17).

The elevation means and ranges of the vineyard areas for each DO (Fig. 5) highlight the large spatial variability in which vineyards are grown, ranging from near sea-level (e.g. DO Tavira, #65) to elevations over 850 m (DO Arlanza, #6). Near coastal regions tend to have vineyards at much lower mean elevations, while in the innermost DO regions vineyards show higher mean elevations and ranges (Fig. 2a). The largest vineyard areas are indeed located at mean elevations ranging from 600 to 800 m, with the most heterogeneous being DO Valencia, which is also one of largest in terms of surface (Table S3). As expected, small DO regions, such as Tavira (#65), Pla i Llevant (#51), Lafões (#34), Monterrei (#43), Mondéjar (#42) and Arlanza (#6), tend to show much lower spatial variability. It should be noticed that La Mancha (with the largest vineyard area; Table S3) shows relatively low variability in elevation (relatively flat area) when compared to other smaller regions in mountainous areas (e.g. Douro, Table S3).

When analysing the geographical aspect of the vineyards within each DO (Fig. 6), it is clear that terrains with an S-SW aspect are preferred for viticultural activities. This outcome mainly reflects the largest vineyard area located at 600–800 m elevations (Fig. 6a). Despite this fact, vineyards in 800–1000 m elevation tend to have S-SE aspects (Fig. 6b), while vineyards at lower elevations (0–200 m) have N-NW aspects (Fig. 6b). For elevations in the 200–600 m range, no clear distinction on aspect preference can be made. Regarding the solar radiation, while most DO regions in Portugal present a high solar radiation, DO regions in Spain are usually located in areas with lower radiations (with the exception of some regions in the south).

In Table 3, the 81 DO are defined according to their predominant categories of CatI, SoilT and EVIc. The cool, dry, with cool nights DO regions (CatI-1) of Tierra del Vino de Zamora (#70) and Toro (#71) exhibit similar clay soils (SoilT-3) with low vegetative growth (EVIc-1), while the also CatI-1 DO Rueda (#61) exhibits sandy-loam soils (SoilT-11) and high vegetative growth (EVIc-2). This effect is less visible in the equally cool, but more humid, CatI-3 regions of Ribera del Duero (#57), Arlanza (#6), Cigales (#22), Chacolí de Álava (#19), Chacolí de Guetaria (#20), Tierra de León (#69) and Valdeorras (#76). While in the first two regions, SoilT-3 (clay) is predominant, each one shows a different EVIc (1 and 2; low and high vigour respectively). The other CatI-3 regions depict SoilT-9 (loam) and EVIc-2, with the exception of Cigales that exhibits EVIc-1.

As previously mentioned, CatI-5 (Temperate, dry, with cool nights) was the 4th leading category (regarding all Iberia). Nonetheless, it is the dominant category when only the area under vineyards is considered, being present in 19 DO regions (Bullas (#14), Conca de Barberà (#24), Manchuela (#39), Ribera del Júcar (#59), Mondéjar (#42), Penedès (#49), Alicante (#4), Almansa (#5), Cariñena (#18), Douro (#28), Priorat (#53), Trás-os-Montes (#73), Uclés (#74), Utiel-Requena (#75), Yecla (#81), Calatayud (#15), Távora-Varosa (#66), Arribes (#7) and Dão (#27)). These regions have SoilT-3, 9 and 11. For CatI-5, only

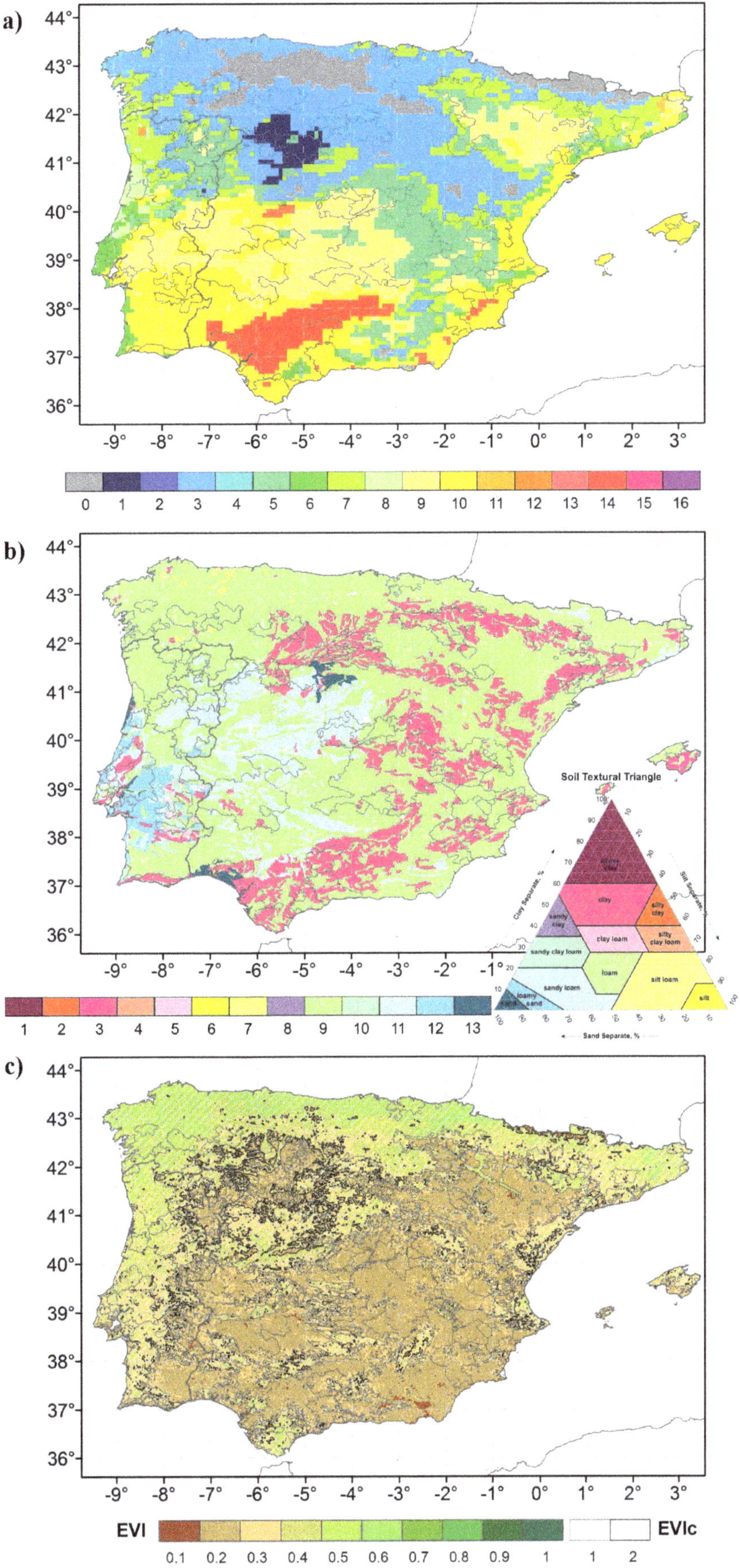
a)
b)
c)
Soil Textural Triangle
EVI
EVIc

Figure 3. Climate, soil and vegetative growth in Iberia. a) CatI over Iberia calculated according to Table 1, for the period of 1989-2012 using WRF simulations. b) SoilT according to Table 2 using HWSD data. c) Mean EVI and EVIc for the grapevine growth period (April-October) in 2012, using MODIS data. The spatial-average of the vineyard areas corresponds to 0.23. Below this value the EVIc equals 1 (transparent overlay), above this value EVIc equals 2 (hatched overlay).

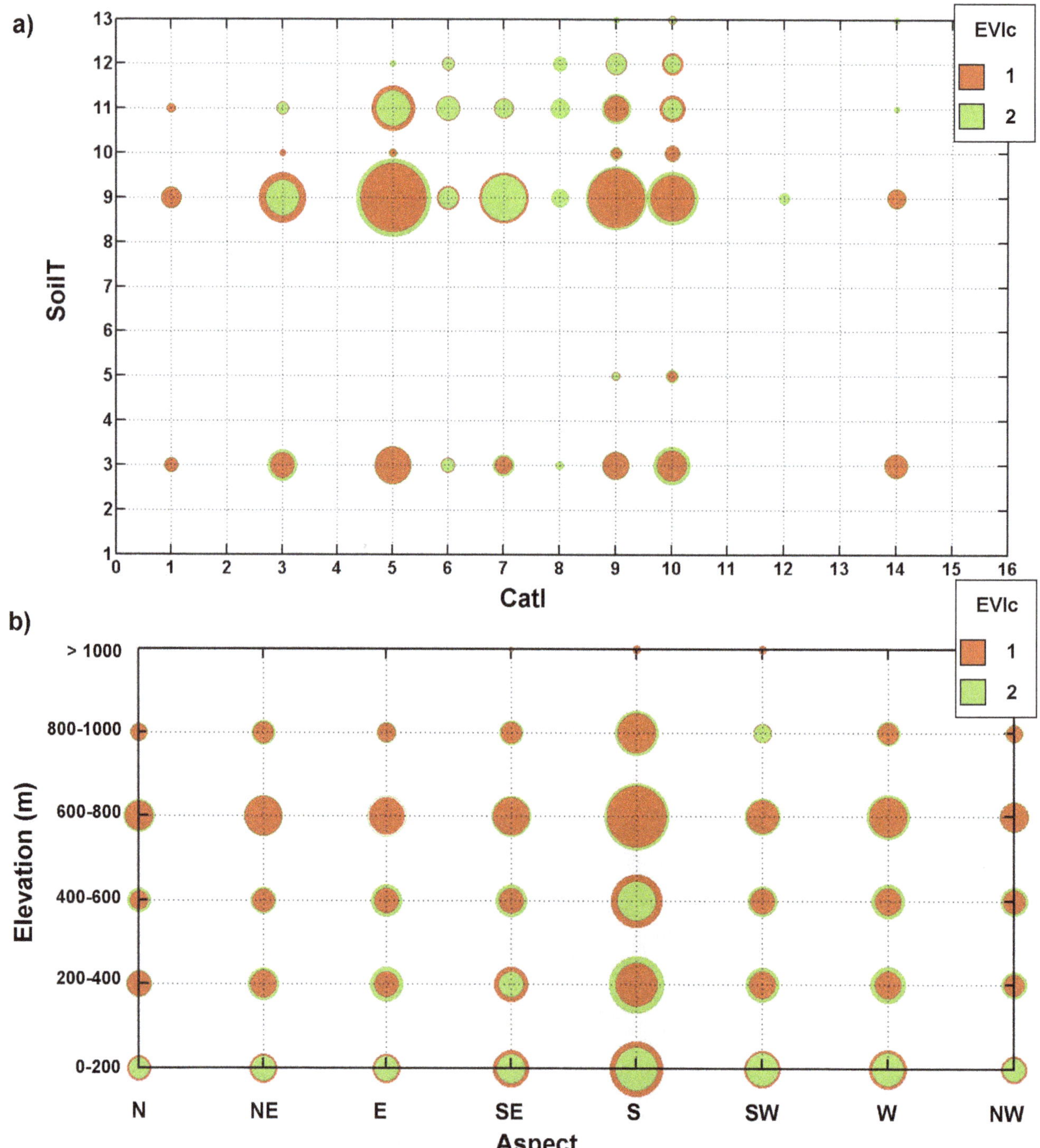

Figure 4. Integrated analysis. a) Circular accumulated EVIc (1 – red, 2 – green) as a function of the SoilT and CatI for all vineyards in Iberia. b) Circular accumulated EVIc as a function of the elevation and aspect for all vineyards in Iberia. The size of each circular chart depicts the accumulated vineyard area belonging to that EVI class, and the inner (outer) circular class depicts the largest (smallest) EVI class.

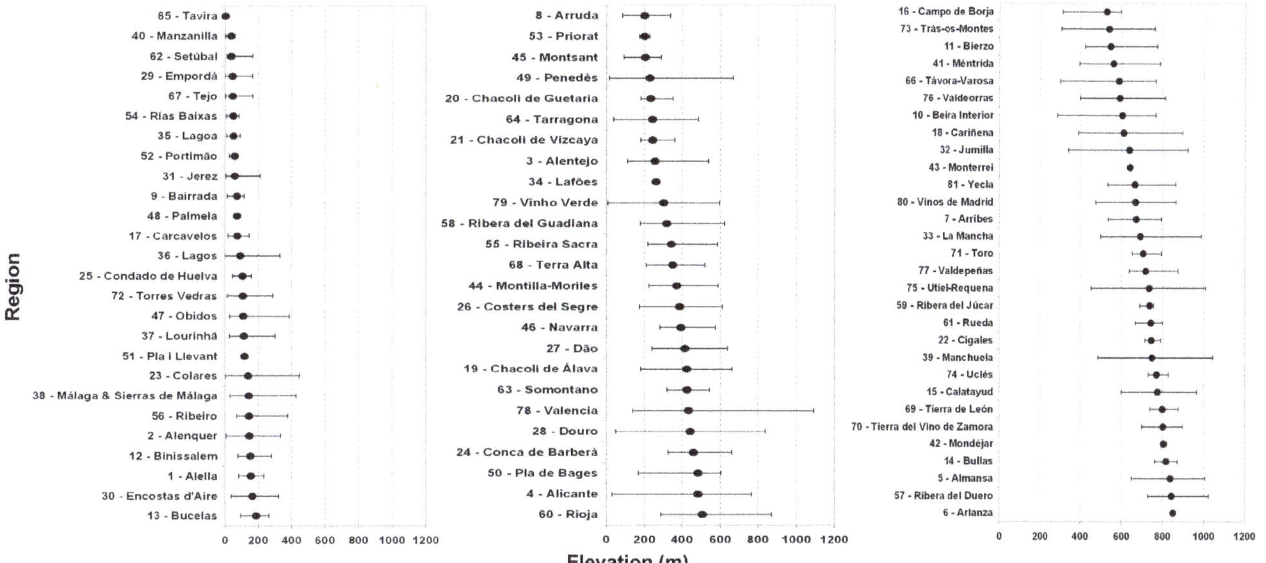

Figure 5. Elevation (m) of the vineyards in each DO/DOCa in Iberia. The inner circle represents the mean elevation and the horizontal bars represent the minimum and maximum, of the locations of the vineyards inside the DO.

SoilT-11 seems to present higher vegetative growth (in more than 50% of these DO regions). However, for CatI-6 regions (same as CatI-5 but with warm nights), all SoilT-9, 11 and 12 regions are keyed to EVIc-2 (Alella (#1), Alenquer (#2), Arruda (#8), Colares (#23), Montsant (#45), Bucelas (#13), Obidos (#47), Torres Vedras (#72) and Lourinha (#37).

The CatI-7 (Temperate, humid with cool nights) regions show the same relationship with soil as previously seen for CatI-3. Since these regions present humid conditions, SoilT-9 (loam) is clearly beneficial for vegetative growth when compared to SoilT-3 (clay).

CatI-8 (Temperate, humid with warm nights) regions (Encostas d'Aire (#30) and Bairrada (#9)) are the only regions where EVIc-2 (high vigour) is present, regardless of SoilT. CatI-9 and CatI-10 (warm, dry with cool or warm nights respectively) regions display similar characteristics to those already reported, higher vegetative growths in SoilT-9, 11 and 12 than in SoilT-3. The only DO regions in Iberia that currently present very warm, dry, with warm nights climate (CatI-14) is the DO Jerez (#31), that also exhibits clay soils (SoilT-3) and consequently low vegetative growth (EVIc-1).

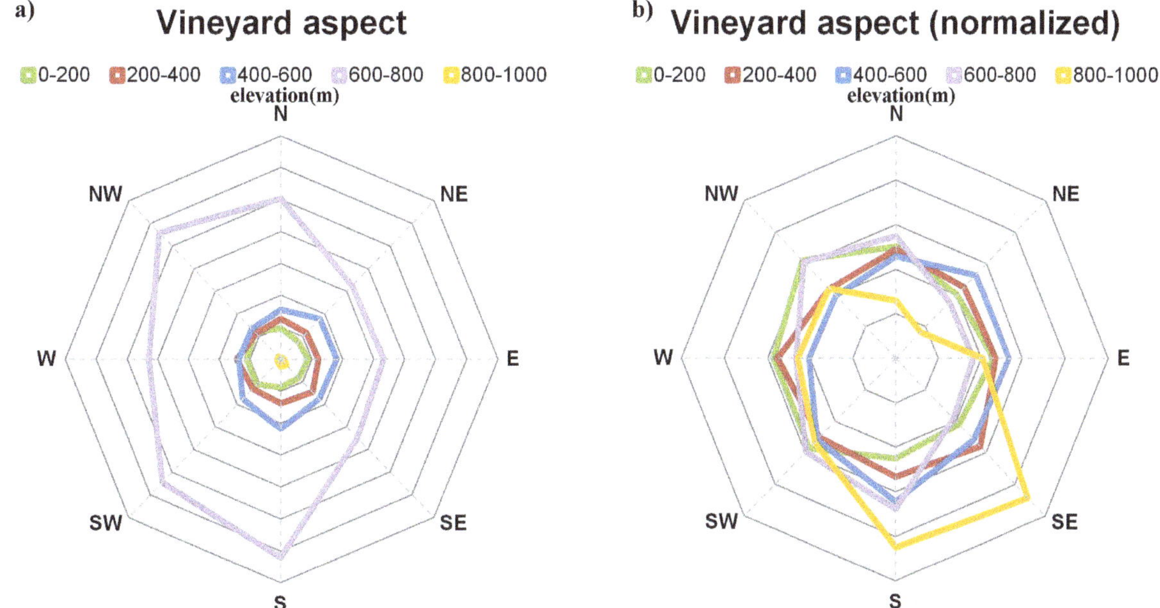

Figure 6. Vineyard aspect in Iberia. a) Geographical aspect (orientation) vineyards in Iberia according to the mean elevation of each DO. b) as in (a) but normalized using the vineyard area.

Table 3. CatI, SoilT and EVI class for each viticultural region in Iberia.

#	Region	CatI	SoilT	EVIc	#	Region	CatI	SoilT	EVIc	#	Region	CatI	SoilT	EVIc
70	Tierra del Vino de Zamora	1	3	1	1	Alella	6	9	2	10	Beira Interior	9	11	2
71	Toro	1	3	1	2	Alenquer	6	9	2	67	Tejo	9	12	1
61	Rueda	1	11	2	8	Arruda	6	9	2	48	Palmela	9	12	2
57	Ribera del Duero	3	3	1	23	Colares	6	9	2	35	Lagoa	10	3	1
6	Arlanza	3	3	2	45	Montsant	6	9	2	44	Montilla-Moriles	10	3	1
22	Cigales	3	9	1	13	Bucelas	6	11	2	68	Terra Alta	10	3	1
19	Chacolí de Álava	3	9	2	47	Obidos	6	11	2	38	Málaga & Sierras de Málaga	10	3	2
20	Chacolí de Guetaria	3	9	2	72	Torres Vedras	6	11	2	65	Tavira	10	3	2
69	Tierra de León	3	9	2	37	Lourinhã	6	12	2	3	Alentejo	10	5	2
76	Valdeorras	3	9	2	60	Rioja	7	3	1	12	Binissalem	10	9	1
14	Bullas	5	3	1	63	Somontano	7	3	1	52	Portimão	10	9	1
24	Conca de Barberà	5	3	1	16	Campo de Borja	7	9	1	78	Valencia	10	9	1
39	Manchuela	5	3	1	11	Bierzo	7	9	2	29	Empordà	10	9	2
59	Ribera del Júcar	5	3	1	21	Chacolí de Vizcaya	7	9	2	51	Pla i Llevant	10	9	2
42	Mondéjar	5	3	2	34	Lafões	7	9	2	64	Tarragona	10	9	2
49	Penedès	5	3	2	43	Monterrei	7	9	2	62	Setúbal	10	12	2
4	Alicante	5	9	1	54	Rías Baixas	7	9	2	40	Manzanilla	10	13	1
5	Almansa	5	9	1	55	Ribeira Sacra	7	9	2	25	Condado de Huelva	10	13	2
18	Cariñena	5	9	1	56	Ribeiro	7	9	2	31	Jerez	14	3	1
28	Douro	5	9	1	79	Vinho Verde	7	9	2					
53	Priorat	5	9	1	30	Encostas d'Aire	8	11	2					
73	Trás-os-Montes	5	9	1	9	Bairrada	8	12	2					
74	Uclés	5	9	1	33	La Mancha	9	3	1					
75	Utiel-Requena	5	9	1	32	Jumilla	9	9	1					
81	Yecla	5	9	1	46	Navarra	9	9	1					
15	Calatayud	5	9	2	58	Ribera del Guadiana	9	9	1					
66	Távora-Varosa	5	9	2	77	Valdepeñas	9	9	1					
7	Arribes	5	11	1	80	Vinos de Madrid	9	9	1					
27	Dão	5	11	2	26	Costers del Segre	9	9	2					
17	Carcavelos	6	3	1	50	Pla de Bages	9	9	2					
36	Lagos	6	3	1	41	Méntrida	9	11	1					

Only the predominant categories are shown.

Discussion and Conclusions

In the current study, an integrated analysis of the climate, soil, topography and vegetative growth was undertaken for the Iberian viticultural areas, using state-of-the-art datasets. Until present, studies on viticultural zoning were mainly focused on climatic conditions [15,62–64], while the combination of the *terroir* composing elements was still underexplored. While in the last decades great advances have been made regarding the quality and availability of these spatial datasets [65], few studies have been devoted to integrate these factors into viticultural zoning [11,66,67], but none for Iberia. Therefore, understanding the spatial variability of these factors provides the basis for a viable characterization of each viticultural region. To our knowledge this is the first study in which climate, soil, topography and vegetative growth, were jointly studied to analyse the viticultural regions in the Iberian Peninsula.

Overall, temperate dry climate with cool nights (CatI-5) is the dominant climatic category for vineyards in Iberia. Further, the results showed that vineyards in DO regions with CatI-5 tend to present lower vigour. Nevertheless, these conditions of moderate water restriction and low nocturnal temperatures during ripening are often beneficial for the production of high quality wines [14,24,68], possibly explaining the higher vineyard density in these regions. Other temperate climate types (CatI-6, 7 and 8) tend to present higher levels of vigour, regardless of soil type, topography or dryness levels (CatI component). This may be partially explained by the fact that grapevines tend to be less exposed to heat and water stress under these climatic conditions, thus experiencing significantly fewer restrictions to its development and growth. Reversely, in warmer and cooler climates, abiotic constrains significantly increase, enhancing the importance of other factors, such as precipitation and soil type, for vine performance [31,69–72].

The results suggested that soil also plays a key role in viticulture, as regions with similar climatic conditions, but with different soil types, can indeed present different vigour attributes. Soils with higher clay content are associated with lower vigour. This can be explained by water uptake restrictions, due to lower root penetration and stronger soil-water retention [73]. Depending on their structure, clay soils can also promote high vigour [74], though not apparent in the present study. To further investigate this, a more detailed soil study including soil structure, porosity and depth would be required, which was not the object of this study. On the other hand, soils with lower clay and higher sand contents seem to promote higher vigour. Although sandy soils retain less water than clay soils, they are better drained, providing a better growing structure for roots and higher absorption capacity for water and mineral nutrients. The results also showed that loamy soils are the most common type in Iberian vineyards, where precipitation plays a leading role in governing the vigour patterns, since more humid (dry) regions tend to present higher (lower) vigour (Fig. 4a). These outcomes enhanced the importance of soil texture in determining vine performance in each region.

Regarding the topographic features, a clear distinction was found between low and high elevations, with the former showing higher vigour vineyards. However, this could be an indirect manifestation of climate-elevation relationships, such as the temperature lapse rate, the Atlantic influence in the Iberian Mediterranean-like climates (higher precipitation amounts over northern and western Iberia than over central and eastern Iberia [14]) and/or the presence of deeper soils with higher water holding capacity in low-elevation areas. For the geographical aspect and considering the peninsula as a whole, no clear relationship with vegetative growth was found, despite some regional differences.

Based on previous research some considerations can still be made on the effects of the main components of the *terroir* in wine quality. First, it has been shown that climate plays a key role on wine quality [2,5]. Regions with temperate/temperate-warm dry climates with cool nights (in this study CatI-5/9) tend to produce a more balanced maturation, by e.g. promoting the synthesis of anthocyanins, resulting in higher quality. Second, soil type is also an important factor for quality, but its suitability largely depends on the targeted wine typicity and attributes. Previous studies showed that wines from clay soils (SoilT-3) showed high sugar accumulation and anthocyanin concentration [66,73], but lower pH [7], than those from sandy soils (SoilT-11/12) [66], while other studies suggest that loamy soils (SoilT-9) may also be favourable to high quality wines [75,76]. Third, studies have shown that lower vigour vineyards (EVIc-1) are generally expected to produce higher grape and wine quality, with higher total soluble solids, total phenolics and anthocyanins and lower titratable acidity [34], but are commonly associated with lower yield [77,78]. From the present study results, some considerations can also be made regarding the regional quality attributes. As an example, the renowned Douro/Porto DO shows climate (CatI-5; Temperate, dry with cool nights), soil (SoilT-9; Loamy) and vigour (EVIc-1; low) conditions particularly suitable for high quality wine production [2,5]. These conditions are also shared by other regions, such as Alicante, Almansa, Cariñena, Priorat, Trás-os-Montes, Uclés, Utiel-Requena and Yecla, highlighting the potential for the future viticultural development of these regions. However, the Iberian Peninsula presents a wide range of *terroirs* (Table 3), from e.g. Toro (CatI-1, SoilT-3 and EVIc-1) to Jerez (CatI-14, SoilT-3 and EVIc-1) or Setúbal (CatI-10, SoilT-12 and EVIc-2), and in all cases these regions are known to produce high quality wines but with different typicity. Nonetheless, it should be noted that the present study did not directly include other important *terroir* elements, such as viticultural and oenological practices, which are also key for the wine attributes produced in each region.

The assessments provided herein may be of great value to viticulturists and may also play a key role when including vineyards into a given DO. Usual methods for DO delineation rely on onsite analysis of climate, soil and topographic attributes. Although these empirical approaches are extremely useful, they are often based on erratic, insufficient or unreliable data (e.g. outdated land cover, soils surveys and topographic maps, assessments made on nearby weather station records), making comparison between regions rather difficult. Furthermore, taking into account the climate change projections for Iberia [15,17,29,79,80], these delineations may require and more continuous update. Our fully integrated approach provides a feasible method for DO comparison on a mesoscale basis. This may allow growers to identify new management practices and grapevine varieties that can be easily adapted to other regions that share the same *terroir* characteristics. Additionally, the implemented methodologies can be extended to other viticultural regions of the world.

Supporting Information

Table S1 The Cool Nigh, Dryness and Huglin indices, along with their mathematical definition, units and classes.

Table S2 Spearman ranked correlation coefficient between the EVI, CatI, SoilT, elevation, aspect and solar radiation in all of Iberia.

Table S3 Area (km^2) of each DO region in Iberia.

Acknowledgments

This study was supported by national (FCT – Fundação para a Ciência e a Tecnologia) and European Union Funds (FEDER/COMPETE - Operational Competitiveness Programme), under the projects PTDC/AGR-ALI/110877/2009, FCOMP-01-0124-FEDER-022692, PEst-OE/AGR/UI4033/2014, by PRODER – Inovação (GreenVitis project IF0018), by ON2 ENOEXEL (NORTE-07-0124-FEDER-000032), and by EUPORIAS - 7th Framework Programme for Research, Grant Agreement 308291.

Author Contributions

Conceived and designed the experiments: HF JAS. Performed the experiments: HF ACM JM-P RMC PMMS JJC JGP JAS. Analyzed the data: HF ACM JM-P RMC PMMS JJC JGP JAS. Contributed reagents/materials/analysis tools: HF ACM JM-P RMC PMMS JJC JGP JAS. Wrote the paper: HF ACM JM-P RMC PMMS JJC JGP JAS.

References

1. OIV (2012) Statistical Report on World Vitiviniculture. OIV, Paris, France, 32 pp.
2. Magalhães N (2008) Tratado de viticultura: a videira, a vinha e o terroir. Lisboa, Portugal: Chaves Ferreira. 605 p.
3. IVV (2011) Vinhos e Aguardentes de Portugal, Anuário 2010/2011. Ministério da Agricultura, do Desenvolvimento Rural e das Pescas: Instituto da Vinha e do Vinho, Lisboa.
4. OIV (2010) RESOLUTION OIV/VITI 333/2010, DEFINITION OF VITIVINICULTURAL "TERROIR", Tbilisi, 25th June 2010.
5. Winkler AJ (1974) General viticulture. California, USA: University of California Press.
6. Morlat R, Jacquet A (2003) Grapevine root system and soil characteristics in a vineyard maintained long-term with or without interrow sward. American Journal of Enology and Viticulture 54: 1–7.
7. Mackenzie DE, Christy AG (2005) The role of soil chemistry in wine grape quality and sustainable soil management in vineyards. Water Science and Technology 51: 27–37.
8. Jackson DI, Lombard PB (1993) Environmental and Management-Practices Affecting Grape Composition and Wine Quality - a Review. American Journal of Enology and Viticulture 44: 409–430.
9. Keller M (2010) The Science of Grapevines: Anatomy and Physiology. Elsevier, Inc. 400pp.
10. Morlat R, Jacquet A (1993) The Soil Effects on the Grapevine Root-System in Several Vineyards of the Loire Valley (France). Vitis 32: 35–42.
11. Yau IH, Davenport JR, Rupp RA (2013) Characterizing Inland Pacific Northwest American Viticultural Areas with Geospatial Data. Plos One 8.
12. Field SK, Smith JP, Holzapfel BP, Hardie WJ, Emery RJN (2009) Grapevine Response to Soil Temperature: Xylem Cytokinins and Carbohydrate Reserve Mobilization from Budbreak to Anthesis. American Journal of Enology and Viticulture 60: 164–172.
13. Flexas J, Galmes J, Galle A, Gulias J, Pou A, et al. (2010) Improving water use efficiency in grapevines: potential physiological targets for biotechnological improvement. Australian Journal of Grape and Wine Research 16: 106–121.
14. Fraga H, Malheiro AC, Moutinho-Pereira J, Santos JA (2014) Climate factors driving wine production in the Portuguese Minho region. Agricultural and Forest Meteorology 185: 26–36.
15. Fraga H, Malheiro AC, Moutinho-Pereira J, Santos JA (2013) Future scenarios for viticultural zoning in Europe: ensemble projections and uncertainties. International Journal of Biometeorology 57: 909–925.
16. Keller M (2010) Managing grapevines to optimise fruit development in a challenging environment: a climate change primer for viticulturists. Australian Journal of Grape and Wine Research 16: 56–69.
17. Malheiro AC, Santos JA, Fraga H, Pinto JG (2010) Climate change scenarios applied to viticultural zoning in Europe. Climate Research 43: 163–177.
18. Jones GV, Davis RE (2000) Climate influences on grapevine phenology, grape composition, and wine production and quality for Bordeaux, France. American Journal of Enology and Viticulture 51: 249–261.
19. Malheiro AC, Campos R, Fraga H, Eiras-Dias J, Silvestre J, et al. (2013) Winegrape phenology and temperature relationships in the Lisbon Wine Region, Portugal. journal International des Sciences de la Vigne et du Vin 47: 287–299.
20. Santos JA, Malheiro AC, Karremann MK, Pinto JG (2011) Statistical modelling of grapevine yield in the Port Wine region under present and future climate conditions. International Journal of Biometeorology 55: 119–131.
21. Bindi M, Fibbi L, Gozzini B, Orlandini S, Miglietta F (1996) Modelling the impact of future climate scenarios on yield and yield variability of grapevine. Climate Research 7: 213–224.
22. Santos JA, Grätsch SD, Karremann MK, Jones GV, Pinto JG (2013) Ensemble projections for wine production in the Douro Valley of Portugal. Climatic Change 117: 211–225.
23. Camps JO, Ramos MC (2012) Grape harvest and yield responses to inter-annual changes in temperature and precipitation in an area of north-east Spain with a Mediterranean climate. International Journal of Biometeorology 56: 853–864.
24. Jones GV, White MA, Cooper O (2004) Climate change and global wine quality. Bulletin of the American Meteorological Society 85: 504–504.
25. Jones GV, Goodrich GB (2008) Influence of climate variability on wine regions in the western USA and on wine quality in the Napa Valley. Climate Research 35: 241–254.
26. Jones GV, White MA, Cooper OR, Storchmann K (2005) Climate Change and Global Wine Quality. Climatic Change 73: 319–343.
27. Jones GV (2006) Climate and Terroir: Impacts of Climate Variability and Change on Wine In Fine Wine and Terroir - The Geoscience Perspective. Macqueen, R.W., and Meinert, L.D., (eds.). Newfoundland, Canada: Geoscience Canada, Geological Association of Canada.
28. Carbonneau A (2003) Ecophysiologie de la vigne et terroir. Terroir, zonazione, viticoltura Trattato internazionale: Phytoline. pp.61–102.
29. Fraga H, Malheiro AC, Moutinho-Pereira J, Jones GV, Alves F, et al. (2014) Very high resolution bioclimatic zoning of Portuguese wine regions: present and future scenarios. Regional Environmental Change 14: 295–306.
30. Jones GV, Snead N, Nelson P (2004) Geology and wine 8. Modeling viticultural landscapes: A GIS analysis of the terroir potential in the Umpqua Valley of Oregon. Geoscience Canada 31: 167–178.
31. Zsofi Z, Toth E, Rusjan D, Balo B (2011) Terroir aspects of grape quality in a cool climate wine region: Relationship between water deficit, vegetative growth and berry sugar concentration. Scientia Horticulturae 127: 494–499.
32. Failla O, Mariani L, Brancadoro L, Minelli R, Scienza A, et al. (2004) Spatial distribution of solar radiation and its effects on vine phenology and grape ripening in an alpine environment. American Journal of Enology and Viticulture 55: 128–138.
33. Hall A, Louis JP, Lamb DW (2008) Low-resolution remotely sensed images of winegrape vineyards map spatial variability in planimetric canopy area instead of leaf area index. Australian Journal of Grape and Wine Research 14: 9–17.
34. Johnson LF, Bosch DF, Williams DC, Lobitz BM (2001) Remote sensing of vineyard management zones: Implications for wine quality. Applied Engineering in Agriculture 17: 557–560.
35. Gouveia C, Liberato MLR, DaCamara CC, Trigo RM, Ramos AM (2011) Modelling past and future wine production in the Portuguese Douro Valley. Climate Research 48: 349–362.
36. Cunha M, Marcal ARS, Silva L (2010) Very early prediction of wine yield based on satellite data from VEGETATION. International Journal of Remote Sensing 31: 3125–3142.
37. Cunha M, Marcal ARS, Rodrigues A (2010) A comparative study of satellite and ground-based vineyard phenology Symposium of the European Association of Remote Sensing Laboratories; Imagin Europe 68–77.
38. AEMET/IMP (2011) Iberian Climate Atlas. Madrid, Spain: Closas-Orcoyen S. L. 79 p.
39. Santos JA, Malheiro AC, Pinto JG, Jones GV (2012) Macroclimate and viticultural zoning in Europe: observed trends and atmospheric forcing. Climate Research 51: 89–103.
40. Nascimbene J, Marini L, Ivan D, Zottini M (2013) Management Intensity and Topography Determined Plant Diversity in Vineyards. Plos One 8: 7.
41. Böhm J (2010) Portugal vitícola: o grande livro das castas. Lisbon, Portugal: C. Ferreira. 234 p.
42. Fraga H, Santos JA, Malheiro AC, Moutinho-Pereira J (2012) Climate Change Projections for the Portuguese Viticulture Using a Multi-Model Ensemble. Ciencia E Tecnica Vitivinicola 27: 39–48.
43. Malheiro AC, Santos JA (2011) Macrozonagem vitivinícola da Península Ibérica face a cenário de alteração climática. Atlas das castas da Península Ibérica - história, terroir, ampelografia: Coordenador Jorge Böhm, Dinalivro. pp.155 – 159.
44. EEA (2002) CORINE Land Cover update, I&CLC2000 project, Technical Guidelines.
45. Büttner G, Feranec G, Jaffrain G (2006) EEA CORINE Land Cover Nomenclature Illustrated Guide - Addendum 2006.
46. Caetano M, Mata F, Freire S (2006) Accuracy assessment of the Portuguese CORINE land cover map. Global Developments in Environmental Earth Observation from Space: 459–467.
47. Genovese G, Vignolles C, Negre T, Passera G (2001) A methodology for a combined use of normalised difference vegetation index and CORINE land

cover data for crop yield monitoring and forecasting. A case study on Spain. Agronomie 21: 91–111.
48. Huglin P (1978) Nouveau mode d'évaluation des possibilités héliothermiques d'un milieu viticole. Comptes Rendus de l'Académie d'Agriculture. Paris, France: Académie d'agriculture de France.
49. Riou C, Carbonneau A, Becker N, Caló A, Costacurta A, et al. (1994) Le determinisme climatique de la maturation du raisin: Application au zonage de la teneur en sucre dans la Communauté Européenne: Office des Publications Officielles des Communautés Européennes, Luxembourg. 319 p.
50. Tonietto J, Carbonneau A (2004) A multicriteria climatic classification system for grape-growing regions worldwide. Agricultural and Forest Meteorology 124: 81–97.
51. Skamarock WC, Klemp JB, Dudhia J, Gill DO, Barker M, et al. (2008) A description of the Advanced Research WRF Version 3. 1–113 p.
52. Soares PM, Cardoso R, Miranda PA, Medeiros J, Belo-Pereira M, et al. (2012) WRF high resolution dynamical downscaling of ERA-Interim for Portugal. Climate Dynamics 39: 2497–2522.
53. Cardoso RM, Soares PMM, Miranda PMA, Belo-Pereira M (2013) WRF high resolution simulation of Iberian mean and extreme precipitation climate. International Journal of Climatology 33: 2591–2608.
54. USDA (2006) United States Department of Agriculture, Soil Mechanics Level 1. Module 3-USDA Textural Soil Classification Study Guide.
55. Juma NG (1999) Introduction to soil science and soil resources, Salman Productions, vol 1. 335pp.
56. McKenzie N, Grundy M, Webster R, Ringrose-Voase A (2008) Guidelines for Surveying Soil and Land Resources, CSIRO PUBLISHING 2ed, 576pp.
57. FAO/IIASA/ISRIC/ISSCAS/JRC (2012) Harmonized World Soil Database (version 1.2). FAO, Rome, Italy and IIASA, Laxenburg, Austria.
58. Huete A, Didan K, Miura T, Rodriguez EP, Gao X, et al. (2002) Overview of the radiometric and biophysical performance of the MODIS vegetation indices. Remote Sensing of Environment 83: 195–213.
59. Pennec A, Gond V, Sabatier D (2011) Tropical forest phenology in French Guiana from MODIS time series. Remote Sensing Letters 2: 337–345.
60. Huete AR, Liu HQ, Batchily K, vanLeeuwen W (1997) A comparison of vegetation indices global set of TM images for EOS-MODIS. Remote Sensing of Environment 59: 440–451.
61. Huete A, Justice C, Liu H (1994) Development of Vegetation and Soil Indexes for Modis-Eos. Remote Sensing of Environment 49: 224–234.
62. Jones GV, Davis RE (2000) Using a synoptic climatological approach to understand climate-viticulture relationships. International Journal of Climatology 20: 813–837.
63. Duchene E, Schneider C (2005) Grapevine and climatic changes: a glance at the situation in Alsace. Agronomy for Sustainable Development 25: 93–99.
64. Webb LB, Whetton PH, Barlow EWR (2008) Modelling the relationship between climate, winegrape price and winegrape quality in Australia. Climate Research 36: 89–98.
65. Jones GV, Duff AA, Hall A, Myers JW (2010) Spatial Analysis of Climate in Winegrape Growing Regions in the Western United States. American Journal of Enology and Viticulture 61: 313–326.
66. van Leeuwen C, Friant P, Choné X, Tregoat O, Koundouras S, et al. (2004) Influence of climate, soil, and cultivar on terroir. American Journal of Enology and Viticulture 55: 207–217.
67. Taylor JA (2004) Digital Terroirs and Precision Viticulture: Investigations Into the Application of Information Technology in Australian Vineyards. Sydney, Australia: Australian Centre for Precision, Agriculture Faculty of Agriculture, Food and Natural Resources, University of Sydney.
68. van Leeuwen C, Tregoat O, Chone X, Bois B, Pernet D, et al. (2009) Vine Water Status Is a Key Factor in Grape Ripening and Vintage Quality for Red Bordeaux Wine. How Can It Be Assessed for Vineyard Management Purposes? Journal International Des Sciences De La Vigne Et Du Vin 43: 121–134.
69. Chaves MM, Santos TP, Souza CR, Ortuno MF, Rodrigues ML, et al. (2007) Deficit irrigation in grapevine improves water-use efficiency while controlling vigour and production quality. Annals of Applied Biology 150: 237–252.
70. Chaves MM, Zarrouk O, Francisco R, Costa JM, Santos T, et al. (2010) Grapevine under deficit irrigation: hints from physiological and molecular data. Ann Bot 105: 661–676.
71. Lopes CM, Santos TP, Monteiro A, Rodrigues ML, Costa JM, et al. (2011) Combining cover cropping with deficit irrigation in a Mediterranean low vigor vineyard. Scientia Horticulturae 129: 603–612.
72. Koundouras S, Van Leeuwen C, Seguin G, Glories Y (1999) Influence of water status on vine vegetative growth, berry ripening and wine characteristics in mediterranean zone (example of Nemea, Greece, variety Saint-George, 1997). Journal International des Sciences de la Vigne et du Vin 33: 149–160.
73. Tramontini S, van Leeuwen C, Domec JC, Destrac-Irvine A, Basteau C, et al. (2013) Impact of soil texture and water availability on the hydraulic control of plant and grape-berry development. Plant and Soil 368: 215–230.
74. Magalhães N (2008) Tratado de viticultura: a videira, a vinha e o terroir: Chaves Ferreira.
75. Carey VA, Archer E, Barbeau G, Saayman D (2008) Viticultural Terroirs in Stellenbosch, South Africa. Ii. The Interaction of Cabernet-Sauvignon and Sauvignon Blanc with Environment. Journal International Des Sciences De La Vigne Et Du Vin 42: 185–201.
76. Winkel T, Rambal S, Bariac T (1995) SPATIAL VARIATION AND TEMPORAL PERSISTENCE OF GRAPEVINE RESPONSE TO A SOIL TEXTURE GRADIENT. Geoderma 68: 67–78.
77. Song JQ, Smart RE, Dambergs RG, Sparrow AM, Wells RB, et al. (2014) Pinot Noir wine composition from different vine vigour zones classified by remote imaging technology. Food Chemistry 153: 52–59.
78. Bramley RGV, Hamilton RP (2007) Terroir and precision viticulture: Are they compatible? Journal International Des Sciences De La Vigne Et Du Vin 41: 1–8.
79. Jones GV, Duchêne E, Tomasi D, Yuste J, Braslavska O, et al. (2005) Changes in European winegrape phenology and relationships with climate. Proc XIV GESCO Symposium. Geisenheim, Germany.
80. Andrade C, Fraga H, Santos JA (2014) Climate change multi-model projections for temperature extremes in Portugal. Atmospheric Science Letters 15: 149–156.

17

The Effect of Leaf Litter Cover on Surface Runoff and Soil Erosion in Northern China

Xiang Li, Jianzhi Niu*, Baoyuan Xie

Key Laboratory of Soil and Water Conservation and Desertification Combating of Education Ministry, Beijing Forestry University, Beijing, China

Abstract

The role of leaf litter in hydrological processes and soil erosion of forest ecosystems is poorly understood. A field experiment was conducted under simulated rainfall in runoff plots with a slope of 10%. Two common types of litter in North China (from *Quercus variabilis*, representing broadleaf litter, and *Pinus tabulaeformis*, representing needle leaf litter), four amounts of litter, and five rainfall intensities were tested. Results revealed that the litter reduced runoff and delayed the beginning of runoff, but significantly reduced soil loss ($p<0.05$). Average runoff yield was 29.5% and 31.3% less than bare-soil plot, and for *Q. variabilis* and *P. tabulaeformis*, respectively, and average sediment yield was 85.1% and 79.9% lower. Rainfall intensity significantly affected runoff ($R=0.99$, $p<0.05$), and the efficiency in runoff reduction by litter decreased considerably. Runoff yield and the runoff coefficient increased dramatically by 72.9 and 5.4 times, respectively. The period of time before runoff appeared decreased approximately 96.7% when rainfall intensity increased from 5.7 to 75.6 mm h^{-1}. Broadleaf and needle leaf litter showed similarly relevant effects on runoff and soil erosion control, since no significant differences ($p\leq0.05$) were observed in runoff and sediment variables between two litter-covered plots. In contrast, litter mass was probably not a main factor in determining runoff and sediment because a significant correlation was found only with sediment in *Q. variabilis* litter plot. Finally, runoff yield was significantly correlated ($p<0.05$) with sediment yield. These results suggest that the protective role of leaf litter in runoff and erosion processes was crucial, and both rainfall intensity and litter characteristics had an impact on these processes.

Editor: Ben Bond-Lamberty, DOE Pacific Northwest National Laboratory, United States of America

Funding: This study was supported by the Fundamental Research Funds for the Central Universities (No.BLYJ201406); National Natural Science Fund of China (41171028); National Advanced Project of the 12th Plan, China (2011BAD38B05); and National Forestry Public Welfare Industry Research Project (201104005). The funders had no role in study design, data collection and analysis, decision to publish, or preparation of the manuscript.

Competing Interests: The authors have declared that no competing interests exist.

* Email: nexk@bjfu.edu.cn

Introduction

Soil erosion has become a serious problem worldwide, causing to decrease productivity of agricultural and forest land, environmental and ecological degradations, and natural disasters such as mudflow that threaten to human safety and infrastructure [1]. Approximately five billion tons of soil is lost annually in China [2,3]. Soil erosion often damages forest ecosystems, including reduction of soil organic matter content and water-holding capacity [4], loss of valuable soil nutrients, and biota, also declines in biodiversity, and collectively lead to ecosystem instability [1,3]. While vegetation litter is often considered as an effective cover above soil surface that prevents soil erosion, it is often burned or removed for fuel by local population in many forests of Northern China. These activities may increase detachment of soil aggregates, so raising the sediment generation and transportation via runoff [5].

Although it is widely recognized that vegetation canopy is important in hindering soil erosion [6], by comparison, the role of vegetation litter layers in modulating surface runoff and soil erosion remains poorly understood. Litter layers are known to protect soil from raindrop splashes by intercepting rainfall, preventing surface sealing and crusting of soil, extend the time of soil infiltration, and enhance sediment deposition by increasing soil surface roughness [7–9]. Previous studies have mainly evaluated the effectiveness of various surface covers in reducing surface runoff and soil loss, including rock fragments [10–13], crop residues [14,15], grass [16–18], geo-textiles [19], post-fire ash and needle cover [20–22], and combined cover such as rock and litter [23]. Nonetheless, few leaf litter materials have been tested [24,25], with variable findings [26–34]. The effects of litter layers on surface runoff and soil loss were not consistent across different types of litter, soil, or different environments. In general, runoff volume and sediment yield were reduced in the presence of litter cover [26–30]. However, some disputable results have shown that runoff and soil erosion were accelerated by impervious plastic mulch covers [31–34].

Moreover, the early studies have mainly examined the percentage of litter cover as a dominant influential factor in surface runoff and soil erosion [20,23,28]. Rainfall intensity and litter characteristics such as litter type and areal litter mass (i.e., litter mass per unit area, kg m^{-2}), have rarely been regarded as key factors in hydrologic and erosion processes in forest ecosystems. These factors were proved to be effective in rainfall interception [35–38], thus possibly exerting strong effects on soil infiltration, surface runoff, and soil erosion.

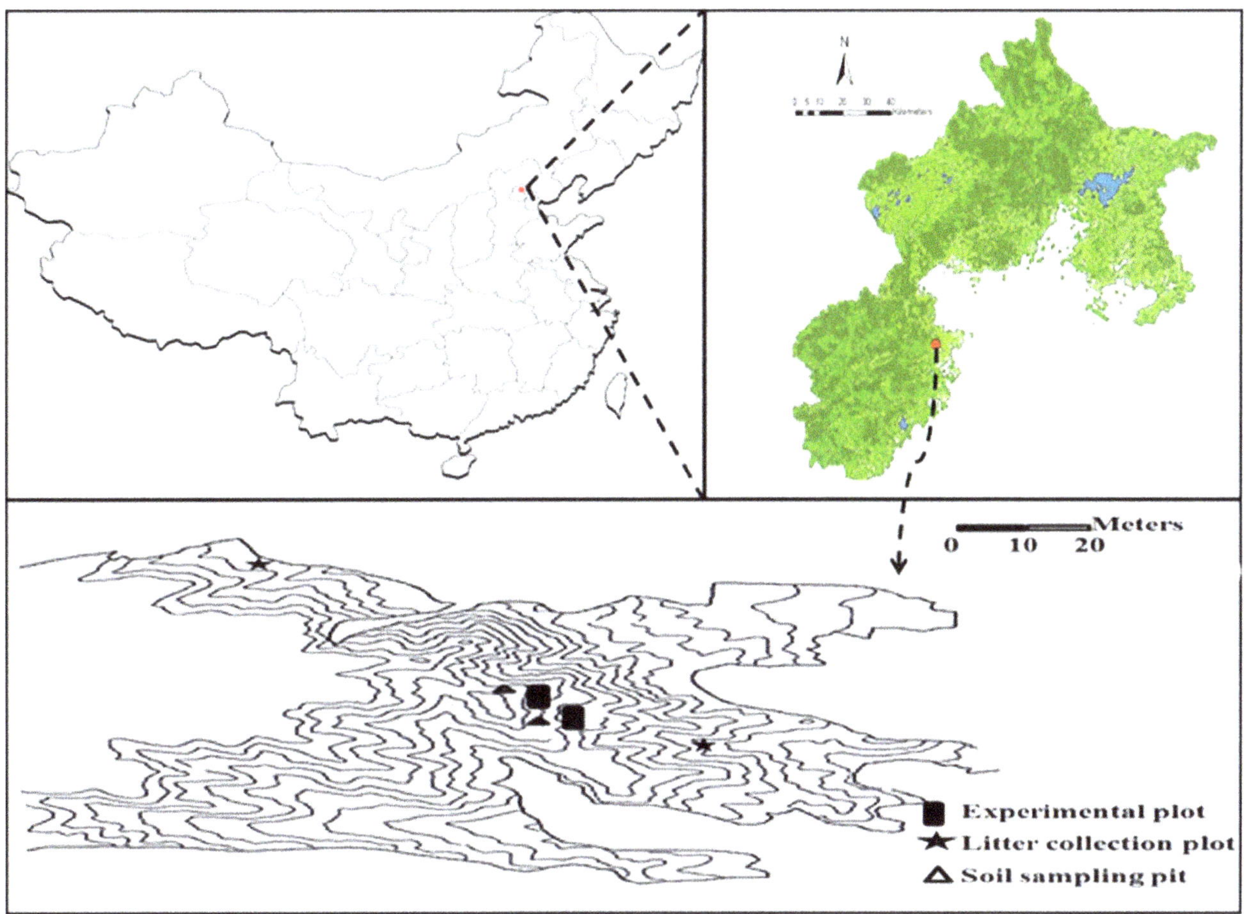

Figure 1. The location of the study area.

In this study, the effects of litter cover (i.e., little type and areal litter mass) and rainfall intensity on surface runoff and soil erosion were quantified through rainfall simulation experiments on sloped field plots with the presence or absence of litter covers. The purpose of the present work was to clarify the role of litter layers on surface runoff and soil erosion, which have caused serious environmental and ecological problems and sometimes even catastrophic events such as debris or mudslides in Northern China [39,40].

Materials and Methods

Ethics Statement

The experimental site, Jiu Feng National Forestry Park is managed by the Forestry Committee of Beijing Forestry University and is available for teaching and research of the university. This field study did not involve any endangered or protected species, and the tree species we selected were common in Northern China.

Study site

This study was performed at Jiu Feng National Forest Park, which is overseen by the Forestry Committee of Beijing Forestry University and available for the university's research and teaching activities. The park is located northwest of Beijing, China (116°28′E, 39°34′N) (Fig. 1). In its warm temperate climate, summers are hot and wet, and winters are cold and dry. Mean daily temperatures are in the range of 23–28°C from May to mid-October, and −5–15°C throughout the winter, resulting in mean annual temperature of 12°C. Mean annual precipitation was approximately 630 mm, most of which occurs as rainfall between June and September, i.e., the period with very intense and erosive rainfall events, usually following the summer drought period. In

Table 1. Forest stands and litter characteristics [38].

Litter species	Plot area (m^2)	Density (trees ha^{-1})	DBH* (cm)	Height (m)	Leaf litter length (cm)	Leaf litter width (cm)
Q. variabilis	25×15	1225	11.2	9.8	11.3–13.7	3.5–4.3
P. tabulaeformis	25×15	1748	7.6	5.2	9.5–12.5	3.3–5.1 (twigs + needles)

*Diameter at breast height.

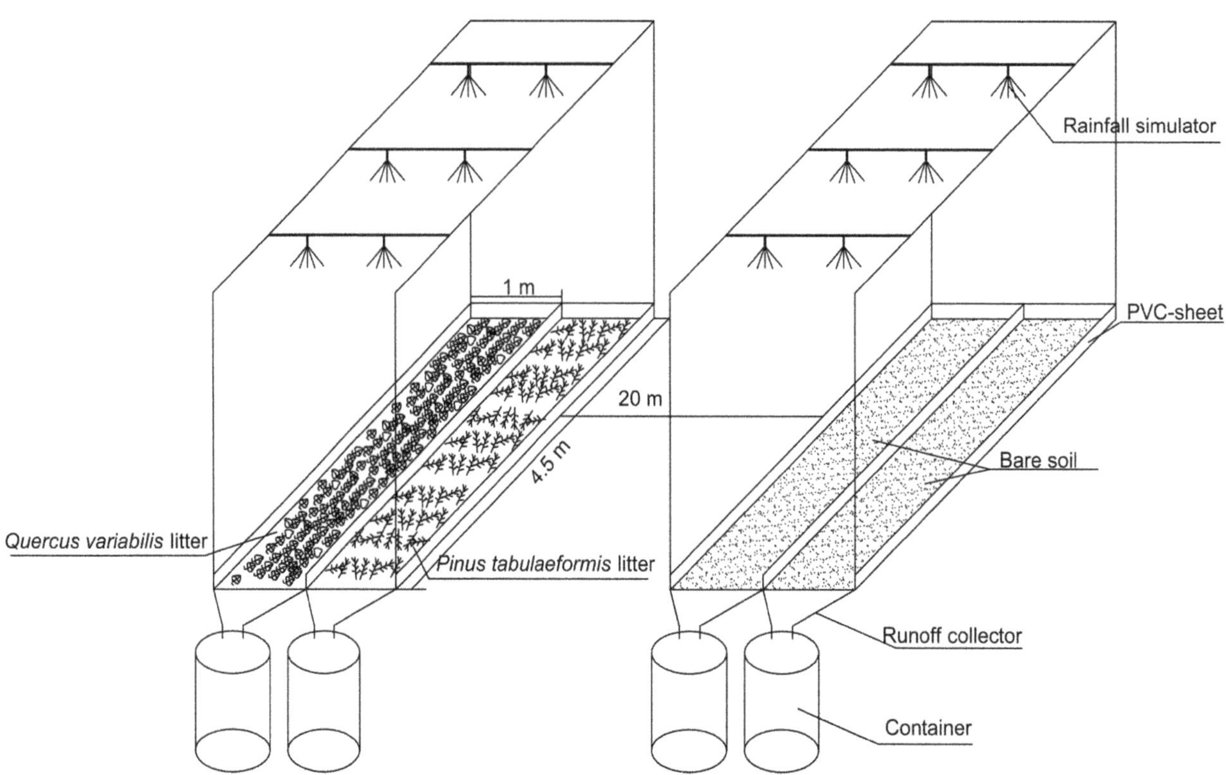

Figure 2. A single set of schematic diagram of experimental runoff plots.

the park Pinaceae *Pinus tabulaeformis*, Fagaceae *Quercus variabilis*, and Cupressaceae *Platycladus orientalis* are dominant tree species, most of which were planted in the 1950s and 1960s. At an elevation of 70–900 m, soils are mainly consisted of cinnamon soils, changing to brown soils at elevations above 900 m. The average slope is approximately 10% with a northeastern aspect.

Leaf litter

Collection and characterization of leaf litter used in this study have been described in detail in Li *et al.* [38]. Briefly, because the tree species of *Pinus tabulaeformis* and *Quercus variabilis* are broadly planted not only in the park but also across the Northern China to prevent wind and water erosion, two dominant and typical types of leaf litter was collected and tested: *Q. variabilis* litter (*QVL*), which here represents broadleaf litter, and *P. tabulaeformis* litter (*PTL*), which here represents needle-leaf litter. Regulators and cleaners remove and burn the litter for fuel almost every year. Prior to this study, the most recent removal and burning occurred in April 2010. In order to investigate the natural litter layer distribution, a 25×15 m^2 experimental stand for each type of litter, respectively. These were further divided into 1×1 m^2 sub-plots, where human activities were not allowed during the collection period from May 2010 to April 2013. Afterwards, in order to consistently investigate the variation of litter mass to ensure that the measured mass values were representative of field conditions, 10 sub-plots were randomly chosen and litter mass was measured and recorded every four months to cover seasonal variation. A 2-cm-high aluminum rectangular frame was placed over the ground to facilitate observation and collection of fallen leaves in each sub-plot. No cover was set in the sub-plots so that the litter was not isolated from the topsoil, also the biological decomposition of litter was not affected. Because the litter was moved around by wind, the mass of collected litter varied from one

Table 2. Soil properties (0–40 cm deep) in experimental and control plots.

	Bare-soil plot	Litter-covered plot
Soil moisture (%)	21.2	17.5
Porosity (%)	57.4	55.4
Bulk density (g cm^{-3})	1.57	1.43
pH	5.6	5.7
Sand (%)	79.5	73.6
Silt (%)	10.9	12.5
Clay (%)	9.6	13.9

Table 3. Runoff yield and runoff coefficient data corresponding to each treatment and simulated rainfall event.

Type of cover	Litter mass (kg m^{-2})	Rainfall intensity (mm h^{-1})				
		5.8	**11.8**	**25.2**	**49.8**	**75.6**
		Runoff yield (mm), runoff coefficient (%)				
Bare soil	0	2.09, 36.7	6.58, 56.2	21.3, 84.5	43.98, 88.3	68.42, 90.5
Q. variabilis	0.3	0.81, 14.2	4.64, 39.7	13.64, 54.1	26.71, 53.6	51.05, 67.5
	0.5	1.28, 22.5	3.66, 31.3	13.24, 52.5	26.53, 53.3	54.50, 72.1
	0.8	0.98, 17.2	3.66, 31.3	13.43, 53.3	33.26, 66.8	52.10, 68.9
	1.0	0.28, 4.9	2.75, 23.5	13.79, 54.7	33.14, 66.6	51.90, 68.7
P. tabulaeformis	0.3	0.71, 12.5	2.98, 25.5	12.31, 48.9	35.74, 71.8	52.06, 68.9
	0.5	0.65, 11.4	3.06, 26.2	11.58, 46.0	29.10, 58.4	43.54, 57.8
	0.8	0.69, 12.1	3.29, 28.1	9.60, 38.1	28.82, 57.9	54.24, 71.8
	1.0	0.35, 6.1	3.16, 27.0	12.76, 50.6	36.35, 73.0	50.30, 66.5

sub-plot to another. Because the interface between the litter and soil was clearly distinguishable, the uncomposed litter was selected as the experimental material instead of the fresh leaves, twigs and decomposed litter, only litter with intact shapes were collected. These samples were manually transferred into plastic bags, and transported to the laboratory, where they were allowed to air-dry.

The lengths and widths of the collected needle-leaf litter and broadleaf litter were measured (Table 1). Because the twigs and needles were difficult to separate, they were weighed together, and decomposed needles and broadleaves were excluded from the measurements. The areal litter mass in kg per m^2 was then calculated by dividing the weighed litter mass over the sub-plot areas. The undecomposed litter was chosen for two main reasons: first, it was the dominant component of the litter (approximately 85%) and was easy to identify; second, the undecomposed litter was located in the upper layer where it would directly intercept the raindrops. For this reason, that the undecomposed litter was assumed to play a more significant role in modulating surface runoff and soil erosion processes than decomposed litter and half-decomposed litter. Accordingly, the mass of this component of litter ranged from 0.33 to 1.24 kg m^{-2} for *QVL* and from 0.18 to 0.77 kg m^{-2} for *PTL* in the sub-plots (1×1 m^2). Based on the measurements mentioned above, to represent the litter variation characteristics and erosion in the slope accurately and to effectively illustrate and compare the hydrological and erosive responses of the two types of litter, four areal litter masses of 0.3, 0.5, 0.8, and 1.0 kg m^{-2} were selected for testing in rainfall simulation studies. This may explain the role of litter in runoff and sediment depletion.

Experimental plot design

Based on initial site survey, two northeastern-facing open runoff plots (4.5×2 m^2 each) with an approximate slope of 10% were selected for rainfall simulation experiments. These served as the bare plot and the litter-covered plot, respectively. The plots were approximately 20 m apart, the soil moisture and soil texture composition were almost the same, so they were not decisive factors in the infiltration and runoff generation in further analysis of runoff yields. All large stones were removed from both plots, and the remaining decomposed litter was removed from the litter-covered plot. No trees or other shrubs were present in either plot. Both plots were divided in half with each part measuring 4.5×1 m^2. The two halves of the divided plots were then covered with *Q. variabilis* litter (*QVL*) and *P. tabulaeformis* litter (*PTL*). These served as litter-covered plots in comparisons of the effects of the two different types of litter on runoff and erosion control (Fig. 2). Before each test run, leaf litter was brought out from plastic bags, then carefully distributed by hand in the litter-covered plot at litter masses of 0.3, 0.5, 0.8, and 1.0 kg m^{-2}. Litter thickness was measured in the upper, middle, and lower part of the plot to test whether the litter covered soil surface uniformly. Several adjustments were made if the litter thickness varied by more than 0.5 cm. The remaining two plots (4.5×1 m^2 each) were regarded as the control (or bare-soil) plot with no litter cover. Within almost 30 min of the cessation of rainfall, leaf litter was re-collected in plastic bags and then oven-dried at 80°C, and evenly returned to the plots before the next rainfall simulation.

Soil properties

The soil properties of each plot are presented in Table 2 (A common Chinese Soil Taxonomy was reported by CRGCST [41] and Shi *et al.* [42]). To comprehensively represent the soil characteristics of the experimental plot, bulk soil samples were collected (at a depth of 0–40 cm) from three different points outside each plot (2–5 m away from the plot) to determine soil properties from May to June, 2013. At each point, soil samples were separated into four sub-samples based on soil depth of 10 cm (0–10 cm, 10–20 cm, 20–30 cm, 30–40 cm) and stored in aluminum specimen boxes and trays. In order to determine soil bulk density (D_b) a weight basis of a 100 cm^3 soil core sample was collected under field-moisture conditions, oven-dried at 105°C for 24 h, and measured for the mass of oven-dried solids, V_t refer to the bulk volume of the soil, which includes the volume of the soil (m_s). Then D_b was calculated as m_s/V_t and the pore space between the soil particles [43], here $V_t = 100$ cm^3. Total porosity S_t can be calculated from the particle density (D_p) and bulk density (D_b) as follows:

$$S_t = 1 - \frac{D_b}{D_P}$$

Here, particle density ($D_p = \frac{m_s}{V_s}$) refers to the mass (m_s) of a unit volume of solid soil particles (V_s) [44]. Gravimetric soil water content (water mass/dry soil mass) was also measured after the

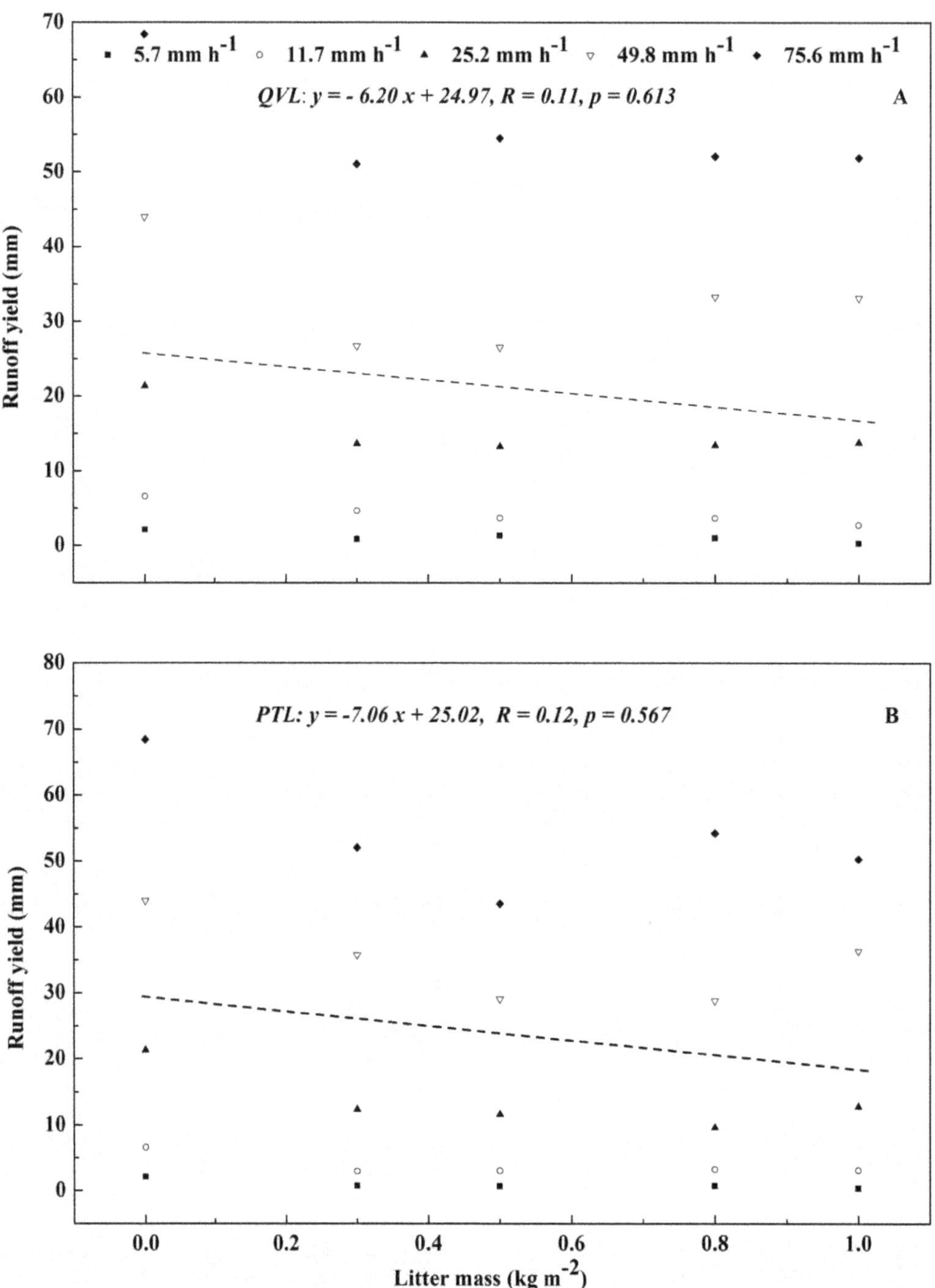

Figure 3. Runoff yield versus litter mass for five rainfall intensities for bare-soil plots (Litter mass equals zero represents 'bare'), ***Quercus variabilis*** **litter (A), and** ***Pinus tabulaeformis*** **litter (B).**

samples were oven dried [45]. Soil particle size distribution was measured after the soil samples were treated with H_2O_2 and dispersed in sodium hexametaphosphate solution using the Bouyoucos densimeter method [46,47]. Soil pH was determined with an electrode pH-meter on the saturated soil paste of 1:2.5 soil to distilled water ratio [48,49].

Rainfall simulation

The relatively high variability of the rainfall intensity in Northern China makes it difficult to examine its role in rainfall-runoff and soil erosion processes that take place under litter cover [50,51]. Here, simulated rainfall produced by an artificial rainfall simulator jointly developed by Beijing Normal University and Beijing Jiaotong University in 2006 was used [52,53]. A detailed description on the rainfall simulator was provided in Li *et al* [38]. The simulator delivered water at a height of 4.5 m and at varying speeds, spraying an area 2.2 m long ×1.5 m wide under a range of rainfall intensities. This rainfall simulator was designed to allow raindrops to reach terminal velocity and allow the flow rates to be easily controlled, resulting in simulated stable rainfall events with

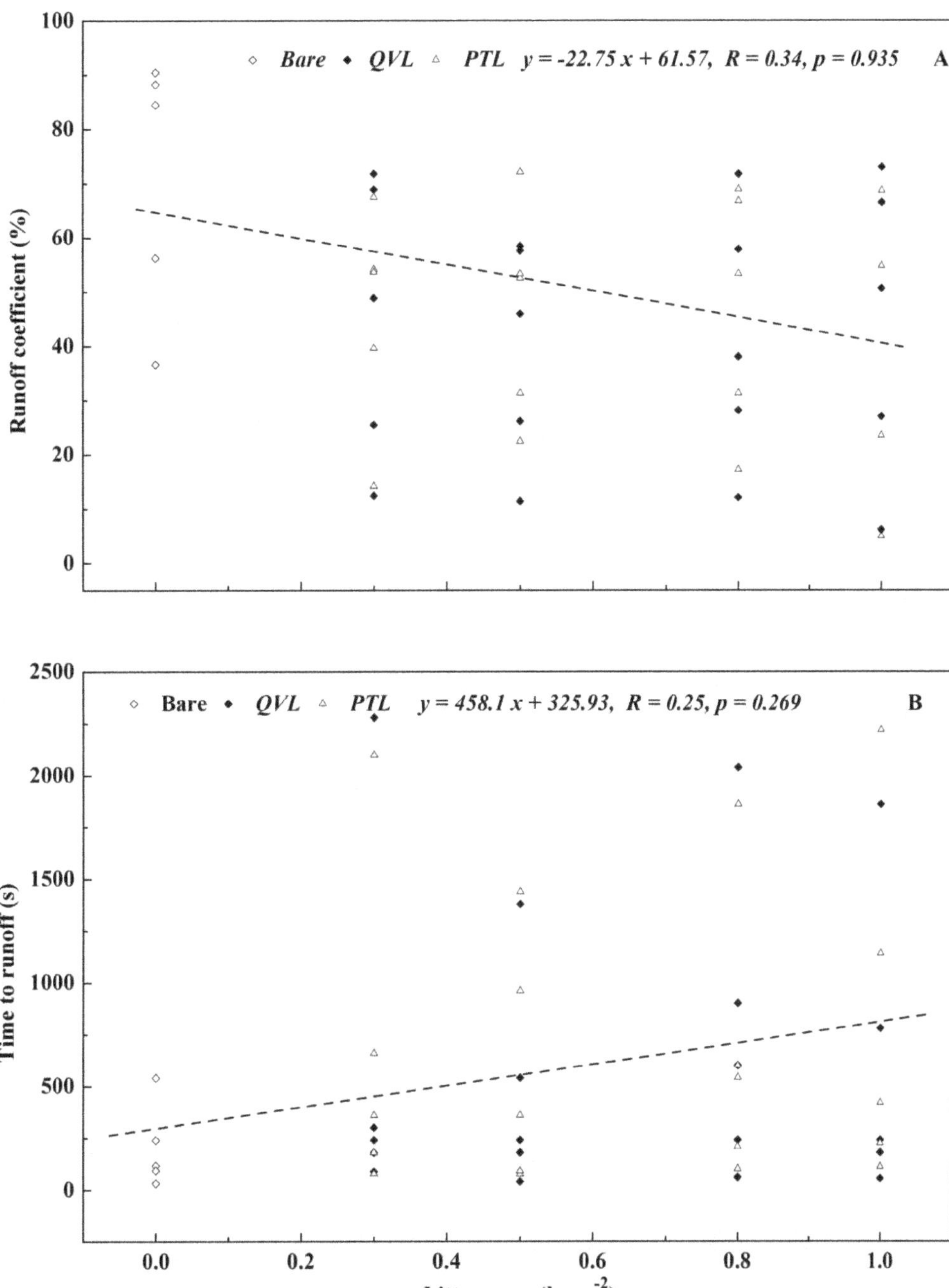

Figure 4. Runoff coefficient and time to runoff versus litter mass for bare-soil plots (open diamonds), *Quercus variabilis* litter (closed diamonds), and *Pinus tabulaeformis* litter (open triangles) for five rainfall intensities.

80% uniformity and raindrop sizes of 2.3±0.3 mm. Simulated rainfall chemistry is also an important factor to be considered in the simulation experiments. Water with high electrical conductivity tends to flocculate soil particles, and water with low electrical conductivity typical of natural rainfall may disperse and readily erode the same soil particles [54,55]. Distilled water was modified using NH_4Cl and NaCl to produce a solution with a pH of 5.5 and electrical conductivity ranged from 12.2 to 20.7 μS cm^{-1} (a detailed description of the electrical conductivity measurement was presented in the National Standard of China [56]), which was similar to the reference value of 14.8 μS cm^{-1} [57].

According to the precipitation data in the study region over 53 years from 1956 to 2008, the maximum rainfall intensity between June and September ranged from 2.6 to 82 mm h^{-1} [58], each plot was subjected to rainfall at intensities of 5.7, 11.7, 25.2, 49.8, and 75.6 mm h^{-1} for 1 h each to represent natural rainfall conditions and to measure the hydrological and erosive response to rainfall accurately. The duration of simulated rainfall was 1 h, which was consistent with the duration of the stable natural rainfall in the study region since 1984 [59,60]. Runoff and sediment samples were collected using a metal runoff collector, which was placed at the bottom of each plot to capture the runoff during the

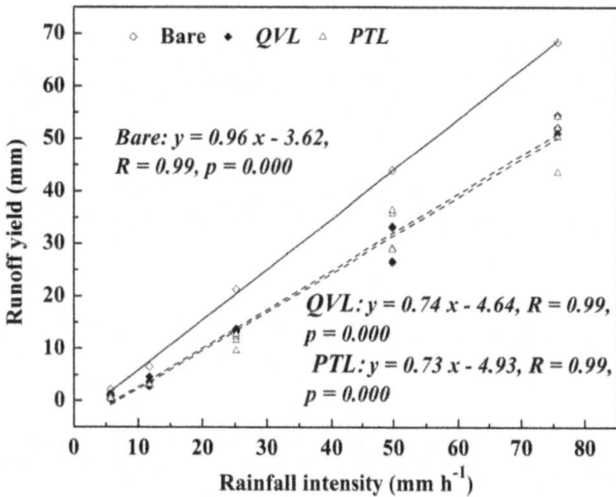

Figure 5. Runoff yield versus rainfall intensity for bare-soil plots (open diamonds), *Quercus variabilis* litter (closed diamonds), and *Pinus tabulaeformis* litter (open triangles) for four litter masses.

test run at 1-minute intervals, and the runoff sample was weighed using an electronic balance. Sediment was settled, separated from the water, and oven-dried at 105°C for 24 h, after which it was weighed to determine the sediment yield. Time to runoff (T_r) was measured as the time when runoff started to develop on the plot surface after the rainfall began [21,61]. Within 30 min after the cessation of rainfall, by which the gravitational water had drained out of the litter, the litter was removed from the soil manually and collected in two plastic bags.

Runoff coefficient (R_c) defined as the proportion of total rainfall that becomes runoff during a storm event, is used to describe the variation of runoff and water resource development [62]. R_c was calculated for each rainfall event. Sediment concentration (SC) was determined by the equation $SC = \frac{Y}{Q}$, where Y (g) represents sediment yield, and Q (L) represents runoff volume. In addition, infiltration and the antecedent soil water content affected runoff process very visibly [63]. Given the wide range of simulated rainfall intensities, the simulated rainfall intensity (RI) order 5.7, 25.2, 49.8, 11.7, and 75.6 mm h^{-1} was set to avoid that two high-intensity of rainfalls were simulated consecutively. To lessen the impact of infiltration on runoff and sediment generation, changes in soil water content in the soil profiles were measured by inserting time-domain reflectometry probes to a depth of 20 cm before and after the experiment at 2-day (48-h) intervals to determine whether the soil water content had reached 20%±2% (Table 2). If the soil water content reached or exceeded 25%, the rainfall simulation would be delayed for another 24 h until the initial water content before every experimental run was 20%±2%. All test runs were carried out from August to September 2013, during which no intensive rain or significant wind occurred. Each test was repeated once.

Statistical analysis

Linear, polynomial and non-linear regressions were used to determine the relationship between the litter mass (or RI) and runoff parameters (or sediment parameters). Correlation analysis among runoff (e.g. runoff yield, R_c, T_r), and sediment (e.g. sediment yield, SC) as dependent variables and litter mass and of RI as an independent variable was performed to evaluate possible relationships among them, and to facilitate understanding of the hydrological and erosion processes. Parameters were considered to be significantly correlated when they were at or above the 95% confidence level ($p \leq 0.05$). A test of normality was conducted by comparing the Sig. value in Kolmogorov-Smirnov and Shapiro-Wilk. Homoscedasticity was carried out to determine the whether the data was homogeneous. Sig>0.05 was defined as homoscedasticity in the Levene Statistic. One-way analysis of variance (ANOVA) was carried out to determine if there were any differences in runoff and sediment yield among three plot treatments (one bare-soil, two litter-covered plots), specifically differences between the broadleaf litter-covered plot and needle-leaf litter covered plot. The Fisher LSD (Least Significant Difference) test at $p \leq 0.05$ was used to test for significant differences. Data were grouped by treatment (bare, *Q. variabilis* litter cover and *P. tabulaeformis* litter cover). All statistical analyses were performed using IBM SPSS Statistics 20.0 software.

Results and Discussion

Runoff

Runoff yield and runoff coefficient (R_c) data are presented in Table 3. Generally, the most runoff took place in the control plot (bare soil). The mean amount of runoff in the QVL and PTL cover plots was lowered by 29.5% and 31.3% compared to the bare plot, respectively. In agreement, similar results were reported by Singer and Blackard [28], Pannkuk and Robichaud [20], Findeling *et al.* [15], and Cerdà and Doerr [21], with a reduction that ranged from 0.12–55.6% in contrast with the bare soil. This was likely due to the protection provided by the litter cover, which absorbed the energy of raindrops and also increased the roughness of soil surface to increase infiltration rate, and delayed and reduced runoff [49]. However, the opposite result was observed for straw mulch cover: runoff yield was slightly higher (by 0.09%) than that in a bare-soil plot [28,33]. Similar results were obtained for rock fragments cover [12,13,64] and plastic mulch cover [34], presumably because the physical shapes or impervious nature of these surface covers concentrated the flow of water to decrease the infiltration rate and increase the flow velocity [25,65]. However, results of LSD test in ANOVA revealed no significant differences ($p \leq 0.05$) in runoff yield between the control and two litter-covered plots ($p = 0.417$ and 0.390 respectively). In contrast, significant differences ($p \leq 0.05$) were found in R_c ($p = 0.000$ and 0.001) and T_r ($p = 0.028$ and 0.012) between bare and litter-covered plots. The results implied that litter cover had an impact on runoff, but other factors, such as precipitation and rainfall intensity should be taken into account as well.

The manner in which litter type (broadleaf *vs* needle-leaf), specially differences in physical leaf shape, might affect runoff was studied. However, no significant differences in runoff yield ($p = 0.696$), R_c ($p = 0.677$) and T_r ($p = 0.681$) were found between QVL and PTL plots, indicating that litter type was a the major influencing factor in controlling runoff in the present study as it was supposed to be. Very few studies concentrated on the role of litter type in runoff reduction, particularly on the physical differences in leaf shape. In a previous work, Neris *et al.* argued that runoff from the pine needle litter cover was twice of the rainforest litter cover, which does not match the current results [47]. The discrepancy suggests that further research is required.

Litter mass was treated as an influencing element in runoff process by covering the topsoil, but its effect on runoff remains unclear. In general, no significant linear correlations were found between litter mass and runoff yield for QVL ($R = 0.11$, $p = 0.923$)

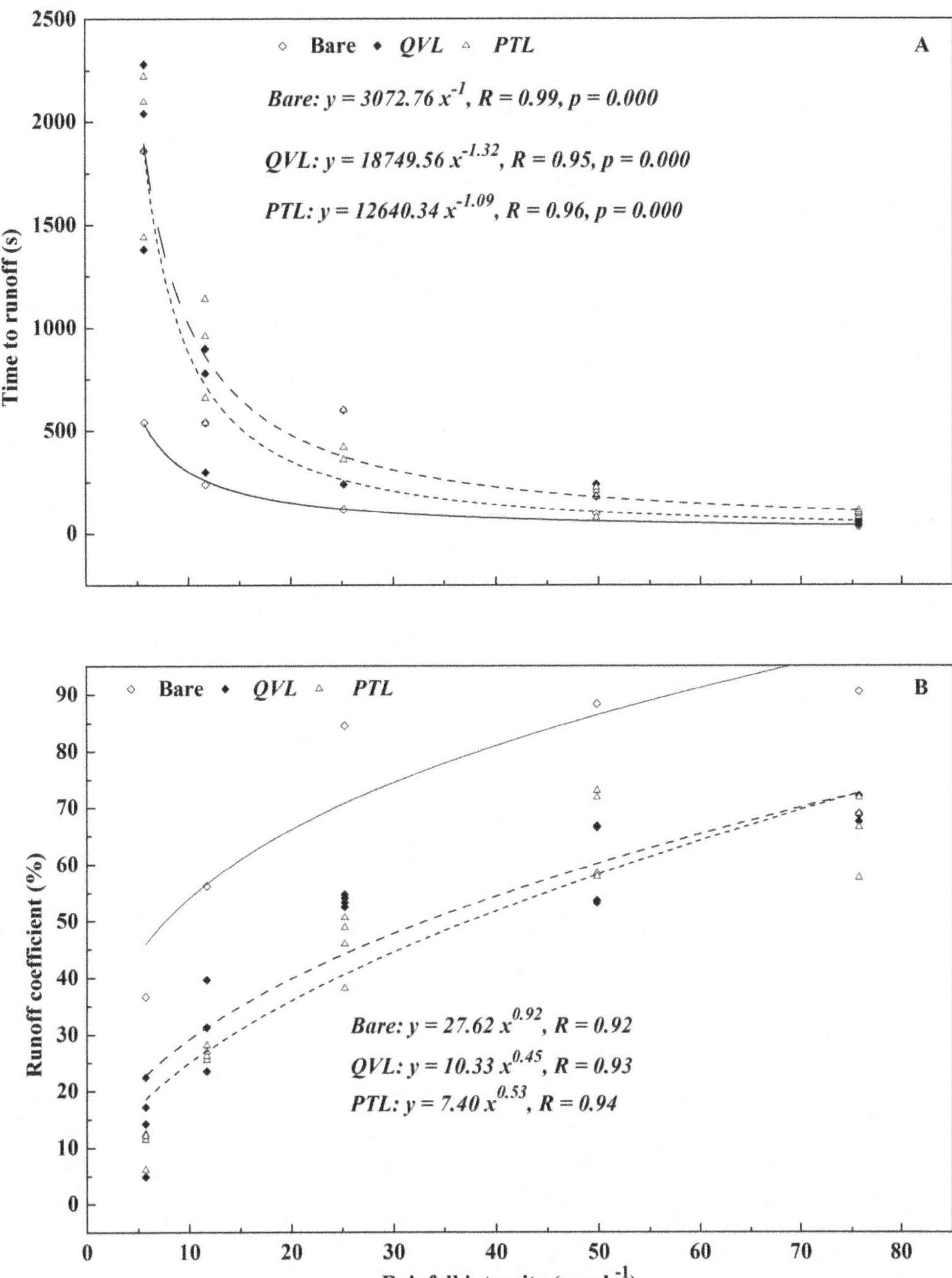

Figure 6. Rainfall intensity versus time to runoff and runoff coefficient for bare-soil plots (open diamonds), *Quercus variabilis* litter (closed diamonds), and *Pinus tabulaeformis* litter (open triangles) for four litter masses.

and *PTL* ($R=0.12$, $p=0.964$) (Fig. 3), and no non-linear correlations were observed. Runoff yield decreased dramatically when litter mass increased from 0 to 0.3 kg m^{-2} but then remained steady regardless of the increasing litter mass afterwards, which proved that litter cover had a strong impact on runoff reduction as stated above. R_c ($R=0.34$, $p=0.114$) and T_r ($R=0.25$, $p=0.078$) showed a similar correlations with litter mass (Fig. 4). One likely explanation was that the increased litter mass might intercept and store more rainwater. However, on the other hand, it may provide a flow channel for runoff by covering more bare topsoil so that runoff flowed on rather than through the leaf litter, particularly under rainfall intensities of 49.8 and 75.6 mm h^{-1}. For this reason, more runoff was observed with higher litter masses of 0.8 and 1.0 kg m^{-2}. While Findeling *et al.* compared the efficiency of different corn residue masses (0, 0.15 and 0.45 kg m^{-2}) in cutting runoff down, and reported that mean values of R_c were 0.44, 0.16, and 0.05, respectively [15]. They confirmed that surface cover contributed notably to the decrease of runoff in contrast with the bare soil, but because they tested only a narrow range of masses,, the slight increase in mass was not sufficient to facilitate extrapolation that runoff would decrease as the ground

Table 4. Sediment yield and sediment concentration data corresponding to each treatment and simulated rainfall event.

Type of cover	Litter mass (kg m^{-2})	Rainfall intensity (mm h^{-1})				
		5.8	11.8	25.2	49.8	75.6
		Sediment yield (kg), sediment concentration (g L^{-1})				
Bare soil	0	0.17, 18.29	0.23, 7.77	1.65, 17.26	4.70, 23.73	7.02, 22.79
Q. variabilis	0.3	0.07, 17.89	0.13, 6.19	0.79, 12.91	0.58, 4.79	2.30, 10.02
	0.5	0.07, 12.00	0.14, 8.42	0.51, 8.53	0.55, 4.64	1.24, 5.06
	0.8	0.05, 10.88	0.08, 4.98	0.18, 2.96	0.31, 2.09	0.42, 1.80
	1.0	0.06, 49.37	0.14, 11.39	0.16, 2.63	0.16, 1.10	0.22, 0.96
P. tabulaeformis	0.3	0.10, 30.14	0.12, 8.61	0.21, 3.77	0.85, 5.29	3.38, 14.44
	0.5	0.06, 19.97	0.15, 10.65	0.32, 6.14	0.68, 5.21	2.94, 15.03
	0.8	0.06, 17.97	0.09, 5.89	0.17, 3.92	0.50, 3.86	0.52, 2.11
	1.0	0.04, 24.44	0.13, 9.17	0.18, 3.08	0.26, 1.58	0.32, 1.40

cover mass increased, which suggested that a broader range of mass should be studied in further works.

Significant linear correlations were found between rainfall intensity (*RI*) and runoff yield for all the plots in Figure 5 ($R = 0.99$, $p<0.05$). In all plots, runoff yield increased with increasing *RI*. Runoff yield at the maximum *RI* of 75.6 mm h^{-1} was 32.7, 62.4, and 83.4 times larger than that at the minimum *RI* of 5.7 mm h^{-1} in bare, *QVL* and *PTL* plots respectively. Based on regression analysis, R_c and time to runoff (T_r) showed a significant power ($R>0.9$, $p<0.05$) and reverse-power ($R>0.95$, $p<0.05$) relationships with *RI* (Figure 6A and 6B). As expected, mean R_c increased from 12.6% ($RI = 5.7$ mm h^{-1}) to 67.8% ($RI = 75.6$ mm h^{-1}) in litter-covered plots (Figure 6B), but T_r decreased apparently from nearly 30 min to 1~1.5 min with increasing *RI*. The findings were in accordance with those reported by Gholami *et al.* [65], who noted that mulch cover was more effective in controlling runoff at a low *RI* of 30 mm h^{-1} ($R_c = 49.9\%$, $T_r = 81.83$ s) than at a high *RI* of 90 mm h^{-1} ($R_c = 66.6\%$, $T_r = 55.22$ s). The phenomena in the present study also proved that *RI* (rather than litter mass) was a dominant factor in determining runoff irrespective of litter cover. This was probably because when *RI* was high, leaf litter was pressed close to the soil surface, blocking the channels between leaves. The accumulative precipitation was assumed to surpass infiltration rate, which resulted in the generation of more runoff.

Sediment

The magnitude of the sediment yield and sediment concentration (*SC*) are shown in Table 4. Generally, bare-soil plot generated 2.75 kg sediment, which was almost 6.7 and 5 times larger than that in *QVL* (0.41 kg) and *PTL* (0.55 kg) plots, respectively. Further, ANOVA analysis indicated there were significant differences ($p = 0.000$ and $p = 0.001$) between bare plot and two litter-covered plots. Similar results were obtained in terms of *SC* in bare and litter-covered plots ($p = 0.001$ and 0.003). In this way, litter cover was able to retard erosion. The observed outcomes were similar to those presented in previous studies. For example, Pannkuk and Robichaud reported that rill and inter-rill erosion were reduced by 70% when plots were covered with Douglas fir needles and 30% when they were covered with Ponderosa pine needles, as compared with the erosion in a bare-soil plot [20]. Cerdà and Doerr reached similar conclusions in their study on soil erosion from the plots covered with ash and pine needles [21]. Benkobi *et al.* suggested that soil loss was reduced by about 87% in 100% litter cover relative to bare soil [23]. Miyata *et al.* reported that annual soil erosion in uncovered plots was 3.7 times greater than in covered plots [30]. However, Singer and Blackard found that more soil loss was observed when mulch covers <40% in contrast with bare soil [28], this was mainly because they excluded the splash erosion off the inter-rill plot, and overall sediment was reduced under mulch cover.

In the present study, because the litter cover prevented raindrops from hitting the soil surface directly by intercepting and storing the rainfall, the detachment of soil aggregates and splash erosion was lessened [66,67]. Less soil was lost in the litter-covered plots than in the bare plot.

Additionally, mean sediment yield and *SC* in the *QVL* plot were slightly lower than in the *PTL* plot, but no statistical differences were found in sediment yield ($p = 0.544$) and *SC* ($p = 0.810$) between the two plots. The findings indicated that litter type (broadleaf or needle-leaf) was may not be a defining predictor in reducing soil loss. This was probably because even though more topsoil was exposed and soil aggregations were detached under needle-leaf litter cover compared with broadleaf litter, needle-leaf litter tended to form mini-debris dams to trap down the soil particles in the runoff and return them to the soil surface [20]. The results were consistent with Neris *et al.*, who also reported that no significant differences were observed between pine forest floor (including needle leaf litter) and a rainforest floor (including broadleaf litter) [47].

Linear relationships between litter mass and sediment yield in *QVL* ($R = 0.50$, $p = 0.05$) and *PTL* plot ($R = 0.52$, $p = 0.112$) are shown in Figure 7. In general, total sediment yield decreased from 13.77 to 0.75 kg (*QVL*) and 0.92 kg (*PTL*) when the litter mass increased from 0 (bare) to 1 kg m^{-2}, but the largest decreases of 71.9% and 66.2% were revealed when litter mass increased from 0 to 0.3 kg m^{-2}. Similar relationships are presented in Figure 8 in terms of litter mass and *SC*. *SC* decreased as litter mass increased. Although no significant correlations were observed, the effect of litter mass on reducing sediment was found to be relevant, as the increasing litter mass covered the remaining bare soil, which indirectly protected soil from raindrop impact, and reduced the velocity of runoff [15,67]. Similarly, Lal reported that soil loss decreased from 52.4 to 0.04 kg m^{-2} when the maize grain yields increased from 0.06 to 0.68 kg m^{-2} [68].

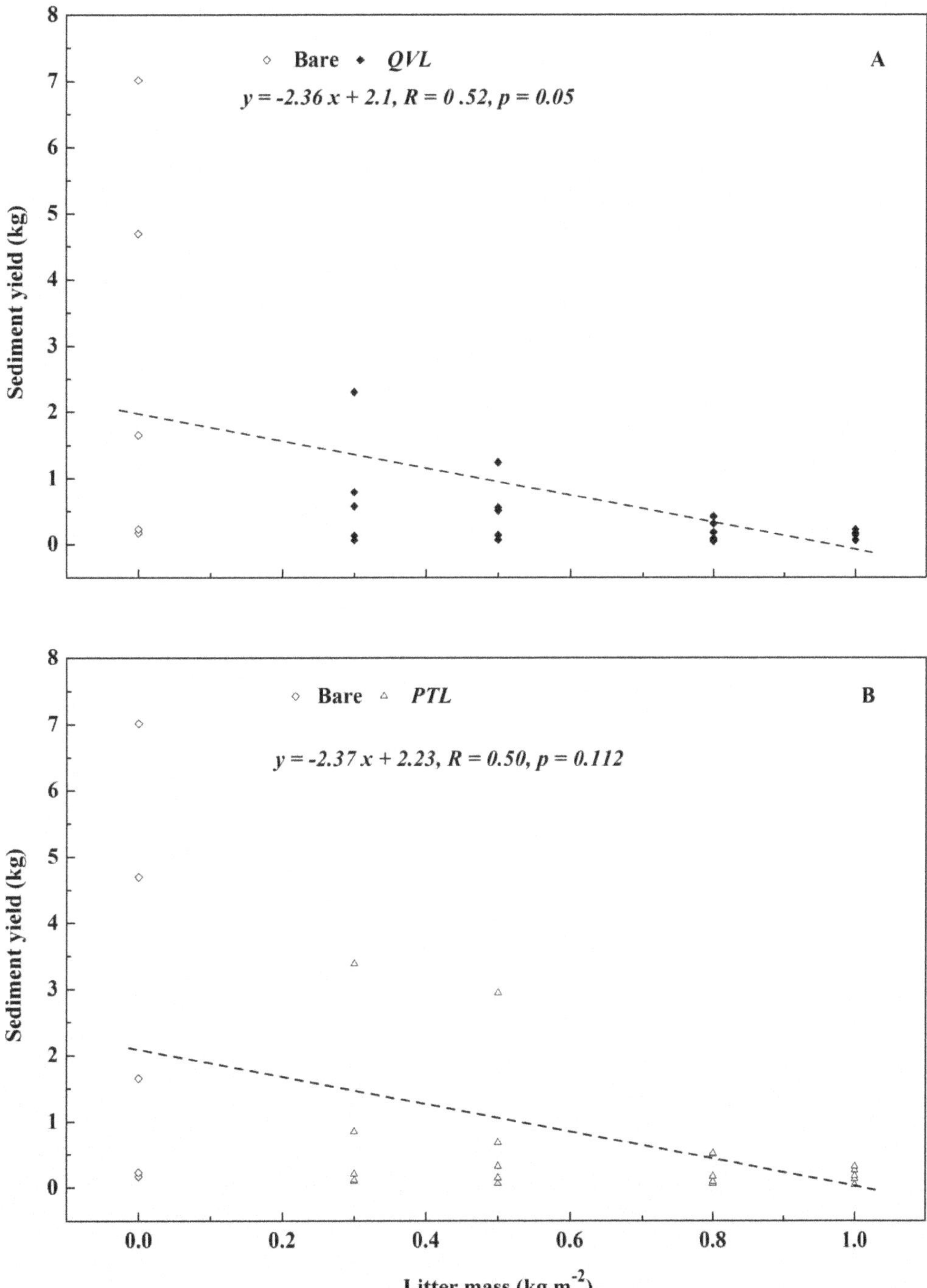

Figure 7. Relationship between litter mass and sediment yield for bare-soil plots (open diamonds), *Quercus variabilis* litter (closed diamonds), and *Pinus tabulaeformis* litter (open triangles) for five rainfall intensities.

RI also had an impact on sediment generation. Both linear (bare soil: $R = 0.99$) and polynomial relationships (*QVL*: $R = 0.63$, $p = 0.055$; *PTL*: $R = 0.71$, $p = 0.022$) between *RI* and sediment yield are presented in Figure 9A. When *RI* increased from 5.7 to 75.6 mm h^{-1}, sediment yield increased 41.3, 17.5, and 29.8 times in bare, *QVL* and *PTL* plots, respectively. *SC* showed a reverse-power decrease trend with *RI* in *QVL* and *PTL* but increased in bare plots. These results verified the protective role of the leaf litter layer, which was in line with the previous studies [24,25,46]. This was mainly because on one hand, leaf litter reduced the detachment of soil aggregation by covering on the topsoil, particularly, high-intensity rainfall (49.8 and 75.6 mm h^{-1}) with intense striking force pressed leaf litter closer to the soil surface [17]; on the other hand, the coverage by leaf litter increased the surface roughness, which facilitated sediment deposition in the runoff [30,67].

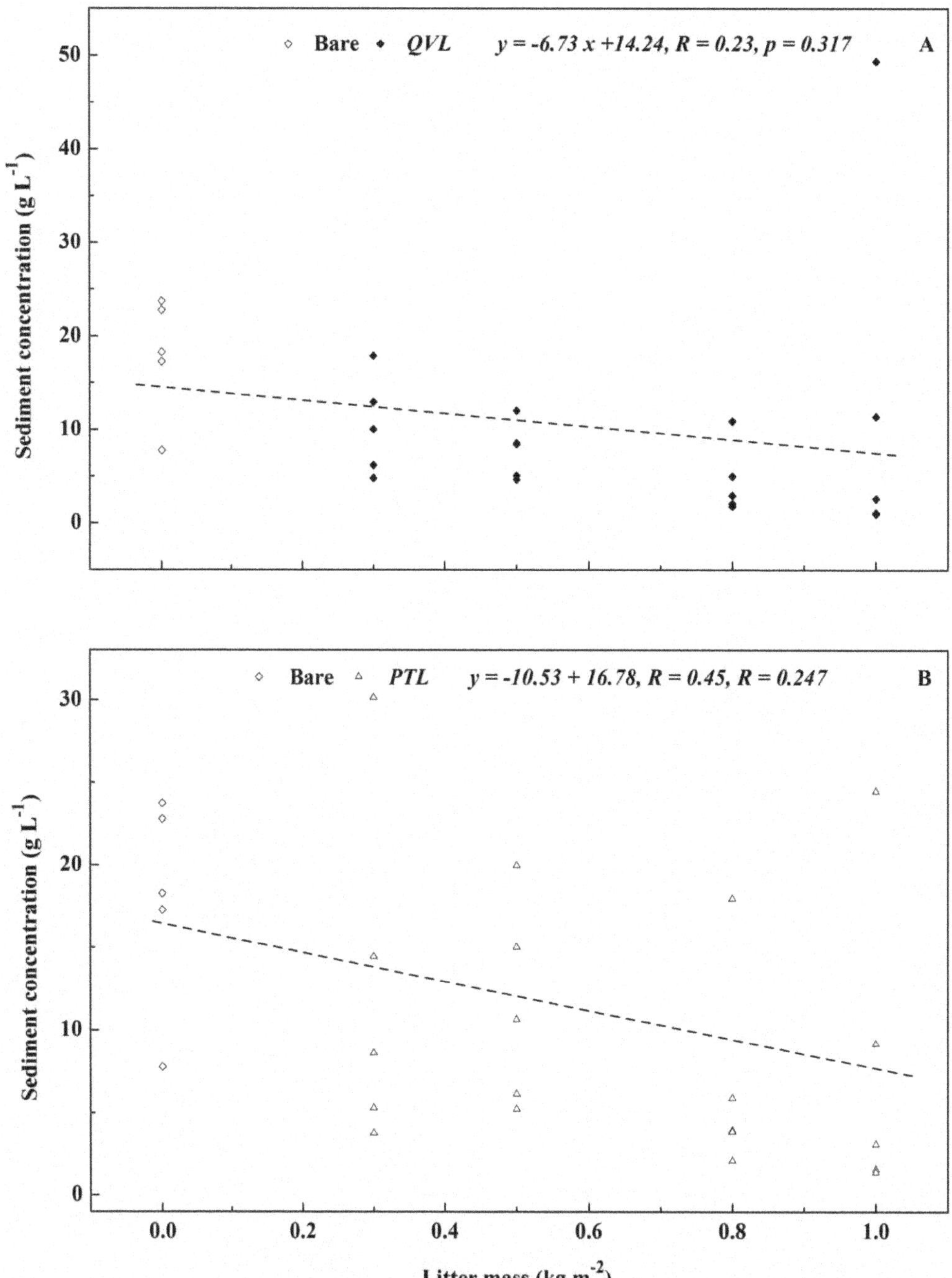

Figure 8. Relationship between litter mass and sediment concentration for bare-soil plots (open diamonds), *Quercus variabilis* litter (closed diamonds), and *Pinus tabulaeformis* litter (open triangles) for four litter masses.

Relationship between runoff and sediment

As analyzed above, runoff yield in the litter-covered plots was not significantly different ($p = 0.226$ and 0.220) from that in the bare plot, but significant differences ($p = 0.000$ and 0.000) were shown in sediment between the bare plot and litter-covered plots. This can be explained briefly by the different forms of interaction between litter type and runoff-erosion process. For broadleaf litter, runoff generated and flowed on the leaf surface, while the presence of needle-leaf litter increased the soil surface roughness and trapped down soil particles in the flow as a barrier [15,21,67,69]. Significant correlations ($p \leq 0.05$) between runoff and sediment were observed for plots in Figure 10A and 10B. Apparently, sediment yield increased with the increasing runoff. This is similar with the previous studies (e.g. [30,65,70]). In addition, an excellent linear regression was observed in bare plot ($R = 0.99$). The greatest runoff, which showed large amounts of energy, was capable of detaching soil aggregation and moving the soil particles. However, leaf litter reduced runoff and sediment by covering soil to prevent splash erosion and largely increasing surface roughness which in turn increased infiltration. The phenomenon confirmed again that litter cover played an irreplaceable role in runoff and soil erosion control.

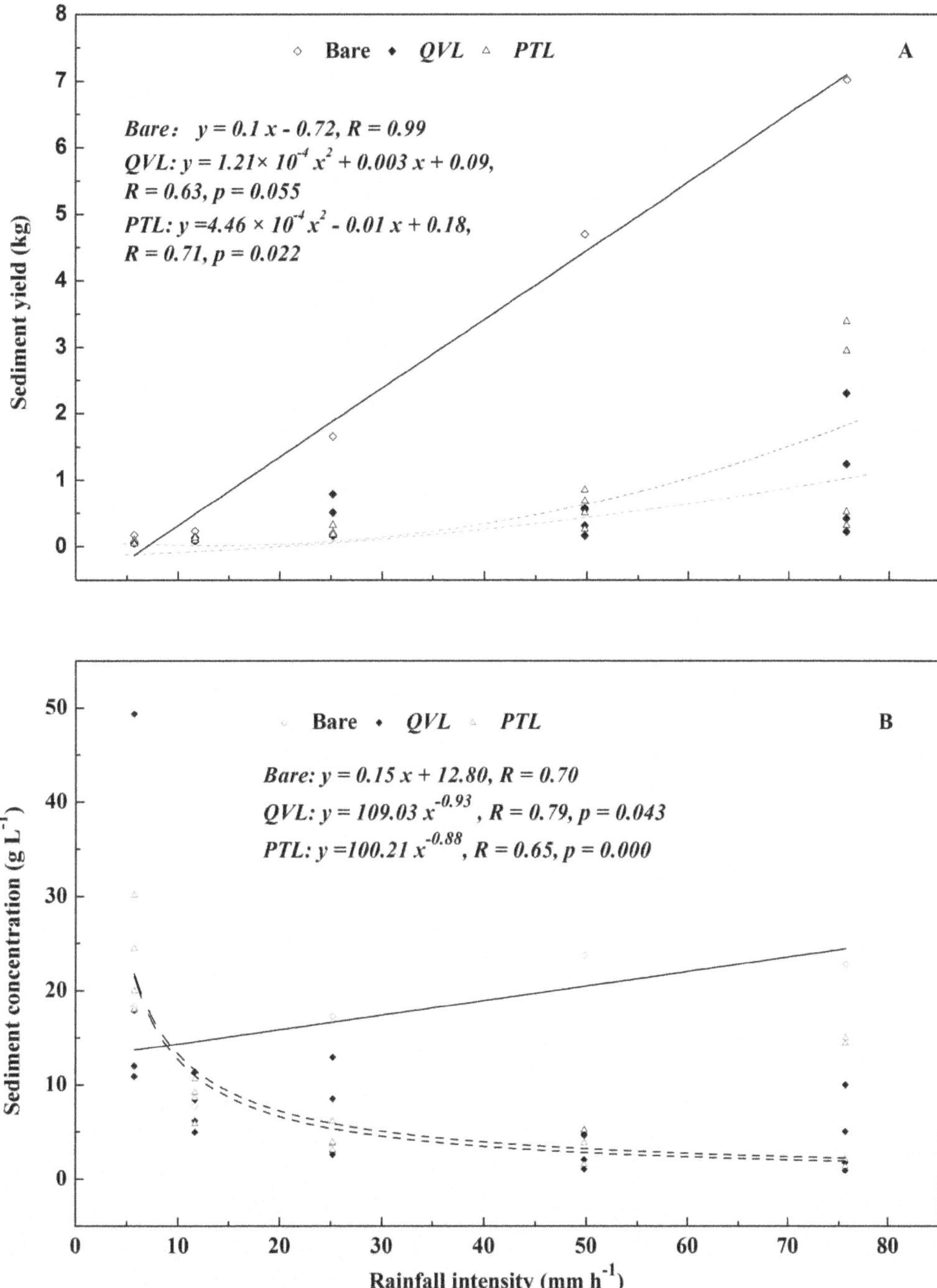

Figure 9. Relationships between sediment yield (A) and sediment concentration (B) versus rainfall intensity for bare-soil plots (open diamonds), *Quercus variabilis* litter (closed diamonds), and *Pinus tabulaeformis* litter (open triangles) for four litter masses.

Conclusion

The research effort was conducted to examine the role of leaf litter in runoff and soil erosion. Results showed that runoff and sediment yield were lower in litter-covered plots than in bare plot, litter cover had a relevant impact on runoff and sediment control. A detailed analysis of how runoff and sediment interacted with litter type, litter mass and rainfall intensity showed that rainfall intensity rather than litter type (broadleaf or needle-leaf) or litter mass was the decisive factor in generating runoff. The correlation between runoff yield and sediment yield was significantly positive ($p<0.05$) in all the plots, which confirmed that increasing runoff with larger energy was able to detach soil aggregation and cause severe erosion afterwards.

These results improved our understanding of litter as a protective layer and rainfall re-distributor. The data may be used to establish hydrological models to predict changes in runoff and soil erosion, and to provide scientific support for managing water resources in Northern China. Future studies should focus on the effects of slope, plot length, other litter types, and hydraulic characteristics (e.g., flow speed, erosive power, and shearing force)

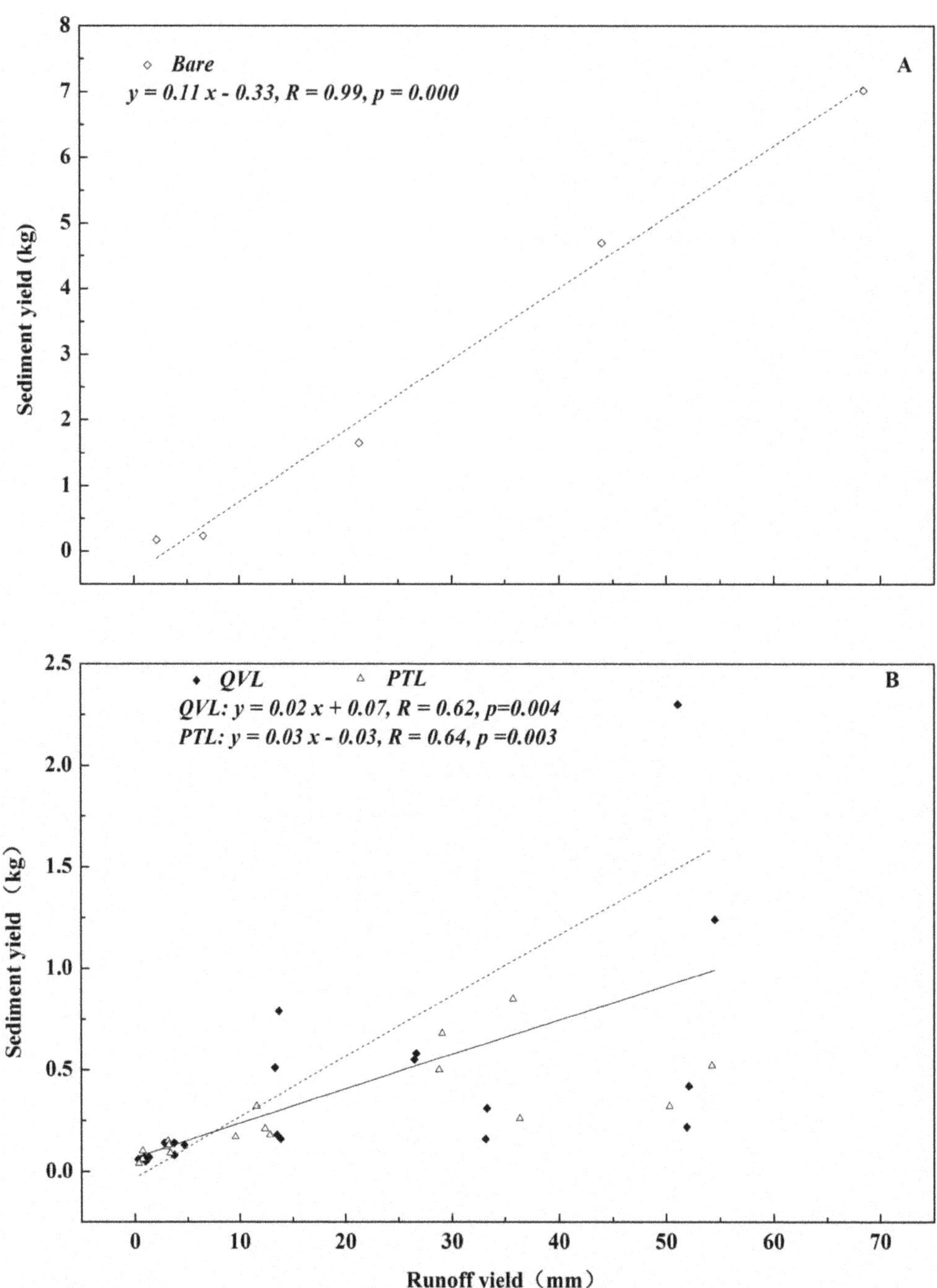

Figure 10. Relationships between runoff yield and sediment yield for bare-soil plots (open diamonds), ***Quercus variabilis*** **litter (closed diamonds), and** ***Pinus tabulaeformis*** **litter (open triangles).**

on runoff and soil erosion under litter cover to draw comprehensive conclusions.

Acknowledgments

We would like to thank the Jiu Feng National Forest Park Administration for their help and support. Further, we deeply appreciate the efficient and effective assistance provided by graduate students Shangjie Chen, Xiaoqing Du, Mingxing Wang, Bingpeng Qu, and Zhentai Pan during the field experiment.

Author Contributions

Conceived and designed the experiments: XL JN BX. Performed the experiments: XL. Analyzed the data: XL JN BX. Contributed reagents/materials/analysis tools: JN. Wrote the paper: XL JN.

References

1. Morgan RPC (2005) Soil and Conservation, 3rd edition. London: Blackwell Publishing Ltd. 304 p.
2. Hugo V, Pleguezuelo C (2008) Soil-erosion and runoff prevention by plant covers. A review. Agron J 28: 65–86. doi: 10.1007/978-90-481-2666-8_48.
3. Li ZG, Cao Y, Liu BZ, Luo ZD (2008) Current status and developing trend of soil erosion in China. Science of Soil and Water Conservation 6(1): 57–62. (In Chinese with English abstract)
4. Zhang HJ (2008) Soil erosion theory, 2nd edition. Beijing: China Forestry Publishing House. 355 p.
5. Kimoto A, Uchida T, Mizuyama T, Li CH (2002) Influences of human activities on sediment discharge from devastated weathered granite hills of southern China: effects of 4-year elimination of human activities. Catena 48: 217–233. doi: 10.1016/S0341-8162(02)00029-2.
6. Bochet E, Poesen J, Rubio JL (2006) Runoff and soil loss under individual plants of a semi-arid Mediterranean shrubland: influence of plant morphology and rainfall intensity. Earth Surf Proc Land 31: 536–549. doi: 10.1002/esp.1351.
7. Walsh RPD, Voight PJ (1977) Vegetation litter: an underestimated variable in hydrology and geomorphology. J Bio 4: 253–277.
8. Geddes N, Dunkerley D (1999) The influence of organic litter on the erosive effects of raindrops and of gravity drops released from desert shrubs. Catena 36: 303–313. Doi: 10.1016/S0341-8162(99)00050-8.
9. Sayer EJ (2006) Using experimental manipulation to assess the roles of leaf litter in the functioning of forest ecosystems. Biol Rev 81: 1–31. Doi: 10.1017/S1464793105006846.
10. Abraham AD, Parsons AJ (1991) Relation between infiltration and stone cover on a semiarid hillslope Arizona. J Hydrol 122: 45–59. doi: 10.1016/0022-1694(91)90171-D.
11. Valentin C, Casenave A (1992) Infiltration into sealed soils as influenced by gravel cover. Soil Sci Soc Am J 56: 1667–1673. doi:10.2136/sssaj1992.03615995005600060002x.
12. Peosen JW, Torri D, Bunte K (1994) Effects of rock fragments on soil erosion by water at different spatial scales: a review. Catena 23: 141–166. Doi: 10.1016/0341-8162(94)90058-2.
13. Jean JS, Ai KF, Shih K, Hung CC (2000) Stone cover and slope factors influencing hillside surface runoff and infiltration: laboratory in investigation. Hydrol Process 14: 1829–1849. doi: 10.1002/1099-1085(200007)14:10<1829::AID-HYP66>3.0.CO;2-#.
14. Dickey EC, Shelton DP, Jasa PJ, Perterson TR (1985) Soil erosion from tillage systems used in soybean and corn residues. T ASAE 28: 1124–1129.
15. Findering A, Ruy S, Scopel E (2003) Modeling the effects of a partial residue mulch on runoff using a physically based approach. J Hydrol 275: 49–66. doi: 10.1016/S0022-1694(03)00021-0.
16. Pan CZ, Shangguan ZP (2006) Runoff hydraulic characteristics and sediment generation in sloped grassplots under simulated rainfall conditions. J Hydrol 331: 178–185. doi: 10.1016/j.jhydrol.2006.05.011.
17. Pan CZ, Ma L, Shangguan ZP (2010) Effectiveness of grass strips in trapping suspended sediments from runoff. Earth Surf Proc Land 35: 1006–1013. doi: 10.1002/esp.1997.
18. Adekalu KO, Olorunfemi IA, Osunbitan JA (2007) Grass mulching effect on infiltration, surface runoff and soil loss of three agricultural soils in Nigeria. Bioresource Technol 98(4): 912–917. doi: 10.1016/j.biortech.2006.02.044.
19. Bhattacharyya R, Smets T, Fullen MA, Poesen J, Booth CA (2010) Effectiveness of geotextiles in reducing runoff and soil loss. A ynthesis. Catena 81: 184–195. doi: 10.1016/j.catena.2010.03.003.
20. Pannkuk CD, Robichaud PR (2003) Effectiveness of needle cast at reducing erosion after forest fires. Water Resour Res 39 (12): 1–9. doi: 10.1029/2003WR002318.
21. Cerdà A, Doerr SH (2008) The effect of ash and needle cover on surface runoff and erosion in the immediate post-fire period. Catena 74: 256–263. doi: 10.1016/j.catena.2008.03.010.
22. Robichaud PR, Lewis SA, Wagenbrenner JW, Ashmun EL, Brown RE (2013) Post-fire mulching for runoff and erosion mitigation Part 1: Effectiveness at reducing hillslope erosion rates. Catena 105: 75–92. doi: 10.1016/j.catena.2012.11.015.
23. Benkobi L, Trlica MJ, Smith JL (1993) Soil loss as affected by different combinations of surface litter and rock. J Environ Qual 4: 657–661. doi:10.2134/jeq1993.00472425002200040003x.
24. Smets T, Poesen J, Knapen (2008) Spatial scale effects on the effectiveness of organic mulches in reducing soil erosion by water. Earth-Sci Rev 89: 1–12. doi: 10.1016/j.earscirev.2008.04.001.
25. Smets T, Poesen J, Bochet E (2008) Impact of plot length on the effectiveness of different soil-surface covers in reducing runoff and soil loss by water. Prog Phys Geog 32 (6): 654–677.
26. Lowdermilk WC (1930) Influence of forest litter on runoff, percolation, and erosion. J Forest 28: 474–491.
27. Adams JE (1966) Influence of mulches on runoff, erosion, and soil moisture depletion. Soil Sci Soc Am J 30 (1): 110–114. doi:10.2136/sssaj1966.03615995003000010036x.
28. Singer MJ, Blackard J (1978) Effect of mulching on sediment in runoff from simulated rainfall. Soil Sci Soc Am J 42 (3): 481–486. doi: 10.2136/sssaj1978.03615995004200030022x.
29. Netto ALC (1987) Overlandflow production in a tropical rainforest catchment: The role of litter cover. Catena 14: 213–231. doi: 10.1016/S0341-8162(87)80019-x.
30. Miyata S, Kosugi K, Gomi T, Mizuyama T (2009) Effect of forest floor coverage on overland flow and soil erosion on hillslopes in Japanese cypress plantation forests. Water Resour Res 45: 1–17. doi: 10.1029/2008WR007270.
31. Pierce SP (1967) Evidence of overland flow on forest watersheds. Inter Symp Forest Hydrol. 25: 247–253.
32. Ruxton BP (1967) Slopewash under primary rainforest in Northern Papua. In: Jennings JN, Mabbutt JA, editors. Landform Studies from Australia and New Guinea. Cambridge University Press. pp. 85–94.
33. McGregor KC, Bengtson RL, Muchler CK (1988) Effects of surface straw on interrill runoff and erosion of Grenada Silt Loam soil. T ASAE 31: 111–116.
34. Y . Wan, SA . El-Swaify (1999) Runoff and soil erosion as affected by plastic mulch in a Hawaiian pineapple field. Soil Till Res 52: 29–35. doi: 10.1016/S0167-1987(99)00055-0.
35. Putuhena W, Cordery I (1996) Estimation of interception capacity of the forest floor. J Hydrol 180: 283–299. doi: 10.1016/0022-1694(95)02883-8.
36. Sato Y, Kumagai T, Kume A, Otsuki K, Ogawa S (2004) Experimental analysis of moisture dynamics of litter layers – the effect of rainfall conditions and leaf shapes. Hydrol Process18: 3007–3018. doi: 10.1002/hyp.5746.
37. Guevara-Escobar, Gonzalez-Sosa, Ramos-Salinas, Hernandez-Delgado (2007) Experimental analysis of drainage and water storage of litter layers. Hydro Earth Syst Sci 11 (5): 1703–1716. doi:10.5194/hess-11-1703-2007.
38. Li X, Niu JZ, Xie BY (2013) Study on Hydrological Functions of Litter Layers in North China. Plos One 8 (7): 1–11. doi: 10.1371/journal.pone.0070328.
39. Wu ZH (2001) The mud-rock flow disaster and their touch off condition by rainfall in Beijing Area. Research Soil Water Conserv 8 (1): 67–72. (In Chinese with English abstract).
40. Wang HZ (2008) A study of the critical rainfall volume for mudflows based on historical data in the mountainous area of Beijing. Analysis Research 3 (1): 18–21. (In Chinese with English abstract).
41. Cooperative Research Group on Chinese Soil Taxonomy (CRGCST) (2001) Chinese Soil Taxonomy, Beijing: Beijing Science Press.
42. Shi XZ, Yu DS, Yang GX, Yang GX, Wang HJ, et al. (2006) Cross-Reference Benchmarks for Translating the Genetic Soil Classification of China into the Chinese Soil Taxonomy. Pedosphere 16(2): 147–153.
43. Blake GR, Hartge KH (1986) Bulk density. In: Klute AK, editor. Methods of soil analysis. Part I: Physical and mineralogical method. American Society of Agronomy-Soil Science Society of America, Madison, pp. 363–375.
44. Carter MC, Gregorich EG (2008) Soil Samping and Methods of Analysis. Canadian Society of Soil Science, pp. 743–760.
45. Carter MC, Gregorich EG (2008) Soil Samping and Methods of Analysis. Canadian Society of Soil Science, pp. 911–938.
46. Gee GW, Bauder JW (1986) In: Klute AK, editor. Methods of soil analysis. Part I: Physical and mineralogical method. American Society of Agronomy-Soil Science Society of America, Madison, pp. 383–411.
47. Neris J, Tejedor M, Rodriguez M, Fuentes J, Jimenez C (2013) Effect of forest floor characteristics on water repellency, infiltration, runoff and soil loss in Andisols of Tenerife (Canary Islands, Spain). Catena 108: 50–57. doi: 10.1016/j.catena.2012.04.011.
48. Carter MC, Gregorich EG (2008) Soil Sampling and Methods of Analysis. Canadian Society of Soil Science, pp. 173–175.
49. Ayed GM, Mohammad AA (2010) The impact of vegetative cover type on runoff and soil erosion under different land uses. Catena 81: 97–103. doi: 10.1016/j.catena.2010.01.008.
50. Jiali W, Renhe Z, Yingchun W (2012) Characteristics of precipitation in Beijing and the precipitation representativeness of Beijing Weather Observatory. Journal of Applied Meteo Sci 23: 265–273. (In Chinese with English abstract).
51. Ma JJ, Li SY, Wang J (2012) The study of Beijing's rainstorm division and its returning periods. Meteo Monthly 38: 569–576. (In Chinese with English abstract).
52. Zhang GH, Liu BY, Li PK (2007) Principles and properties of artificial trough rainfall simulator. Bulletin of Soil and Water Conservation 27 (6): 56–60. (In Chinese with English abstract).
53. Xie Y, Lin XJ, Liu YN, Zhen YZ, Liu BY, et al. (2008) Calibration of simulated rainfall intensity and its special distribution for trough rainfall simulator. Bulletin of Soil and Water Conservation 28 (4): 1–6. (In Chinese with English abstract).
54. Battany MC, Grismer ME (2000) Rainfall runoff and erosion in Napa Vally vineyards: effects of slope, cover and surface roughness. Hydrol Process 14: 1289–1304. doi: 10.1002/(SICI)1099-1085(200005)14:7<1289::AID-HYP43>3.0.CO;2-R.
55. Kim KH, Miller WP (1996) Effect of rainfall electrolyte concentration and slope on infiltration and erosion. J Soil Tech 9: 173–185. doi: 10.1016/S0933-3630(96)00011-6.
56. Ministry of Environmental Protection of the P.R. China (1993) Determination of specific conductance in the wet precipitation. http://www.mep.gov.cn/image20010518/1881.pdf. Accessed 2014 Aug 29.
57. Tang J, Xue HS, Yu XL Cheng HB, Xu XB, et al. (2000) The preliminary study on chemical characteristics of precipitation at Mt. Waliguan. Acta Sci Circums 20(4): 420–425. (In Chinese with English abstract).

58. Zhong YD, Jia YW, Li ZW (2013) Spatial and Temporal Changes of Maximum 1h Precipitation Intensity in Beijing Region in Last 53 Years. J China Hydrol 33 (1): 32–37. (In Chinese with English Abstract).
59. Jiali W, Renhe Z, Yingchun W (2012) Characteristics of precipitation in Beijing and the precipitation representativeness of Beijing Weather Observatory. J Applied Meteo Sci 23: 265–273. (In Chinese with English abstract).
60. Ma JJ, Li SY, Wang J (2012) The study of Beijing's rainstorm division and its returning periods. Meteo Monthly 38: 569–576. (In Chinese with English abstract).
61. Imeson AC (1983) Studies of erosion thresholds in semi-arid areas:field measurements of soil loss and infiltration in northern Morocco. Catena Sup 4: 79–89.
62. Savenije HHG (1996) The runoff coefficient as the key to moisture recycling. J Hydrol 176: 219–225. doi: 10.1016/0022-1694(95)02776-9.
63. Uttam KM, Rao KV, Mishra PK, Vittal KPR, Sharma KL, et al. (2005) Soil infiltration, runoff and sediment yield from a shallow soil with varied stone cover and intensity of rain. Eur J Soil Sci 56: 435–443. doi: 10.1111/j.1365-2389.2004.00687.x.
64. Abrahams AD, Parsons AJ (1991) Relation between infiltration and stone cover on a semiarid hillslope, southern Arizona. J Hydrol 122: 45–59. doi: 10.1016/0022-1694(91)90171-D.
65. Gholami L, Sadeghi SH, Homaee M (2013) Straw mulching effect on splash erosion, runoff, and sediment yield from eroded plots. Soil Sci Soc Am J 77: 268–278. doi:10.2136/sssaj2012.0271.
66. Ghahramani A, Ishikawa Y, Gomi T (2011) Downslope soil detachment–transport on steep slopes via rain splash. Hydrol. Processes 25: 2471–2480. doi: 10.1002/hyp.8086.
67. Ghahramani A, Ishikawa Y, Gomi T (2011) Effect of ground cover on splash and sheetwash erosion over a steep forested hillslope: A plot-scale study. Catena 85: 34–47. doi: 10.1016/j.catena.2010.11.005.
68. Lal R (1981) Soil erosion problems on alfisols in Western Nigeria, VI. Effects of erosion on experimental plots. Geoderma 25: 215–230. doi: 10.1016/0016-7061(81)90037-9.
69. Miura S, Yoshinaga S, Yamada T (2003) Protective effect of floor cover against soil erosion on steep slopes forested with Chamaecyparis obtusa (hinoki) and other species. J Forest Res-Japan 8: 27–35. doi: 10.1007/s103100300003.
70. Dunjó G, Pardini G, Gispert M (2004) The role of land use–land cover on runoff generation and sediment yield at a microplot scale, in a small Mediterranean catchment. Journal of Arid Environments 57: 99–116. doi: 10.1016/S0140-1963(03)00097-1.

18

Density and Stability of Soil Organic Carbon beneath Impervious Surfaces in Urban Areas

Zongqiang Wei[1,2], Shaohua Wu[1]*, Xiao Yan[2], Shenglu Zhou[1]

1 School of Geographic and Oceanographic Science, Nanjing University, Nanjing, China, **2** School of Environmental and Land Resource Management, Jiangxi Agricultural University, Nanchang, China

Abstract

Installation of impervious surfaces in urban areas has attracted increasing attention due to its potential hazard to urban ecosystems. Urban soils are suggested to have robust carbon (C) sequestration capacity; however, the C stocks and dynamics in the soils covered by impervious surfaces that dominate urban areas are still not well characterized. We compared soil organic C (SOC) densities and their stabilities under impervious surface, determined by a 28-d incubation experiment, with those in open areas in Yixing City, China. The SOC density (0–20 cm) under impervious surfaces was, on average, 68% lower than that in open areas. Furthermore, there was a significantly ($P<0.05$) positive correlation between the densities of SOC and total nitrogen (N) in the open soils, whereas the correlation was not apparent for the impervious-covered soils, suggesting that the artificial soil sealing in urban areas decoupled the cycle of C and N. Cumulative CO_2-C evolved during the 28-d incubation was lower from the impervious-covered soils than from the open soils, and agreed well with a first-order decay model ($C_t = C_1 + C_0(1\text{-}e^{-kt})$). The model results indicated that the SOC underlying capped surfaces had weaker decomposability and lower turnover rate. Our results confirm the unique character of urban SOC, especially that beneath impervious surface, and suggest that scientific and management views on regional SOC assessment may need to consider the role of urban carbon stocks.

Editor: Dafeng Hui, Tennessee State University, United States of America

Funding: This work was funded by the National Natural Science Foundation of China (41001047). The funders had no role in study design, data collection and analysis, decision to publish, or preparation of the manuscript.

Competing Interests: The authors have declared that no competing interests exist.

* Email: shaohuawu@126.com

Introduction

At present more than half of the world's population resides in cities and towns, and the percentage of urban population is projected to increase to 70% by 2050 [1]. As a result, urban areas are increasing in extent at a greater pace than any other land use type [2]. The rapid expanding of urban areas have caused large areas of agricultural, pasture, or forest soils to be changed to urban soils [3,4]. To date, urban land is estimated to cover 9% of the continent [5], in which the impervious surface (e.g., buildings, roads, and other pavements) is estimated to cover nearly 580 000 km^2 globally, an area larger than France [6]. Presently, the resulting impacts of the impervious surfaces on urban soil function, including soil organic carbon (SOC) storage and dynamics, remain largely unknown.

Yet a growing body of literature suggest that urban soils still have robust C storage capacity [7–10], especially in the areas covered by green vegetation (e.g., meadow, forest) which could provide fairly C sequestration [11,12]. Generally, the SOC storage capacity of urban open soils (without impervious surfaces) is comparable to that of adjacent agricultural soils and varies highly amongst different cities/regions, which is might be controlled by several factors, such as urbanization histories, land use types, soil parent materials, topography, and climate (Table 1). However, the storage and turnover of SOC beneath impervious surfaces are still poorly characterized due to their inaccessibility. The impervious surface in urban area is still rapidly expanding due to urbanization, thus it is critical to investigate the SOC stocks and dynamics beneath impervious, to provide accurate inventories in estimates of the entire SOC storage in urban areas, and to promote our understanding of the net impact of urbanization on terrestrial C pools. The relatively limited previous studies suggest that, in urban areas, the SOC density, an important parameter to calculate SOC storage, is significantly lower under impervious surfaces than in open sites [8,15]. These studies mainly focus on the amount or stocks of SOC in urban areas, whereas few attempts are made to study the SOC dynamics in the impervious-covered soils.

Here we collected some urban soil samples from typical impervious-covered and open areas in Yixing City, China, an area experienced rapid urbanization these decades. The main objectives of this study were to investigate the SOC density underlying impervious surfaces in urban areas, and to further study the stability or decomposability of SOC in the sealing environment.

Materials and Methods

Ethics Statement

This study was conducted in Yixing, China. The impervious-covered and open soils were collected from part of Yixing urban area (31°20′–31°25′N, 119°45′–119°50′E). The selected sampling

Table 1. Soil organic carbon densities (kg m^{-2} at a depth where indicated) for urban soils located in different cities.

City	Latitude/longitude	Urban SOC density	Suburban SOC density	Adjacent agricultural SOC density	Literature cited
		kg m^{-2}			
Liverpool, UK	53.4° N/2.98° W	4[a]	–	–	[13]
Boston, U.S.A.	42.35° N/71.06° W	4.02–4.24[b]	3.33–3.99	2.83–4.29	[14]
New York, U.S.A.	42.34° N/75.19° W	5.67[a]	–	–	[15]
New York, U.S.A.	42.34° N/75.19° W	5.1[c]	3.5	3.4	[7]
Baltimore, U.S.A.	39.28° N/76.62° W	4–7[a]	–	–	[7]
Chuncheon, South Korea	37.87° N/127.73° E	2.48[d]	–	3.16	[16]
Phoenix, U.S.A.	33.45° N/112.06° W	0.5–1.1[b]	–	0.75	[17]
Nanjing, China	32.05° N/118.77° E	4.52[e]	–	–	[18]
Chongqing, China	29.56° N/106.57° E	2–3.6[e]	–	–	[19]

a, 0–15 cm; b, 0–10 cm; c, 0–100 cm; d, 0–60 cm; e, 0–20 cm; –, not determined.
Cities were ranked by latitude.

areas did not involve endangered or protected species, and no specific permissions were required for the soil sampling.

Study area

The soils in this study were sampled in January 2008 from Yixing City (31°07′–31°37′N, 119°31′–120°03′E), which has an urban area of 13.4 km^2 in 2000 [3]. Yixing which locates on the plain of the lower reaches of the Yangtze River has had rapid urbanization in recent decades mainly at the expense of agricultural land. Average annual rainfall in Yixing is 1177 mm, and average annual temperature is 15.7°C. The soils in the studied area were formed on the alluvium of the Yangtze River, and the dominant soil type in the agricultural lands around Yixing City is Hydragric Anthrosols [20].

Soil sampling and analysis

Seven sites were selected for impervious-covered soils, and six sites with similar soil parent materials were selected for urban open soils (Figure 1). Soil in our study area had SOC content of 1.48 (±0.19 SE, $n = 9$) g kg^{-1}, TN content of 1.38 (±0.11 SE, $n = 9$) g kg^{-1}, and bulk density of 1.3 (±0.13 SE, $n = 61$) g cm^{-3} in 1982, which these parameters had relatively small variation as their CVs ranged from 8% to 13% (Table 2) [21]. The impervious sites consisted of road pavements, and paved residential squares and alleys, whereas the open area reference sites mainly comprised residential and commercial lawns and gardens, and public greenspaces. Outline description (0–20 cm) for the impervious-covered and open soils selected in our study is shown in Table 3. The soils sampled in our study area primarily had a slit loam texture but varied in the contents of soil admixtures. Compared with the soils in the open areas, the impervious-covered soils had little root penetration and greater amounts of artifacts, including asphalt, cement, brick, tiles, and gypsum. Generally, soil physico-chemical properties were more variable in impervious areas than in the grass areas, and more severely affected by human activities. The mean soil pH and particle fraction larger than 2 mm beneath impervious surfaces (0–20 cm) were 7.72 (range, 7.08–8.33) and 4.28% (range, 6.4%–7.74%), respectively (Table 3).

Each sample, consisting of 3 separate soil cores (5 cm in diameter), was taken from 0–20 cm depth, as SOC storage was supposed to be mainly allocated in the upper soil layers. To make the impervious-covered and open soils more comparable, in the impervious areas, only soils having roots penetrations and little mixture of artifacts were selected. We assumed that the selected soil horizon having these properties is of significance in sequestrating SOC. Specific soil horizon from which the soil was sampled was not determined, but the sampled soils had similar soil texture and chemical properties (Table 3). At the impervious sites, we punched holes on the hard ground near the central position of the selected sealing area (2 m×2 m) and removed the padding to make the covered soil accessible for sampling. After collection, soil samples were transported to the laboratory and air dried at room temperature. Then, the air-dried soils were grounded and sieved through a 2-mm nylon mesh; stones, artifacts, and coarse roots greater than 2 mm in size were weighed.

Soil pH was measured in water (1 soil: 2.5 water, w/v) using a glass electrode, and soil particle size analysis was determined using the hydrometer method [22]. Soil organic C concentrations in the samples were determined using the potassium dichromate sulfuric acid oxidation method [23]. Total nitrogen (N) concentrations were measured by Kjeldahl digestion [24]. Soil bulk density was measured by automated three-dimensional laser scanning [25] (NextEngine Desktop 3D Scanner, NextEngine, Inc., US). The densities of SOC and total N in a horizon of unit area (1 m^2) in each site were calculated as:

$$c = c_c \times BD \times (1 - \delta_{2mm}) \times H \quad (1)$$

where c (kg m^{-2}) represents SOC density or total N density, c_c (g kg^{-1}) represents SOC concentration or total N concentration, BD (g cm^{-3}) represents soil bulk density, δ_{2mm} (g g^{-1}) represents the fraction of material lager than 2 mm diameter, and H (m) represents the soil sampling depth. In our study, the densities of SOC and total N in each site were calculated based on 1 m^2 square to a 0.2 m depth.

To investigate the stability of SOC in urban areas, a thermostatic soil incubation experiment was conducted. Before the incubation, we first determined soil full water holding capacity. The sieved soil (<2 mm) was packed to the same bulk density (~1.5 g cm^3) in cutting ring (100 cm^3), saturated with water, weighed and then dried at 105°C for 48 hours to determine soil water content. Soil full water holding capacity was the water

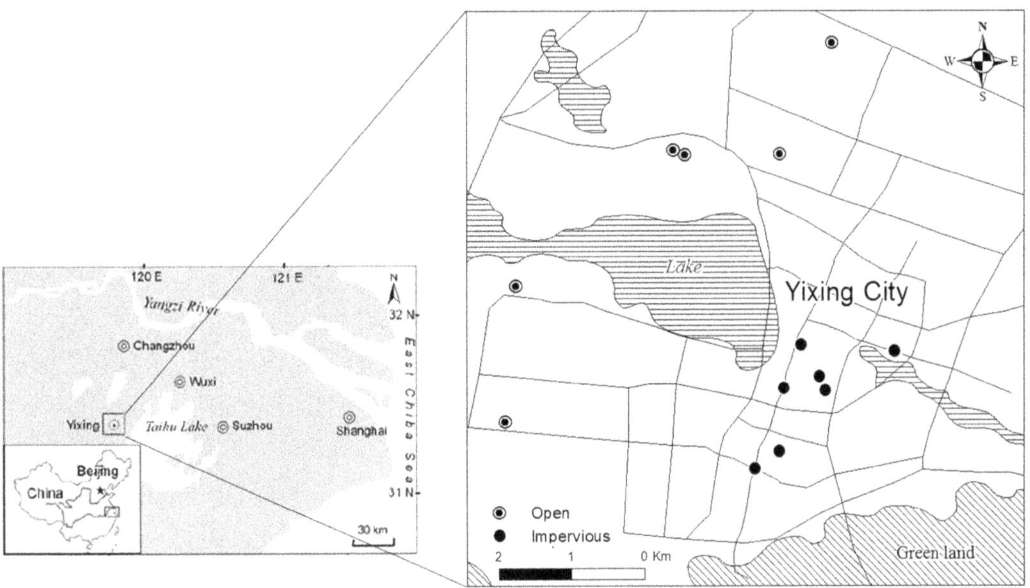

Figure 1. Soil sampling points in Yixing city. Seven sites were selected for impervious-covered soils, and six sites with similar soil parent materials were selected for open soils. Soil samples were collected at 0–20 cm depth.

content of the saturated soil. Fifty grams of each urban soil (< 2 mm) were placed into 500-mL capacity plastic bottles and incubated simultaneously in one incubator (SP-300B, Hengyu, China) for 28 days. The soil water content of each sample in the bottle was then adjusted to 60% of full water holding capacity, which the actual amount of water needed was calculated based on the preliminary test. Each plastic bottle was sealed airtight and incubated in the dark at 25°C. Each sample was pre-incubated without CO_2 absorption for 5 days [18]. Then, a beaker with 5 mL of 0.6 mol L^{-1} NaOH was placed in each jar to absorb the evolved CO_2 during the incubation. Control jars (without soil) were used to measure the background CO_2 concentration. The NaOH was renewed after 1, 3, 5, 7, 14, 21, and 28 days. Any loss of water from the cylinders (checked by weighing) was corrected by a mist sprayer. The CO_2 trapped in the NaOH was determined by back-titration of excess NaOH with 1.5 mol L^{-1} H_2SO_4 after precipitation with 1 mol L^{-1} $BaCl_2$ [26].

A first-order decay equation was used to describe the organic C mineralization in the samples studied [27]:

$$C_t = C_1 + C_0(1 - e^{-kt}) \quad (2)$$

where C_t (mg C g^{-1} C) is the cumulative amount of SOC mineralized during time t (day), C_1 (mg C g^{-1} C) is the rapidly mineralizable SOC pool, C_0 (mg C g^{-1} C) is the potentially mineralizable SOC pool, and k (mg C g^{-1} C d^{-1}) is the corresponding mineralization rate constant.

The SOC mineralization half-time (i.e., time required to mineralize half of the potentially mineralizable SOC) was calculated as follows [28]:

Table 2. Initial properties of the urban soil in Yixing obtained from the second Chinese soil survey [a] ($n = 9$ except where noted).

Soil property	0–14 cm		14–25 cm	
	Mean ±SE	CV, %	Mean ±SE	CV, %
Bulk density [b], g cm^{-3}	1.3±0.13	10	1.44±0.15	10.4
pH	6.6±0.3	4.5	6.9±0.6	8.7
TOC [c], %	1.48±0.19	13.3	1.14±0.30	26.4
N, g kg^{-1}	1.38±0.11	8.0	1.13±0.28	24.8
C/N	10.6±1.2	11.3	9.9±1.5	15.2
P, %	0.095±0.01	10.5	0.09±0.125	16.7
K, %	1.67±0.09	5.4	1.68±0.11	6.7
Olsen P, mg kg^{-1}	5.8±2.8	48.3	5±3.8	76.0
Available K [d], mg kg^{-1}	85±30	35.3	96±30	31.2
CEC, cmol kg^{-1}	19.51±2.36	12.1	20.11±2.02	10.2

a, Office of Soil Survey in Yixing County, conducted in 1979–1982; b, $n = 61$; c, determined using the dichromate oxidation method; d, CH_3COONH_4 extractable K.

Table 3. Outline description for the urban soils (0–20 cm) selected in our study.

Soil sample	Texture	pH	artifacts, rock fragments, roots	> 2 mm fraction
				%
Impervious soils				
1	Silt loam	7.08	coarse angular weathered brick, roots	1.60
2	Silt	7.42	coarse angular freshly weathered rock fragment, roots	9.01
3	Silt loam	7.9	plastic, roots	5.19
4	Silt loam	8.15	coarse angular highly weathered brick and tiles	2.67
5	Silt loam	7.56		8.53
6	Silt	7.58	gypsum, roots	1.44
7	Silt loam	8.33		1.51
Open soils				
8	Silt loam	7.21	heavily roots	5.61
9	Silt loam	6.4		0.72
10	Silt loam	7.12	coarse angular weathered rock fragment, roots	4.25
11	Silt loam	7.58	heavily roots	2.78
12	Silt loam	7.74	heavily roots	2.33
13	Silt loam	6.94	heavily roots	2.35

$$t_{1/2} = \ln 2/k \qquad (3)$$

Statistical analysis

Model fit was conducted using the Global curve fit wizard in SigmaPlot 12.0 software package (Systat Software, Inc., Chicago, IL, USA). Linear regression analysis and the comparisons in the densities of SOC and total N and the model parameters between the impervious and open land uses (group t-test) were conducted using SAS 8.2 software (SAS Institute, Cary, NC).

Results and Discussion

SOC density under impervious surfaces

The impervious-covered sites had lower SOC and TN concentrations (0–20 cm) compared with the open sites and those in 1982 (Figure 2a, b). The SOC and TN concentrations of the open soils were comparable with those in 1982. The mean SOC density beneath impervious surfaces was 68% (±7.7% SE, $P<0.05$) lower than that in open areas (2.46 versus 7.59 kg m^{-2}, respectively, Figure 2c). Similar results were also reported by Pouyat *et al.* (2006) and Raciti *et al.* (2012) that SOC densities under impermeable surfaces were lower than that of open soils [8,15]. It is interesting to note that the SOC densities observed at the impervious sites were comparable to those in New York City (Raciti *et al.* 2012) although the samples were distributed at different cities [15], suggesting that there might be an equilibrium value for the depletion of SOC densities under impervious surfaces. The total N densities were 0.25 kg m^{-2} and 0.32 kg m^{-2}, respectively, for the impervious-covered and open soils in Yixing City, but no significant difference was found between them (Figure 2d), probably due to the relatively small sample size in our study.

The exact mechanisms of SOC loss from the impervious areas are unknown, but likely possibilities include gaseous losses, aqueous losses as dissolved organic and inorganic C, and physical removal of topsoil during the construction process. The removal of surface soil may readily result in a significant depletion of organic C stock due to enhanced mineralization and re-use. The situation could be worse when topsoil is not re-used and is left slowly to decompose. If the potential loss of SOC inferred by this study hold true for other impervious-covered soils, this would suggest that roughly 0.45 Pg (1 Pg = 10^{15} g) of SOC would be potentially lost in China associated with 87128 km^2 of impervious surface [6]. The potential loss of SOC stock could be more significant in magnitude when taking into account the large vegetable C losses in response to initial land use change to urban.

Although the SOC densities beneath impermeable surfaces were relatively lower, it is not adequate to exclude these C pools for estimates of ecosystem C stocks in national scale as required by Kyoto Protocol signatories [10]. In fact, investigation of SOC stocks and dynamics in urban impervious areas are getting more important to the C study, since there is ongoing significant expanding of constructed impervious surfaces around the world [5,6]. In our study, soil samples were collected at 0–20 cm soil layer and therefore SOC densities were calculated to 0.2 m depth. Urban subsoil horizons may also contain considerable amounts of SOC since physical disturbance (e.g., mixing, burying) can result in translocation of topsoil that is rich in OC to deeper soil layer [29]. Thus, the SOC storage underlying capped surfaces in urban areas could be larger than we estimated when it was calculated to 1 m soil depth as usually did. The SOC inventory involving subsoil horizons beneath impervious surfaces in urban areas will strengthen our understanding on the impact of urbanization on the ecosystem. Furthermore, the SOC inventory in urban area should consider the definition and built-up age of the urban used by the investigators. Urban areas, regardless of definition, are rapidly expanding at unprecedented rates; inconsistent definitions of 'urban' will result in different conclusions about the size of urban C stocks [30]. Moreover, it was suggested that the built-up age could modify urban C stocks [31].

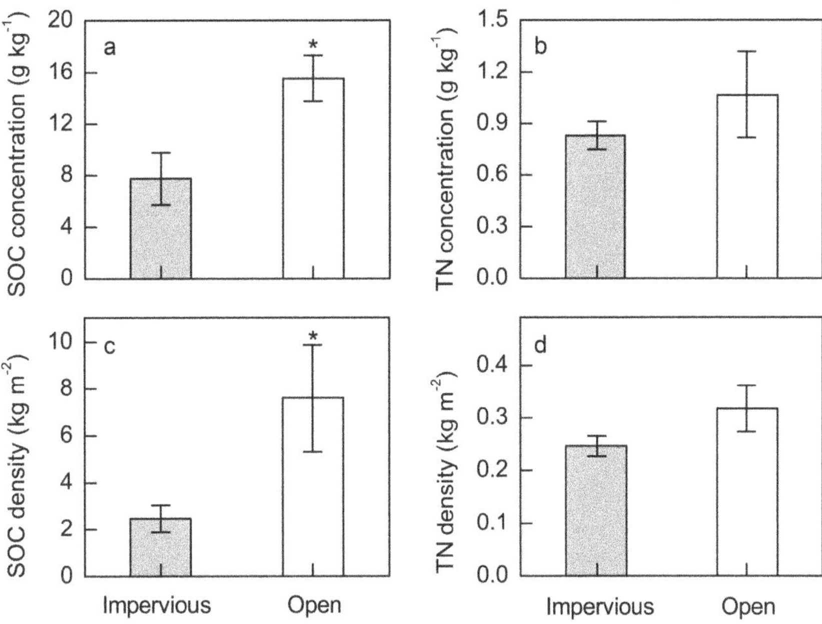

Figure 2. The concentrations and densities of SOC and TN for the impervious-covered and open soils in Yixing city. Values are means ± SE, SOC represents soil organic carbon, and TN represents total nitrogen, * $P<0.05$ ($n=7$ for urban impervious-covered soils, and $n=6$ for urban open soils).

In most soils, there is a tight coupling between the stocks and fluxes of C and N. The mean C/N ratio in the soils covered by impervious surfaces was significantly lower than the ratio in open soils (10.8 versus 22.1, respectively, $P<0.05$, data not shown), indicating that the microorganisms had a relative C deficit state in the impervious-covered soils. The higher C/N ratios in the open soils were probably due to the abundant input of organic materials (e.g., leaves, branches) from plant as these materials always had relatively higher C/N ratios [32]. Regression of SOC density as a function of total N density in the open soils revealed a strong linear relationship between the two variables at the 0–20 cm depth ($r^2=0.73$, $P<0.05$, Figure 3). In contrast, there was no clear relationship between SOC and total N for the soils beneath impervious surfaces, suggesting that the paving in urban areas decoupled the cycle of C and N in the soils [15].

Installation of impervious surfaces in urban areas had negative impact on urban ecosystem, which was indicated by large amount of SOC loss that can offset the C stored in trees or other green vegetation in urban areas, and perturbed soil C and N cycling in impervious area. More greenspaces or semi-pervious pavement systems were, therefore, recommended in future urban construction to mitigate the negative consequences of urban artificial soil sealing [5]. In addition, soil management (e.g., fertilization, tillage, irrigation) for the open sites could be optimized to enhance the SOC sequestration in urban soils.

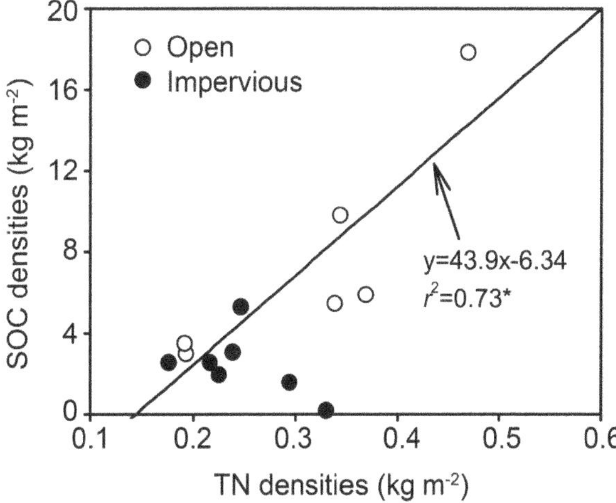

Figure 3. The correlations between the densities of SOC and TN for urban soils in Yixing city. SOC represents soil organic carbon, and TN represents total nitrogen, * $P<0.05$ ($n=7$ for urban impervious-covered soils, and $n=6$ for urban open soils).

SOC stability under impervious surfaces

The laboratory incubation can provide an insight of SOC dynamics (e.g., [33]). The amount of SOC mineralized (mg C g^{-1} C) over the 28-d incubation was calculated based on the total (whole soil) SOC content in each sample to enable us directly compare the potential C loss (i.e., CO_2 emissions) from the impervious-covered and open soils, because the portion of SOC is comparable amongst urban soils although the initial SOC content for the urban soils were unequal (Figure 4). It was found that less CO_2 was emitted from the soils underlying capped surfaces than from the open soils during the incubation, indicating that the SOC was more stable beneath impervious surfaces than in open areas. The lower transformation of organic C in the covered soils can be partially explained by a low microbial activity in these soils [34].

The first-order decay model ($C_t=C_1+C_0(1-e^{-kt})$) used in this study fitted the data well (r^2 were 0.97 and 0.96, Table 4). The easily decomposable SOC pool (C_1) was small relative to the potentially decomposable SOC pool (C_0) in all samples. The C_1 and C_0 of SOC under impervious surfaces were 47% and 27%, respectively, lower than those in open areas, indicating that there is a more severe depletion of readily decomposable SOC pool after

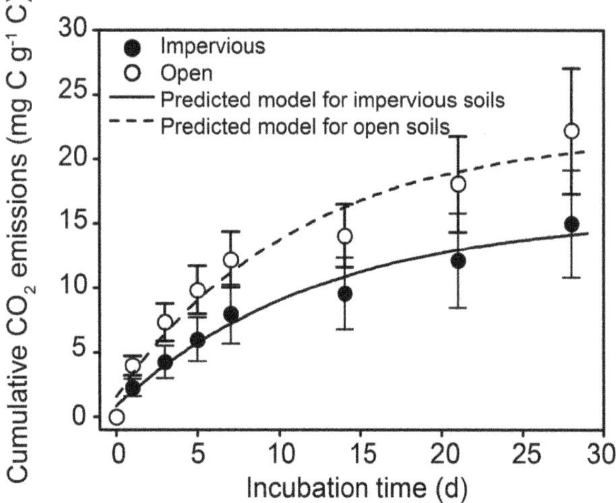

Figure 4. Cumulative carbon release (as CO_2) from the impervious-covered and open soils during the 28-d incubation. Data were fitted by the first-order decay model. The bars indicate standard errors ($n=7$ for urban impervious-covered soils, and $n=6$ for urban open soils).

paving. The impervious-covered and open soils had similar mineralization rate constants (k) of the potentially decomposable organic C. However, the lower value of C_0k, a parameter that could be comparable with the initial potential rate of SOC mineralization [35], together with the longer mineralization half-life time ($t_{1/2}$) of SOC under impervious surfaces revealed these soils had weaker organic C decomposability and lower turnover rate.

Edmondson *et al.* (2012) suggested that the turnover of SOC beneath impervious surface likely depend upon the type of capping and the extent of impervious surface [10]. In our study, the covered soils were collected conformably from areas with more than 95% impervious surface, in which the exchanges of gas and water between the soil and atmosphere were supposed to be rare. In fact, there are some urban impervious-covered soils (e.g., patio, garden path) which were distributed in areas dominated by vegetation. The soils underlying capped surfaces in these areas could be colonized by the root systems of lawn grasses and garden trees and shrubs. Thus, it is likely that below these smaller patches of impervious surfaces the soil remains active potentially accumulating SOC and has more robust organic C transformation. Accordingly, more systematic research and soil samples including different land use types and sealing degrees in urban areas are required to better understand the SOC turnover under impervious surfaces.

Additionally, some impervious surfaces in urban area may be removed due to urban land use change. The differences in soil biochemical properties between the impervious and open sites are supposed to be minimized, since the removal of impervious surface. Investigation on the temporal dynamic of SOC after removal of sealing will be beneficial for comprehensively understanding the ecological effects of urbanization. Therefore, management and time, which were needed to rebuild organic C sequestration capacity for the soil after removal of sealing, need to be studied.

Conclusions

Our data demonstrated that the SOC density decreased at 0–20 cm depth in urban areas (regardless of the definition) after the installation of impervious surfaces, although the precise mechanism of organic C loss from these soils was uncertain. The artificial soil sealing in urban areas also decoupled the cycle of C and N, since SOC density correlated positively with total N density in the open soils ($P<0.05$), but they did not exhibit an apparent relation in the impervious-covered soils. The SOC underlying capped surfaces had weaker decomposability and lower turnover rate compared with that in open areas, which was indicated by a smaller readily decomposable SOC pool, a longer half-time, and a smaller amount of CO_2-C emission during the 28-d incubation. More greenspaces or semi-pervious pavement systems in future urban construction will mitigate the negative consequences of urban artificial soil sealing.

Acknowledgments

We are grateful to the Center of Modern Analysis Nanjing University for supporting soil analysis in our study. We thank the three anonymous reviewers for their very helpful comments and revisions of the manuscript.

Author Contributions

Conceived and designed the experiments: ZQW SHW SLZ. Performed the experiments: ZQW SHW XY. Analyzed the data: ZQW SHW. Contributed reagents/materials/analysis tools: SHW XY. Wrote the paper: ZQW SHW.

Table 4. The first-order decay model (Eq. 2) parameters and coefficients of determination (r^2) for carbon mineralization in urban impervious-covered and open soils.

Parameters	Impervious	Open	
C_1 (mg C g^{-1} C)	0.84	1.59	n.s.
C_0 (mg C g^{-1} C)	14.9	20.53	n.s.
k (mg C g^{-1} C d^{-1})	0.08	0.09	n.s.
C_0k	1.17	1.87	n.s.
$t_{1/2}$ (d)	8.84	7.62	n.s.
r^2	0.97	0.96	n.s.
p	0.01	0.01	n.s.

C_1, rapidly mineralizable SOC pool; C_0, potentially mineralizable SOC pool; k, SOC mineralization rate constant; C_0k, a parameter that could be comparable with the initial potential rate of SOC mineralization [35]; $t_{1/2}$, SOC mineralization half-time.
n.s., not significant.

References

1. United Nations (2008) World urbanization prospects: the 2007 revision.
2. Hansen AJ, Knight RL, Marzluff JM, Powell S, Brown K, et al. (2005) Effects of exurban development on biodiversity: patterns, mechanisms, and research needs. Ecol Appl 15: 1893–1905. doi:10.1890/05-5221.
3. Pan XZ, Zhao QG (2007) Measurement of urbanization process and the paddy soil loss in Yixing city, China between 1949 and 2000. Catena 69: 65–73. doi:10.1016/j.catena.2006.04.016.
4. Su SL, Zhang Q, Zhang ZH, Zhi JJ, Wu JP (2011) Rural settlement expansion and paddy soil loss across an ex-urbanizing watershed in eastern coastal China during market transition. Reg Environ Change 11: 651–662. doi:10.1007/s10113-010-0197-2.
5. Scalenghe R, Ajmone-Marsan F (2009) The anthropogenic sealing of soils in urban areas. Lanscape Urban Plan 90: 1–10. doi:10.1016/j.landurbplan.2008.10.011.
6. Elvidge CD, Tuttle BT, Sutton PC, Baugh PC, Howard AT, et al. (2007) Global distribution and density of constructed impervious surfaces. Sensors 7: 1962–1979. doi:10.3390/s7091962.
7. Pouyat R, Groffman P, Yesilonis I, Hernandez L (2002) Soil carbon pools and fluxes in urban ecosystems. Environ Pollut 116: S107–S118. doi:10.1016/S0269-7491(01)00263-9.
8. Pouyat RV, Yesilonis ID, Nowak DJ (2006) Carbon storage by urban soils in the United States. J Environ Qual 35: 1566–1575. doi:10.2134/jeq2005.0215.
9. Churkina G, Brown DG, Keoleian G (2010) Carbon stored in human settlement: the conterminous United States. Global Change Biol 16: 135–143. doi:10.1111/j.1365-2486.2009.02002.x.
10. Edmondson JL, Davies ZG, McHugh N, Gaston KJ, Leake JR (2012) Organic carbon hidden in urban ecosystems. Sci Rep 2: 963. doi:10.1038/srep00963.
11. Townsend-Small A, Czimczik C (2010) Carbon sequestration and greenhouse gas emissions in urban turf. Geophys Res Lett 37: L02707. doi:10.1029/2009GL041675.
12. Raciti SM, Groffman PM, Jenkins JC, Pouyat RV, Fahey TJ, et al. (2011) Accumulation of carbon and nitrogen in residential soils with different land-use histories. Ecosystems 14: 287–297. doi:10.1007/s10021-010-9409-3.
13. Beesley L (2012) Carbon storage and fluxes in existing and newly created urban soils. J Environ Manage 104: 158–165. doi:10.1016/j.jenvman.
14. Rao P, Hutyra LR, Raciti SM, Finzi AC (2013) Field and remotely sensed measures of soil and vegetation carbon and nitrogen across an urbanization gradient in the Boston metropolitan area. Urban Ecosystems 16: 593–616. doi:10.1007/s11252-013-0291-6.
15. Raciti SM, Hutyra LR, Finzi AC (2012) Depleted soil carbon and nitrogen pools beneath impervious surfaces. Environ Pollut 164: 248–251. doi:10.1016/j.envpol.2012.01.046.
16. Jo HK (2002) Impacts of urban greenspace on offsetting carbon emissions for Middle Korea. J Environ Manage 64: 115–126. doi:10.1006/jema.2001.0491.
17. Kaye JP, Majumdar A, Gries C, Buyantuyev A, Grimm NB, et al. (2008) Hierarchical Bayesian scaling of soil properties across urban, agricultural, and desert ecosystems. Ecol Appl 18: 132–145. doi:10.1890/06-1952.1.
18. Wei ZQ, Wu SH, Zhou SL, Li JT, Zhao QG (2014) Soil organic carbon transformation and related properties in urban soil under impervious surfaces. Pedosphere 24: 56–64. doi:10.1016/S1002-0160(13)60080-6.
19. Liu Y, Wang C, Yue W, Hu Y (2013) Storage and density of soil organic carbon in urban topsoil of hilly cities: a case study of Chongqing Municipality of China. Chin Geogra Sci 23: 26–34. doi:10.1007/s11769-013-0585-x.
20. IUSS Working Group WRB (2007) World reference base for soil resources 2006, first update 2007. World Soil Resources Reports No. 103. FAO, Rome.
21. Office of Soil Survey in Yixing County (1988) The soils of Yixing County. Office of Soil Survey in Jiangsu Province, Nanjing. (in Chinese).
22. Gee GW, Or D (2002) Particle-size analysis. In: Dame JH, Topp GC, editors. Methods of Soil Analysis, Part 4. Physical Methods.pp.255–293.
23. Nelson DE, Sommers LE (1982) Total carbon, organic carbon, and organic matter. In: Page AL, Miller RH, Keeney DR, editors. Methods of soil analysis, Part 2. Chemical and microbiological properties. pp.539–580.
24. Bremner JM, Mulvaney CS (1982) Nitrogen total. In: Page AL, Miller RH, Keeney DR, editors. Methods of soil analysis. Part 2. Chemical and microbiological properties. pp.595–624.
25. Rossi AM, Hirmas DR, Graham RC, Sternberg PD (2008) Bulk density determination by automated three-dimensional laser scanning. Soil Sci Sco Am J 72: 1591–1593. doi:10.2136/sssaj2008.0072N.
26. Goyal S, Chander K, Mundra MC, Kapoor KK (1999) Influence of inorganic fertilizers and organic amendments on soil organic matter and soil microbial properties under tropical conditions. Biol Fertil Soils 29: 196–200. doi:10.1007/s003740050544.
27. Jones CA (1984) Estimation of an active fraction of soil nitrogen. Commun Soil Sci Plant Anal 15: 23–32. doi:10.1080/00103628409367451.
28. Zhang ZD, Yang XM, Drury CF, Reynolds WD, Zhao LP (2010) Mineralization of active soil organic carbon in particle size fractions of a Brookston clay soil under no-tillage and mouldboard plough tillage. Can J Soil Sci 90: 551–557. doi:10.4141/cjss09081.
29. Lorenz K, Kandeler E (2005) Biochemical characterization of urban soil profiles from Stuttgart, Germany. Soil Biol Biochem 37: 1373–1385. doi:10.1016/j.soilbio.2004.12.009.
30. Raciti SM, Hutyra LR, Rao P, Finzi AC (2012) Inconsistent definitions of "urban" result in different conclusions about the size of urban carbon and nitrogen stocks. Ecol Appl 22: 1015–1035. doi:10.1890/11-1250.1.
31. Scalenghe R, Malucelli F, Ungaro F, Perazzone L, Filippi N, et al. (2011) Influence of 150 years of land use on anthropogenic and natural carbon stocks in Emilia-Romagna region (Italy). Environ Sci Technol 45: 5112–5117. doi:10.1021/es1039437.
32. McGroddy ME, Daufresne T, Hedin LO (2004) Scaling of C:N:P stoichiometry in forests worldwide: implications of terrestrial Redfield-type ratios. Ecology 85: 2390–2401. doi:10.1890/03-0351.
33. Haile-Mariam S, Collins HP, Wright S, Paul EA (2008) Fractionation and long-term laboratory incubation to measure soil organic matter dynamics. Soil Sci Soc Am J 72: 370–378. doi:10.2136/sssaj2007.0126.
34. Wei ZQ, Wu SH, Zhou SL, Lin C (2013) Installation of impervious surfaces in urban areas affects microbial biomass, activity (potential C mineralization), and functional diversity of the fine earth. Soil Res 51: 59–67. doi:10.1071/SR12089
35. Fernández JM, Plaza C, Hernández D, Polo A (2007) Carbon mineralization in an arid soil amended with thermally-dried and composted sewage sludges. Geoderma 137: 497–503. doi:10.1016/j.geoderma.2006.10.013.

Effects of Winter Cover Crops Straws Incorporation on CH_4 and N_2O Emission from Double-Cropping Paddy Fields in Southern China

Hai-Ming Tang*, Xiao-Ping Xiao*, Wen-Guang Tang, Ke Wang, Ji-Min Sun, Wei-Yan Li, Guang-Li Yang

Hunan Soil and Fertilizer Institute, Changsha, PR China

Abstract

Residue management in cropping systems is believed to improve soil quality. However, the effects of residue management on methane (CH_4) and nitrous oxide (N_2O) emissions from paddy field in Southern China have not been well researched. The emissions of CH_4 and N_2O were investigated in double cropping rice (*Oryza sativa* L.) systems with straw returning of different winter cover crops by using the static chamber-gas chromatography technique. A randomized block experiment with three replications was established in 2004 in Hunan Province, China, including rice–rice–ryegrass (*Lolium multiflorum* L.) (Ry-R-R), rice–rice–Chinese milk vetch (*Astragalus sinicus* L.) (Mv-R-R) and rice–rice with winter fallow (Fa-R-R). The results showed that straw returning of winter crops significantly increased the CH_4 emission during both rice growing seasons when compared with Fa-R-R. Ry-R-R plots had the largest CH_4 emissions during the early rice growing season with 14.235 and 15.906 g m^{-2} in 2012 and 2013, respectively, when Ry-R-R plots had the largest CH_4 emission during the later rice growing season with 35.673 and 38.606 g m^{-2} in 2012 and 2013, respectively. The Ry-R-R and Mv-R-R also had larger N_2O emissions than Fa-R-R in both rice seasons. When compared to Fa-R-R, total N_2O emissions in the early rice growing season were increased by 0.05 g m^{-2} in Ry-R-R and 0.063 g m^{-2} in Mv-R-R in 2012, and by 0.058 g m^{-2} in Ry-R-R and 0.068 g m^{-2} in Mv-R-R in 2013, respectively. Similar result were obtained in the late rice growing season, and the total N_2O emissions were increased by 0.104 g m^{-2} in Ry-R-R and 0.073 g m^{-2} in Mv-R-R in 2012, and by 0.108 g m^{-2} in Ry-R-R and 0.076 g m^{-2} in Mv-R-R in 2013, respectively. The global warming potentials (GWPs) from paddy fields were ranked as Ry-R-R>Mv-R-R>Fa-R-R. As a result, straw returning of winter cover crops has significant effects on increase of CH_4 and N_2O emission from paddy field in double cropping rice system.

Editor: Dafeng Hui, Tennessee State University, United States of America

Funding: This study was supported by the National Natural Science Foundation of China (No. 31201178), and the Public Research Funds Projects of Agriculture, Ministry of Agriculture of the P.R. China (No. 201103001). The funders had no role in study design, data collection and analysis, decision to publish, or preparation of the manuscript.

Competing Interests: The authors have declared that no competing interests exist.

* Email: hntfsxxping@163.com (XPX); tanghaiming66@163.com (HMT)

Introduction

With the current rise in global temperatures, numerous studies have focused on greenhouse gases (GHG) emissions [1–3]. Agriculture production is an important source of GHG emission [4]. In addition to carbon dioxide (CO_2), methane (CH_4) and nitrous oxide (N_2O) play important roles in global warming. The global warming potentials (GWPs) of CH_4 and N_2O are 25 and 298 times that of CO_2 in a time horizon of 100 years, respectively [5]. The concentrations of CH_4 and N_2O in the atmosphere are estimated to be increasing at the rates of 1% and 0.2–0.3% per year [6]. In addition to industrial emissions, farmland is another important source of atmospheric GHG [7–10]. Numerous results indicate that rice (*Oryza sativa* L.) paddy field is a significant source of CH_4 and N_2O emissions [10,11]. The anaerobic conditions in wetland rice field are favorable for fostering CH_4 emission [12]. Thus, the characteristics of CH_4 and N_2O emissions from paddy field and the reduction of emission have received attentions from scientists.

A considerable number of studies have shown that some farm operations can influence CH_4 and N_2O emission. For example, cropping system, crop type, water and nitrogen (N) management, organic matter application and tillage can regulate CH_4 and N_2O emission [13–15]. Tillage and crop straws retention have a great influence on CH_4 and N_2O emission through the changes of soil properties (e.g., soil porosity, soil temperature and soil moisture, etc.) [16–17]. In paddy soils, CH_4 is produced by archaea bacteria during the anaerobic degradation of organic matter and oxidized by methanotrophic bacteria [18]. Incorporation of organic material into soil can enhance the number and activity of archaea bacteria [19] and provide large amounts of active organic substrate for CH_4 production [20]. Soil amendment with organic material, such as crop straw [21] and green manure incorporation [22], has been well estimated to promote CH_4 emission in paddy fields. Biogenic N_2O production originates from nitrification and denitrification [23], which are processes involving microorganisms in the soil. N_2O flux in paddy fields was small in flooding condition, but peaked after drainage [24]. Some studies have indicated that the cropping system of winter fallow with cover crops has advantages of promoting soil quality, enhancing nutrient utilization, increasing crop yield, reducing soil erosion and chemical runoff, and inhibiting weed growth in paddy field [25–26].

Winter cover crops, which are grown during an otherwise fallow period, are a possible means of improving nutrient dynamics in the surface layer of intensively managed cropping systems. Chinese milk vetch (*Astragalus sinicus* L.) and ryegrass (*Lolium multiflorum* L.) are the main winter cover crops in Southern China. Growing these cover crops with straw mulching in the winter season after late rice harvest and incorporating them into soil as green manure before early rice transplanting next year is a traditional practice as well as rice straw incorporation. Hermawan and Bomke [26] suggested that growing winter cover crops such as annual ryegrass may protect aggregate breakdown during winter and result in a better soil structure after spring tillage, as opposed to leaving soil bare. Other potential benefits of winter cover crops are the prevention of nitrate leaching [27]; weed infestation [28]; and improvement of soil water retention, soil organic matter content and microbial activity [29]. Returning of crop straws have been suggested to improve overall soil conditions, reduce the requirement for N fertilizers and support sustainable rice productivity.

In recent years, many researches have studied the effects of winter cover crops on soil physical properties and crop productivity, methane emission, N availability and nitrogen surplus [30–32]. However, relatively few studies related to CH_4 and N_2O emissions and yields under different double cropping rice systems with different winter cover crops have been conducted in double–cropping paddy field in Southern China. Monitoring CH_4 and N_2O emissions of different winter cover crops–double cropping rice cultivation modes is important to maintain soil productivity, increase carbon (C) storage, and regulate the greenhouse effects. Therefore, the objectives of this research were: (1) to quantify CH_4 and N_2O emissions from paddy field and grain yield under different winter cover crops and double cropping rice systems, (2) to evaluate the GWPs of different winter cover crops–double cropping rice treatments in southern China.

Materials and Methods

Experimental site

The experiment was initiated in winter 2004 at the experimental station of the Institute of Soil and Fertilizer Research, Hunan Academy of Agricultural Sciences, China (28°11′58″ N, 113°04′47″ E). The typical cropping system in this area is double cropping rice. The soil type is a Fe–accumuli–Stagnic Anthrosol derived from Quaternary red clay (clay loam). The characteristics of the surface soil (0–20 cm) in 2004 are as follows: pH 5.40, soil organic carbon (SOC) 13.30 g kg^{-1}, total N 1.46 g kg^{-1}, available N 154.5 mg kg^{-1}, total phosphorous (P) 0.81 g kg^{-1}, available P 39.2 mg kg^{-1}, total potassium (K) 13.0 g kg^{-1}, and available K 57.0 mg kg^{-1}. All these data were tested before the experiment in 2004. This region has the subtropical monsoonal humid climate with a long hot period and short cold period. The average annual precipitation is approximately 1500 mm and the annual mean temperature is 17.1°C, the annual frost-free period is approximately from 270 days to 310 days. The daily precipitation and mean temperature data during the early and late rice growing season during 2012–2013 are presented in Fig. 1. The cropping system was that the early rice rotated with the late rice, and then planted winter cover crops till the next year's early rice transplanting.

Experimental design and field management

A randomized block experiment with three replications was established in 2004, and this study was conducted from 2012 to 2013. The experiment included three cropping systems: rice–rice–ryegrass (Ry-R-R), rice–rice–Chinese milk vetch (Mv-R-R), and rice–rice with winter fallow (Fa-R-R). The plot area was 1.1 m^2 (1 m × 1.1 m). After winter cover crops harvested, a moldboard plow was used to incorporate part of the crop straw into soil: both the ryegrass and Chinese milk vetch straw returned was 22500.0 kg ha^{-1}. All the plots were plowed once to a depth of 20 cm by using a moldboard plow 15 d before rice seedling transplanting. The early rice variety (*Oryza sativa* L.) Lingliangyou 211 and late rice variety (*Oryza sativa* L.) Fengyuanyou 299 were used as the materials in 2012 and 2013. One-month-old seedlings were transplanted with a density of 150,000 plants ha^{-1} (one seed per 16 cm × 16 cm) and 2–3 plants per hill. Gramoxone (paraquat) was applied to control weeds at 2 d before rice transplantation. The basal fertilizer of the early and late rice was applied at the rate of 150.0 kg N ha^{-1} and 180.0 kg N ha^{-1} as urea (60% for basal; 40% for top–dressed at the tillering stage), 75.0 kg P_2O_5 ha^{-1} as diammonium phosphate and 120.0 kg K_2O ha^{-1} as potassium sulfate. The different treatments during early and late rice season and field management were presented in Table 1.

Collection and measurement of CH_4 and N_2O

CH_4 and N_2O emitted from paddy field were collected using the static chamber–GC technique at 9:00–11:00 in the morning during the early and late rice growing season. The chamber (50 cm × 50 cm × 120 cm) was made of 5 mm PVC board with a PVC base. The base had a groove in the collar, in which the chamber could be settled. The chamber base was inserted into soil about 5 cm in depth with rice plant growing inside the base. The groove was 1 cm below flooded water, and the chamber was settled into the groove of the collar with water to prevent leakage and gas exchange. The chamber contained a small fan for stirring air, a thermometer sensor, and a trinal–venthole. From the second day after transplanting of early or late rice, gases were sampled weekly. Before sampling, the fan in the chamber started working to allow an even mix of air before extracting the air with a 50 ml injector at 0, 10, 20, and 30 min after closing the box. The air samples were transferred into 0.5 L sealed sample bags by rotating trinal venthole.

The quantities of CH_4 and N_2O emission were measured with a gas chromatograph (Agilent 7890A) equipped with flame ionization detector (FID) and electron capture detector (ECD). Methane was separated using 2 m stainless-steel column with an inner diameter of 2 mm 13XMS column (60/80 mesh), with FID at 200°C. Nitrous oxide was separated using a 1 m stainless-steel column with an inner diameter 2 mm Porapak Q (80/100 mesh) and ECD at 330°C.

Data analysis

Fluxes of CH_4 and N_2O were calculated with the following equation [33]:

$$F = ph \times \frac{273}{273+t} \times \frac{dc}{dt}$$

Where, F is the CH_4 flux (mg m^{-2} h^{-1}) or N_2O flux (μg m^{-2} h^{-1}); T is the air temperature (°C) inside the chamber; ρ is the CH_4 or N_2O density at standard state (0.714 kg m^{-3} for CH_4 and

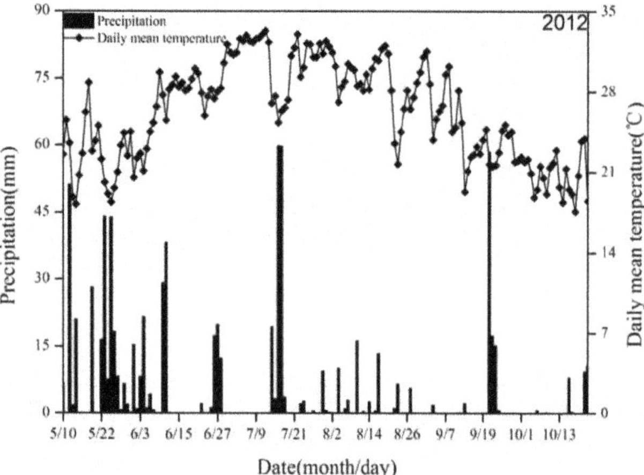

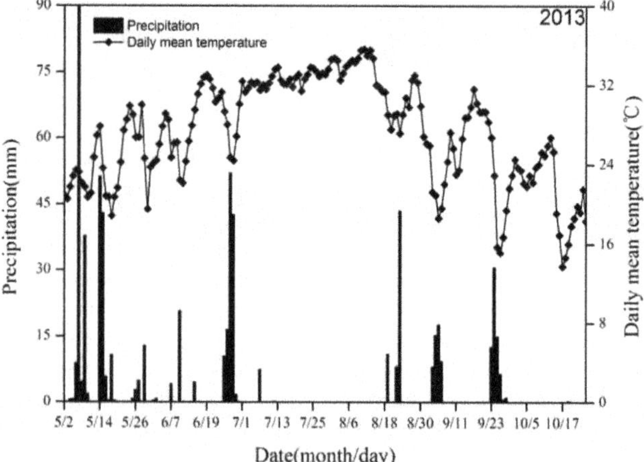

Figure 1. Daily precipitation and mean temperature at the study site in 2012 and 2013.

1.964 kg m^{-3} for N_2O); h is the headspace height of the chamber (m); and dc/dt is the slope of the curve of gas concentration variation with time.

The total emissions of CH_4 and N_2O were sequentially computed from the emissions between every 2 adjacent intervals of the measurements, based on a non–linear, least–squares method of analysis [34,35].

GWPs is defined as the cumulative radiative forcing both direct and indirect effects integrated over a period of time from the emission of a unit mass of gas relative to some reference gas. Carbon dioxide was chosen as this reference gas. The GWPs conversion parameters of CH_4 and N_2O (over 100 years) were adopted with 25 and 298 kg ha^{-1} CO_2-equivalent [5].

Statistical analysis

Data presented herein are means of 3 replicates in each treatment. All data were expressed as mean ± standard error. The data were analyzed as a randomized complete block, using the PROC ANOVA procedure of SAS [36]. Mean values were compared using the least significant difference (LSD) test, and a probability value of 0.05 was considered to indicate statistical significance.

Results

Characteristics of CH_4 emission flux from early and late rice fields

In the early rice season, the curve of CH_4 flux was low when early rice was newly transplanted, but increased quickly until the first peak about 2 weeks after transplanting, and then dramatically declined to a low level with relative stability with the second small peak appeared at 36 and 35 d after transplanting in 2012 and 2013, respectively (Fig. 2). The gradual increase of CH_4 emission after transplanting resulted from the decomposition of organic matter and the growth of rice. The second peak was mainly because of the continuous decomposition of organic matter under

Table 1. Management practices of different cropping systems.

Crop	Date (month/day)		Field management
	2012	2013	
Early rice	4/12	4/5	Sowing and seedling raising
	5/9	5/1	Paddy tillage
	5/10	5/2	Transplanting (16 cm×16 cm)
	5/18	5/10	Urea were applied at 130.0 kg ha^{-1} for top-dressed at tillering
	6/7–6/15	5/27–6/5	Drained out water and dried the soil at maximum tillering stage
	6/16–7/13	6/6–7/13	Wetting-drying alternation irrigation
	7/18	7/18	Grains were harvested
Late rice	6/25	6/27	Sowing and seedling raising
	7/21	7/19	Paddy tillage (The rate of early rice straw returning was 4 500.0 kg ha^{-1})
	7/22	7/20	Transplanting (16 cm×16 cm)
	7/30	7/28	Urea were applied at 156.5 kg ha^{-1} for top-dressed at tillering
	8/20–8/27	8/16–8/26	Drained out water and dried the soil at maximal tillering stage
	8/28–10/17	8/27–10/19	Wetting-drying alternation irrigation
	10/22	10/25	Grains were harvested

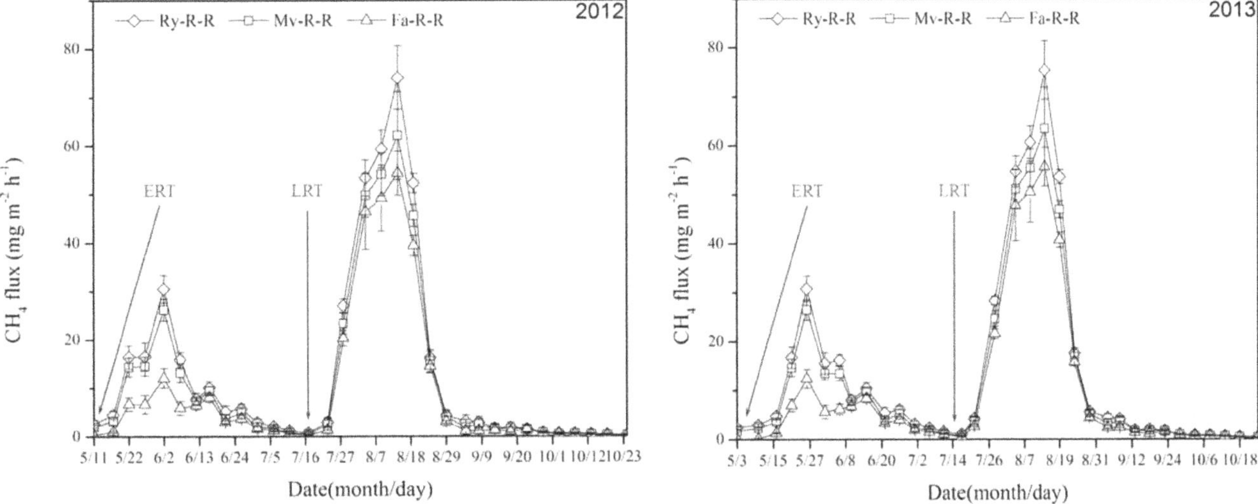

Figure 2. Effects of winter cover crops on CH_4 flux in early and late rice fields in 2012 and 2013. Ry-R-R: rice–rice–ryegrass cropping system; Mv-R-R: rice–rice–Chinese milk vetch cropping system; Fa-R-R: rice–rice cropping system with winter fallow. ERT: early rice transplanting; LRT: late rice transplanting. CH_4 emission rate is the mean of values measured within each treatment (n = 3).

high temperature. In the early rice season, the CH_4 flux values were significantly different among treatments with the order of Ry-R-R>Mv-R-R>Fa-R-R ($P<0.05$) (Fig. 2).

Methane emission in the late rice growing season mainly focused at tillering stage, and the peak value of CH_4 flux was observed at 23 and 24 d after transplanting in all treatments in 2012 and 2013, respectively. Then, the emission rate dramatically decreased to a low and stable level, especially from field drainage to harvest. The order of treatments in CH_4 emission was Ry-R-R>Mv-R-R>Fa-R-R (Fig. 2).

Characteristics of N_2O emission flux from early and late rice fields

The peak flux N_2O was emitted when the field was drained. Meanwhile, part of N_2O was emitted during wetting–drying alternation irrigation period. The first peak value of N_2O flux appeared at 7 and 15 d after transplanting in all treatments in 2012 and 2013, respectively, and then decreased. The order among treatments was Mv-R-R>Ry-R-R>Fa-R-R during the period from transplanting to field drainage, and Ry-R-R>Mv-R-R>Fa-R-R during wetting–drying alternation period. The N_2O flux in early rice paddy reached the highest peak at 32 and 35 d after transplanting in 2012 and 2013, respectively (Fig. 3).

In the late rice growing season, N_2O emission increased from field drainage to full heading stage, and mainly focused at booting stage. The order of N_2O emission fluxes among different treatments was Ry-R-R>Mv-R-R>Fa-R-R in the late rice growing season. In 2012, the average N_2O fluxes in the late rice growing season were 78.718 μg m^{-2} h^{-1} in Ry-R-R, 64.928 μg m^{-2} h^{-1} in Mv-R-R, and 32.275 μg m^{-2} h^{-1} in Fa-R-R. In 2013, the average N_2O fluxes in the late rice growing season were 81.453 μg m^{-2} h^{-1} in Ry-R-R, 67.662 μg m^{-2} h^{-1} in Mv-R-R, and 34.623 μg m^{-2} h^{-1} in Fa-R-R (Fig. 3).

Total CH_4 and N_2O emission from paddy fields in the growing durations of early and late rice

In the early rice growing season, the total CH_4 emissions of Ry-R-R and Mv-R-R were significantly higher than Fa-R-R ($P<0.05$), and the order of treatments was Ry-R-R>Mv-R-R>Fa-R-R (Table 2). The straws of winter cover crops incorporated into soil provided favorable soil condition and sufficient substance to be decomposed in the early rice season; therefore, the CH_4 emission quantities in straw returning treatments were higher than Fa-R-R ($P<0.05$). In 2012, the total CH_4 emissions from paddy fields during late rice entire growing season were 35.673 g m^{-2} in Ry-R-R, 31.542 g m^{-2} in Mv-R-R, 27.874 g m^{-2} in Fa-R-R. In 2013, the total CH_4 emissions from paddy fields during late rice whole growing season were 38.606 g m^{-2} in Ry-R-R, 34.358 g m^{-2} in Mv-R-R, 30.550 g m^{-2} in Fa-R-R. The order of treatments in total CH_4 emission was Ry-R-R>Mv-R-R>Fa-R-R (Table 2).

Compared to Fa-R-R, the other treatments increased total N_2O emissions in the early rice growing season, and the N_2O emissions increased by 0.05 g m^{-2} (131.58%) in Ry-R-R and 0.063 g m^{-2} (165.79%) in Mv-R-R in 2012, and by 0.058 g m^{-2} (138.1%) in Ry-R-R and 0.068 g m^{-2} (161.90%) in Mv-R-R in 2013, respectively. Similar results were observed in the late rice growing season in 2012, the total N_2O emissions increased by 0.104 g m^{-2} (144.44%) in Ry-R-R and 0.073 g m^{-2} (101.39%) in Mv-R-R. And the total N_2O emissions increased by 0.108 g m^{-2} (135.00%) in Ry-R-R and 0.076 g m^{-2} (95.00%) in Mv-R-R in 2013 (Table 2).

The emissions of CH_4 and N_2O were closely related to farming system, soil type, climate, and field management practices. Ry-R-R and Mv-R-R had larger total CH_4 emissions than Fa-R-R in the double rice growing season ($P<0.05$). Ry-R-R had the largest total N_2O emissions in the double rice growing season with the quantities of 0.264 g m^{-2} in 2012, and 0.288 g m^{-2} in 2013, respectively (Table 3).

Global warming potentials of CH_4 and N_2O

GWPs is an indicator to reflect the relative radioactive effect of a greenhouse gas, and the GWPs of CO_2 is defined as 1. In this study, the GWPs of CH_4 and N_2O from double cropping paddy fields varied with different winter cover crops, and the trend showed as Ry-R-R>Mv-R-R>Fa-R-R. In 2012, Ry-R-R had the largest GWPs (13281.79 kg CO_2–eq ha^{-1}) of total CH_4 and N_2O from double cropping paddy fields, followed by Mv-R-R (11657.44 kg CO_2–eq ha^{-1}), and Fa-R-R had the lowest GWPs

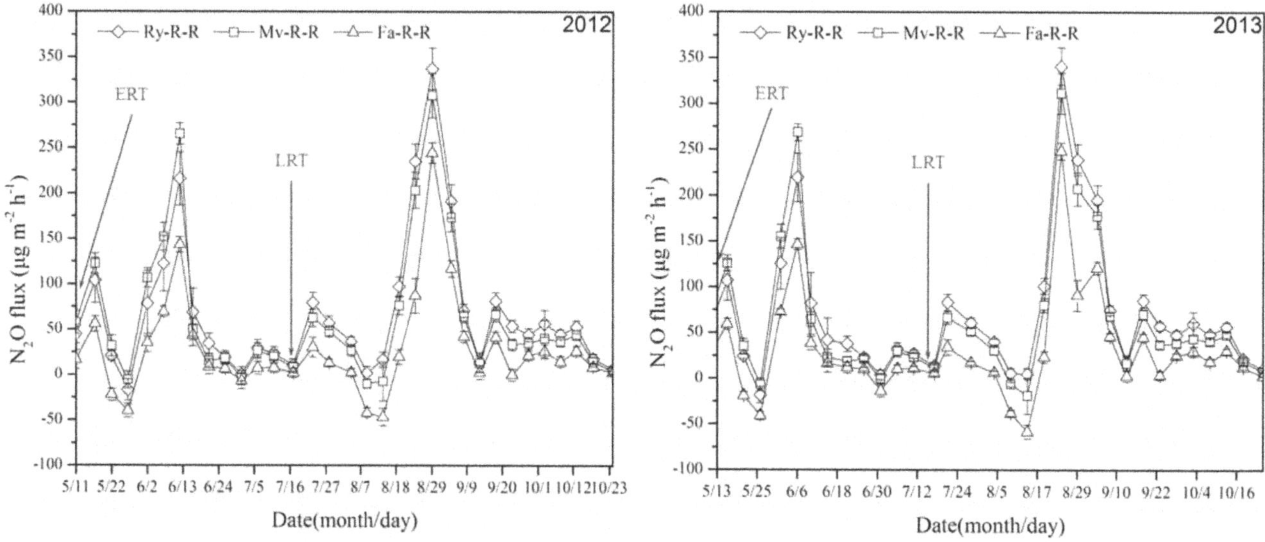

Figure 3. Effects of winter cover crops on N_2O flux in early and late rice fields in 2012 and 2013. Ry-R-R: rice–rice–ryegrass cropping system; Mv-R-R: rice–rice–Chinese milk vetch cropping system; Fa-R-R: rice–rice cropping system with winter fallow. ERT: early rice transplanting; LRT: late rice transplanting. N_2O emission rate is the mean of values measured within each treatment (n = 3).

of total CH_4 and N_2O (8993.12 kg CO_2–eq ha^{-1}). In 2013, Ry-R-R had the largest GWPs (14506.80 kg CO_2–eq ha^{-1}) of total CH_4 and N_2O from double cropping paddy fields, followed by Mv-R-R (12780.73 kg CO_2–eq ha^{-1}), and Fa-R-R had the lowest GWPs of total CH_4 and N_2O (9899.22 kg CO_2–eq ha^{-1}). According to GWPs, CH_4 from double cropping paddy fields had greater contribution to global warming than N_2O (Table 3).

Double rice grain yield of Mv-R-R was the highest, the lowest was Fa-R-R (Table 3). We also estimated per yield GWPs which was calculated as GWPs divided by rice grain yield. As is shown in Table 3, per yield GWPs of Ry-R-R was significantly higher than Mv-R-R and Fa-R-R ($P<0.05$), and the lowest was Fa-R-R.

Discussion

CH_4 emission

Methane emission is complex processes including production, oxidation, and emission. Chidthaisong et al. [37] reported that the highest CH_4 peaks were observed at flowering and heading stages, which could be related to the development of intense reducing conditions in the rice rhizosphere. In this study, we found that CH_4 emission was low in paddy fields after transplanting during early rice–growing season, and increased with the decomposition of organic matters and growth of rice. In addition, CH_4 emission was influenced by soil temperature and soil redox potential (Eh). Yu et al. [38] reported that CH_4 emission showed an exponential decrease by an Eh increase. In this study, the CH_4 flux and total CH_4 emission from paddy fields during the early and late rice growing season were much larger in Ry-R-R and Mv-R-R compared to Fa-R-R, which was similar to the result by Lee et al. [22]. The reasons for above result may be: first, microbial activities were improved after returning straws of winter cover crops into the soil due to the supplements of carbon source and energy for microbial activities to accelerate consumption of soil oxygen and decrease of soil Eh; second, methanogens became active due to the large quantities of C source, which provided reactive substrate for CH_4 emission from paddy fields. In the early rice growing season, the order of CH_4 flux and total CH_4 emission from paddy fields varied among treatments, which were highly related to the returning straw type, and straw decomposition rate. During the late rice growing season, the CH_4 emission increased gradually with the decomposition of organic matters and growth of rice after transplanting, and reach the peak value at tillering stage in all treatments. However, CH_4 emissions in both rice seasons were reduced in a large extent after field drying, because (1) soil aeration was improved during this period, and the activities of methanogens were restricted; and (2) the physiological activity of rice plant decreased, thereby limiting the ability for transportation and emission of CH_4 [39].

Although straw returning helps to maintain soil fertility and protect environment, but it enhances CH_4 emission simultaneously. Pandey et al. [12] showed that CH_4 emission was positively related to straw returning amount under permanent flooding condition, whereas N_2O emission had a reverse relationship with the amount of straw returning. In this study, we found that CH_4 flux in the late rice growing season was much higher than that in the early rice growing season, and peak appeared earlier. As straws of early rice (4500 kg ha^{-1}) returned to field before transplanting of late rice, the paddy soil of late rice was under anoxic condition after transplanting, which was favorable for CH_4 production and emission. Temperature was the major reason for the differences in the CH_4 emission pattern between the early and the late rice season. Soil temperature had a predictive functional relationship with CH_4 emission. Zhang et al. [40] reported that there was a strong positive correlation between CH_4 emission and soil temperature. In this experimental area, the late rice season was the hottest time in summer (Fig. 1). Therefore, high temperatures enhanced the decomposition of crop straws in the moist environment. In contrast to the warm temperatures of the late rice season, the air temperatures of the early rice season were lower, which resulted in slower crop straws decomposition and little CH_4–substrate. Hence, these differences in weather factors (e.g., temperature) resulted in the different characteristics of CH_4 between the early and the late rice seasons. However, there were significantly differences among treatments although they had similar trends. This indicated that CH_4 flux and emission from paddy fields were affected by different winter cover crops.

Table 2. Effects of winter cover crops on CH_4 and N_2O emission from rice fields during whole growing season of early and late rice (g m^{-2}).

Year	Treatment	CH_4			N_2O		
		Early rice	Late rice	Total	Early rice	Late rice	Total
2012	Ry-R-R	14.235±0.411a	35.673±1.030a	49.908±1.441a	0.088±0.003a	0.176±0.05a	0.264±0.008a
	Mv-R-R	12.092±0.349b	31.542±0.912b	43.634±1.260b	0.101±0.003b	0.145±0.04b	0.246±0.007a
	Fa-R-R	6.732±0.194c	27.874±0.805c	34.606±0.999c	0.038±0.001c	0.072±0.02c	0.110±0.003b
2013	Ry-R-R	15.906±0.459a	38.606±1.115a	54.512±1.574a	0.100±0.003b	0.188±0.006a	0.288±0.008a
	Mv-R-R	13.523±0.390b	34.358±0.992b	47.882±1.382b	0.110±0.003a	0.156±0.005b	0.266±0.008a
	Fa-R-R	7.535±0.218c	30.550±0.882c	38.085±1.099c	0.042±0.002c	0.080±0.002c	0.122±0.004b

Ry-R-R: rice–rice–ryegrass cropping system; Mv-R-R: rice–rice–Chinese milk vetch cropping system; Fa-R-R: rice–rice cropping system with winter fallow.
Values are presented as mean ± SE (n = 3). Means in each column with different letters are significantly different at the $P < 0.05$ level.

Table 3. Double rice grain yield, global warming potentials (GWPs) of CH_4 and N_2O and per yield GWPs from rice fields under different cropping patterns.

Year	Treatment	CH_4 emission	N_2O emission	GWPs of CH_4	GWPs of N_2O	GWPs of CH_4 and N_2O	Double rice grain yield	Per yield GWPs CO_2
		(g m^{-2})	(g m^{-2})	(kg CO_2-eq ha^{-1})	(kg CO_2-eq ha^{-1})	(kg CO_2-eq ha^{-1})	(kg ha^{-1})	(kg kg^{-1})
2012	Ry-R-R	49.908±1.441a	0.264±0.008a	12494.38±360.68a	787.41±22.73a	13281.79±383.41a	13800.23±398.38a	0.96±0.03a
	Mv-R-R	43.634±1.260b	0.246±0.007a	10923.85±315.34b	733.58±21.18a	11657.44±336.52b	15089.30±435.59a	0.77±0.02b
	Fa-R-R	34.606±0.999c	0.110±0.003b	8663.66±250.10c	329.46±9.51b	8993.12±259.61c	14359.00±414.51a	0.63±0.02c
2013	Ry-R-R	54.512±1.574a	0.288±0.008a	13646.99±393.95a	859.81±24.82a	14506.80±418.76a	14738.87±425.47a	0.98±0.03a
	Mv-R-R	47.882±1.382b	0.266±0.008a	11987.20±346.04b	793.53±22.91a	12780.73±368.95b	14896.57±430.03a	0.86±0.02b
	Fa-R-R	38.085±1.099c	0.122±0.004b	9534.57±275.24c	364.64±10.53b	9899.22±285.77c	13625.16±322.60a	0.73±0.02c

Ry-R-R: rice–rice–ryegrass cropping system; Mv-R-R: rice–rice–Chinese milk vetch cropping system; Fa-R-R: rice–rice cropping system with winter fallow.
Values are presented as mean ± SE (n = 3). Means in each column with different letters are significantly different at the $P<0.05$ level.

N_2O emission

The emissions of N_2O are closely related to soil moisture, oxygen, temperature, content of soil organic matter and pH [4,11,17]. Great positive interaction has been reported between N_2O emission and green manure or chemical nitrogen fertilizer in early rice growing season [41]. In this study, we found that N_2O emission in the early rice growing season focused in the period of field drainage, and the Ry-R-R and Mv-R-R with winter cover crops had more N_2O emissions than Fa-R-R in both rice growing seasons (Fig. 3). N_2O emission from paddy field is promoted with the amount of straw returning via increasing soil denitrification, which provides the soil microbial substrates and energy for soil nitrification and denitrification process [42]. Different ranking of treatments in N_2O flux and total N_2O emission might be related to the decomposition rates of winter crop species during the rice growing season. In the late rice growing season, the total N_2O emissions of treatments Ry-R-R and Mv-R-R were significantly higher than Fa-R-R ($P<0.05$). This possibly results from that soil nitrification and denitrification process has been facilitated after the early rice straw returning through carbon and energy resource regulation (Table 1); a small amount of winter crop straw remains in the soil until the growing season of late rice; and tillage practice before late rice transplanting helps the incorporation of straws into soil, which may improve the soil nitrification and denitrification process.

Global warming potentials of CH_4 and N_2O

Global warming potential can be used as an index to estimate the potential effects of different greenhouse gases on the global climate system. Bhatia et al. [5] estimated that GWPs of rice–wheat system increased by 28% on full substitution of organic N by chemical N. Zhu et al. [43] reported that the highest GWPs was found in Chinese milk vetch incorporation in double cropping rice system, which was 21–325% higher than the other three treatments. In this study, the GWPs of CH_4, N_2O or both had different orders. For a comprehensive consideration, GWPs of both CH_4 and N_2O is more important to assess the effect of a farming system on climate warming. Therefore, it is necessary to make a combined estimate of global warming effects of CH_4 and N_2O emitted from each treatment. Thus, we introduced the GWPs and per yield GWPs into this study for global warming calculations. Although the global warming effect of N_2O is 12 times as large as that of CH_4, CH_4 emissions were nearly 370 times that of N_2O, resulting in the majority of GWPs originating from CH_4 (Table 3). Therefore, it is certain that the GWPs and per yield GWPs values for Ry-R-R and Mv-R-R were larger than Fa-R-R ($P<0.05$), due to their greater CH_4 emissions. But the GWPs of CH_4 and N_2O and per yield GWPs of Mv-R-R was significantly lower than Ry-R-R ($P<0.05$). It should be mentioned that, the cultivation of ryegrass, Chinese milk vetch and its incorporation is a process involving C accumulation from the atmosphere to the soil, while the production of synthetic nitrogen fertilizer consumes fossil fuels that release C and contribute to greenhouse gas emissions. Therefore, we recommend Mv-R-R pattern in double cropping rice areas in the Middle and Lower reaches of Yangtze River in China, which correspond to Chinese milk vetch as winter cover crop + double rice.

Conclusions

The emissions of CH_4 and N_2O from double cropping paddy fields were significantly enhanced by returning different winter cover crops. The effects on CH_4 and N_2O fluxes and emissions were different among treatments, and the emission characteristics varied greatly between early and late rice growing season. The orders of treatments were Ry-R-R>Mv-R-R>Fa-R-R for total emissions of CH_4 and N_2O during double rice seasons, and Ry-R-R>Mv-R-R>Fa-R-R for GWPs of total CH_4 and N_2O from double cropping paddy fields. Compared with Ry-R-R, Mv-R-R and Fa-R-R reduced CH_4 emission during rice growing seasons. The GWPs (based on CH_4 emission) under Mv-R-R and Fa-R-R was significantly ($P<0.05$) lower than Ry-R-R. Although the cumulative N_2O emission under Ry-R-R and Mv-R-R were higher than that from Fa-R-R ($P<0.05$), GWPs of N_2O was relatively low compared to that of CH_4. The GWPs (based on CH_4 and N_2O) of Mv-R-R and Fa-R-R is lower than that of Ry-R-R ($P<0.05$). Meanwhile, the GWPs of CH_4 and N_2O and per yield GWPs of Mv-R-R was significantly lower than Ry-R-R ($P<0.05$). Thus, Mv-R-R is beneficial in GHG mitigation and it can be extended as an excellent cropping pattern in double rice cropped regions.

Author Contributions

Conceived and designed the experiments: XPX GLY. Performed the experiments: HMT. Analyzed the data: HMT WGT. Contributed reagents/materials/analysis tools: JMS KW WYL. Wrote the paper: HMT.

References

1. Levy PE, Mobbs DC, Jones SK, Milne R, Campbell C, et al. (2007) Simulation of fluxes of greenhouse gases from European grasslands using the DNDC model. Agric Ecosyst Environ 121: 186–192.
2. Saggar S, Hedley CB, Giltrap DL, Lambie SM (2007) Measured and modeled estimates of nitrous oxide emission and methane consumption from a sheep-grazed pasture. Agric Ecosyst Environ 122: 357–365.
3. Hernandez-Ramirez G, Brouder SM, Smith DR, Van Scoyoc GE (2009) Greenhouse gas fluxes in an eastern corn belt soil: Weather, nitrogen source, and rotation. J Environ Qual 38: 841–854.
4. Wassmann R, Neue HU, Ladha JK, Aulakh MS (2004) Mitigating greenhouse gas emissions from rice-wheat cropping systems in Asia. Environ Devel Sustain 6: 65–90.
5. Bhatia A, Pathak H, Jain N, Singh PK, Singh AK (2005) Global warming potential of manure amended soils under rice-wheat system in the Indo-Gangetic plains. Atmospheric Environ, 39(37): 6976–6984.
6. Verge XPC, Kimp CD, Desjardins RL (2007) Agricultural production, greenhouse gas emissions and mitigation potential. Agric Forest Meteorol 142: 255–269.
7. Lokupitiya E, Paustian K (2006) Agricultural soil greenhouse gas emissions: A review of national inventory methods. J Environ Qual 35: 1413–1427.
8. Verma A, Tyagi L, Yadav S, Singh SN (2006) Temporal changes in N_2O efflux from cropped and fallow agricultural fields. Agric Ecosyst Environ 116: 209–215.
9. Liu H, Zhao P, Lu P, Wang YS, Lin YB, et al. (2008) Greenhouse gas fluxes from soils of different land-use types in a hilly area of South China. Agric Ecosyst Environ 124: 125–135.
10. Tan Z, Liu S, Tieszen LL, Tachie-Obeng E (2009) Simulated dynamics of carbon stocks driven by changes in land use, management and climate in a tropical moist ecosystem of Ghana. Agric Ecosyst Environ 130: 171–176.
11. Kallenbach CM, Rolston DE, Horwath WR (2010) Cover cropping affects soil N_2O and CO_2 emissions differently depending on type of irrigation. Agric Ecosyst Environ 137: 251–260.
12. Pandey D, Agrawal M, Bohra JS (2012) Greenhouse gas emissions from rice crop with different tillage permutations in rice-wheat system. Agric Ecosyst Environ 159: 133–144.
13. Yagi K, Minami K (1990) Effect of organic matter application on methane emission from some Japanese paddy fields. Soil Sci Plant Nutr 36: 599–610.
14. Yagi K, Tsuruta H, Kanda KI, Minami K (1996) Effect of water management on methane emission from a Japanese rice paddy field: Automated methane monitoring. Global Biogeochem Cycles 10: 255–267.
15. Nishimura S, Sawamoto T, Akiyama H, Sudo S, Yagi K (2004) Methane and nitrous oxide emissions from a paddy field with Japanese conventional water

management and fertilizer application. Global Biogeochem Cycles 18, GB2017, doi:10.1029/2003GB002207
16. Al-Kaisi MM, Yin X (2005) Tillage and crop residue effects on soil carbon and carbon dioxide emission in corn-soybean rotations. J Environ Qual 34: 437–445.
17. Yao Z, Zheng X, Xie B, Mei B, Wang R, et al. (2009) Tillage and crop residue management significantly affects N-trace gas emissions during the non-rice season of a subtropical rice-wheat rotation. Soil Biol Biochem 41: 2131–2140.
18. Groot TT, VanBodegom PM, Harren FJM, Meijer HAJ (2003) Quantification of methane oxidation in the rice rhizosphere using ^{13}C-labelled methane. Biogeochemistry 64: 355–372.
19. Yue J, Shi Y, Liang W, Wu J, Wang CR, et al. (2005) Methane and nitrous oxide emissions from rice field and related microorganism in black soil, northeast China. Nutr Cy Agroecosyst 73: 293–301.
20. Sethunathan N, Kumaraswamy S, Rath AK, Ramakrishnan B, Satpathy SN, et al. (2000) Methane production, oxidation, and emission from Indian rice soils. Nutr Cy Agroecosyst 58: 377–388.
21. Ma J, Xu H, Yagi K, Cai ZC (2008) Methane emission from paddy soils as affected by wheat straw returning mode. Plant Soil 313: 167–174.
22. Lee CH, Park KD, Jung KY, Ali MA, Lee D, et al. (2010) Effect of Chinese milk vetch (*Astragalus sinicus* L.) as a green manure on rice productivity and methane emission in paddy soil. Agric Ecosyst Environ 138: 343–347.
23. Bouwman AF (1998) Nitrous oxides and tropical agriculture. Nature 392: 866–867.
24. Cai ZC, Lanughlin RJ, Stevens RJ (2001) Nitrous oxide and dinitrogen emissions from soil under different water regimes and straw amendment. Chemosphere 42: 113–121.
25. Rittera WF, Scarborough RW, Chirnside AEM (1998) Winter cover crops as a best management practice for reducing nitrogen leaching. J Contam Hydrol 34: 1–15.
26. Hermawan B, Bomke AA (1997) Effects of winter cover crops and successive spring tillage on soil aggregation. Soil Tillage Res 44: 109–120.
27. McCracken DV, Smith MS, Grove JH, MacKown CT, Blevins RL (1994) Nitrate leaching as influenced by cover cropping and nitrogen source. Soil Sci Soc Am J 58: 1476–1483.
28. Barnes JP, Putnam AR (1983) Rye residues contribute weed suppression in no-tillage cropping systems. J Chem Ecol 9: 1045–1057.
29. Powlson DS, Prookes PC, Christensen BT (1987) Measurement of soil microbial biomass provides an early indication of changes in total soil organic matter due to straw incorporation. Soil Biol Biochem 19(2): 159–164.
30. Mitchell JP, Shennan C, Singer MJ, Peters DW, Miller RO, et al. (2000) Impacts of gypsum and winter cover crops on soil physical properties and crop productivity when irrigated with saline water. Agr Water Manag 45: 55–71.
31. Chang HL, Ki DP, Ki YJ, Muhammad AA, Dokyoung L, et al. (2010) Effect of Chinese milk vetch (*Astragalus sinicus* L.) as a green manure on rice productivity and methane emission in paddy soil. Agr Ecosyst Environ 138: 343–347.
32. Salmeróna M, Isla R, Cavero J (2011) Effect of winter cover crop species and planting methods on maize yield and N availability under irrigated Mediterranean conditions. Field Crops Res 123: 89–99.
33. Zheng X, Wang M, Wang Y, Shen R, Li J, et al. (1998) Comparison of manual and automatic methods for measurement of methane emission from rice paddy fields. Adv Atmos Sci 15: 569–579.
34. Parashar DC, Gupta PK, Rai J, Sharma RC, Singh N (1993) Effect of soil temperature on methane emission from paddy field. Chemosphere 26: 247–250.
35. Singh JS, Singh S, Raghubanshi AS, Saranath S, Kashyap AK (1996) Methane flux from rice/wheat agroecosystem as affected by crop phenology, fertilization and water lever. Plant Soil 183: 323–327.
36. SAS Institute (2003) SAS Version 9.1.2 2002–2003. SAS Institute Inc., Cary, NC.
37. Chidthaisong A, Obata H, Watanabe I (1999) Methane formation and substrate utilization in anaerobic rice soils as affected by fertilization. Soil Biol Biochem 31: 135–143.
38. Yu K, Bohme F, Rinklebe J, Neue HU, DeLaune RD (2007) Major biogeochemical processes in soils-A microcosm incubation from reducing to oxidizing conditions. Soil Sci Soc Am J 71: 1406–1417.
39. Yang X, Shang Q, Wu P, Liu J, Shen Q, et al. (2010) Methane emissions from double rice agriculture under long-term fertilizing systems in Hunan, China. Agric Ecosyst Environ 137: 308–316.
40. Zhang HL, Bai XL, Xue JF, Chen ZD, Tang HM, et al. (2013) Emissions of CH_4 and N_2O under different tillage systems from double-cropped paddy fields in Southern China. PLoS ONE 8(6): e65277. doi:10.1371/journal.pone.0065277.
41. Petersen SO, Mutegi JK, Hansen EM, Munkholm LJ (2011) Tillage effects on N_2O emissions as influenced by a winter cover crop. Soil Biol Biochem 43: 1509–1517.
42. Huang Y, Zou JW, Zheng XH, Wang YS, Xu XK (2004) Nitrous oxide emissions as influenced by amendment of plant residues with different C: N ratios. Soil Biol Biochem 36: 973–981.
43. Zhu B, Yi LX, Hu YG, Zeng ZH, Tang HM, et al. (2012) Effects of Chinese Milk Vetch (*Astragalus sinicus* L.) residue incorporation on CH_4 and N_2O emission from a double-rice paddy soil. J Integrative Agric 11(9): 1537–1544.

20

Biochar-Induced Changes in Soil Hydraulic Conductivity and Dissolved Nutrient Fluxes Constrained by Laboratory Experiments

Rebecca T. Barnes*¤a, Morgan E. Gallagher¤b, Caroline A. Masiello, Zuolin Liu, Brandon Dugan

Department of Earth Science, Rice University, Houston, Texas, United States of America

Abstract

The addition of charcoal (or biochar) to soil has significant carbon sequestration and agronomic potential, making it important to determine how this potentially large anthropogenic carbon influx will alter ecosystem functions. We used column experiments to quantify how hydrologic and nutrient-retention characteristics of three soil materials differed with biochar amendment. We compared three homogeneous soil materials (sand, organic-rich topsoil, and clay-rich Hapludert) to provide a basic understanding of biochar-soil-water interactions. On average, biochar amendment decreased saturated hydraulic conductivity (K) by 92% in sand and 67% in organic soil, but increased K by 328% in clay-rich soil. The change in K for sand was not predicted by the accompanying physical changes to the soil mixture; the sand-biochar mixture was less dense and more porous than sand without biochar. We propose two hydrologic pathways that are potential drivers for this behavior: one through the interstitial biochar-sand space and a second through pores within the biochar grains themselves. This second pathway adds to the porosity of the soil mixture; however, it likely does not add to the effective soil K due to its tortuosity and smaller pore size. Therefore, the addition of biochar can increase or decrease soil drainage, and suggests that any potential improvement of water delivery to plants is dependent on soil type, biochar amendment rate, and biochar properties. Changes in dissolved carbon (C) and nitrogen (N) fluxes also differed; with biochar increasing the C flux from organic-poor sand, decreasing it from organic-rich soils, and retaining small amounts of soil-derived N. The aromaticity of C lost from sand and clay increased, suggesting lost C was biochar-derived; though the loss accounts for only 0.05% of added biochar-C. Thus, the direction and magnitude of hydraulic, C, and N changes associated with biochar amendments are soil type (composition and particle size) dependent.

Editor: Zhi Zhou, Purdue University, United States of America

Funding: This work was supported by National Science Foundation (NSF-EAR-0911685) to CAM & BD (www.nsf.gov), US Department of Agriculture (2011-67009-20074) to CAM (www.usda.gov), Rice University's Shell Center for Sustainability to CAM (shellcenter.rice.edu). The funders had no role in study design, data collection and analysis, decision to publish, or preparation of the manuscript.

Competing Interests: The authors have declared that no competing interests exist.

* Email: becca.barnes@coloradocollege.edu

¤a Current address: Environmental Program, Colorado College, Colorado Springs, Colorado, United States of America
¤b Current address: Center on Global Change, Duke University, Durham, North Carolina, United States of America

Introduction

Woolf et al. [1] estimate that 1.8 Pg CO_2-carbon equivalents can be sequestered each year through the sustainable production and application of 0.9 Pg of biochar to agricultural land which sequesters carbon (C), reduces CH_4 and N_2O emissions, and results in avoided CO_2 emissions. Biochar, charcoal intentionally produced by humans through pyrolysis for soil amendment, is a type of black carbon, like soot or charcoal [2,3]. Adding 0.9 Pg of biochar to the agricultural landscape would correspond to a 4–20 fold increase in global black carbon production (0.04 to 0.194 Pg yr−1; [4]). Recent work illustrates the likely mobility of biochar added to soil: charcoal in soils can be released into rivers [5,6], where, given its aromatic structure, it can be photo-oxidized [7] and/or transported downstream where it has the potential to alter ecosystem processes [6]. Given the potential magnitude of ecosystem perturbations from full-scale implementation of bio-char-C sequestration, it is critical to determine the effects of biochar soil amendment on water and biogeochemical cycling.

Biochar soil amendment can increase crop productivity [3,8], potentially by improving the hydrologic properties of the soils [9]. Biochar can also increase soil water-holding capacity, and available water content [8,10–13], plant available water [14–16], alter soil hydrophobicity [17], and change soil hydraulic conductivity [18–22]. Biochar is predicted to cause sandy soils to drain more slowly [23] and clay-rich soils to drain more rapidly [24]. However, past results have not been consistent, likely due to confounding factors such as biochar characteristics (i.e. feedstock and pyrolysis temperatures), application rates, and soil characteristics. Given the importance of hydraulic conductivity in determining the partitioning of precipitation between infiltration and overland flow (i.e. infiltration rates; [25]), which impacts water storage in the subsurface and thus plant available water, it is necessary to understand the effects of biochar on the hydraulic properties of different soil types.

The C and nitrogen (N) content of biochar varies with feedstock and production conditions [26]. These conditions and the C:N ratio of biochar influence its stability [27,28] as well as possible soil C and N losses [29]. While dependent on production conditions, biochar tends to have a high cation exchange capacity [30] and anion sorption ability [31], allowing for adsorption of dissolved organic matter (DOM) [32] and N [33], and can alter greenhouse gas emissions [34–36]. While biochar amendment adds C and N to soils (which may be available for leaching), it is also able to sequester additional C and nutrients in the soil due to its sorptive properties. Therefore the addition of biochar to soil could result in a net increase or decrease in dissolved C and N losses [37]. Worldwide analysis of dissolved black carbon (DBC), which includes derivatives of charcoal and biochar, exported from rivers indicates that, on average, DBC contributes 10% of the global total dissolved organic carbon (DOC) flux [6]. In addition, analysis of the bioavailability of biochar extracts in natural stream water suggests that some biochar-C molecules have turnover rates on the order of days to a month [38], indicating that at least a portion of biochar-C is not recalcitrant. In addition to the dissolution of biochar within soils, particles are also transported. Recent studies illustrate that the movement of biochar particles is related to particle size and surface chemistry, as well as pore water salt content and pH [39,40]. Collectively this research points to the uncertainty in the fate of biochar and biochar-C and their down-gradient effects on aquatic ecosystems.

Using column experiments, we quantify the effects of a 10% (by mass) biochar amendment on the saturated hydraulic conductivity (K) of sand, clay-rich, and organic-rich soil materials as well as report the effects of this soil amendment on C and N leaching. Using simple, homogeneous soil materials with different grain sizes and surface chemistry allowed us to examine biochar-soil interactions and to compare our results to established soil hydrology models. These data begin to address an important knowledge gap by providing new quantitative constraints on how biochar amendments change K and the chemistry of soil leachate; this work points to the need for more mechanistic studies to examine biochar-soil-water interactions.

Materials and Methods

Soil and Biochar

Three soil materials were used to gauge soil property responses to biochar amendment. Sand (Pavestone Natural Play Sand) and organic-rich soil (Micro-Gro Organic Rich Garden Soil, with no N fertilizer) were purchased from Home Depot. The organic-rich soil was texturally similar to a sandy loam. A clay-loam Hapludert, characterized by its poor drainage [41], was collected from Rice University's campus. All materials were oven dried at 60°C to remove any moisture prior to dry sieving. Dry materials were mixed, and then oven dried at 100°C for 24 hours to create homogenous mixtures with initial water content of zero; 100°C facilitates water loss but minimizes chemical impacts as it is significantly lower than our pyrolysis temperature [42]. We determined the grain size of the three soils and the biochar using seven sieves (38 μm to 500 μm).

We produced biochar from mesquite wood (*Prosopis* sp.), ground to smaller than 20 mesh (850 μm). Batches of mesquite (70–80 g) were pyrolyzed using the reactor design described in Kinney et al. [17], by heating in a muffle furnace at 6°C min^{-1} and holding at 400°C for 4 hours. On average these pyrolysis conditions provided a biochar mass yield of 40.4%. The produced biochar had a pH of 6.5±0.1, ash content of 3.33%±0.04%, and liming equivalent of 4%, determined using protocols outlined by the International Biochar Initiative [43–45].

Column Experiments

To test the response of K and dissolved nutrient fluxes to biochar amendment we conducted falling head experiments [46] across six materials: sand, sand+biochar, organic soil, organic soil+biochar, clay, and clay+biochar. We packed 150 g of each mixture into three replicate columns, 50 g per column, with 54 μm polyester mesh screen (Small Parts Inc.) at the bottom. Materials were packed with a consistent force into columns in four equal increments to achieve uniform bulk density [47] and the initial soil length was recorded. Bulk density for soil materials and soil+biochar mixtures was determined using the dry mass and column dimensions (height of soil materials, diameter of column) at the start of the experiment. Grain size distributions of soil+biochar mixtures were estimated using the proportional masses of each material (i.e. 10% biochar, 90% soil material) and appropriate grain size data (Table 1).

Biochar constituted 10% of the total mass in the columns that contained biochar. This represents a 133 tons ha^{-1} (95 Mg C ha^{-1}) application rate with a 10 cm tillage depth. We chose a high amendment rate to ensure that we altered the soil-water system in a way that would allow us to detect any effects across the three contrasting soil materials. Specifically, we were concerned about the potential of soil biochar amendments increasing C exports to surface waters, as well as any unforeseen consequences on soil hydrologic properties. This amendment rate is above what is likely to be added to an agricultural field, though within the range reported for positive or neutral productivity effects (up to 140 Mg C ha−1; [3]), similar to the application made by Chan et al. [48], and well below what was shown to be an upper limit to biochar-induced benefits to plants (200 tons C ha−1; [15]).

To saturate the soils, we capped the column bottoms, added 150 mL of 18 MΩ-cm MilliQ water, and allowed the columns to sit for 48 h before drainage. Six consecutive falling head experiments [i.e. flushing events; [46] were conducted on saturated soils using 150 mL of MilliQ water, with leachate collected at the end of each experiment for all columns. The same experimental set up was used for all six materials to allow for inter-comparison. The leachate was weighed, filtered through a pre-combusted glass fiber filter (Whatman GF/F) and kept at 4°C until analysis. This process was repeated without allowing columns to dry between flushing events. Evaporation was monitored daily (via the net change in the mass of water in a beaker) and water throughput (leachate volume) was corrected for evaporation by adding the product of the daily evaporative loss (mg hr^{-1}) and duration of flushing event (hrs).

Saturated hydraulic conductivity (K) from falling head data was calculated using equation 1 [49]:

$$K = (L/\Delta t) * ln(h_2/h_1) \tag{1}$$

where L is soil sample length (m), Δt is time elapsed (s), and h_1 and h_2 are the initial and final water heights (m), respectively (data available in Table S1). Separate experiments on sand-only systems confirm consistent measures of K from top-saturated and bottom-saturated falling head and constant head experiments, suggesting full saturation and steady-state conditions were achieved using this falling-head technique.

Table 1. Physical and elemental properties of soil materials and biochar.

Material	Grain Size			Bulk Density	%C	%N
	d_{10} (μm)	d_{50} (μm)	d_{90} (μm)	ρ_d (g cm^{-3})		
sand	70	160	380	1.68±0.18	0.4	0.01
organic-rich	95	400	480	0.43±0.002	37.9	0.54
clay-rich	45	115	460	1.72±0.04	0.9	0.03
biochar (mesquite)	75	320	470	0.36±0.03	71.6	0.84

Soil and sample analysis

We used a Shimadzu TOC-VCN to determine the DOC and total dissolved nitrogen (TDN) concentrations of the filtered leachate. Sample replicates indicate a 0.08 mg L^{-1} and 0.04 mg L^{-1} precision for DOC and TDN, respectively. Mass loss of C and N (dissolved flux) were determined by multiplying the DOC and TDN concentrations by the water volume of the sample, respectively. In contrast, water throughput was calculated using the evaporation-corrected water volumes. The aromaticity of the dissolved C was determined by calculating the specific UV absorbance at 254 nm (SUVA254; [50]). The UV absorption (m^{-1}) of filtered leachate was measured on a Cary UV-Vis spectrophotometer and divided by the DOC concentration (mg C L^{-1}) following the protocol developed by Weishaar et al. [50] to calculate $SUVA_{254}$ (L mg C^{-1} m^{-1}).

After six flushing events, soil mixtures were removed from the columns, weighed, dried at 100°C for at least 24 hours, and reweighed to determine water content at field capacity by mass. We measured C and N content on original soils, biochar, and dried, post-experiment soils using a Costech 4010 CHNS/O Elemental Analyzer. Replicate analysis shows a precision of 0.6% and 0.02% for C and N measurements, respectively.

Statistical analyses

We used two-sample t-tests to determine statistical differences in the soil and leachate characteristics between treatments and controls. Paired t-tests were used to determine the statistical differences of *K* over the course of the experiment. Finally, a general linear model with flush number as a covariate, was used to determine if *K* changed significantly over the course of the experiment as a result of the biochar amendment. All statistical tests were done in the RStudio environment (v0.98.507, 2014 RStudio, Inc.) and results were considered significant when $p < 0.05$.

Results and Discussion

Soil physical characteristics

The addition of biochar to the soil materials changed a number of physical properties, e.g. grain size distribution and bulk density, which likely affected water movement. The grain size distribution of biochar differed from that of the three soils and was most similar to the organic soil (Table 1). When biochar was added to sand and clay soils, the d_{50} of the mixtures increased; however, when biochar was added to the organic soil the d_{50} decreased. Given the similarity in d_{10} of sand and biochar, the addition of biochar did not appreciably change the proportion of fines in the sand versus the sand+biochar mixture. The addition of 10% biochar changed soil bulk density (ρ_b) (Table 2), though these changes were not always significant; the addition of biochar decreased the ρ_b of sand and clay by 17% ($p = 0.056$) and 20% ($p = 0.052$), respectively. Biochar addition to clay lowered ρ_b enough to bring it within the range recommended by the National Soil Conservation Service to allow for adequate root growth (<1.47 g cm^{-3}, USDA, 2008). In contrast, when biochar was added to organic soil the mixture ρ_b increased 10% ($p = 0.018$), despite the biochar having a lower ρ_b (organic: 0.43±0.002 g cm^{-3}; biochar: 0.36±0.03 g cm^{-3}; Table 1); this is likely related to the smaller relative grain size of biochar and grain arrangement during packing of the columns (Table 1).

Soil hydraulic characteristics

Saturated hydraulic conductivity (*K*) describes the ease of fluid flow through saturated porous media and it can be directly measured with flow through experiments [46,51] or estimated using theoretical or empirical models. Theoretical models, such as the Kozeny equation [52,53] require significant knowledge of porosity, tortuosity, pore shape, grain density, and specific surface area of solid grains. Because of the difficulty constraining all these parameters, others have developed empirical models that relate porosity and *K* [54] or grain size and *K* [55]. Here we compare our *K* results with these empirical relationships using our bulk density, grain size, and porosity data. Changes in *K* accompanying biochar amendment for both clay- and organic-rich soils follow the empirical relationships discussed above; specifically the change in *K* is inversely related to the change in bulk density (or positively related to the change in porosity) caused by the biochar amendment. The addition of biochar to clay-rich soil resulted in a ρ_b decrease of 20% (a porosity increase), an estimated d_{50} increase of 18%, and an increase in *K* of over 300% (3.26×10^{-8} m s^{-1} to 1.16×10^{-7} m s^{-1}; Figure 1c, Table 2). In contrast, when biochar was added to organic-rich soil, the mixture was 10% denser (lower porosity) with a 2% smaller d_{50}, and *K* decreased by 67% (2.23×10^{-6} m s^{-1} to 7.79×10^{-7} m s^{-1}; Figure 1b, Table 2). Similarly, experiments using silt loams reported an increase in *K* with biochar amendment, though there were not always corresponding decreases in ρ_b (Table 3, [20]). Previous work in organic-rich soils has not documented changes in *K* in response to biochar amendment (0.5 to 2% by mass) (Table 3; [56]). Differences in results are attributable to different amendment rates, biochar grain size, soil properties, and/or threshold effects of amendment rate or grain size.

Several studies have documented threshold effects on porosity and permeability when fine-grained particles are added to a soil. Boadu [57] found that porosity decreased with increasing fines up to a threshold of ~5% (by mass) at which point fine grain additions increased the porosity. Similarly, Crawford et al. [58] noted that porosity is lowest when the fine particle volume equals the pore space of the coarse grains; however, if fine grains are removed or added, the porosity increases.

Table 2. The mean (and standard deviation) of physical, hydraulic, and nutrient properties of the three replicates of each soil and soil+biochar treatment.

Soil	Hydraulic conductivity (K) (m s^{-1})		Bulk Density (ρ_d) (g cm^{-3})		Water Content at Field Capacity (fraction water)		Cumulative DOC loss (mg)		Cumulative TDN loss (mg)		SUVA$_{254}$ (L mg C^{-1} m^{-1})	
	Soil	+biochar	Soil	+biochar	Soil	+biochar	Soil	+biochar	Soil	+biochar	Soil	+biochar
sand	2.9×10^{-6} (6.3×10^{-7})	2.3×10^{-7} (5.9×10^{-8})	1.69 (0.18)	1.39 (0.06)	0.15 (0.02)	0.30 (0.03)	1.71 (0.08)	3.29 (0.35)	0.25 (0.04)	0.19 (0.01)	1.86 (0.83)	2.75 (0.41)
p-value	*<0.001*		*0.056*		*0.002*		*0.017*		*0.081*		*0.043*	
organic-Rich	2.1×10^{-6} (1.9×10^{-6})	7.8×10^{-7} (7.0×10^{-7})	0.43 (0.002)	0.47 (0.008)	0.65 (0.01)	0.66 (0.02)	105.45 (3.74)	95.6 (3.07)	5.52 (0.26)	5.99 (1.25)	3.48 (0.96)	3.38 (0.73)
p-value	*0.038*		*0.018*		*0.757*		*0.039*		*0.565*		*0.442*	
clay-rich	3.2×10^{-8} (1.9×10^{-8})	1.2×10^{-7} (1.2×10^{-8})	1.72 (0.04)	1.38 (0.14)	0.27 (0.01)	0.33 (0.004)	4.86 (2.22)	4.03 (0.13)	0.66 (0.30)	0.25 (0.01)	2.54 (0.70)	3.75 (0.74)
p-value	*<0.001*		*0.052*		*0.003*		*0.587*		*0.078*		*<0.001*	

Two-tailed t-tests were conducted to determine statistical differences between control and +biochar treatments; p-values are shown in italics below mean and standard deviation values for each treatment.

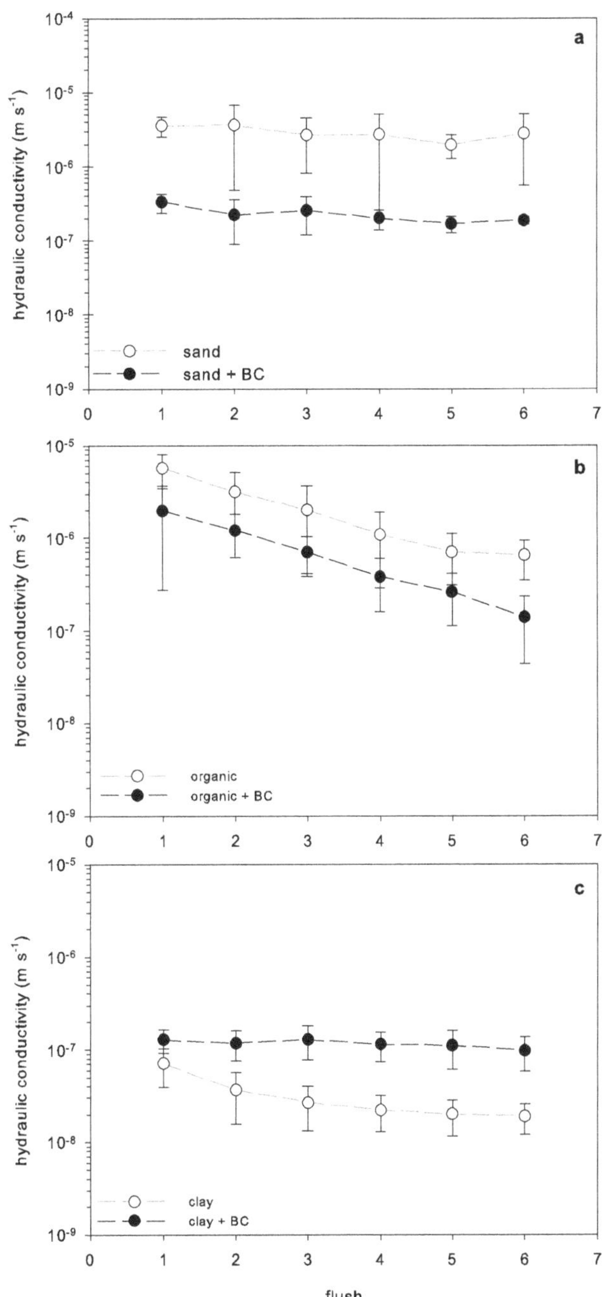

Figure 1. Impact of biochar amendment on saturated soil hydraulic conductivity. The saturated hydraulic conductivity (K), as measured using falling head experiments, for six soil treatments over subsequent flushing events: (a) sand and sand+biochar, (b) organic and organic+biochar, and (c) clay and clay+biochar. Note: the different soil treatment flushing events varied in duration with the clay (c) taking up to 10× longer to drain than the sand (a) or organic soil (b). Saturated hydraulic conductivity data and flushing duration for each flushing experiment available in Table S1.

Models relating K to porosity and grain size suggest that amending sand with biochar should increase K given the greater porosity, overall grain size, and decreased ρ_b. A review of research reveals that this inverse relationship between ρ_b and K is preserved in most cases (Table 3). However, in our experiments biochar amendment to sand decreased the ρ_b and decreased K by 92%

Table 3. A comparison of studies that examined the impact of biochar amendments on soil saturated hydraulic conductivity (*K*) and bulk density (ρ_d).

Feedstock	Temperature (°C)	Application rate[a]	Application rate (tons biochar ha^{-1})[b]	Experiment duration (d)	Soil type and/or % sand/silt/clay	Response of soil bulk density to biochar addition	Response of *K* to biochar addition	Reference
mixed hardwood lump charcoal (e.g. oak & hickory)	NR	5, 10, 20 g kg^{-1}	5.5, 11, 22 t ha^{-1}	500	Mesic Typic Hapludolls	decreased	no effect detected	Laird et al. 2010
mesquite	*400*	*10% by dry weight mass*	*133 t ha^{-1}*	*1–20*	*organic rich*	*increased*	*decreased*	*this study*
commercial biochar, wood residue (e.g. teak and rosewood)	NR	4, 8, 16 t ha^{-1}	4, 8, 16 t ha^{-1}	60	18/34/48	NR	increased	Asai et al. 2009
					27/45/28	NR	no significant change	
corn stover	350 & 550	7.18 t C ha^{-1}	350°C = 11.3 t ha^{-1}	295	Alfisol (silt loam)	decreased	increased with both biochars	Herath et al. 2013
			550°C = 10.0 t ha^{-1}		Andisol (silt loam)	no change	increased with 350°C, no change for 550°C	
mesquite	*400*	*10% by dry weight mass*	*133 t ha^{-1}*	*3–23*	*clay-loam Hapludert*	*decreased*	*increased*	*this study*
dairy manure	300, 500, 700	5% by dry weight mass	61 t ha^{-1}	180	loamy, 40/35/25	decreased	increased, greater increases for biochars at higher temperatures	Lei & Zhang 2013
woodchip						decreased	increased, greater increases for biochars at higher temperatures	
charcoal produced in kilns	~300–500	ambient levels - beneath charcoal kilns	NA	NA	Haplic Acrisols	no significant change	increased	Oguntude et al. 2008
black locust	300, 400, 500	10, 20 Mg ha^{-1}	10, 20 t ha^{-1}	27	sandy	no change	decreased, greatest decrease seen for 500°C biochar & higher application rate	Uzoma et al. 2011
powdered wood charcoal	NR	0.5%, 1.5%, 2.5%, 5% by dry weight mass	6.3, 18.8, 31.3, 62.5 t ha^{-1}	60	sandy loam	decreased	decreased with increasing biochar concentration	Devereaux et al. 2012
acacia green waste	NR	47 Mg ha^{-1}	47 t ha^{-1}	900	Planosol, 72.8/16.8/10.4	decreased	increased near saturated K, no effect on unsaturated K	Hardie et al. 2014
mesquite	*400*	*10% by dry weight mass*	*133 t ha^{-1}*	*1–4*	*sand*	*decreased*	*decreased*	*this study*

[a] Biochar application rate reported in paper.
[b] Biochar application rate converted using bulk density of soils or column materials provided in the paper and assuming tillage depth of 10 cm.
NR: not reported.
NA: not applicable.
Biochar amendment rates are provided two ways: the units provided by the study and in tons biochar ha^{-1}. The conversion assumed a tillage depth of 10 cm and the bulk density of the soil or column materials provided in the paper. Studies are organized by soil type, top to bottom: organic-rich soils, clay- and silt-rich soils, and sandy soils. Results from this study are in italics.

(2.88×10^{-6} m s^{-1} to 2.28×10^{-7} m s^{-1}; Figure 1a, Table 2). Similar changes in K and ρ_b were reported by Deveraux et al. ([22]; Table 3), though this may be due to a decrease in porosity attributed to the relatively smaller d_{50} of the biochar. In our experiments, the median grain size of biochar was larger than the sand and therefore the decrease in ρ_b was accompanied by an increase in porosity.

In addition to grain size, how the biochar is mixed into the soils is also important. The incorporation of charcoal into sandy, Haplic Acrisols underlying kilns, resulted in a non-significant change (9%) in ρ_b and an 88% increase in K ([18]; Table 3). It is likely that the location of the charcoal within the soil column (only in the surface layer), the soil-charcoal layering, and the charcoal properties, significantly affected infiltration rates and other hydrologic properties. Furthermore, in cases where biochar amendment results in bulk density decreases and/or porosity increases, concomitant with a decrease in K (e.g. our results and [21]; (Table 3), changes in physical soil properties are not sufficient to explain the observed hydrologic changes.

The observed decreases in K despite the increased porosity and decreased ρ_b are likely due to the internal structure of biochar. The biochar had an average pore volume of 1.18 cm^3 g^{-1}, porosity of 0.62, and surface area of 6×10^5 cm^2 g^{-1} [59]. Thus the biochar has a greater porosity (typical sand has a porosity between 0.17 and 0.33 [60]) and surface area than our sand (based on d_{50}, sand = 140 cm^2 g^{-1}). The highly porous structure of biochar [59,61] creates two theoretical flow pathways, one in the interstitial space within the biochar-sand matrix and a second connecting the pores within the biochar. According to measurements made by Brewer et al. [59] the biochar we used is dominated (99%) by macropores (0.05 to 1000 μm), and therefore includes many pores larger than the diameter of a water molecule (0.28 nm). However, this second pathway likely has greater tortuosity and smaller median pore throat size due to the size of the smallest pores as well as their lack of complete connectivity [59]. While these pores contribute to the bulk density and total porosity of the mixture, they may not contribute to the effective porosity. In addition, the biochar grains likely create torturous interstitial space between the sand and biochar grains, further decreasing K. This mechanism assumes that the internal pores and surface of biochar are not hydrophobic. While we did not make hydrophobicity measurements on the mesquite biochar, multiple studies have reported that biochar amendments do not result in greater soil hydrophobicity [17,20]. Furthermore, Briggs et al. [12] found that laboratory-produced charcoals leached with distilled water and naturally-produced charcoal collected beneath leaf litter were less water repellent than non-leached and surficial charcoal, respectively; suggesting that the hydrophobic surface compounds may be easily removed.

A second mechanism driving decreases in K could be related to the high field capacity of biochar [8]; i.e. water may have continued to sorb to biochar particles, contributing to the apparent decrease in K for some of the soil mixtures (Figure 1a). While partial saturation of column materials is possible, the measured K of sand-only systems from constant-head and falling-head experiments were similar, suggesting full saturation and equilibrium conditions for both experiments. Saturated hydraulic conductivity changed with repeated flushing events in four of the six soil treatments ($p<0.01$) (Figure 1). The change in K with flushing events was only dependent on the presence of biochar in the case of the organic-rich soil ($F = 7.366$, $p = 0.01$), i.e. while K changed over time in the unamended organic soil, the change was greater in organic+biochar columns (Figure 1b). These shifts in K could be related to physical mechanisms, such as swelling and grain segregation, leading to the clogging of pores, decrease in pore radii, and possibly a variation in bulk density and sample heterogeneity over the course of the experiment. Comparing soil column heights from before and after flushing events revealed that all materials swelled between 6% (sand) and 21% (organic+biochar) suggesting that bulk density changed over the course of the experiments; but biochar addition did not result in statistically different amounts of swelling. Visual observations indicate that biochar particles moved upwards within the columns during the experiment, consistent with observations made by Wang et al. [62]; however, these changes were not quantified and require further study to determine how this movement is related to flow in soil systems and how it impacts hydraulic properties.

The addition of biochar significantly ($p<0.01$) increased the water content at field capacity of the sand and clay. On average biochar amendment doubled the water content at field capacity in the sand and increased it by 20% in clay (Table 2). Several studies have reported increased water content in biochar-amended soils: e.g. organic-rich Mollisols [56], Amazonian *terra preta* [8], sandy loams [13,16,21,22], and silt loams [20]. The increased water holding capacity is likely the result of the internal porosity of the biochar and grain-to-grain interactions; interactions that are highly dependent on soil type and biochar production conditions and are likely the reason that other studies point to mixed results [63]. The organic soil's water content at field capacity did not change with biochar addition, likely due to the already high water holding capacity of the organic soil (Table 2).

Soil carbon and nitrogen dynamics

The addition of biochar significantly ($p<0.05$) increased the %C, by mass in all soil materials: sand increased from 0.40±0.35 to 6.98±0.5%C, clay-rich from 0.90±0.06 to 8.33±0.21%C, and organic-rich from 37.85±3.25 to 42.47±0.30%C. Similar patterns were seen for the N content of the soil materials, though not all changes were significant: %N, by mass, increased from 0.01±0.0 to 0.06±0.03%N for sand, 0.03±0.0 to 0.11±0.02%N for clay, and 0.51±0.01 to 0.54±0.06%N for organic-rich soil.

While biochar addition increased the amount of C in all soils, it did not have a universal effect on the C loss as soil leachate (i.e. DOC flux), suggesting that the release of DOC from biochar-amended soils varies with soil type. When added to the C-poor sand, biochar amendments significantly increased the aromaticity and cumulative loss of DOC (Table 2). The increase in aromaticity of the leached DOC from the biochar treatments suggests that biochar-derived C, and not native C, was lost. DOC fluxes decreased with each subsequent flushing event, suggesting that the easily leachable biochar-C was quickly depleted.

In contrast to sand, the addition of biochar to the organic- and clay-rich soils did not increase the cumulative DOC loss (Table 2). When added to C-rich organic soil, biochar significantly decreased the cumulative DOC loss and did not change the aromaticity (Table 2). This suggests that while the biochar added leachable C to the soil (as observed in the sand+biochar treatment); it is also capable of sorbing soil C. While there was no significant change in the magnitude of DOC lost when biochar was added to clay-rich soil, the aromaticity of the leachate was significantly greater when biochar was present, suggesting that while biochar-derived leachable DOC was lost, additional soil-derived C was retained within the soil-biochar matrix.

The partitioning of biochar-C between soil sequestration and leachate is partially dependent on soil type. Approximately 0.05% of the added biochar-C was lost from the sand+biochar mixture, an amount on the low end of the cumulative losses (0.1 to 1.7% of added C) reported by Mukherjee and Zimmerman [37] for similar short-term laboratory experiments. Furthermore, the $SUVA_{254}$ measurements indicate that while biochar-C is preferentially lost with water flow-through in sandy soils, there was also a net sorption of soil-DOC in both the clay and organic-rich soils. Thus the risk of a net increase of DOC export associated with biochar soil amendments may be minimal in soils with moderate amounts of clay, silt or organic material, a conclusion that mirrors field trials and modeling results [24,64,65].

Given the potential use of biochar to improve soil fertility in arid, drought-prone environments with C-poor sandy soils (e.g. [21] and references therein) it is important to quantify the partitioning of C between soil and leachate in sandy soils. In particular, while our results suggest that the majority of biochar-C remains in the sandy soil, other studies have shown that this partitioning is dependent on pore water and biochar characteristics [37,66]. For example, Bruun and others [66] found that there is a tradeoff between increased water holding capacity and increased biochar-C leachate in sandy soils; the amount of C lost was dependent on production conditions, with greater losses seen when fast- versus slow-pyrolysis biochar was added to soils. Given the results from these experiments and recent analyses of DBC in rivers [5,6] biochar amendments could change the chemical composition of DOC exported to downstream aquatic systems. The ecosystem impacts of this change in the chemical composition of the DOC pool remain unclear [6] and require further study.

Many studies have shown that biochar is able to retain inorganic N, reducing the nitrate flux from soils [29,33,67–69], and biochar-amended sand and clay retained more N in our experiments, resulting in a 24% ($p = 0.081$) and 62% ($p = 0.078$) reduction in N losses (Table 2); though these differences were not statistically significant. The well documented sorptive properties of biochar are attributed to surficial carboxylic and other acid functional groups that provide the cation and anion exchange capacity of biochar [32,70,71] and are dependent on production temperatures and feedstock [30]. The lack of statistical difference with biochar amendment in our study is likely due to the relatively low N content of the soil materials and biochar (<0.8%, Table 1), the only sources of N within our columns. Thus while the results of our experiments are in line with past research, it is important to note that they do not represent the full potential of biochar to mitigate N leaching from agricultural fields.

Conclusions

Biochar-associated changes in K and field capacity have implications for infiltration rates and plant water availability. As shown by our experiments, the addition of biochar to coarser soils decreases K, indicating the potential to decrease crop water stress and reduce nutrient loss below the rooting zone [29]. Conversely, biochar is able to increase porosity and permeability in fine-grained clay soils, making them more suitable for crop growth by increasing infiltration rates. Our results, combined with those of other studies, strongly support the argument that biochar addition increases the water holding capacity in coarse-grained soils, likely improving plant water availability. The saturated hydraulic conductivity (K) results of these laboratory and field experiments should be considered short-term, as reported changes in soil structure and hydrology are likely to evolve with time, impacting K. Brodowksi et al. [72] showed that biochar degrades into silt-sized particles over time likely changing the porosity and K of the amended soil. Other mechanisms may additional act to alter K on longer timescales, including increased microbial activity and increased bioturbation. Increased microbial activity and the addition of OM associated with biochar amendments has been shown to increase soil aggregation and macropore volume, thereby increasing K [73]. Increased bioturbation associated with biochar amendment can also increase K [74]. These examples provide further impetus for examining the effects of biochar particle size on soil hydraulic properties in the near- and long-term and in both laboratory/greenhouse and field settings.

Our experiments also illustrate that biochar addition to soils can add and sorb leachable DOC, and potentially add aromatic DOC to rivers [5]. Furthermore, biochar has the potential to reduce TDN in leachates, mitigating environmental impacts of agricultural N pollution. These changes in biogeochemical cycling, as well as alterations in greenhouse gas fluxes, plant productivity, and microbial activity are inherently linked to soil hydraulic properties. It is therefore crucial that future research addresses the complex interactions between biochar amendment and soil hydrology, C and N cycling, as well as how different soil types, biochars (varying feedstock, particle size, and production conditions) and amendment rates control these processes.

Supporting Information

Table S1 Average saturated hydraulic conductivity results for soil materials with and without biochar. The average and standard deviation saturated hydraulic conductivity (K) for each flushing experiment (n = 3 for each soil material) in m/s. The duration of the experiment (in hours) is also provided.

Acknowledgments

We thank F. Liljestrand, A. Woda, A. Hilton, and D. Leahy for help with experiments, soil processing and data collection; and V. Chuang for biochar production.

Author Contributions

Conceived and designed the experiments: RTB MEG CAM BD. Performed the experiments: RTB MEG ZL. Analyzed the data: RTB MEG. Contributed reagents/materials/analysis tools: CAM BD. Wrote the paper: RTB. Significant help revising the manuscript: MEG CAM BD.

References

1. Woolf D, Amonette JE, Street-Perrott FA, Lehmann J, Joseph S (2010) Sustainable biochar to mitigate global climate change. Nature Communications 1: 1–9.
2. Masiello CA (2004) New directions in black carbon organic geochemistry. Marine Chemistry 92: 201–213.
3. Lehmann J, Gaunt J, Rondon M (2006) Bio-char sequestration in terrestrial ecosystems - A review. Mitigation and Adaptation Strategies for Global Change 11: 403–427.
4. Kuhlbusch TAJ, Crutzen PJ (1996) Black Carbon, the Global Carbon Cycle, and Atmospheric Carbon Dioxide. In: Levine JS, editor. Biomass Burning and Global Change: Remote Sensing, Modeling and Inventory Development, and Biomass Burning in Africa. Cambridge, MA: Massachusetts Institute of Technology. pp. 160–169.
5. Dittmar T, de Rezende CE, Manecki M, Niggemann J, Ovalle ARC, et al. (2012) Continuous flux of dissolved black carbon from a vanishing tropical forest biome. Nature Geoscience 5: 618–622.
6. Jaffe R, Ding Y, Niggemann J, Vahatalo AV, Stubbins A, et al. (2013) Global Charcoal Mobilization from Soils via Dissolution and Riverine Transport to the Oceans. Science 340: 345–347.
7. Stubbins A, Spencer RGM, Chen H, Hatcher PG, Mopper K, et al. (2010) Illuminated darkness: Molecular signatures of Congo River dissolved organic matter and its photochemical alteration as revealed by ultrahigh precision mass spectrometry. Limnology and Oceanography 55: 1467–1477.
8. Glaser B, Lehmann J, Zech W (2002) Ameliorating physical and chemical properties of highly weathered soils in the tropics with charcoal - a review. Biology and Fertility of Soils 35: 219–230.

9. Kammann C, Ratering S, Eckhard C, Müller C (2012) Biochar And Hydrochar Effects On Greenhouse Gas (carbon Dioxide, Nitrous Oxide, And Methane) Fluxes From Soils. J Environ Qual 41: 1052–1066.
10. Novak JM, Busscher WJ, Watts DW, Amonette JE, Ippolito JA, et al. (2012) Biochars Impact on Soil-Moisture Storage in an Ultisol and Two-Aridisols. Soil Science 177: 310–320.
11. Gaskin JW, Speir A, Morris LM, Ogden L, Harris K, et al. (2007) Potential for Pyrolysis Char to Affect Soil Moisture and Nutrient Status of a Loamy Sand Soil. March 27–29; University of Georgia.
12. Briggs C, Breiner JM, Graham RC (2012) Physical and chemical properties of *Pinus ponderosa* charcoal: implications for soil modification. Soil Science 177: 263–268.
13. Tryon EH (1948) Effects of Charcoal on Certain Physical, Chemical, and Biological Properties of Forest Soils. Ecological Monographs 18: 81–115.
14. Baronti S, Vaccari FP, Miglietta F, Calzolari C, Lugato E, et al. (2014) Impact of biochar application on plant water relations in *Vitis vinifera* (L.). European Journal of Agronomy 53: 38–44.
15. Kammann CI, Linsel S, Gößling JW, Koyro H-W (2011) Influence of biochar on drought tolerance of *Chenopodium quinoa* Willd and on soil-plant relations. Plant and Soil.
16. Bruun EW, Petersen CT, Hansen E, Holm JK, Hauggaard-Nielsen H (2014) Biochar amendment to coarse sandy subsoil improves root growth and increases water retention. Soil Use and Management 30: 109–118.
17. Kinney TJ, Masiello CA, Dugan B, Hockaday WC, Dean MR, et al. (2012) Hydrologic properties of biochars produced at different temperatures. Biomass and Bioenergy 41: 34–43.
18. Oguntunde PG, Abiodun BJ, Ajayi AE, van de Giesen N (2008) Effects of charcoal production on soil physical properties in Ghana. Journal of Plant Nutrition and Soil Science 171: 591–596.
19. Asai H, Samson BK, Stephan HM, Songyikhangsuthor K, Homma K, et al. (2009) Biochar amendment techniques for upland rice production in Northern Laos 1. Soil physical properties, leaf SPAD and grain yield. Field Crops Research 111: 81–84.
20. Herath HMSK, Camps-Arbestain M, Hedley M (2013) Effect of biochar on soil physical properties in two contrasting soils: An Alfisol and an Andisol. Geoderma 209–210: 188–197.
21. Uzoma KC, Inoue M, Andry H, Zahoor A, Nishihara E (2011) Influence of biochar application on sandy soil hydraulic properties and nutrient retention. Journal of Food, Agriculture & Environment 9: 1137–1143.
22. Deveraux RC, Sturrock CJ, Mooney SJ (2012) The effects of biochar on soil physical properties and winter wheat growth. Earth and Environmental Science Transactions of the Royal Society of Edinburgh 103: 13–18.
23. Atkinson CJ, Fitzgerald JD, Hipps NA (2010) Potential mechanisms for achieving agricultural benefits from biochar application to temperate soils: a review. Plant and Soil 337: 1–18.
24. Major J, Lehmann J, Rondon M, Goodale CL (2010) Fate of soil-applied black carbon: downward migration, leaching and soil respiration. Global Change Biology 16: 1366–1379.
25. Mishera SK, Tyagi JV, Singh VP (2003) Comparison of infiltration models. Hydrological Processes 17: 2629–2652.
26. Krull ES, Baldock JA, Skjemstad JO, Smernik RJ (2009) Characteristics of biochar: organo-chemical properties. In: Lehmann J, editor. Biochar for environmental management science and technology. London: Earthscan. pp. 53–65.
27. Baldock JA, Smernik RJ (2002) Chemical composition and bioavailability of thermally altered *Pinus resinosa* (Red pine) wood. Organic Geochemistry 33: 1093–1109.
28. Schneider MPW, Hilf M, Vogt UF, Schmidt MWI (2010) The benzene polycarboxylic acid (BPCA) pattern of wood pyrolyzed between 200°C and 1000°C. Organic Geochemistry 41: 1082–1088.
29. Major J, Steiner C, Downie A, Lehmann J (2009) Biochar effects on nutrient leaching. In: Lehmann J, Joseph S, editors. Biochar for environmental management, science, and technology. London: Earthscan. pp. 271–287.
30. Lehmann J (2007) Bio-energy in the black. Frontiers in Ecology and the Environment 5: 381–387.
31. Cheng C-H, Lehmann J, Engelhard MH (2008) Natural oxidation of black carbon in soils: Changes in molecular form and surface charge along a climosequence. Geochimica et Cosmochimica Acta 72: 1598–1610.
32. Liang B, Lehmann J, Solomon D, Kinyangi J, Grossman J, et al. (2006) Black carbon increases cation exchange capacity in soils. Soil Science Society of America Journal 70: 1719–1730.
33. Steiner C, Glaser B, Teixeira WG, Lehmann J, Zech W (2008) Nitrogen retention and plant uptake on a highly weathered Amazonian Ferralsol amended with compost and charcoal. Journal of Plant Nutrition and Soil Science 171: 893–899.
34. Yu L, Tang J, Zhang R, Wu Q, Gong M (2013) Effects of biochar application on soil methane emission at different soil moisture levels. Biology and Fertility of Soils 49: 119–128.
35. Mukome FND, Six J, Parikh SJ (2013) The effects of walnut shell and wood feedstock biochar amendments on greenhouse gas emissions from a fertile soil. Geoderma 200–201: 90–98.
36. Angst TE, Patterson CJ, Reay DS, Anderson P, Peshkur TA, et al. (2013) Biochar Diminishes Nitrous Oxide and Nitrate Leaching from Diverse Nutrient Sources. Journal of Environmental Quality 42: 672–682.
37. Mukherjee A, Zimmerman AR (2013) Organic carbon and nutrient release from a range of laboratory-produced biochars and biochar-soil mixtures. Geoderma 193–194: 122–130.
38. Norwood MJ, Louchouarn P, Kuo L-J, Harvey OR (2013) Characterization and biodegradation of water-soluble biomarkers and organic carbon extracted from low temperature chars. Organic Geochemistry 56: 111–119.
39. Zhang W, Niu J, Morales VL, Chen X, Hay AG, et al. (2010) Transport and retention of biochar particles in porous media: effect of pH, ionic strength, and particle size. Ecohydrology 3: 497–508.
40. Wang D, Zhang W, Hao X, Zhou D (2013) Transport of Biochar Particles in Saturate Granular Media: Effects of Pyrolysis Temperature and Particle Size. Environmental Science & Technology 47: 821–828.
41. Wheeler FF (1976) Soil Survey of Harris County, Texas. United States Department of Agriculture, Soil Conservation Service. pp. 152.
42. ASTM International (2010) D2216-10, Standard Test Methods for Laboratory Determination of Water (Moisture) Content of Soil and Rock by Mass.
43. ASTM International (2007) D1762-84, Standard Test Method for Chemical Analysis of Wood Charcoal.
44. Rayment GE, Lyons DJ (2010) Soil Chemical Methods. Melbourne: CSIRO.
45. Rajkovich S, Enders A, Hanley K, Hyland C, Zimmerman A, et al. (2011) Corn growth and nitrogen nutrition after additions of biochars with varying properties to a temperate soil. Biology and Fertility of Soils.
46. ASTM International (2010) D5084-10 Standard Test Methods for Measurement of Hydraulic Conductivity of Saturated Porous Materials Using a Flexible Wall Perimeter.
47. Oliveira IB, Demond AH, Salehzadeh A (1996) Packing of sands for the production of homogeneous porous media. Soil Science Society of America Journal 60: 49–53.
48. Chan KY, Van Zwieten L, Meszaros I, Downie A, Joseph S (2007) Agronomic values of greenwaste biochar as a soil amendment. Australian Journal of Soil Research 45: 629–634.
49. Klute A (1986) Methods of soil analysis. Part 1. Physical and mineralogical methods. Madison, WI: American Society of Agronomy.
50. Weishaar JL, Aiken GR, Bergamaschi BA, Fram MS, Fujii R, et al. (2003) Evaluation of specific ultraviolet absorbance as an indicator of the chemical composition and reactivity of dissolved organic carbon. Environmental Science & Technology 37: 4702–4708.
51. ASTM International (2006) D2434, Standard Test Method for Permeability of Granular Soils (Constant Head) In: Materials ASfTa, editor. West Conshohocken, PA.
52. Kozeny J (1927) Über kapillare Leitung des Wassers im Boden – Aufstieg, Versickerung und Anwendung auf die Bewässerung. Sitzungsberichte Akad Wiss Wien 136: 271–306.
53. Carmen PC (1937) Flow through a granular bed. Transactions of Institute of Chemical Engineers, London 15: 150–156.
54. Neuzil CE (1994) How permeable are clays and shales? Water Resources Research 30: 145–150.
55. Alyamani MS, Sen Z (1993) Determination of hydraulic conductivity from complete grain-size distribution curves. Groundwater 31: 551–555.
56. Laird DA, Fleming P, Davis DD, Horton R, Wang B, et al. (2010) Impact of biochar amendments on the quality of a typical Midwestern agricultural soil. Geoderma 158: 443–449.
57. Boadu FK (2000) Hydraulic conductivity of soils from grain-size distribution: New models. Journal of Geotechnical and Geoenvironmental Engineering 126: 739–746.
58. Crawford BR, Faulkner DR, Rutter EH (2008) Strength, porosity, and permeability development during hydrostatic and shear loading of synthetic quartz-clay fault gouge. Journal of Geophysical Research 113.
59. Brewer CE, Chuang VJ, Masiello CA, Gonnermann H, Gao X, et al. (2014) New approaches to measuring biochar density and porosity. Biomass and Bioenergy 66: 176–185.
60. Holtz RD, Kovacs WD (1981) An Introduction to Geotechnical Engineering. Upper Saddle River, NJ: Prentice Hall.
61. Sun H, Hockaday WC, Masiello CA, Zygourakis K (2012) Multiple Controls on the Chemical and Physical Structure of Biochars. Industrial and Engineering Chemistry Research 51: 3587–3597.
62. Wang C, Walter MT, Parlange J-Y (2013) Modeling simple experiments of biochar erosion from soil. Journal of Hydrology 499: 140–145.
63. Novak JM, Lima IM, Xing B, Gaskin JW, Steiner C, et al. (2009) Characterization of designer biochar produced at different tempeartures and their effects on a loamy sand. Annals of Environmental Science 3: 195–2006.
64. Jones DL, Rousk J, Edwards-Jones G, DeLuca TH, Murphy DV (2012) Biochar-mediated changes in soil quality and plant growh in a three year field trial. Soil Biology and Biochemistry 45: 113–124.
65. Foereid B, Lehmann J, Major J (2011) Modeling black carbon degradation and movement in soil. Plant and Soil 345: 223–236.
66. Bruun EW, Petersen CT, Strobel BW, Hauggaard-Nielsen H (2012) Nitrogen and carbon leaching in repacked sandy soil with added fine particulate biochar. Soil Science Society of America Journal 76: 1142–1148.
67. Novak JM, Busscher WJ, Watts DW, Laird DA, Ahmedna MA, et al. (2010) Short-term CO_2 mineralization after additions of biochar and switchgrass to a Typic Kandiudult. Geoderma 154: 281–288.

68. Bell MJ, Worrall F (2011) Charcoal addition to soils in NE England: A carbon sink with environmental co-benefits? Science of the Total Environment 409: 1704–1714.
69. Clough TJ, Condron LM (2010) Biochar and the Nitrogen Cycle: Introduction. Journal of Environmental Quality 39: 1218–1223.
70. Singh BP, Hatton BJ, Singh B, Cowie AL, Kathuria A (2010) Influence of biochars on nitrous oxide emission and nitrogen leaching from two contrasting soils. Journal of Environmental Quality 39: 1224–1235.
71. Asada T, Ishihara S, Yamane S, Toba T, Yamada A, et al. (2002) Science of bamboo charcoal: Study of carbonizing temperature of bamboo charcoal and removal capability of harmful gases. Journal of Health Sciences 48: 473–479.
72. Brodowski S, Amelung W, Haumaier L, Zech W (2007) Black carbon contribution to stable humus in German arable soils. Geoderma 139: 220–228.
73. Lei O, Zhang R (2013) Effects of biochars derived from different feedstocks and pyrolysis temperatures on soil physical and hydraulic properties. Journal of Soils and Sediments.
74. Hardie M, Clothier B, Bound S, Oliver G, Close D (2014) Does biochar influence soil physical properties and soil water availability? Plant and Soil 376: 347–361.

21

Spatial Variation in Carbon and Nitrogen in Cultivated Soils in Henan Province, China: Potential Effect on Crop Yield

Xuelin Zhang[1]*, Qun Wang[1], Frank S. Gilliam[2], Yilun Wang[1], Feina Cha[3], Chaohai Li[1]

1 The Incubation Base of the National Key Laboratory for Physiological Ecology and Genetic Improvement of Food Crops in Henan Province, Zhengzhou, China; Agronomy College of Henan Agricultural University, Zhengzhou, China, **2** Department of Biological Sciences, Marshall University, Huntington, West Virginia, United States of America, **3** Meteorological Bureau of Zhengzhou, Zhengzhou, China

Abstract

Improved management of soil carbon (C) and nitrogen (N) storage in agro-ecosystems represents an important strategy for ensuring food security and sustainable agricultural development in China. Accurate estimates of the distribution of soil C and N stores and their relationship to crop yield are crucial to developing appropriate cropland management policies. The current study examined the spatial variation of soil organic C (SOC), total soil N (TSN), and associated variables in the surface layer (0–40 cm) of soils from intensive agricultural systems in 19 counties within Henan Province, China, and compared these patterns with crop yield. Mean soil C and N concentrations were 14.9 g kg^{-1} and 1.37 g kg^{-1}, respectively, whereas soil C and N stores were 4.1 kg m^{-2} and 0.4 kg m^{-2}, respectively. Total crop production of each county was significantly, positively related to SOC, TSN, soil C and N store, and soil C and N stock. Soil C and N were positively correlated with soil bulk density but negatively correlated with soil porosity. These results indicate that variations in soil C could regulate crop yield in intensive agricultural systems, and that spatial patterns of C and N levels in soils may be regulated by both climatic factors and agro-ecosystem management. When developing suitable management programs, the importance of soil C and N stores and their effects on crop yield should be considered.

Editor: Dafeng Hui, Tennessee State University, United States of America

Funding: This study was supported by grants from Henan Science and Technology Department of China under the Key Research Project (30200051). The funder had no role in study design, data collection and analysis, decision to publish, or preparation of the manuscript.

Competing Interests: The authors have declared that no competing interests exist.

* Email: xuelinzhang1998@163.com

Introduction

Safeguarding food security and ensuring sustainable development are two fundamental goals of intensive agriculture in China [1,2]. Increasing soil C and N sequestration while reducing C and N emissions from agricultural fields are important aspects of sustainable farming and these goals can be achieved through improvement in soil quality [1,3]. This requires a better understanding of the functional relationship between crop yield and soil organic C and N stores.

Indeed, variations in soil C and N stores may closely regulate crop yield, although published data on the relationship between these parameters are inconsistent. Some studies have reported a positive correlation between soil C and N and crop yield [4,5], whereas other studies have found no significant relationship between these parameters [6,7]. Lal (2006) reported that the relationship between soil organic C and crop yield may vary between patterns that are sigmoidal, linear, or exponential [8]. Clearly, the existence of such variability warrants further investigation.

Soil C and N stores in crop lands, especially in the topsoil layer, are potentially greatly affected by human activity; thus, understanding the spatial pattern of soil C and N stores on a regional scale is crucial to developing a management strategy for improving soil fertility [1,2]. Spatial variation in soil C and N stores in agroecosystems has been widely reported [9,10,11], including from the northern [12,13], eastern [14], and southern [15,16] regions of China. Since these reports from China were based on two national surveys from 1960 and 1983, such data may have limited use in helping to develop management strategies based on current practices [17]. Therefore, in order to better understand the spatial patterns and their relationship to crop yield, it is necessary to update regional soil organic C and N information with contemporary measurements, especially for intensively-used crop land.

Henan Province is the second largest area of crop production in China (China National Bureau of Statistics). To produce an adequate supply of food for the domestic population, unsustainable production methods have often been used in this province. Historically, intensive production based on an annual wheat-maize system has been used to achieve high crop yield. This practice, however, has resulted in badly degraded agricultural soils, causing erosion and a loss of good soil structure. More than 600 kg N ha^{-1} annually has been applied in this production area, resulting in an increase in soil acidity [18]. Based on the determination that crop yields in China will need to increase from 50 billion in 2010 to 65 billion kg in 2020, the provincial crop lands in Henan Province will continue to play an important role in food production. Such goals create the challenge of improving soil quality, enhancing soil fertility, and mitigating C and N loss, while achieving food security and practicing sustainable agriculture. A better understanding of the spatial variability of soil organic C and N, and their relationship to crop yield, should help to develop management practices that are designed to meet this challenge [1,19].

The objective of the present study was to characterize the spatial distribution of C and N stores in intensively cultivated counties within the Henan Province of China and to determine the relationship between crop yield and soil organic C and N.

Materials and Methods

Statement: We have field permits for sampling soil in each of the field sites within each county of Henan Province, China. All of the sampling sites are privately owned, and there was no potential impact on any endangered or protected species among these sampling sites.

Study site

The study was carried out in 19 counties within Henan Province, located in central China (Figure 1). Map data were obtained from the National Geomatics Center of China (http://ngcc.sbsm.gov.cn/) using ArcGIS software. As of 2009, the human population of Henan was about 9.9×10^7 persons. The Province is

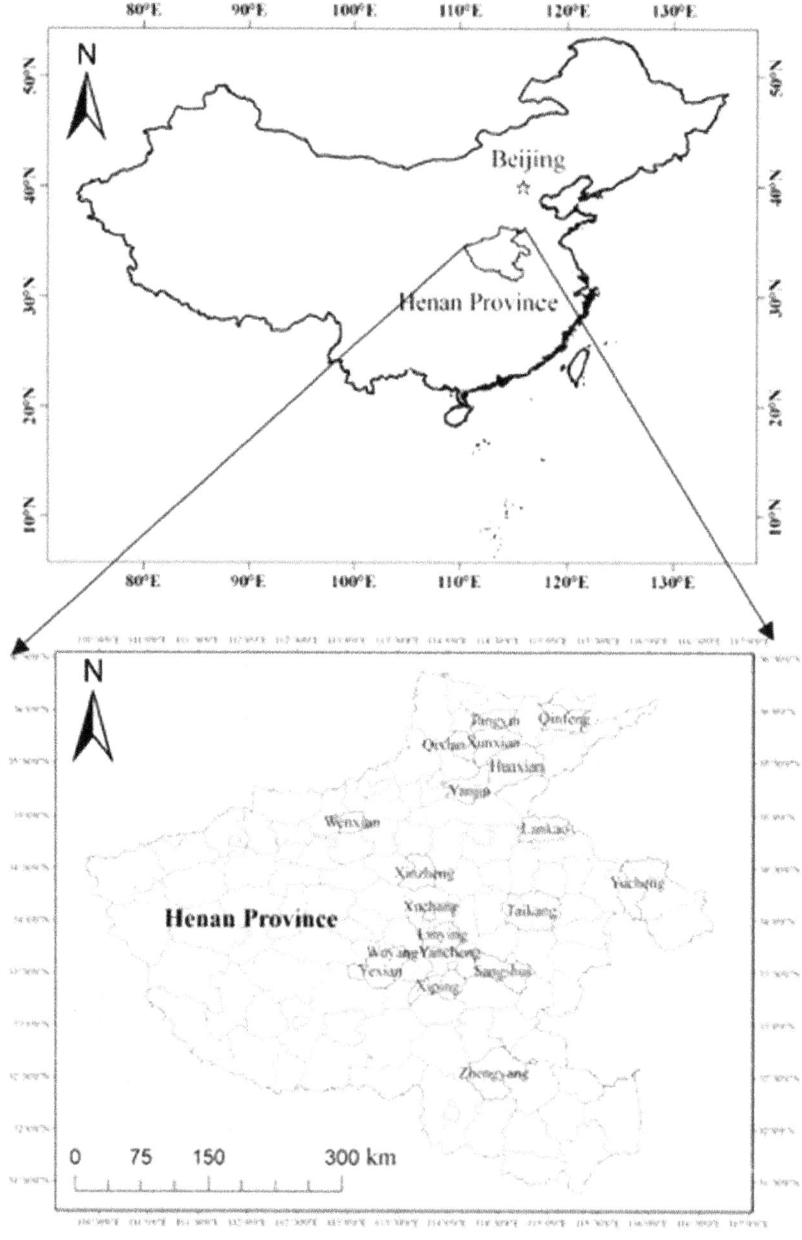

Figure 1. Map of China (top) showing location of Henan Province and counties (bottom) within Henan Province used in this study.

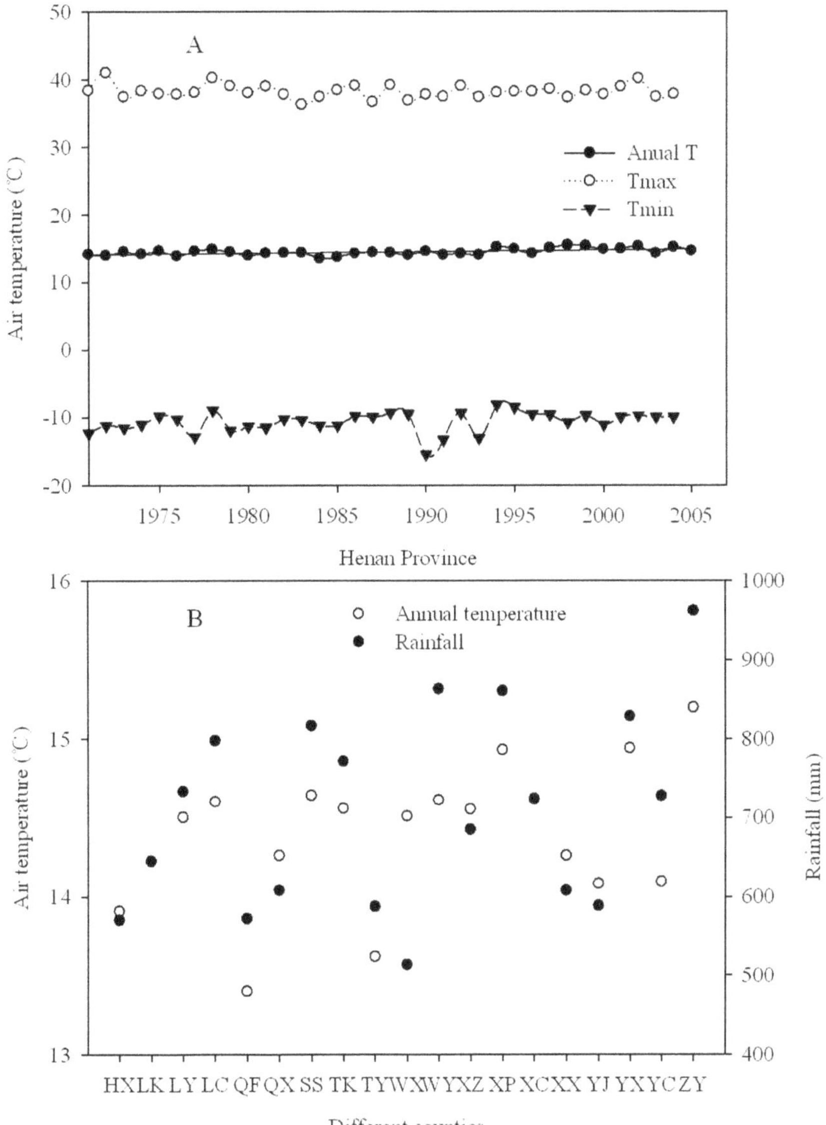

Figure 2. Average annual, maximum, and minimum temperature from 1971 to 2004 in Henan Province, China (A), and (B) average temperature and rainfall from 1975 to 2006 in different counties within Henan Province, China (B). See Table 1 for key to county name abbreviations. All these counties were arranged in English alphabetical order.

approximately 167,000 km^2 in land area, lying within the monsoonal temperate zone. It has a cultivated land area of 79, 260 km^2 for the production of wheat and maize. There are three dominant soil types in Henan Province: Yellow-cinnamon soil (Eutric Cambisols in FAO taxonomy), Sajiang black soil (Eutric Vertisols/Gleyic Cambisol), and Fluvo-aquic soil (Fluvisols in FAO taxonomy) [20]. Mean annual precipitation ranges from 400 to 1000 mm among the counties of the study, with ~70% of it occurring from July to September; mean annual temperature ranges from 13.6 to 15°C (Figure 2). Cultivated agricultural fields are the predominant land use, representing 60% of the total land area in Henan Province. A double cropping system of winter wheat (early October-early June) and maize (mid-June–later September) is the most common planting system used in this region.

Collection of crop yield and soil sampling and analysis

Data on total crop production (including wheat, maize and millet) and wheat yield from 1978–2009 (Figure. 3A) were obtained from the Henan Statistical Yearbook 2010 (13–17) (http://www.ha.stats.gov.cn/hntj/index.htm). Annual yield data for winter wheat and total crop production in 2009 were also obtained from Henan Statistical Yearbook 2010 (29-7) and the Agricultural Bureau of each of the 19 counties in which soil sampling took place (Figure. 3B). These counties, along with basic climatic information, are listed in Table 1. Climatic data of each county were obtained from Meteorological Bureau of Zhengzhou. All counties will be referred to by the two-letter codes presented in Table 1.

The 19 counties were selected as representative of the main agro-ecosystems of Henan Province. Soil samples were collected during June 1–15, 2009 following the wheat harvest but prior to the sowing of maize. Six representative, replicate field plots,

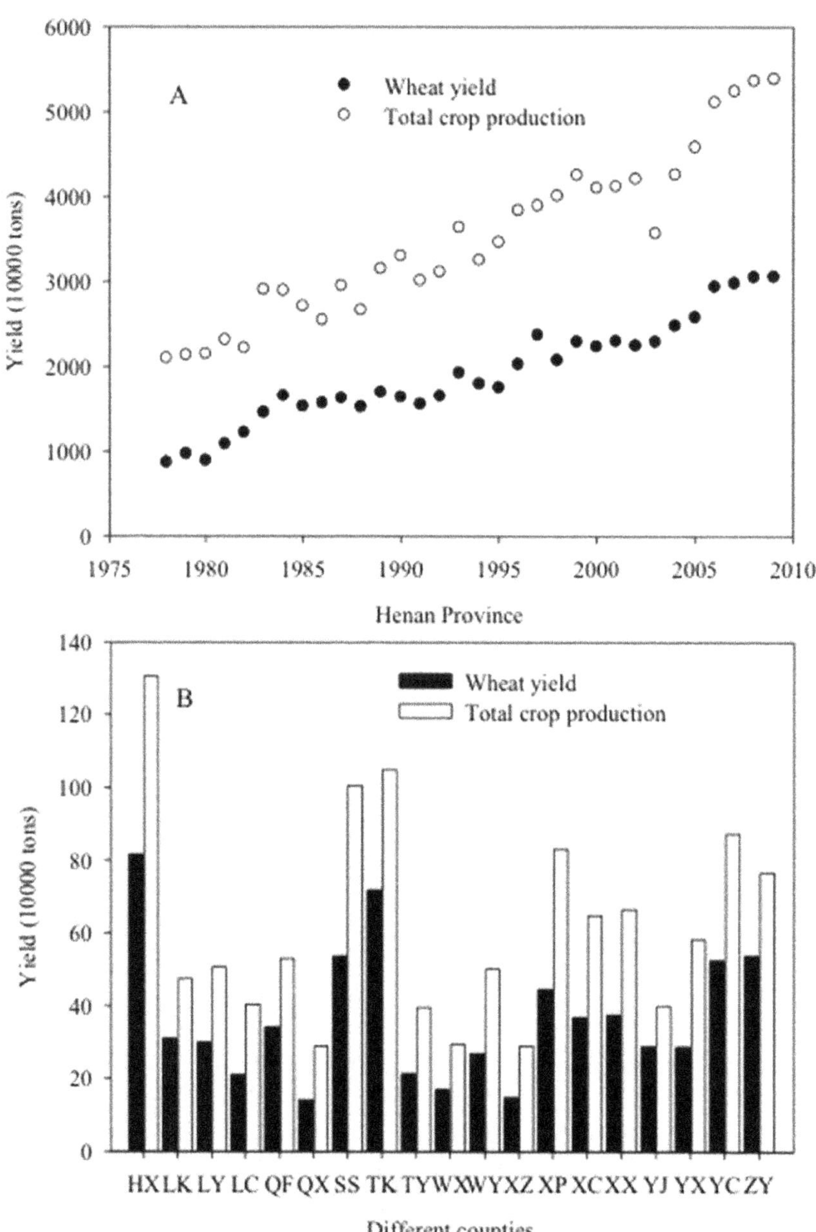

Figure 3. Wheat yield and total crop production (including wheat, maize, millet,) in Henan Province from 1978–2009 (A), and wheat yield and total crop production in different counties within Henan Province in 2009 (B). See Table 1 for key to county name abbreviations.

located at least 6 km apart, were selected within each county based on four criteria: (1) the field plots had been continuously cultivated for at least 30 yr with a native variety, (2) the cropland area was located within 5 km of native vegetation with a similar landscape, soil type and texture, and a relatively flat terrain, and (3) all of the sampling sites are privately owned, and (4) there was no potential impact on any endangered or protected species in the sampling site. Geographic coordinates of each sampling site was recorded by handed GPS of Magellan eXplorist 210(USA), and all of these data were attached in the supporting information.

Sample areas of ~1300 m^2 were established in each plot, with sixteen sampling points taken at random in each of two layers (0–20 cm and 20–40 cm) using a 70 mm - diameter auger. All of the soil samples taken at each layer within a sample plot were mixed together and treated as one sample to represent the value of the plot, yielding 114 soil samples at each layer.

Residual plant material was removed from the soil samples after the samples were air-dried at room temperature. The soil samples were then ground to pass a 2 mm sieve, and a portion of the ground sample was subsequently ground again in a porcelain mortar in order to pass through a 0.15–mm sieve. Organic C and total N measurements were obtained from the twice-ground soil samples. Soil organic C (SOC) was measured using a modified Mebius method. Briefly, 0.1 g soil samples were digested for 5 min

Table 1. Basic geographic coordinates for each county, along with climate data for 19 counties within Henan Province, China.

County	Latitude	Longitude	Sea level (m)	Average Temp (°C)	Rainfall (mm)	Sunshine (h)
Huaxian (HX)	35°44′	114°28′	68	13.9	570.0	2060.9
Lankao (LK)	34°55′	114°46′	70	14.2	644.5	2183.2
Linying (LY)	33°55′	113°55′	63	14.5	732.9	2141.3
Luoheyancheng (LC)	33°35′	114°02′	65	14.6	797.2	2273.0
Qingfeng (QF)	35°53′	115°06′	51	13.4	571.9	2209.1
Qixian (QX)	35°35′	114°12′	72	14.3	607.5	2133.8
Shangshui (SS)	33°39′	114°34′	52	14.6	815.8	1902.0
Taikang (TK)	34°05′	114°50′	53	14.6	770.9	1998.4
Tangyin (TY)	36°03′	114°19′	103	13.6	587.1	2159.3
Wenxian (WX)	35°01′	113°03′	109	14.5	513.2	2302.2
Wuyang (WY)	33°36′	113°32′	77	14.6	862.3	2060.4
Xinzheng (XZ)	34°30′	113°39′	159	14.6	684.6	2058.7
Xiping (XP)	33°29′	113°59′	65	14.9	859.8	2084.7
Xuchang (XC)	34°04′	113°52′	72	14.6	722.7	1959.8
Xunxian (XX)	35°40′	114°32′	59	14.3	607.5	2133.8
Yanjin (YJ)	35°13′	114°11′	69	14.1	588.0	2287.8
Yexiang (YX)	33°38′	113°21′	88	14.9	827.8	1972.4
Yucheng (YC)	34°25′	115°52′	46	14.1	727.3	2244.6
Zhengyang (ZY)	32°37′	114°24′	70	15.2	961.8	2004.4

Note: Counties are arranged in English alphabetical order.

with 5 mL of 1N $K_2Cr_2O_7$ and 10 mL of concentrated H_2SO_4 at 150°C, followed by titration of the digests with standardized $FeSO_4$. Total soil N (TSN) was measured using a modified Kjeldahl wet digestion procedure and a Tector Kjeltec System 1026 distilling unit. Soil available N was analyzed using a micro-diffusion technique after alkaline hydrolysis (1.8 mol L^{-1} NaOH). The Olsen method was used to determine available soil phosphorus (P), and available soil potassium (K) was measured in 1 mol L^{-1} NH_4OAc extracts by flame photometry (Table 2).

Three sampling points were used to determine soil bulk density in each plot. Samples were collected separately from four layers within a depth of 0–40 cm in each sampling point. Soil bulk density was measured using 100-cm^3 soil cores obtained from the four layers. Soil porosity was calculated from soil bulk density and specific gravity, with any stone material removed and not considered in bulk density calculations.

Calculation of soil organic C and N stores and SOC and TSN

Total soil organic C store (TSOCS) and total soil N stores (TSNS) at 0–40 cm depth were calculated as follows:

$$\mathrm{TSOC}(\mathrm{g.m^{-2}}) =$$
$$\text{Soil organic C}(\mathrm{g.kg^{-1}}) \times \text{soil bulk density}(\mathrm{g.cm^{-3}}) \times \text{sampling depth}(\mathrm{cm})$$
$$\mathrm{TSN}(\mathrm{g.m^{-2}}) =$$
$$\text{Soil total N}(\mathrm{g.kg^{-1}}) \times \text{soil bulk density}(\mathrm{g.cm^{-3}}) \times \text{sampling depth}(\mathrm{cm})$$

Given the cultivated area, the total cultivated topsoil (0–40 cm) C and N stocks of each county were estimated by the equation:

$$\mathrm{CS} = \sum area_i \times \mathrm{TSOC}$$
$$\mathrm{NS} = \sum area_i \times \mathrm{TSN}$$

where *area* is the given total cultivated area of each county, and CS and NS are C and N stocks, respectively. SOC and TSN were means of six sampling sites of each county.

Statistics

Analysis of variance was used to assess the significance of location (county) on soil C and N concentration and storage; means were compared using Duncan's multi-range test at $\alpha = 0.05$. Linear regression was used to determine the relationships between C and N stock versus wheat and total crop production. Principle components analysis was used to assess patterns of similarity/dissimilarity among counties with respect to several environmental variables [21]. All statistical analyses were performed using SPSS 10.0 (Chicago IL, USA).

Results

Wheat yields increased more than 250% from 1978 to 2009 while total annual crop production in Henan Province increased from 21 to 54 million tons over the same time period (Figure 3A). Wheat yield varied from 143 to 729 thousand tons among the different counties in 2009 (Figure 3B).

The absolute value of SOC concentration in the top 40 cm of soil varied from 8.13 to 27.89 g kg^{-1} among the 19 counties in 2009 (Table 2) while TSN concentration varied from 0.84 to 2.2 g kg^{-1}. Soil C/N varied from 6.4 to 20 (Table 2). Soil organic C stores (TSOCS) in the 0–40 cm soil layer varied from 2,322 g m^{-2} to 8,038 g m^{-2}, whereas total N stores (TSNS) varied from 221 to

Table 2. Spatial variation in soil (0–40 cm depth) properties, soil organic C (SOC), total soil N (TSN) concentration (g kg^{-1}), and C/N in the 0–40 cm soil layer in 19 counties within Henan province, China.

	Alkaline-extractable N (mg kg^{-1})	Olsen-extractable P (mg kg^{-1})	NH_4OAc-extractable K (mg kg^{-1})	Bulk density (g cm^{-3})	Soil porosity (%)	SOC (g kg^{-1})	TSN (g kg^{-1})	C/N
HX	48.9±3.2abc	1.8±0.7a	80.1±9.2abc	1.44±0.03de	38.3±1.4abcd	12.4±0.9abc	1.4±0.05abcd	8.8±0.7abc
LK	56.5±2.4abcd	7.6±1.9ab	71.9±11.9abc	1.42±0.02bcde	40.7±1.0bcdef	11.2±0.7ab	1.4±0.09abcd	7.9±0.5a
LY	49.9±3.1abc	4.2±0.7a	145.4±24.1ef	1.36±0.02abc	41.7±1.4cdefg	15.5±1.0bcd	1.1±0.07a	14.2±0.2efg
LC	49.0±1.6abc	11. 5±2.1abc	103.6±8.9abcde	1.39±0.02bcd	38.7±1.6abcd	14.6±1.4abcd	1.4±0.14bcd	10.5±1.1abcd
QF	47.5±2.8abc	10.9±4.9abc	71.7±6.3abc	1.39±0.01bcd	41.8±0.4cdefg	11.8±0.6ab	1.4±0.08abcd	8.7±0.8abc
QX	51.9±3.7abc	6.3±2.2ab	82.1±10.8abc	1.44±0.01de	38.7±0.3abcd	21.1±1.8f	1.5±0.21cd	16.2±3.1g
SS	45.1±1.9abc	11.7±3.7abc	169.3±33.9f	1.35±0.02abc	37.9±1.3abcd	14.5±0.9abcd	1.3±0.08abc	11.4±0.8bcde
TK	59.3±4.5cd	17.7±8.2bcd	140.6±23.9def	1.35±0.02ab	41.5±1.5cdef	13.4±1.0abcd	1.1±0.05ab	11.9±0.7cde
TY	59.2±2.4cd	6.4±2.6ab	110.9±12.8bcde	1.45±0.03de	38.6±1.3abcd	15.0±0.4abcd	1.7±0.09de	8.9±0.5abc
WX	56.8±3.5abcd	11.0±1.9abc	82.1±7.7abc	1.3±0.03a	43.6±1.5fg	17.1±1.9de	1.5±0.11bcd	11.5±0.6bcde
WY	47.9±1.1abc	10.3±2.9ab	84.2±10.5abc	1.38±0.02bcd	36.8±1.2ab	14.9±1.6abcd	1.6±0.09cd	9.5±0.6abc
XZ	72.9±7.4e	7.8±2.1ab	95.8±18.9abcd	1.47±0.02e	38.3±0.7abcd	16.1±1.6cd	1.1±0.12ab	14.5±1.1efg
XP	72.2±4.2e	27.3±5.8d	117.5±17.7cde	1.43±0.01cde	37.8±1.2abc	19.9±1.9ef	1.3±0.07abc	15.4±1.3fg
XC	43.2±3.4a	17.2±5.8bcd	66.2±9.6ab	1.42±0.03bcde	42.9±0.9efg	16.6±2.3cde	1.3±0.18abc	12.7±0.7def
XX	53.1±4.2abcd	5.1±0.5ab	89.9±3.0abc	1.41±0.02bcde	35.3±1.0a	14.8±0.6abcd	1.9±0.11e	7.9±0.3a
YJ	49.5±2.9abc	12.8±4abc	77.0±9.6abc	1.39±0.02bcd	38.2±1.0abcd	10.8±0.6a	1.3±0.07abc	8.2±0.4ab
YX	44.5±2.1ab	22.9±5.9cd	89.5±11.4abc	1.3±0.04a	45.5±1.9g	15.4±0.9bcd	1.2±0.07ab	13.6±1.1defg
YC	66.4±11.3de	3.5±1.1a	59.9±8.3a	1.44±0.02de	39±1.3abcde	14.8±0.9abcd	1.1±0.08ab	13.1±0.2defg
ZY	58.7±2.4bcd	6.3±1.4ab	73.8±5.7abc	1.35±0.02ab	41.9±0.9defg	12.7±0.3abc	1.5±0.06bcd	8.8±0.5abc

Different letters indicate significant differences ($p = 0.05$) among the 19 counties. Counties are arranged in English alphabetical order.

Table 3. Total C (TSOCS) and N (TSNS) stores in the surface soil layer (0–40 cm) of soils in 19 counties in Henan Province, China.

	C store (g m^{-2})	N store (g m^{-2})
HX	3541±261.8 abcd	410.1±18.9 cde
LK	3118.9±189.3 ab	399.7±23.6 bcde
LY	4106.7±294.1 abcd	290.4±22.2 a
LC	4023.8±372 abcd	398.6±35.6 bcde
QF	3229.7±140 abc	381.8±22.3 abcd
QX	5977.9±524.3 e	429.1±63.1 de
SS	3881.8±219 abcd	348.3±24.0 abcd
TK	3605.1±328.1 abcd	303.4±19.2 ab
TY	4300.4±98.5 bcd	494.3±24.9 ef
WX	4396.2±451.9 cd	379.4±23.4 abcd
WY	4081.3±455.2 abcd	429.3±25.3 de
XZ	4528.2±516.5 d	320.3±40.1 abc
XP	5709.2±582.9 e	369.8±21.9 abcd
XC	4614.5±609.9 d	366.3±47.7 abcd
XX	4323.8±224.7 bcd	558.2±35.1 f
YJ	3072.5±178.3 a	378.7±19.2 abcd
YX	4190.7±241.3 abcd	315.4±17.4 abc
YC	3926.2±304.8 abcd	299.3±26.3 ab
ZY	3413.4±104.3 abcd	398.1±21.7 bcde

Counties are arranged in English alphabetical order.

659 g m^{-2}. The highest value was in XX County and the lowest in LY County in N reserves (Table 3).

Linear regression analysis indicated that total crop production was significantly and positively correlated with SOC and TSN (Figure 4A), soil C and N store (Figure 4B), and soil C and N stocks (Figure 4C). Soil bulk density was significantly and positively correlated with soil N concentration ($r=0.25$, $p=0.008$, $n=114$), soil C ($r=0.21$, $p=0.03$, $n=114$) and N store ($r=0.43$, $p=0.001$, $n=114$). While soil porosity was significantly and negatively correlated with soil N concentration ($r=-0.19$, $p=0.05$, $n=114$), soil C ($r=-0.25$, $p=0.007$, $n=114$) and N store ($r=-0.32$, $p=0.001$, $n=114$).

Principle components analysis revealed that Axis 1, which explained 98% of the variation in all data (eigenvalue = 0.98), was highly correlated with soil C, whereas Axis 2, explaining 1% of the variation (eigenvalue = 0.09), was highly correlated with soil N. Thus, counties such as QX and XP located highly positive on Axis 1 with high levels of soil C, but other counties, such as LK, YJ, and QF, occupied positions toward the negative end of Axis 1 with low soil C (Figure 5).

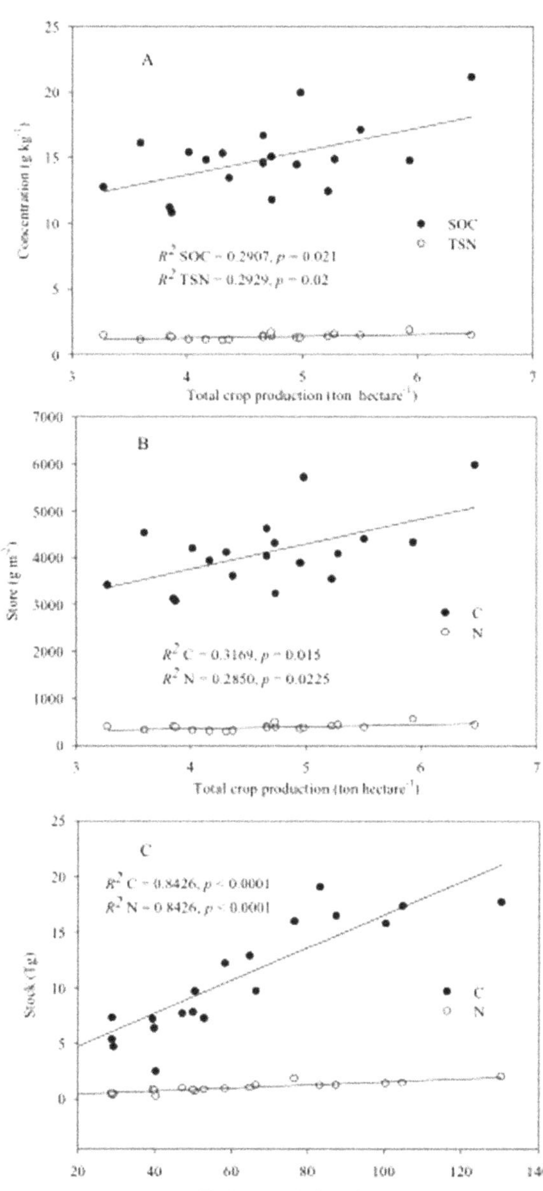

Figure 4. Linear regression analysis of total crop production in each county (ton ha^{-1}) with SOC and TSN (A) and with soil C and N store (0–40 cm) (B), and the total crop production of each county (10000 ton) with their soil C and N stock (C) (n = 19).

Discussion

Potential influences on crop yield

It is notable that 14 environmental (e.g., mean annual temperature and precipitation -Table 1) and soil variables (including extractable nutrients-Table 2) examined in our analysis of the data from the 19 counties in Henan Province were correlated with either wheat or total crop yield (data not shown), and total crop production were significantly, positively related to SOC and TSN, soil C and N store, and soil C and N stock (Figure 4). Part of this is likely related to the highly integrated nature of the measures of C and N stocks, i.e., their calculations combine soil concentrations of C and N, soil bulk density, sampling depth, and area of cultivation. However, all of these have been shown to directly influence crop performance. For example, increases in soil C have been shown to increase crop yield in other studies. Lal (2004, 2006) reported increases in yield from 20 to 70 kg ha^{-1} and 10 to 300 kg ha^{-1} for wheat and maize, respectively, following increases of 1 MT of C in agricultural soils in Africa [1,8]. Similarly, loss of soil C has been shown to decrease yield in agricultural soils of Canada and the U.S. [4,5].

Soil C-mediated increases of crop yields also may arise from improvements in soil structure and available water-holding

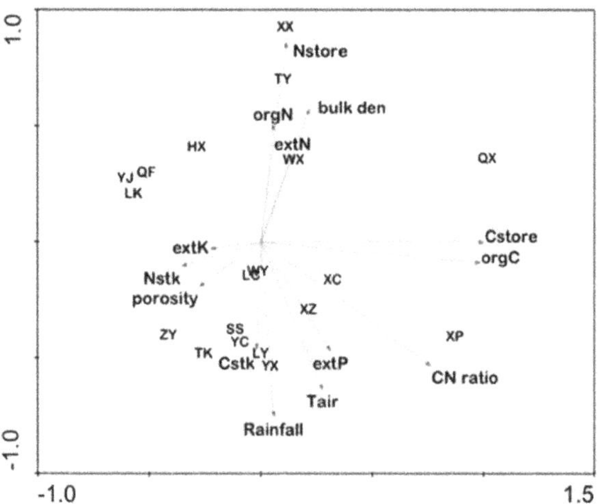

Figure 5. Principle components analysis of environmental and soil variables for agricultural soils in 19 counties within Henan Province. Length of arrows is directly proportional to their importance in explaining spatial patterns in the counties. Direction of the arrows indicates increasing values. Thus, the x-axis is primarily a gradient in soil C, whereas the y-axis is primarily a gradient in soil N and rainfall and secondarily a gradient in soil N. See Table 1 for key to county name abbreviations.

capacity. Enhanced soil structure, via increased soil C, generally arises from several processes, including increasing stability of soil aggregates [22,23,24]. As a result of the increased stability of the aggregates, soils become less prone to crusting, compaction, and erosion [25,26,28]. Emerson (1995) demonstrated that an increase of 1 g of soil organic matter (~50% of which is C) can increase available soil moisture by up to 10 g [27], which is enough to maintain crop growth between periods of rainfall of 5 to 10 days [8].

Spatial variation in cultivated soils

In this study, soil organic C concentration averaged 14.9 g kg^{-1} and total N averaged 1.4 g kg^{-1} in the 0–40 cm layer across all sites, while soil C and N stores averaged 4.1 kg C m^{-2} and 0.38 kg N m^{-2}, respectively. These values are comparable to published values from other regions of China, including 9–15 g C kg^{-1} and 1.2–1.8 g N kg^{-1} in northern China [12,29], and 16.1 g C kg^{-1} and 1.04 g N kg^{-1} in eastern and southern China [14,16,30]. Liu et al. (2011) reported soil C stores of 4.57 kg C m^{-2} in the Loess Plateau region in northwestern China [13].

Principal components analysis separated the 19 counties primarily along a gradient in soil C, with counties LK, YJ, QF, ZY, HX, and TK (mean soil C = 12.1 g C kg^{-1}) toward the lower end and XP and QX (mean soil C = 20.5 g C kg^{-1}) toward the upper end of Axis 1, which accounted for nearly 80% of the variation in soil and environmental data (Figure 5). Spatial variation in soil organic C in agricultural systems can be influenced by several factors, including microclimate, soil type, topography, and especially human activity [31].

Spatial variation in soil N was essentially orthogonal to that of soil C. This was surprising since typically, the two are highly correlated in terrestrial ecosystems [32]. As a result, the secondary gradient (i.e., Axis 2) was one of soil N, with counties TK, YC, SS, LY, YX, and XP (mean soil N = 1.15 g N kg^{-1}) located toward the lower end of Axis 2 (accounting for <10% of variation) and XX and TY (mean soil N = 1.81 g N kg^{-1}) located toward the upper end of Axis 2 (Figure 5). Although C and N are often correlated through their organic forms in plant detritus, spatial variation of N in soils of agro-ecosystems can also be greatly influenced by the extensive use of N fertilizers.

Management methods used in crop production systems, including tillage practices and fertilizer use, can affect soil C and N on broad spatial scales, including that of an entire Province [33]. Over the course of repeated seasons of crop growth in Henan Province, agricultural fields are repeatedly subjected to soil tillage, planting, fertilization, irrigation, and harvest, all of which potentially influence soil C and N stores [30,34]. In contrast, Zhang et al. (2012) reported that raised-bed planting, a viable alternative to conventional tillage, can significantly enhance the yield of summer maize while simultaneously improving soil structure, as well as the structure and function of microbial communities essential to the quality of agricultural soils [22].

Results presented in the current study underscore the complexity of factors that can impact agricultural soils and their ability to produce crops to meet the ever-increasing demand in China resulting from population growth. Some of the spatial pattern exhibited in ordination space (Figure 5) is clearly related to regional factors, such as microclimate. For example, WY and LC are adjacent to each other in Henan Province (Figure 1) and are also closely clustered in ordination space, indicating that they are very similar with respect to environmental and soil characteristics. XP and SS, however, are also adjacent counties; yet occur distant from each other in ordination space, indicating great dissimilarity in environmental and soil factors. Agronomists should take into account the large spatial variability in important components of the soils in Henan Province, especially in the variation of soil C and N, when considering appropriate agronomic management practices.

Author Contributions

Conceived and designed the experiments: XLZ QW CHL. Performed the experiments: XLZ QW YLW. Analyzed the data: XLZ QW FSG. Contributed reagents/materials/analysis tools: XLZ QW YLW FNC. Contributed to the writing of the manuscript: XLZ FSG CHL.

References

1. Lal R (2004) Soil carbon sequestration impacts on global climate change and food security. Science 304: 1623–1627.
2. Liu DW, Wang ZM, Zhang B, Song KS, Li XY, et al. (2006) Spatial distribution of soil organic carbon and analysis of related factors in croplands of the black soil region, Northeast China. Agriculture, Ecosystems and Environment 113: 73–81.
3. Smith WN, Desjardins RL, Pattey E (2000) The net flux of carbon from agricultural soils in Canada 1970–2010. Global Change Biology 6: 557–568.
4. Bauer A, Black AL (1994) Quantification of the effect of soil organic matter content on soil productivity. Soil Science Society of America Journal 58: 185–193.
5. Larney FJ, Janzen HH, Olson BM, Lindwall CW (2000) Soil quality and productivity response to simulated erosion and restorative amendments. Canadian Journal of Soil Science 80: 515–522.
6. Hairiah K, Van Noordwijk M, Cadisch G (2000) Crop yield, C and N balance of the three types of cropping systems on an Ultisol in northern Lampung. Netherland Journal of Agricultural Science 48: 3–17.
7. Duxbury JM (2001) Long-term yield trends in the rice-wheat cropping system: results from experiments in Northwest India. Journal of Crop Production 3: 27–52.
8. Lal R (2006) Enhancing crop yields in the developing countries through restoration of the soil organic carbon pool in agricultural lands. Land Degradation and Development 17: 197–209.
9. Batjes NH (2002) Carbon and nitrogen stocks in the soils of Central and Eastern Europe. Soil Use and Management 18: 324–329.

10. Maia SMF, Ogle SM, Cerri CC, Cerri CEP (2010) Changes in soil organic carbon storage under different agricultural management systems in the Southwest Amazon Region of Brazil. Soil and Tillage Research 106: 177–184.
11. Piao SL, Fang JY, Ciais P, Peylin P, Huang Y, et al. (2009) The carbon balance of terrestrial ecosystems in China. Nature 458, doi:10.1038/nature 07944.
12. Wang ZM, Zhang B, Song KS, Liu DW, Ren CY (2010) Spatial variability of soil organic carbon under maize monoculture in the Song-Nen plain, Northeast China. Pedosphere 20: 80–89.
13. Liu ZP, Shao MA, Wang YQ (2011) Effect of environmental factors on regional soil organic carbon stocks across the Loess Plateau region, China. Agriculture, Ecosystems and Environment 142: 184–194.
14. Liao QL, Zhang XH, Li ZP, Pan GX, Smith P, et al. (2009) Increase in soil organic carbon stock over the last two decades in China's Jiangsu Province. Global Change Biology 15: 861–875.
15. Zhang HB, Luo YM, Wong MH, Zhao QG, Zhang GL (2007) Soil organic carbon storage and changes with reduction in agricultural activities in Hong Kong. Geoderma 139: 412–419.
16. Feng S, Tan S, Zhang A, Zhang Q, Pan G, et al. (2011) Effect of household land management on cropland topsoil organic carbon storage at plot scale in a red earth soil area of South China. Journal of Agricultural Science 149: 557–566.
17. Harper RJ, Gilkes RJ (1995) Some factors affecting the distribution of carbon in soils of a dry land agricultural system in southwestern Australia. In: Lal R, Kimble JM, Follett RF, Stewart BA (editors). Assessment Methods for Soil Carbon. CRC Press. Boca Raton, FL, USA. PP.577–591.
18. Guo JH, Liu XJ, Zhang Y, Shen JL, Han WX, et al. (2010) Significant Acidification in Major Chinese Croplands. Science 327: 1008–1010.
19. Pan GX, Li LQ, Wu LS, Zhang XH (2003) Storage and sequestration potential of topsoil organic carbon in China's paddy soils. Global Change Biology 10: 79–92.
20. Wu HB, Guo ZT, Gao Q, Peng CH (2009) Distribution of soil inorganic carbon storage and its changes due to agricultural land use activity in China. Agriculture, Ecosystems and Environment 129: 413–421.
21. Gilliam FS, Saunders NE (2003) Making more sense of the order: A review of Canoco for Windows 4.5, PC-ORD version 4 and SYN-TAX 2000. Journal of Vegetation Science 14: 297–304.
22. Zhang XL, Ma L, Gilliam FS, Wang Q, Liu T, et al. (2012) Effects of raised-bed planting for enhanced summer maize yield on soil microbial functional groups and enzyme activity in Henan Province, China. Field Crops Research 130: 28–37.
23. Feller C, Beare MH (1997) Physical control of soil organic matter dynamics in tropics. Geoderma 79: 69–116.
24. Haynes RJ, Naidu R (1998) Influence of lime, fertilizer and manure applications on soil organic matter content and soil physical conditions: a review. Nutrient Cycling in Agroecosystems 51: 123–137.
25. Diaz-Zorita M, Grosso GA (2000) Effect of soil texture, organic carbon and water retention on the compatibility of soils from the Argentinean Pampas. Soil and Tillage Research 54: 121–126.
26. Schertz DL, Moldenhauer WC, Livingston SJ, Weeisies GA, Hintz AE (1989) Effect of past soil erosion on crop productivity in Indiana. Journal of Soil and Water Conservation 44: 604–608.
27. Emerson WW (1995) Water-retention, organic-carbon and soil texture. Australian Journal of Soil Research 33: 241–251.
28. Powlson DS, Hirsch PR, Brookes PC (2001) The role of soil micro-organisms in soil organic matter conservation in the tropics. Nutritional Cycling in Agroecosystems 61: 41–51.
29. Du ZL, Ren TS, Hu CS (2010) Tillage and residue removal effects on soil carbon and nitrogen storage in the North China Plain. Soil Science Society of American Journal 74: 196–202.
30. Pan GX, Li LQ, Zhang Q, Wang XK, Sun XB, et al. (2005) Organic carbon stock in topsoil of Jiangsu Province, China, and the recent trend of carbon sequestration. Journal of Environmental Sciences 17: 1–7.
31. Post WM, Pastor J, Zinke PJ, Stangenberger AG (1985) Global patterns of soil nitrogen storage. Nature 317: 613–616.
32. Gilliam FS, Dick DA, Kerr ML, Adams MB (2004) Effects of silvicultural practices on soil carbon and nitrogen in a nitrogen saturated Central Appalachian (USA) hardwood forest ecosystem. Environmental Management 33: S108–S119.
33. Pan GX, Zhao QG (2005) Study on evolution of organic carbon stock in agricultural soils of China: facing the challenge of global change and food security. Advances in Earth Science 20: 384–393 (in Chinese).
34. Dersch G, Böhm K (2001) Effects of agronomic practices on the soil carbon storage potential in arable farming in Austria. Nutrient Cycling in Agroecosystems 60: 49–55.

22

Contribution for the Derivation of a Soil Screening Value (SSV) for Uranium, Using a Natural Reference Soil

Ana Luisa Caetano[1,2*], Catarina R. Marques[1,2], Ana Gavina[6,2], Fernando Carvalho[3], Fernando Gonçalves[1,2], Eduardo Ferreira da Silva[4], Ruth Pereira[5,6]

1 Department of Biology, University of Aveiro, Campus Universitário de Santiago, Aveiro, Portugal, **2** CESAM, University of Aveiro, Campus Universitário de Santiago, Aveiro, Portugal, **3** Nuclear and Technological Institute (ITN) Department of Radiological Protection and Nuclear Safety, Sacavém, Portugal, **4** Department of Geosciences, University of Aveiro, GeoBioTec Research Center, Campus Universitário de Santiago, Aveiro, Portugal, **5** Department of Biology, Faculty of Sciences of the University of Porto, Porto, Portugal, **6** Interdisciplinary Centre of Marine and Environmental Research (CIIMAR/CIMAR), University of Porto, Porto, Portugal

Abstract

In order to regulate the management of contaminated land, many countries have been deriving soil screening values (SSV). However, the ecotoxicological data available for uranium is still insufficient and incapable to generate SSVs for European soils. In this sense, and so as to make up for this shortcoming, a battery of ecotoxicological assays focusing on soil functions and organisms, and a wide range of endpoints was carried out, using a natural soil artificially spiked with uranium. In terrestrial ecotoxicology, it is widely recognized that soils have different properties that can influence the bioavailability and the toxicity of chemicals. In this context, SSVs derived for artificial soils or for other types of natural soils, may lead to unfeasible environmental risk assessment. Hence, the use of natural regional representative soils is of great importance in the derivation of SSVs. A Portuguese natural reference soil PTRS1, from a granitic region, was thereby applied as test substrate. This study allowed the determination of NOEC, LOEC, EC_{20} and EC_{50} values for uranium. Dehydrogenase and urease enzymes displayed the lowest values (34.9 and <134.5 mg U Kg, respectively). *Eisenia andrei* and *Enchytraeus crypticus* revealed to be more sensitive to uranium than *Folsomia candida*. EC_{50} values of 631.00, 518.65 and 851.64 mg U Kg were recorded for the three species, respectively. Concerning plants, only *Lactuca sativa* was affected by U at concentrations up to 1000 mg U kg^1. The outcomes of the study may in part be constrained by physical and chemical characteristics of soils, hence contributing to the discrepancy between the toxicity data generated in this study and that available in the literature. Following the assessment factor method, a predicted no effect concentration (PNEC) value of 15.5 mg $kg^{-1}{}_{dw}$ was obtained for U. This PNEC value is proposed as a SSV for soils similar to the PTRS1.

Editor: Stephen J. Johnson, University of Kansas, United States of America

Funding: ALC was supported by a PhD grant from Fundação para a Ciência e Tecnologia (FCT) (http://www.fct.pt/). The funders had no role in study design, data collection and analysis, decision to publish, or preparation of the manuscript.

Competing Interests: The authors have declared that no competing interests exist.

* Email: ana.caetano@ua.pt

Introduction

Uranium (U) is a natural soil component, being originated from rocks in the Earth's crust, where it mainly occurs in the form of oxides. Natural processes acting on rocks and soils, such as wind, water erosion, dissolution, precipitation and volcanic activity contribute for U dispersal in the environment [1]. The use of U as fuel in nuclear power plants has driven to its large-scale exploration worldwide. The U exploration became significantly important in the world during the Second World War, and later on during the Cold War, in both cases to supply military needs of the greatest potencies. Recently, the World Nuclear Association estimated worldwide reserves of U at 5.4 million tons in 2009, of which Australia had about 31%, followed by Kazakhstan (12%), Canada and Russia with 9% (http://www.world-nuclear.org/info/inf75.html). The remarkable energy crisis that is currently faced worldwide due to the exhaustion of carbon based energy resources is demanding further extraction of U, as nuclear energy arises as a potential solution. Hence, it is expected that the mining and milling of U will increase in the next decades, contributing for its widespread in the environment [2].

During the last century, Portugal has actively explored radioactive ores and was for some time ranked as one of the main U producers. The extraction of U ore in Portugal started in 1908, first driven by the interest in radium (being U a by-product) and then by the interest in its military applications, till 2001 [3,4]. Most of the old U mines were located in the granitic regions of the Iberian Meseta, in the centre-north of Portugal (Beiras), [5]. Nowadays, although the mining activities ceased, like in several other places in the world, the old U mines represent a serious environmental problem, due to waste accumulation (mainly tailings and sludge) and improper disposal of radioactive material, composed by U and its daughter radionuclides [1,5–16]. Soils and water are the two major environmental matrices affected by U contamination.

U has a long half-life, persisting in nature as different isotopes, with different chemical and radiological characteristics [17]. The toxic effects induced by this metal are caused by both properties.

However, since U isotopes mainly emit alpha particles, with little penetration capacity, the main radiation hazards only occur after ingestion or inhalation of these isotopes and daughter radionuclides [17]. Once in the soil, U interacts with all the components of this matrix, such as clay minerals, aluminum and iron oxides, organic matter and microorganism, in a very complex system, where pH and organic matter seem to have the major role in controlling U mobility (pH 6) and leaching (pH<6) [18]. The high mobility/availability of U will in turn increase the ecological risks posed to soil and water compartments [19–27].

The soil has been recognized as an important compartment that provides crucial ecosystem services (e.g., filtering of contaminants, reservoir of carbon and a bank of genes) and is the support of agro-sylvo-pastoral production [28,29] and of several other human activities. The soil compartment offers raw materials (e.g., peat, clay, ore) and contributes for climate regulation and biodiversity conservation, as well as other cultural services [30,31]. The recognition of the importance of maintaining the provision of such services has increased the necessity to create appropriate legal tools to correctly and effectively protect and manage this resource. In this sense, the Soil Framework Directive proposed by the Commission of the European Communities (CEC), aims to establish a common strategy for the protection and sustainable use of soils [32]. For that end, this proposal defines measures for the identification of the main problems faced by soils, the adoption of strategies to prevent their degradation, as well as for the rehabilitation of contaminated or degraded soils [33]. The Soil Framework Directive will fill in the gap regarding soil protection, since this compartment has never been a target of specific protection policies at the European Community level [32]. Many countries, committed in regulating the management of contaminated land, have adopted generic quality standards, the soil screening values (SSVs) [34]. SSVs are concentration thresholds above which, more site-specific evaluations are required to assess the risks posed by soil contamination [35]. The SSVs should provide a level of protection to terrestrial species and ecological functions of the soil [35–37]. SSVs are particularly useful for the first tier of Ecological Risk Assessment (ERA) processes applied to contaminated sites, supporting the decision-making at this initial stage of assessment [38], which at the end is aimed in setting priorities for remediation and risk reduction measures [39]. In the case of Portugal, SSVs for soils have never been established for metals or organics. Only threshold concentrations of metals on sewage sludge were legally established to regulate the application of this solid waste on agricultural soils [40]. However, they are not appropriate for soil ERA purposes.

The use of natural reference soils in ecotoxicological tests has been recommended by several authors [41–43]. This is because the properties of the OECD artificial soil are not representative of the great majority of natural soils [44]. Different levels of toxicity, for each contaminant, can be expected in soils with different properties [45–48], hence it is important each country derives their own SSVs using natural reference soils representing the main types of soils within their territories. In this context, the main aim of this work was to obtain ecotoxicological data for U, performing soil enzymes activity tests, invertebrates and plant tests, using for that a Portuguese natural reference soil (PTRS1), that represents one of the dominant types of soil from a granitic region (cambisol) of the country [49]. As a result, enough data are gatheredas to make the first proposal of a SSV for this metal.

Materials and Methods

The present study used a natural soil that was collected in a non-protected area, requiring no specific permission for its collection. Further, no work with endangered species was performed, and no vertebrate species were used in the ecotoxicological assays. Only tests with invertebrates and plants were performed. The invertebrates were obtained from laboratorial cultures maintained by the authors of this manuscript and plant seeds were obtained from a local supplier.

1. Test soil

The natural soil (PTRS1) used as test substrate in this study was collected in Ervas Tenras [Pinhel, Guarda, Portugal center; geographical coordinates: 40°44′4.27″N and 7°10′54.3″W), at 655 m altitude, in a granitic region.

A composite soil sample was collected and immediately brought to the laboratory where it was air dried. Another portion of the soil, was immediately sieved through a 2 mm mesh size and the sieved fraction (<2 mm) was stored in polyethylene bags, at −20°C, until further analysis of soil microbial parameters, which were performed within the period of one month. For the tests with soil organisms and plants, the soil was passed through a 4 mm mesh sieve and the sieved fraction (<4 mm) was defaunated through two freeze–thawing cycles (48 h at −20°C followed by 48 h at 25°C), before the beginning of the assays.

The physical and chemical properties (including total metal contents) of the PTRS1 soil were presented in a preliminary study by Caetano et al. [49], aimed in characterizing this soil as a reference substrate for ecotoxicological purposes. The main properties of the PTRS1 are also described in Tables 1 and 2. Briefly, soil-KCl 1 M and soil-deionized water suspensions (1:5 m/v) were used for pH (KCl, 1 M) and pH-H_2O measurements, respectively, according to ISO 17512–1 [50]- After 15 min of magnetic stirring and 1 h resting period, the pH of the suspension was measured using a WTW 330/SET-2 pH meter. A soil water suspension (1:5 w/v) was used for the measurement of soil conductivity [51] Ten grams of PTRS1 were mechanically shaken in polypropylene flasks with 50 ml with deionized water filtered in a Milli-Q equipment (hereinafter referred as deionized water), water for 15 min. The mixture was left to rest overnight for soil bulk settling [51]. The conductivity of the resulting suspension was measured using an LF 330/SET conductivity meter. Soil water content was determined from the loss of weight after drying at 105°C, for 24 h. Organic matter (OM) content was determined by loss of ignition of dried soil samples at 450°C during 8 h [52]. For determination of water holding capacity (WHC) polypropylene flasks were prepared with a filter paper-replaced bottom, which after being filled up with soil samples, were immersed in water for 3 h. After this period, samples were left for water drainage during 2 h and the WHC was determined accounting to the loss of weight after drying at 105°C until weight stabilization [50].

2. Test substance

For all the test organisms, the natural soil was spiked with a stock solution of uranyl nitrate 6-hydrate, $UO_2(NO_3)_26H_2O$ (98%, PANREAC) prepared with deionized water in order to obtain a range of concentrations, which were ascertained by range finding tests performed with the different test species.

For soil enzyme tests, the PTRS1 soil was spiked with the following concentrations: 0.0, 134.6, 161.5, 193.8, 232.5, 279.0, 334.8, 401.8, 482.2, 578.7, 694.4, 833.3, 1000 mg U $kg^{-1}{}_{dw}$. To obtain these concentrations, the stock solution of uranyl nitrate was diluted in the volume of deionized water required to adjust the soil moisture at 80% of its maximum water holding capacity (WHC_{max}).

The following U concentrations were used to expose the earthworms in the reproduction tests: 0.0, 113.1, 124.4, 136.9, 150.5, 165.6, 231.9, 324.6, 454.5, 500.0, 550.0, 605.0, 665.5 mg U $kg^{-1}{}_{dw}$. For potworms, collembolans and terrestrial plant assays the same range of concentrations was tested: 0.0, 167.4, 192.5, 221.4, 254.6, 292.7, 336.6, 420.8, 526.0, 657.5, 756.1, 869.6, 1000 mg U $kg^{-1}{}_{dw}$.

The volume of deionized water required to adjust the WHC of the soil to a given percentage of its maximum value was used to dilute the stock solution for these tests. After spiking the soil was left to rest for equilibration for 48 h before testing.

3. Ecotoxicological assessment

3.1 Soil microbial activity. For testing the effect of increasing concentrations of U on soil microbial parameters, a 30-day exposure was firstly conducted. Ten grams of sieved PTRS1 soil per replicate and concentration were spiked with different U concentrations, a total of three replicates were used per treatment. Six replicates with the same amount of soil only moistened with deionized water were also prepared for the control. The soil was incubated at 20±2°C and a photoperiod of 16 h^L:8 h^D. During the incubation period, the soil moisture was weekly monitored by weighing the pots, and whenever needed it was adjusted to 80% of its WHC_{max} by adding deionized water. At the end, 1 g of each replicate from the control and concentrations tested was stored in individual falcon tubes at −20°C for approximately one month. Thereby, a total of 9 sub-replicates were made for each concentration. The soil was thawed at 4°C before analysis.

The biochemical parameters analyzed were: the activity of arylsulphatase, dehydrogenase, urease, and cellulase enzymes and changes in the nitrogen mineralization (N mineralization) and potential nitrification.

For the determination of arylsulphatase activity, the method proposed by Tabatabai and Bremner [53] and Schinner et al. [54] was followed. After addition of 1 mL of p-nitrophenylsulfate (0.02 M), soil sub-samples were incubated for one hour, at 37°C. The nitrophenyl liberated by the activity of arylsulphatase was extracted and colored with a 4 mL of sodium hydroxide (0.5 M) and determined photometrically at 420 nm. The results were expressed as μg p-nitrophenylsulfate (p-NP) g^{-1} soil $_{dw}$ h^{-1}.

The method proposed by Öhlinger [55] was used to assess the dehydrogenase activity. The samples were suspended in 1 mL of trifeniltetrazol chloride (TTC) (3.5 g L^{-1}) and incubated at 40°C for 24 h. The triphenylformazan (TPF) produced was extracted with acetone and measured spectrophotometrically at 546 nm. The results were expressed as μg TPF g^{-1} $soil_{dw}$ h^{-1}.

The cellulase activity was tested according to the method proposed by Schinner et al. [54] and Schinner and von Mersi [56]. The reducing sugars produced during the incubation period, after addition of 1.5 mL of acetate buffer (2 M), caused the reduction of hexacyanoferrate (III) potassium to hexacyanoferrate (II) potassium in an alkaline solution. This last compound reacts with ferric ammonium sulfate in acid solution to form a ferric complex of hexacyanoferrate (II), of blue colour, which is colorimetrically measured at 690 nm and expressed as μg glucose g^{-1} $soil_{dw}$ 24 h^{-1}.

N mineralization activity was measured according to Schinner et al. [54]. For this purpose the soil samples were incubated for 7 days at 40°C. During this period, the organic forms of N were converted to inorganic forms (mainly ammonium ion, NH_4^+),

Table 1. Physical and chemical properties of PTRS1 soil (retrieved from Caetano et al. [49]).

	pH (H_2O)	pH (KCl, 1 M)	Conductivity (mS cm^{-1})	OM (%)	WHC (%)	Particle-size distribution (%)				[U] (mg Cu kg^{-1} $soil_{dw}$)
						Clay (<4 μm)	Silt (4–63 μm)	Sand (63 μm–2 mm)	Gravel (>2 mm)	
PTRS1	5.9±0.09	4.3±0.02	4.8±0.23	6.5±0.004	23.9±1.84	3.3	22.8	46.9	23.9	9.0

pH (H_2O), pH (KCl, 1 M), OM (organic matter), and WHC (water holding capacity) are represented as average ± STDEV.

Table 2. Pseudo-total concentrations (mg/kg) of metals recorded in PTRS1 soil (average ± standard deviation) extracted with aqua régia, (retrieved from Caetano et al.[49]).

Metal	PTRS1
Ag	0.1±0.0
Al	25628.5±5130.0
B	2.2±0.8
Ba	45.8±8.0
Be	1.2±0.2
Cd	0.1±0.1
Co	5.6±1.1
Cr	10.8±2.1
Cu	9.0±1.8
Fe	24921.4±4534.4
Li	124.4±22.9
Hg	5253.5±1025.5
Mn	386.8±77.9
Mo	0.9±0.2
Na	78.1±14.9
Ni	4.6±0.9
Pb	12.5±2.2
Sb	0.2±0.0
Sn	10.4±1.9
U	7.8±1.7
V	37.8±14.1
Zn	57.1±8.9

which were determined by a modification of the Berthelot reaction, after extraction with 3 mL of potassium chloride (2 M). The reaction of ammonia with sodium salicylate in the presence of sodium dichloroisocyanurate formed a green colored complex in alkaline pH that was measured at 690 nm and expressed as µg N g^{-1} $soil_{dw}$ d^{-1}.

The urease activity was assayed according to the method proposed by Kandeler and Gerber [57] and, Schinner et al. [54]. The samples were incubated for 2 h at 37°C after the addition of 4 mL of a buffered urea solution (720 mM). The ammonia released was extracted with 6 mL of potassium chloride (2 M) and determined by the modified Berthelot reaction. The quantification was based on the reaction of sodium salicylate with ammonia in the presence of chlorinated water. UR was detected at 690 nm and expressed as µg N g^{-1} $soil_{dw}$ 2 h^{-1}.

The quantification of potential nitrification was determined by the method of Kandeler [58], which is a modification of the technique proposed by Berg and Rosswall [59]. The ammonium sulphate (4 mL, 10 mM) was used as substrate, and soil samples were incubated for 5 h, at 25°C. Nitrate released during the incubation period was extracted with 1 mL of potassium chloride (2 mM) and determined colorimetrically at 520 nm. This reaction was expressed as µg nitrite (N) g^{-1} $soil_{dw}$ h^{-1}.

3.2. Invertebrate and plant tests. Test organisms and culture conditions: The earthworm *Eisenia andrei* (Oligochaeta: Lumbricidae), the potworm *Enchytraeus crypticus* (Oligochaeta: Enchytraeidae) and the springtail *Folsomia candida* (Collembola: Isotomidae) were used as invertebrate test organisms. All organisms were obtained from laboratorial cultures, kept under controlled environmental conditions (temperature: 20±2°C; photoperiod: 16 h^L:8 h^D). The earthworms (*E. andrei*) are maintained in plastic boxes (10 to 50 L) containing a substrate composed by peat, dry and defaunated horse manure (through two freeze–thawing cycles (48 h at −20°C followed by 48 h at 65°C), and deionized water. The pH of the culture medium is adjusted to 6.0–7.0 with $CaCO_3$. The organisms are fed, every 2 weeks, with six table spoons of oatmeal previously hydrated with deionized water and cooked for 5 min. The potworms (*E. crypticus*) are cultured in plastic containers (25.5 cm length; 17.4 cm width; 6.5 cm height), which are filled with pot soil moistened to the nearest 60% of its WHC_{max} and with pH adjusted to 6.0±0.5. The organisms are fed twice a week with a tea spoon of macerated oat. The collembolans (*F. candida*) are maintained in plastic containers filled with culture medium composed by moistened Plaster of Paris mixed with activated charcoal 8:1 (w:w). They are fed with half of a tea spoon of granulated dry yeast, twice a week. The food is added in small amounts to avoid spoilage by fungi.

Seeds from four plant species (two dicotyledonous and two monocotyledonous), purchased from a local supplier, were used for seed germination and growth tests: *Avena sativa, Zea mays, Lacuta sativa* and *Lycopersicon esculentum*.

Reproduction tests with invertebrates: Previous studies from our team, at least with earthworms from the same laboratorial cultures, have proved that these organisms were not exposed to meaningful levels of metals (especially U, in laboratorial culture conditions) [60]. The accomplishment of validity criteria, by all the controls of the assays (herein described) with the three invertebrate species, also confirmed that the test animals were not previously exposed to toxic levels of metals through test containers, substrates or food. The reproduction tests with *E. andrei*, *E. albidus* and *F. candida* were carried out according to the ISO guidelines 11268-2 [61], 16387 [62] and 11267 [63], respectively. Each replicate of the invertebrate tests contained 10 individuals in a certain developmental stage: the earthworms had a fully developed clitellum and an individual fresh weight between 250 and 600 mg; the potworms were 12-mm size; and the springtails were 10–12 days old. Five hundred grams of dry soil were weighted per test vessel for earthworms. For the tests with potworms and collembolans 20 g and 30 g of soil were weighted per replicate, respectively. Following an ECx sampling design, which considers more concentrations and less number of replicates, two replicates per concentration and five replicates for the control were prepared in the reproduction tests with *E. andrei*. Adult earthworms were removed from the test containers after 28 days. The produced cocoons persisted in the soil until 56 days have been completed. After this period, the juveniles from each test container were counted. During the test, organisms were fed once a week, with 5 g per box of defaunated horse manure (using the same procedure above described), and the soil moisture content was weekly monitored (following the procedures outlined in ISO guideline 11268-2 [61]).

The *E. albidus* reproduction test was held for 28 days and the adults were left in the vessels until the end of the test. About 2 mg of rolled oats were placed on the soil surface, weekly to feed the animals. At the end of the test, the potworms were killed with alcohol, colored with Bengal red and counted according to the Ludox Flotation Method, as described in ISO 16387 [62]. The reproduction tests with *F. candida* took four weeks to be completed. The collembolans were fed with granulated dry yeast, obtained from a commercial supplier, being weekly added (about 2 mg of yeast per test vessel) to the soil surface. At the end of the test, the containers were filled with water and the juveniles were

Table 3. Toxicity data obtained for copper (mg U kg^{-1} soil$_{dw}$) in PTRS1 soil on soil microbial processes, invertebrates and plants.

Biota	Endpoint	NOEC	LOEC	EC_{20}	EC_{50}
Microbial parameters					
Arylsulphatase		232.5	279	155.3 (84.76–255.87)	295.6 (216.09–375.17)
Dehydrogenase		<134.5	≤134.5	34.9 (20.52–59.35)	110.3 (83.25–137.47)
Nitrogen mineralization	Enzyme activity	694.4	833.3	152.2 (46.66–257.79)	347.0 (211.25–482.91)
Celulase		≤134.5	≥134.5	n.d.	n.d.
Urease		<134.5	≤134.5	<134.5	<134.5
Potencial nitrification		<134.5	≤134.5	429.5 (229.53–629.46)	610.0 (459.07–761.11)
Invertebrates					
Eisenia andrei	Rep. (56 days)	500.0	550.0	474.8 (391.47–558.04)	631.0 (532.78–699.21)
Enchytraeus crypticus	Rep. (28 days)	420.8	526.0	469.7 (355.47–584.04)	518.6 (480.40–556.90)
Folsomia candida	Rep. (28 days)	675.5	756.1	343.4 (172.23–514.60)	851.64 (606.10–1097.18)
Plants					
Avena sativa	Germination	≥1000	>1000	n.d.	n.d.
Zea mays	Germination	≥1000	>1000	n.d.	n.d.
Lactuca sativa	Germination	≥1000	>1000	n.d.	n.d.
Lycopersicon esculentum	Germination	≥1000	>1000	n.d.	n.d.
Avena sativa	Fresh mass	≥1000	>1000	n.d	n.d.
Zea mays	Fresh mass	≥1000	>1000	n.d	n.d.
Lactuca sativa	Fresh mass	≥1000	>1000	n.d	n.d.
Lycopersicon esculentum	Fresh mass	≥1000	>1000	n.d	n.d.
Avena sativa	Dry mass	≥1000	>1000	n.d.	n.d.
Zea mays	Dry mass	≥1000	>1000	n.d.	n.d.
Lactuca sativa	Dry mass	<167.4	≤167.4	n.d.	n.d.
Lycopersicon esculentum	Dry mass	≥1000	>1000	n.d.	n.d.

For ECx point estimates the 95% confidence limits are presented in brackets. n.d.- not determined; Rep. – reproduction.

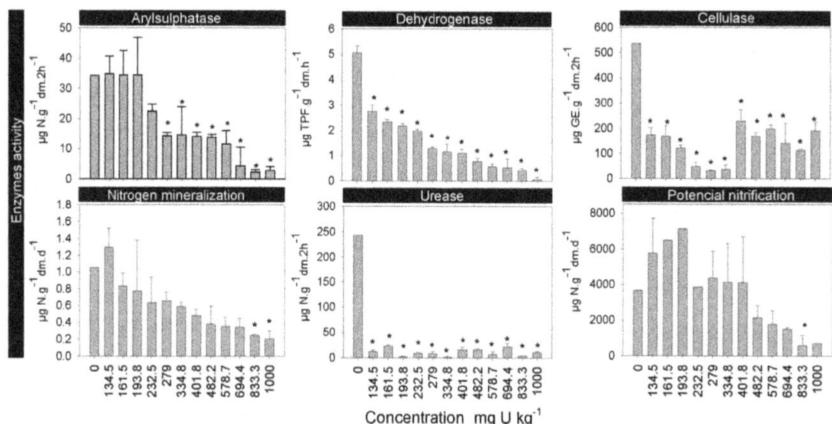

Figure 1. Soil enzyme activities, N mineralization and potential nitrification. Response of the arylsulphatase, dehydrogenase, cellulase urease, activity, N mineralization and potential nitrification to soils spiked with a range of uranium concentrations. The error bars indicate the standard deviation. The asterisks point out significantly differences from the control (P<0.05).

Table 4. Toxicity of copper (mg U $kg^{-1}soil_{dw}$) reported in the literature for the reproduction of soil invertebrates using different soil types with different physical and chemical characteristics.

Species	Endpoint	Soil type	Physical-chemical parameters			Point estimates (mg U $kg^{-1}soil_{dw}$)				Reference
			pH	OM (%)	Clay (%)	NOEC	LOEC	EC_{20}	EC_{50}	
		Canadian soil	6.2	1.0	2.0	n.d.	n.d.	>1000	n.d.	Sheppard and Stephenson [86]
Eisenia fetida	Rep. (56 days)	Canadian soil	6.2	1.0	2.0	n.d.	n.d.	>1120	n.d.	Sheppard and Stephenson [86]
		Canadian soil	7.5	2.2		>838	n.d.	n.d.	n.d.	Sheppard and sheppard [98]
		Canadian soil	7.5	18.4		>994	n.d.	n.d.	n.d.	Sheppard and sheppard [98]
Folsomia Candida	Rep. (28 days)	Canadian soil	7.5	2.2	24.0	n.d.	n.d.	840.0	n.d.	Sheppard and Stephenson [86]
		Canadian soil	7.5	n.d.	n.d.	n.d.	n.d.	>720	n.d.	Sheppard and Sheppard [98]
Elymus lanceolatus	Germination	Canadian soil	6.2	1	2	n.d.	>1000	n.d.	n.d.	Sheppard and Stephenson [86]
Elymus lanceolatus	Germination	Canadian soil	7.5	2.2	24	n.d.	>1001	n.d.	n.d.	Sheppard and Stephenson [86]
Zea mays	Dry mass	European soil	5.2	2.5	n.d.	n.d.	>100	n.d.	n.d.	Stojanović et al., [108]

OM - organic matter, Rep. - reproduction, n.d. - not determined., germ.- germination.

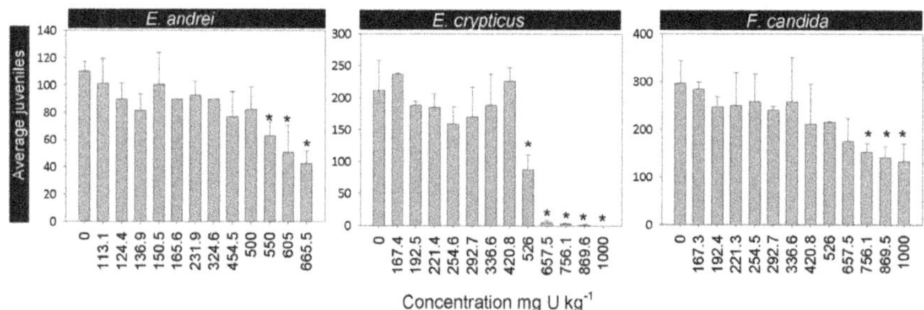

Figure 2. Reproduction of invertebrates. Results obtained exposing *Eisena andrei*, *Enchytraeus crypticus* and *Folsomia candida*, to natural PTRS1 soil, contaminated with different concentrations of U. The error bars indicate the standard deviation. The asterisks point out significantly differences from the control ($P<0.05$).

counted after flotation. The addition of a few dark ink drops provided a higher contrast between the white individuals and the black background. The organisms were then counted through the use of the ImageJ software (online available: http://rsb.info.nih.gov/ij/download.html). The exposure was carried out at 20±2°C and a photoperiod of $16^L:8^D$. For both species five replicates of uncontaminated natural PTRS1 soil were prepared for the control. The same ECx sampling design applied for earthworms was followed. However, in order to reduce the variability of the results, three replicates were prepared per test concentration (instead of two for the earthworms).

Seed germination and plant growth tests: Germination and growth tests with terrestrial plants were performed following standard procedures described by the ISO guideline 11269-2 [64]. For this purpose, 200 g_{dw} of the spiked soil with the concentrations described above was placed in plastic pots (11.7 cm diameter, 6.2 cm height) and tested. In this case, the amount of water required to adjust the WHC_{max} of the soil to 45% was used to dilute the stock solution and to moist the soil at the beginning of the test. The soil was placed in the plastic pots (11.7 cm diameter, 6.2 cm height). In the bottom of each plastic pot a hole was previously made to let a rope passing through, hence allowing communication with the pot below that was filled with deionized water. After soil spiking and soil saturation with water twenty seeds were added to each pot and gently covered with the spiked soil. The level of water in the lower recipient was adjusted whenever needed, as to guarantee the necessary conditions of moisture according to, the recommendations specified in [64]. Five replicates of uncontaminated natural PTRS1 soil were prepared for the control, while three replicates were tested per

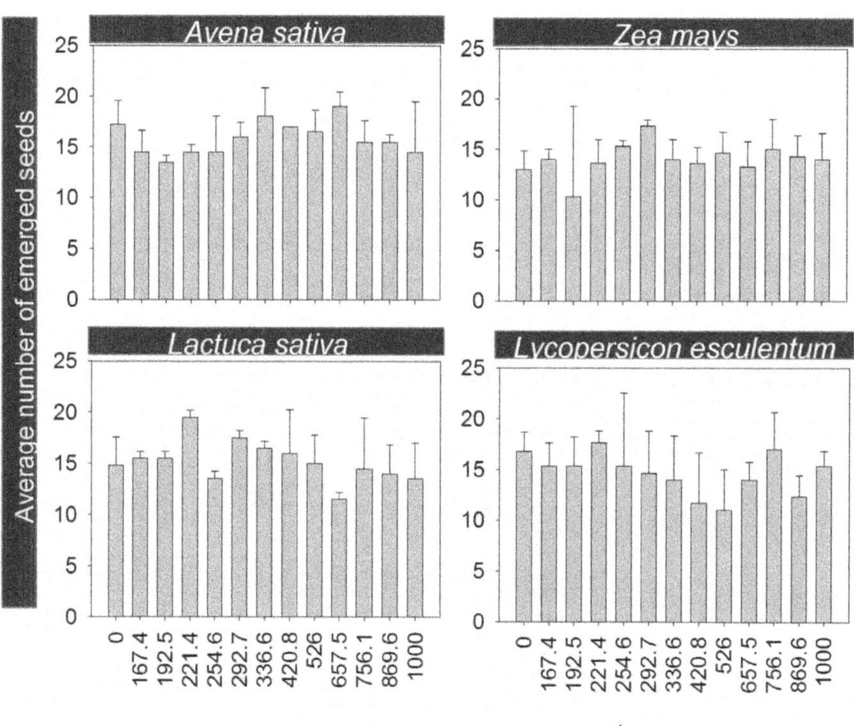

Figure 3. Seed germination of plants. Average number of emerged seeds in monocotyledonous, *Avena sativa and Zea mays* and in dicotyledonous species, *Lycopersicon esculentum and Lactuca sativa,* grown in PTRS1 soil contaminated with U. The error bars indicate the standard deviation.

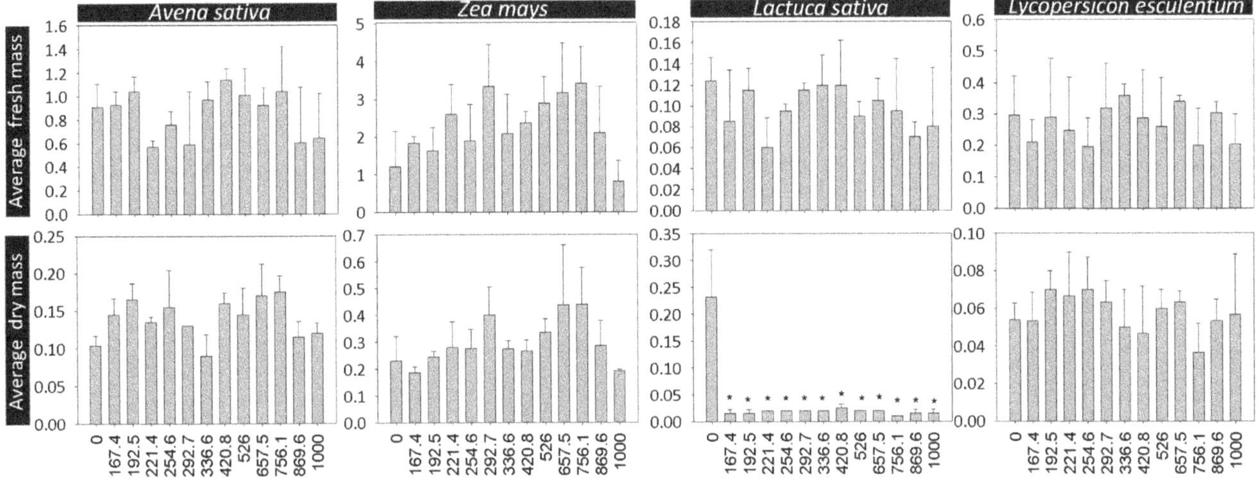

Figure 4. Growth of plants. Average values of fresh mass and dry mass in monocotyledonous, *Avena sativa and Zea mays* and in dicotyledonous species, *Lycopersicon esculentum and Lactuca sativa* grown in PTRS1 soil, contaminated with U. The error bars indicate the standard deviation. The asterisks point out significantly differences from the control (P<0.05).

concentration, in order to minimize the variability of the results, and to follow the ECx sampling design, similarly used for the invertebrate tests.

At the beginning of the test, nutrients (Substral - Plants fertilizer using 1 bottle cap for 2 L of water proportion according to the manufacturer recommendation; Fertilizer NPK: 6-3-6; nitrogen (N): 6%; phosphate (P_2O_5): 3%; potassium (K_2O): 6%; iron (Fe): 0,03%; trace elements: Cu, Mn, Mo and Zn), were added in each lower recipient containing the water. Pots were maintained at constant conditions of temperature (20±2°C), photoperiod (16 hL: 8 hD) and light intensity (25.000 lux). Daily observations were carried out to record the number of emerged seeds. Only the first five emerged seeds were left to grow, the remaining ones were counted and harvested. Fourteen days later, the assay was finished and the fresh and dry biomass above soil was assessed for each test species at the end of the exposure period.

The endpoints seed germination, and fresh and dry biomass, above soil, were assessed for each species at the end of the exposures according to the methods outlined in ISO guideline 11269-2 [64].

For this work, a battery of enzymes involved in different biogeochemical cycles S (sulfur cycle), N (Nitrogen cycle), C (Carbon cycle), as well as enzymes more indicative of the good physiological conditions of the whole microbial community (e.g. dehydrogenases) were selected. The species of invertebrates and plants were selected based on the availability of standard protocols. Since we aimed to obtain data for the derivation of SSVs, for regulatory purposes, this procedure is recommended.

Statistical Analysis

A one-way analysis of variance (one-way ANOVA) was performed to test significant differences between the uranium concentrations tested for each endpoint analyzed: the activity of enzymes, the number of juveniles produced by potworms and collembolans, the number of emerged seeds, and the fresh and dry mass of the plants. The Kolmogorov-Smirnov test was applied to check data normality, whereas homoscedasticity of variances was checked by the Levene's test. When these two assumptions of the one-way ANOVAs were not met, a Kruskal-Wallis analysis was performed. The statistical analysis was run in the SigmaPlot 11.0 software for Windows. When statistical significant differences were recorded, the Dunnett's (for parametric one-way ANOVA) or the Dunn's test (for non-parametric ANOVA) was carried out to perceive which concentrations were significantly different from the respective control. Based on the outcomes of the multiple comparison tests the NOEC (no-observed-effect-concentration) and LOEC (low-observed-effect-concentration) values were deter-

Table 5. Soil quality guideline values derived for copper in Portugal, USA and Canada (mg U Kg^{-1} soildw).

Portugal				Canada	Other reference
Backgound concentrations	PNEC		Proposed SSV[b]	SQG_E[c]	
	NOEC	EC_{20}			
7.8[a]	23.3	15.5	15.5	23	100[d]

[a]Caetano et al.[49];
[b]SSV - soil screening value;
[c]Canadian Soil Quality Guidelines for environmental health (SQGe), Scott-Fordsmand and Pedersen [116].;
[d]Sheppard and Sheppard [98].

mined. The EC_{20} and EC_{50} values for each endpoint were calculated whenever possible, after fitting the data to a log-logistic model using the STATISTICA 7.0 software.

PNEC-Based SSV Derivation

Following the approach suggested by the Technical Guidance Document published by the European Commission [65], a predicted no effect concentration (PNEC) for U in the PTRS1 soil was determined, based on the assessment factor method For that, it was by used the lowest point estimate (i.e., NOEC and EC_{20} values) and applied the appropriate assessment factor based on the criteria of the Guidance Document [65]. The lowest point estimate calculated was for arylsulphatase activity. Considering that more than three NOEC values were obtained in this study, for at least three species, an assessment factor of 10 was applied, The PNEC value was calculated through the application of the following equation:

$$PNEC = \frac{\text{lowest point estimate}}{10}$$

Results and Discussion

1. Soil microbial activity

As far as authors are aware, this study is one of the few studies gathering data regarding the ecotoxicity of spiked soils with U on soil microbial parameters. Only a study from Sheppard et al. [66] has analyzed the effect of U on soil phosphatase activity in eleven different Canadian soils (including an agricultural, a boreal forest and a garden soil). This study recorded a significantly depressed activity only at the highest concentration tested (1000 mg U kg $soil_{dw}^{-1}$) for all the soils. These results suggested that probably, soil phosphatase activity was one of the less sensitive soil microbial parameters to U. In fact, Pereira et al. [7] also reported the low sensitivity of this parameter in mine soils contaminated with metals.

The variation in soil enzyme activities, N mineralization and potential nitrification in the PTRS1 soil, spiked with different U concentrations, is shown in Figure 1, and the Table 3 summarizes the toxicity values obtained for each biochemical parameter.

U had a clear inhibitory effect in almost all functional parameters tested. Overall, dehydrogenase and urease were the most affected soil enzymes by U, being their activity significantly inhibited at concentrations equal or lower than 134.5 mg U kg $soil_{dw}^{-1}$ (Table 3). Dehydrogenases have a relevant role in the oxidation of soil organic matter (SOM), being a good indicator of the active microbial biomass in the soil compartment [67]. As such, U (in the form of uranyl) strongly affected the normal microbial activity in PTRS1 soil. Meyer et al. [68] also observed a significant reduction in respiration rates of a soil exposed to depleted uranium (DU), but only for concentration equal and higher than 500 mg U kg $soil_{dw}^{-1}$. Indeed, the inhibition of urease activities indicates that U had a deleterious effect on soil N-cycle (Figure 1, Table 3). The reduction in the activity of this enzyme may have been caused by a negative effect of U on the overall microbial biomass, which in turn was translated into a reduction in the oxidation rate of organic N into ammonium [58,69]. Arylsulphatase is regularly involved in the S-cycle by catalyzing hydrolysis reactions in the biogeochemical transformation of S [67]. This parameter was significantly affected by U at a LOEC of 279.0 mg U kg $soil_{dw}^{-1}$. On its turn, the cellulase activity was significantly inhibited at intermediate U concentrations. However in the highest concentrations the tendency was reversed and the activity increased, but not for levels significantly different from the control (Figure 1). Thereby, we can conclude that the C-metabolism associated with the degradation of SOM and catalyzed by these extracellular enzymes [70] was constrained by U. N mineralization and potential nitrification are indicators of the functioning of the N-cycle, hence providing an overview of the activity of specific microbial groups (nitrifying bacteria) directly involved in both processes [71]. The general pattern of response observed for these two parameters corresponded to stimulation at the lower U concentrations and inhibition under the highest ones (Figure 1), leading to EC_{50} values of 347.0 and 610.0 mg U kg $soil_{dw}^{-1}$ (Table 3), respectively. It has been stated that N mineralization is normally less sensitive than potential nitrification, since the former is carried out by a wider diversity of microorganisms [71]. However, our data showed the opposite (Figure 1). Meyer et al [68] did not observe effects on nitrogen mineralization of the test soil for U concentrations up to 25000 mg kg $soil_{dw}^{-1}$, however the form of U tested by these authors (schoepite $UO2(OH_2).H_2O$) was less soluble than the one tested in this soil.

The sensitivity of soil microbial parameters to metals has already been demonstrated by several authors, either in metal-polluted or in artificially spiked soils (e.g.,[4,72–77]). Dehydrogenase and urease had generally been referred as the most affected enzymes for different metals (e.g., Cu, Pb, Zn, Cd, Fe, Cr, Ni), (e.g.,[72,73,78,79]). Arylsulphatase and cellulase, however, have shown contradictory responses in different studies. Some authors observed negative correlations between arylsulphatase and cellulase activities and Zn [75] and Cu concentrations, respectively [80,81]; while others observed positive correlations between arylsulphatase and Cd [81], and no changes on cellulase activities in the presence of metals in urban soils was observed [82]. Usually, potential nitrification is negatively influenced by the presence of metals and metalloids such as Pb, Cu and As [7,81]; and the inhibitory effect of some metals (like Zn, Cd and Pb) on N mineralization was also reported by Dai et al. [83]. Antunes et al [81] found negative correlations (based on the Spearman coefficient) between U levels in soils from an abandoned U mine (presenting a mixture of metals) and the activities of urease and cellulase enzymes. For dehydrogenase, potential nitrification and arylsulphatase no significant correlations were detected. Nevertheless, this study analyzed mine contaminated soils, where the mixture of metals, may cause either synergistic or antagonistic effects, and where a well adapted and functional microbial community was likely established.

The inhibition of soil enzyme activities recorded could have been caused by toxicological effects of metals on soil microorganisms with subsequent decrease in their abundance and/or biomass; and/or by the direct inactivation of extracellular enzymes by metals [84]. Notwithstanding, the levels of metals may be not the sole effect on soil microbial activity. Soil properties (e.g., pH, organic matter content, nutrients and soil texture) may also interfere and modulate the bioavailability and toxicityof metals on soil enzymes [74,85]. Clays can retain and protect extracellular hydrolases, namely urease [73]. But the low clay content of PTRS1 soil (3.32%) (Table 1) might have increased U bioavailability, leading to the impairment of soil microbial community through cytotoxic effects, hence reducing their metabolic activity [81]. Additionally, the low pH of PTRS1 soil (Table 1) might have contributed for U availability and impacts on enzyme processes, potential nitrification and N mineralization, particularly at higher

U concentrations, as previously observed by Coppolecchia et al. [75] for arylsulphatase in the presence of Zn and low pH.

The above results illustrated well the effects of U in the performance of soil enzymes, reinforcing the importance of these parameters as bioindicators of soil quality. Indeed, the EC_{20} values calculated for dehydrogenase (34.9 mg Ukg $soil_{dw}^{-1}$), urease (< 135.5 mg Ukg $soil_{dw}^{-1}$), N mineralization (152.2 mg Ukg $soil_{dw}^{-1}$) and arylsulphatase (155.3 mg Ukg $soil_{dw}^{-1}$) are within the environmental concentrations quantified in soils from an abandoned U mine, following extractions with *aqua regia* or with rainwater [8]. In this sense, the data herein generated represent a great asset for the derivation of SSVs, since they have a great ecological representativeness.

2. Uranium toxicity to the reproduction of soil invertebrates

The reproduction tests with the three invertebrate species revealed that *E. andrei, E. crypticus and F. candida* were quite sensitive to U in the PTRS1 soil. Tests fulfilled the validity criteria established by the standard guidelines for control replicates [61–63]. The resulting NOEC, LOEC, EC_{20} and EC_{50} values obtained in this study and toxicity data available in the literature are summarized in the Table 3.

The effects of U in the reproduction of *E. andrei* were evident, since statistical significant differences were found between the control and the highest tested concentrations of U for this organism ($F = 5.218$, d.f. $= 23$, $p = 0.002$) (Figure 2). The tested metal did not significantly affect the reproduction of *E. andrei* at concentrations up to 500.0 mg U kg $soil_{dw}^{-1}$ (NOEC) but compromised this endpoint for concentrations above 550.0 mg U kg $soil_{dw}^{-1}$ (LOEC). EC_{20} and EC_{50} values of U for *E. andrei* reproduction were 474.83 mg U kg $soil_{dw}^{-1}$ and 631.00 mg U kg $soil_{dw}^{-1}$, respectively (Table 3). The results obtained in our study, did not support those of Sheppard and Stephenson [86] (Table 4) that did not record toxic effects for *E. andrei* below 1000 mg U kg $soil_{dw}^{-1}$ (soils (carbonated): pH 7.5, 18% organic matter, 18% clay). However, they found an inhibition of juveniles production in two soils spiked with U, presenting low organic matter (2.2% and 1%) and a pH of 7.5 and 6.2, respectively (Table 4). According to the literature, the adsorption of metals to soil components is dependent on its physical and chemical properties, therefore influencing their toxicity to soil organisms [41,47,48]. Chelinho et al. [87], observed that soils with an organic matter content below 4% reduced or completely inhibited earthworms reproduction. However, the PTRS1 natural soil, had a high organic matter content, 6.2% (according to the classification provided by Murphy et al. [88]). Besides, as previously checked, the intrinsic properties of this soil did not compromise the performance of earthworms [49]. A high organic matter content of soils is usually related with a decrease in the toxicity of the contaminants for the organisms [41,43,89]. However, this was not the case in the study. In fact, Lourenço [60,90] exposed *E. andrei* to a U mine soil with 215.72±8.50 mg U kg $soil_{dw}^{-1}$, a pH of 7.79±0.01, and 7.71±0.60% of organic matter and observed that the bioaccumulation of U and daughter radionuclides was in tandem with loss of DNA integrity of coelomocyte cells, changes in the frequency of cells of immune system and also with histopathological changes (especially of the epidermis and chloragogenous tissue and intestinal epithelium). In fact, some other authors [91] had also suggested that the direct dermal exposure of earthworms to metals in the soil pore water, the ingestion of water, and/or soil particles may strongly favor the bioaccumulation of metals. Since pH is variable in the different compartments of gastrointestinal tract of earthworms, it can increase the mobilization of contaminants from soil after its ingestion [92,93].

Although, other metals were present in the contaminated soil tested by Lourenço et al. [60,90] U likely had a crucial role in the toxic effects observed, because it persisted in the whole body till 56 days. These authors suggested that the changes observed in DNA integrity were likely early warning indicators of effects on the growth and reproduction of the organisms. And in fact, effects on reproduction were observed in our study. Further, Giovanetti et al. [94]. exposed *E. fetida* natural U- and DU-contaminated soil (no information on soil type) for 7 and 28 days. Regarding natural U, no mortality or significant changes in weight were observed for both exposure periods up to 600 mg U $kg^{-1}{}_{dw}$. The chloragogeneous tissue, the main storage tissue of U, presented meaningful changes after 7 days of exposure for 300 mg U Kg^{-1}, while DNA strand breaks increased in a dose dependent manner above 150 mg U kg $soil_{dw}^{-1}$.

Regarding *E. crypticus* reproduction, it was significantly not reduced above 526.0 mg U kg $soil_{dw}^{-1}$ (LOEC) (Table 3) ($F = 31.05$, d.f. $= 12$, $p<0.05$). The EC_{20} and EC_{50} values estimated were respectively 469.7 and 518.6 mg U kg $soil_{dw}^{-1}$. Although no toxicity values are reported for the lowest concentrations tested, enchytraeids showed considerable sensitivity to U, since the number of juveniles was minimal or no juveniles were produced by *E. crypticus* at concentrations above 657.5 mg U kg $soil_{dw}^{-1}$ (Figure 2). Despite enchytraeids are commonly used in standardized toxicity tests, to the best of our knowledge, no data are available in the literature regarding the effects of U on the reproduction of this species. The available information concerns only the toxic effects caused by other metals or by natural soil properties in the reproduction of this species [43,48,95–97]. Thus, taking into account this literature review, pH and CEC were the most important parameters controlling the high sensitivity of enchytraeids to metals. Additionally, and according to Kuperman et al. [43], adults survival and juveniles production by *E. crypticus* can be maximized in natural soils with properties within the following ranges: 4.4–8.2 pH; 1.2–42% OM; 1–29% clay. The PTRS1 natural soil used as test substrate fell into in these ranges (Table 3), and similarly to *E. andrei*, the reproduction of this species was not compromised during the validation of the PTRS1 natural soil as a reference soil [49], meaning that the soil properties did not limit the performance of *E. crypticus*.

Concerning *F. candida*, U affected the production of juveniles, as shown by a significant decrease of this endpoint along the concentrations tested ($F = 11.6$, d.f. $= 12$, $p<0.05$) (Figure 2). The number of juveniles was not significantly affected up to 675.50 mg U kg $soil_{dw}^{-1}$ (NOEC), but it was significantly decreased for U concentrations equal to or greater than 756.10 mg U Kg^{-1} (LOEC) (Table 3). The EC_{20} value estimated for reproduction was 343.41 mg U kg $soil_{dw}^{-1}$ which is considerably lower than the toxicity data reported by Shepard et al. [98], EC_{20}>710 mg U kg $soil_{dw}^{-1}$, in two loam soils with pH 7.5 (Table 4). The low sensitivity of *F. candida* to U was also observed by Sheppard and Stephenson [86] which tested 3 soils amended with a range of U concentrations and aged for 10 years before testing. In this study, the lowest EC_{20} value obtained was 840 mg U kg $soil_{dw}^{-1}$ in a loam soil (pH 7.5, 24% clay, 2.2% OM) (Table 4). Despite this, *F. candida* was more sensitive in the study of Sheppard and Stephenson (since their EC_{20} value was similar to the EC_{50} recorded in our study 851.64 mg U kg $soil_{dw}^{-1}$). When considering the number of juveniles produced, U was less toxic to *F. candida* comparatively to *E. andrei* and *E. crypticus*. The lower sensitivity of *F. candida* is also consistent with other studies, when the effects of other metals in the reproduction of the three species was investigated [97,99], or even when other species of collembolans were analyzed [86]. The exposure of *F. candida* to chemicals in soil is apparently lower than for earthworms, which are exposed both by ingestion of contaminated soil (mineral particles, organic matter and chemicals in the soil solution) and also through direct dermal contact [100]. Despite the widely known influence of soil parameters on the bioavailability of chemicals and their influence on the reproduction of soil organisms, less is known about the intrinsic effects of physico-chemical parameters of the soils in the reproduction of *F. candida*. In general, several authors had reported a high tolerance of *F. candida* reproduction to a wide range of soil textural classes, organic matter contents and soil pH [48,101,102]. Once again the performance of this species was not compromised by the intrinsic properties of the PTRS1 soil. Hence, the effects observed can undoubtedly be attributed to U exposure.

3. Phytotoxicity of uranium

Relatively to terrestrial plants tests, all the validity criteria as described by the standard guidelines were attained [64]. Data obtained showed no significant effects on seeds emergence for all species tested ($p>0.05$). In fact, it was observed a relatively high rate of germination, either in monocotyledonous and dicotyledonous species (Figure 3). This outcome was somewhat expected, based on previous studies (e.g.,[22]). Seed coats form a barrier which protects embryos from a wide range of contaminants, especially metals. Thus, the germination relies almost exclusively on the seed reserves making it a less sensitive endpoint to the toxicity of soil pollutants [103].

An apparent hormetic effect was recorded for the other endpoints measured for almost all plant species. Such occurrence was recorded by other authors and it was attributed to the use of U as uranyl nitrate, which corresponds to a supplementary dose of N given to plants [98].

With regard to production of fresh- and dry-mass, it was possible to perceive that the tested plants displayed different sensitivities to this metal. However, no significant differences were generally observed comparatively to the control, exception for *L. sativa* dry mass ($H = 22.8$, d.f. = 12, $p = 0.029$). Thus, and according to Figure 4, *L. sativa* was the most sensitive terrestrial plant to U. The high sensitivity of *L. sativa* was also found by Hubálek et al. [104] and Soudek [105]. This was probably caused by the high capacity of this species to bioaccumulate high concentrations of metals, including U [22].

The exposure of plants to metals, was already extensively studied, showing that these contaminants can induce biological effects on germination, growth and development, as well as, alterations in the nutrient profile of plants [22,106]. However, only some studies (e.g., [66,107] and others reviewed [98]) have assessed the ecotoxicological effects of U on terrestrial plant species.

Based on our study, once again was proved the diverse ecotoxicological outcomes for U effects on plant species, since no effects were observed, in the range of tested concentrations for the three evaluated endpoints (in three out of four species), in PTRS1 soil. Similar results were obtained by Sheppard and Sheppard [86] in acidic soils (Table 4), when testing the emergence and growth of wheatgrass *Elymus lanceolatus*. Like in our study, these authors did not observe any effect on this species (up to 1000 mg U kg $soil_{dw}^{-1}$). In opposition, Sheppard and Sheppard [81] revised data on U toxicity to terrestrial plants and reported EC_{25} values ranging from 300 to 500 mg U kg $soil_{dw}^{-1}$, considering only the most reliable studies. Stojanović et al. [108] also reported phytotoxic effects of U on *Zea mays* exposed, in different soil types, to 250, 500 and 1000 mg U kg $soil_{dw}^{-1}$, but especially at the highest concentration tested and in the most acidic soil. However, no statistical analysis of the data was performed in this study.

Soil properties are also the factors that most strongly affect U uptake and phytotoxic effects, [18,109–111]. The bivalent uranyl ion (UO_2^{2+}) is sorbed to the negatively charged surfaces of clay minerals and organic compounds. In acidic soils subjected to pH increase, more negatively charged binding sites are available on mineral surfaces due to the progressive reduction of protons occupying these sites. However, pH values close to 6, like the one of PTRS1, favors U availability, since the concentrations of carbonates tends to increase, and U is released to the soil solution in the form of U-carbonate complexes [18]. The natural soil PTRS1, besides being acidic, has a lower clay content, which means lower binding sites for the bivalent uranyl ion (UO_2^{2+}), hence constraining U bioavailability. other soil properties and plant mechanisms may explain the reduced sensitivity of the plants in comparison with soil microbial parameters and invertebrates. Viehweger and Geipel [112] reported an increased U absorption by *Arabidopsis halleri* attributed to Fe deficiency in the medium of hydroponically grown plants. With respect to this metal, in the natural PTRS1 soil, the analyses done by Caetano et al. [49] showed that Fe surpassed the soil benchmark values proposed by two EPA regions (http://rais.ornl.gov/tools/eco_search.php http://rais.ornl.gov/tools/eco_search.php). In this sense, it is hypothesized that the high Fe content of the PTRS1 soil, may have also contributed for reducing the absorption of U by plants. As far as plant mechanisms are considered, in several studies reviewed by Mitchell et al. [113] the transport of U within plants was reduced and higher concentrations were consistently found in the roots. Using X-ray absorption spectroscopy (XAS) and transmission electron microscopy (TEM), Laurette et al. [114] observed that when plants are exposed to U and phosphates, needle-like U-phosphates are formed and precipitate, both outside and inside the cells, or persist in the subsurface of root tissues. The precipitation of U-phosphate complexes acts as a protective mechanism preventing U translocation to the shoots and leaves. This can also occur when the culture medium of the plants has no phosphate, since some plants are able to exudate phosphates. Further, U may be also absorbed like UO_2^{2+} and linked to endogenous organophospate groups [114]. In opposition, when translocation occurs within plants, U has mainly formed U-carboxylated complexes. Plants can also exudate organic acids to the rhizosphere environment or UO_2^{2+} may form complexes with endogenous compounds like malic, citric, oxalic and acetic acid [114]. In summary, the different resistance mechanisms described above could explain the lack of toxic effects observed for *A. sativa*, *Z. mays* and *L. esculentum*, in opposition to *L. sativa*. Most concerning is the fact that the majority of studies testing the phytotoxicity of U, including those performed by us, were made with the addition of nutrients solution, which increased the availability of phosphates to the test soil, likely decreasing the sensitivity of plants to U. Hence, to enhance the protection level of SSVs derived for plants, more assays with different plant species should be performed and the addition of nutrients should be prevented, or at least the tests may include replicates with and without nutrients.

Derivation of a Soil Screening Value (SSV) for Uranium Applying Assessment Factors

The PNEC values obtained for U were based in EC_{20} and NOEC values varied between 15.5 and 23.3 mg kg $soil_{dw}^{-1}$, respectively (Table 5). These values were six to four times lower than the PNEC value suggested by Sheppard and Sheppard [86] (Table 5), which was 100 mg Ukg $soil_{dw}^{-1}$. In opposition, they are close to the lowest Canadian Soil Quality Guideline for both environment and human health (23 mg U kg $soil_{dw}^{-1}$). Thereby, while more ecotoxicological data is being obtained or other methods are being applied to derive soil screening values (SSVs) we prefer to be precautious by proposing a PNEC of 15.5 mg Kg^{-1} $soil_{dw}$ as a SSV for U, in soils similar to the PTRS1. This SSV value is near the background value found in non-contaminated soils [8,48], but not in some areas with naturally occurring U anomalies in soils, where concentrations ranging between 13–724 mg U kg $soil_{dw}^{-1}$ can be found [115].

Conclusion

With the present study it was possible to generate a set of important ecotoxicological data for the derivation of a SSV for U, using a Portuguese natural soil representative of a granitic region, where this type of mine exploration occurred.

Soil enzyme activities were clearly inhibited by U. The obtained results depended not only on the concentrations of U but also on the properties of soil, which were likely responsible for the bioavailability of U and subsequent impairments on soil microbial population and, consequently, in their activity. Dehydrogenase and urease were particularly sensitive to U. Further, and comparatively to the remaining effect concentrations obtained/estimated for invertebrates and plants, the soil microbial parameters were more affected by U contamination[1].

The toxic effects of U in soil invertebrates were also confirmed, but the tested species showed a variable sensitivity to this metal. The increasing order of species sensitivity to U based on EC_{50} values for reproduction was *E. crypticus* > *E. andrei* > *F. candida*. However, if EC_{20} values are considered *F. candida* is the most sensitive invertebrate, since its EC_{20} value was 343.41 mg U kg $soil_{dw}^{-1}$, compared to 474.83 mg U kg $soil_{dw}^{-1}$ and 469.76 mg U kg $soil_{dw}^{-1}$ EC_{20} values estimated for *E. andrei* and *E. crypticus*, respectively. The EC_{20} values estimated were lower than the NOEC values for *E. andrei* and *F. candida*. Thus, the EC_{20} point estimate should be selected for the derivation of more protective SSVs. Relatively to the plants, the tested species showed no adverse effects caused by U in PTRS1, with the exception of *L. sativa* dry mass yield. Considering the results obtained, it was possible to verify a great variability between the EC_x values estimated in this study and those reported in the scientific literature. Multiple factors can contribute to this discordance, but probably at least for some species, soils physical and chemical properties were the main factors responsible for such differences. Although, this reinforces, at least in part, the importance of using natural soils representatives of the main types of soil from each region in ecotoxicological evaluations and for the derivation of SSVs, the data generated suggests that the SSV (15.5 mg Kg^{-1} $soil_{dw}$) derived for U, was six times lower than the PNEC value proposed by other authors. Nevertheless, as mentioned previously, more data should be obtained following standard protocols.

Author Contributions

Conceived and designed the experiments: ALC CRM RP. Performed the experiments: ALC AG. Analyzed the data: ALC CRM RP. Contributed reagents/materials/analysis tools: FC FG EFS. Contributed to the writing of the manuscript: ALC CRM RP.

References

1. Gavrilescu M, Pavel L, Cretescu I (2009) Characterization and remediation of soils contaminated with uranium. J Hazard Mater 163: 475–510. doi:10.1016/j.jhazmat.2008.07.103.
2. Malyshkina N, Niemeier D (2010) Future sustainability forecasting by exchange markets: basic theory and an application. Environ Sci Technol 44: 9134–9142. Available: http://www.ncbi.nlm.nih.gov/pubmed/21058697.
3. Carvalho F, Oliveira J, Faria I (2009) Alpha emitting radionuclides in drainage from Quinta do Bispo and Cunha Baixa uranium mines (Portugal) and associated radiotoxicological risk. Bull Environ Contam Toxicol 83: 668–673. Available: http://www.ncbi.nlm.nih.gov/pubmed/19590808. Accessed 19 November 2012.
4. Pereira R, Barbosa S, Carvalho FP (2014) Uranium mining in Portugal: a review of the environmental legacies of the largest mines and environmental and human health impacts. Environ Geochem Health 36: 285–301. Available: http://www.ncbi.nlm.nih.gov/pubmed/24030454. Accessed 7 March 2014.
5. Carvalho F, Madruga M, Reis M, Alves J, Oliveira J, et al. (2007) Radioactivity in the environment around past radium and uranium mining sites of Portugal. J Environ Radioact 96: 39–46. Available: http://www.ncbi.nlm.nih.gov/pubmed/17433852. Accessed 25 October 2012.
6. Vandenhove H, Sweeck L, Mallants D, Vanmarcke H, Aitkulov A, et al. (2006) Assessment of radiation exposure in the uranium mining and milling area of Mailuu Suu, Kyrgyzstan. J Environ Radioact 88: 118–139. Available: http://www.ncbi.nlm.nih.gov/pubmed/16581165.
7. Pereira R, Sousa J, Ribeiro R, Gonçalves F (2006) Microbial Indicators in Mine Soils (S. Domingos Mine, Portugal). Soil Sediment Contam 15: 147–167. Available: http://www.informaworld.com/openurl?genre=article&doi=10%2e1080%2f15320380500506813&magic=crossref%7c%7cD404A21C5BB053405B1A640AFFD44AE3. Accessed 5 October 2013.
8. Pereira R, Antunes S, Marques S, Gonçalves F (2008) Contribution for tier 1 of the ecological risk assessment of Cunha Baixa uranium mine (Central Portugal): I soil chemical characterization. Sci Total Environ 390: 377–386. Available: http://www.ncbi.nlm.nih.gov/pubmed/17919686. Accessed 6 September 2011.
9. Arogunjo A, Höllriegl V, Giussani A, Leopold K, Gerstmann U, et al. (2009) Uranium and thorium in soils, mineral sands, water and food samples in a tin mining area in Nigeria with elevated activity. J Environ Radioact 100: 232–240. Available: http://www.ncbi.nlm.nih.gov/pubmed/19147259. Accessed 14 October 2013.
10. Momčilović M, Kovačević J, Dragović S (2010) Population doses from terrestrial exposure in the vicinity of abandoned uranium mines in Serbia. Radiat Meas 45: 225–230. Available: http://linkinghub.elsevier.com/retrieve/pii/S1350448710000363. Accessed 25 October 2012.
11. Figueiredo M, Silva T, Batista M, Leote J, Ferreira M, et al. (2011) Uranium in surface soils: An easy-and-quick assay combining X-ray diffraction and X-ray fluorescence qualitative data. J Geochemical Explor 109: 134–138. Available: http://linkinghub.elsevier.com/retrieve/pii/S0375674210001366. Accessed 23 October 2012.
12. Patra A, Sumesh C, Mohapatra S, Sahoo S, Tripathi R, et al. (2011) Long-term leaching of uranium from different waste matrices. J Environ Manage 92: 919–925. Available: http://www.ncbi.nlm.nih.gov/pubmed/21084148. Accessed 14 October 2013.
13. Scheele F (2011) Uranium from Africa: Mitigation of Uranium Mining Impacts on Society and Environment by Industry and Governments. SSRN Electron J. Available: http://papers.ssrn.com/abstract=1892775. Accessed 14 October 2013.
14. Niemeyer J, Moreira-Santos M, Nogueira M, Carvalho G, Ribeiro R, et al. (2010) Environmental risk assessment of a metal-contaminated area in the Tropics. Tier I: screening phase. J Soils Sediments 10: 1557–1571. Available: http://www.springerlink.com/index/10.1007/s11368-010-0255-x. Accessed 29 October 2012.
15. Carvalho F (2011) Environmental Radioactive Impact Associated to Uranium Production Nuclear and Technological Institute Department of Radiological Protection and Nuclear Safety,. Journal, Am Sci Environ Publ Sci 7: 547–553.
16. Wang J, Lu A, Ding A (2007) Effect of cadmium alone and in combination with butachlor on soil enzymes. Heal (San Fr: 395–403. doi:10.1007/s10653-007-9084-2.
17. ASTDR (2011) Agency for Toxic Substances & Disease Registry-Toxicological profile for uranium. Draft. U.S. Department of Health and Human Services. Public Health Service and Agency for Toxic Substances and Disease Registry. 452 + annexes.
18. Vandenhove H, Van Hees M, Wouters K, Wannijn J (2007) Can we predict uranium bioavailability based on soil parameters? Part 1: effect of soil parameters on soil solution uranium concentration. Environ Pollut 145: 587–595. Available: http://www.ncbi.nlm.nih.gov/pubmed/16781802. Accessed 19 November 2012.
19. Gongalsky K (2003) Impact of pollution caused by uranium production on soil macrofauna. Environ Monit Assess 89: 197–219. Available: http://www.ncbi.nlm.nih.gov/pubmed/14632090. Accessed 15 October 2013.
20. Geras'kin S, Evseeva T, Belykh E, Majstrenko T, Michalik B, et al. (2007) Effects on non-human species inhabiting areas with enhanced level of natural

radioactivity in the north of Russia: a review. J Environ Radioact 94: 151–182. Available: http://www.ncbi.nlm.nih.gov/pubmed/17360083. Accessed 26 October 2012.

21. Joner E, Munier-Lamy C, Gouget B (2007) Bioavailability and microbial adaptation to elevated levels of uranium in an acid, organic topsoil forming on an old mine spoil. Environ Toxicol Chem 26: 1644–1648. Available: http://www.ncbi.nlm.nih.gov/pubmed/17702337.
22. Pereira R, Marques CR, Ferreira MJS, Neves MFJV, Caetano AL, et al. (2009) Phytotoxicity and genotoxicity of soils from an abandoned uranium mine area. Appl Soil Ecol 42: 209–220. Available: http://linkinghub.elsevier.com/retrieve/pii/S0929139309000778. Accessed 6 September 2011.
23. Kenarova A, Radeva G, Danova I, Boteva S, Dimitrova I (2010) Soil bacterial abundance and diversity of uranium impacted. Second Balkan conference on Biology: 5–9.
24. Islam E, Sar P (2011) Molecular assessment on impact of uranium ore contamination in soil bacterial diversity. Int Biodeterior Biodegradation 65: 1043–1051. Available: http://linkinghub.elsevier.com/retrieve/pii/S0964830511001697. Accessed 29 October 2012.
25. Geng F, Hu N, Zheng J-F, Wang C-L, Chen X, et al. (2011) Evaluation of the toxic effect on zebrafish (*Danio rerio*) exposed to uranium mill tailings leaching solution. J Radioanal Nucl Chem 292: 453–463. Available: http://link.springer.com/10.1007/s10967-011-1451-x. Accessed 15 October 2013.
26. Lourenço J, Pereira R, Silva A, Carvalho F, Oliveira J, et al. (2012) Evaluation of the sensitivity of genotoxicity and cytotoxicity endpoints in earthworms exposed in situ to uranium mining wastes. Ecotoxicol Environ Saf 75: 46–54. Available: http://www.sciencedirect.com/science/article/pii/S0147651311002685. Accessed 14 October 2013.
27. Islam E, Sar P (2011) Molecular assessment on impact of uranium ore contamination in soil bacterial diversity. Int Biodeterior Biodegradation 65: 1043–1051. Available: http://www.sciencedirect.com/science/article/pii/S0964830511001697. Accessed 25 October 2013.
28. Lavelle P, Decaens T, Aubert M, Barot S, Blouin M, et al. (2006) Soil invertebrates and ecosystem services. Eur J Soil Biol 42: S3–S15. Available: http://linkinghub.elsevier.com/retrieve/pii/S1164556306001038. Accessed 19 July 2011.
29. O'Halloran K (2006) Toxicological Considerations of Contaminants in the Terrestrial Environment for Ecological Risk Assessment. Hum Ecol Risk Assess An Int J 12: 74–83. Available: http://www.tandfonline.com/doi/abs/10.1080/10807030500428603. Accessed 7 September 2011.
30. Barrios E (2007) Soil biota, ecosystem services and land productivity. Ecol Econ 64: 269–285. Available: http://linkinghub.elsevier.com/retrieve/pii/S0921800907001693. Accessed 4 October 2012.
31. Dominati E, Patterson M, Mackay A (2010) A framework for classifying and quantifying the natural capital and ecosystem services of soils. Ecol Econ 69: 1858–1868. Available: http://www.sciencedirect.com/science/article/pii/S0921800910001928. Accessed 14 October 2013.
32. CEC (2006) Comission of the European Cominities, Directive 2004/35/EC. Commissiona of the European Communities, Brussels.
33. Bone J, Head M, Barraclough D, Archer M, Scheib C, et al. (2010) Soil quality assessment under emerging regulatory requirements. Environ Int 36: 609–622. Available: http://www.ncbi.nlm.nih.gov/pubmed/20483160. Accessed 6 October 2012.
34. Jensen J, Mesman M (2006) Ecological Risk Assessment of Contaminated Land. Decision support for site specific investigations. ISBN 90-6960-138-9 978-90-6960-138-0.
35. Fishwick S (2004) Soil screening values for use in UK ecological risk assessment. Soil Quality & Protection, Air Land and Water Group. Science Environment Agency's Project Manager.
36. USEPA (2003) United States Environmental Protection Agency. Guidance for Developing Ecological Soil Screening Levels (Eco-SSLs). Attachment 1-2: Assessment of Whether to Develop Ecological Soil Screening Levels for Microbes and Microbial Processes, OSWER Directi. Environ Prot.
37. Carlon C (2007) Derivation methods of soil screening values in europe. a review and evaluation of national procedures towards harmonisation. European Commission, Joint Research Centre, Ispra, EUR 22805-EN, 306.
38. Provoost J, Reijnders L, Swartjes F, Bronders J, Carlon C, et al. (2008) Parameters causing variation between soil screening values and the effect of harmonization. J Soils Sediments 8: 298–311. Available: http://www.springerlink.com/index/10.1007/s11368-008-0026-0. Accessed 29 October 2012.
39. Van Gestel C (2012) Soil ecotoxicology: state of the art and future directions. Zookeys 296: 275–296. Available: http://www.pubmedcentral.nih.gov/articlerender.fcgi?artid=3335420&tool=pmcentrez&rendertype=abstract. Accessed 29 October 2012.
40. MAOTDR (2006) Ministerio do Ambiente do Ordenamento do Territorio e do Desenvolvimento Regional. Diário da República, 1. a série – N. o 208–27 de Outubro.
41. Römbke J, Jänsch S, Junker T, Pohl B, Scheffczyk A, et al. (2006) Improvement of the applicability of ecotoxicological tests with earthworms, springtails, and plants for the assessment of metals in natural soils. Environ Toxicol Chem 25: 776–787. Available: http://www.ncbi.nlm.nih.gov/pubmed/16566163.
42. Van Assche F, Alonso JL, Kapustka L, Petrie R, Stephenson GL, et al. (2002) Terrestrial plant toxicity tests. In: Fairbrother A., Glazebrock P.W., Van Straalen N.M., Tarazona J.V. (eds) Test methods to determine hazards of sparingly soluble metal compounds in soils. SETAC, Pensacola, Forida, 37–57. Available: https://www.setac.org/store/view_product.asp?id=1038018.
43. Kuperman R, Amorim M, Römbke J, Lanno R, Checkai R, et al. (2006) Adaptation of the enchytraeid toxicity test for use with natural soil types. Eur J Soil Biol 42: S234–S243. Available: http://www.sciencedirect.com/science/article/pii/S1164556306000719. Accessed 14 October 2013.
44. Hofman J, Hovorková I, Machát J (2009) Comparison and Characterization of OECD Artificial Soils. Ecotoxicological Characterization of Waste 2009, 223–229. Available: http://www.rivm.nl/bibliotheek/rapporten/711701047.html.
45. Song J, Zhao F, McGrath S, Luo Y (2006) Influence of soil properties and aging on arsenic phytotoxicity. Environ Toxicol Chem 25: 1663–1670. Available: http://www.ncbi.nlm.nih.gov/pubmed/16764487. Accessed 14 October 2013.
46. Rooney C, Zhao F, McGrath S (2007) Phytotoxicity of nickel in a range of European soils: influence of soil properties, Ni solubility and speciation. Environ Pollut 145: 596–605. Available: http://www.ncbi.nlm.nih.gov/pubmed/16733077. Accessed 14 October 2013.
47. Van Gestel C, Borgman E, Verweij R, Ortiz M (2011) The influence of soil properties on the toxicity of molybdenum to three species of soil invertebrates. Ecotoxicol Environ Saf 74: 1–9. Available: http://www.ncbi.nlm.nih.gov/pubmed/20951431. Accessed 14 October 2013.
48. Domene X, Chelinho S, Campana P, Natal-da-Luz T, Alcañiz JM, et al. (2011) Influence of soil properties on the performance of Folsomia candida: implications for its use in soil ecotoxicology testing. Environ Toxicol Chem 30: 1497–1505. Available: http://www.ncbi.nlm.nih.gov/pubmed/21437938. Accessed 13 March 2013.
49. Caetano A, Gonçalves F, Sousa J, Cachada A, Pereira E, et al. (2012) Characterization and validation of a Portuguese natural reference soil to be used as substrate for ecotoxicological purposes. J Environ Monit 14: 925–936. Available: http://www.ncbi.nlm.nih.gov/pubmed/22297688. Accessed 31 May 2013.
50. ISO 17512-1 (2008) International Organization for Standardization 17512-1: 2008.Soil Quality: Avoidance Test for Testing the Quality of Soils and the Toxicity of Chemicals-Test with Earthworms (*Eisenia Fetida*). Geneva, Switzerland. International Organization for Standard. Geneva, Switzerland.
51. FAOUN (1984) Food and agriculture organization of the United Nations – physical and chemical methods of soil and water analysis. Soils Bull. 10, 1–275.
52. SPAC (2000) Soil and Plant Analysis Council – Handbook of Reference Methods. CRC Press, Boca Raton, Florida.
53. Tabatabai M, Bremner J (1970) Arylsulfatase activity in soils. Soil Science Society of America, 34: 225–9.
54. Schinner F, Kandeler E, Öhlinger R, Margesin R (1996) Methods in soil biology. Germany: Springer-Verlag.
55. Öhlinger R (1996) Soil sampling and sample preparation. In: Schinner F, Öhlinger R, Kandeler E, Margesin R, editors. Methods in Soil Biology. Springer-Verlag.
56. Schinner F, von Mersi W (1990) Xylanase, CM-cellulase and invertase activity in soil, an improved method. Soil Biol Biochem 22: 511–5.
57. Kandeler E, Gerber H (1988) Short-term assay of soil urease activity using colorimetric determination of ammonium. biol Fert Soils 6: 68–72.
58. Kandeler E (1996) Potential nitrification. In: Schinner F, Öhlinger R, Kandeler E, Margesin R, editors. Methods in soil biology. Germany: Springer-Verlag-Berlin-Heidelberg, 146–9.
59. Berg P, Rosswall T (1985) Ammonium oxidizer numbers, potential and actual oxidation rates en two Swedish arable soils. Biology Fertility Soils 1: 131–40.
60. Lourenço JI, Pereira RO, Silva AC, Morgado JM, Carvalho FP, et al. (2011) a) Genotoxic endpoints in the earthworms sub-lethal assay to evaluate natural soils contaminated by metals and radionuclides. J Hazard Mater 186: 788–795. Available: http://www.ncbi.nlm.nih.gov/pubmed/21146299. Accessed 21 September 2013.
61. ISO 11268-2 (1998) International Organization for Standardization ISO 11268-2. Soil quality: effects of pollutants on earthworms (*Eisenia fetida*) - Part 2: Determination of effects on reproduction. ISO 11268-2. Geneva, Switzerland: International Organization for Standardiza. Geneva, Switz.
62. ISO 16387 (2004) International Organization for Standardization ISO 16387. Soil quality: effects of pollutants on Enchytraeidae (*Enchytraeus* sp.)-Determination of effects on reproduction and survival. ISO16387. Geneva, Switzerland: International Organization for Standard. Geneva, Switz.
63. ISO 11267 (1999) International Organization for Standardization ISO 11267. Soil quality: inhibition of reproduction of Collembola (*Folsomia candida*) by soil pollutants. ISO 11267. Geneva, Switzerland: International Organization for Standardization. Geneva, Switz.
64. ISO 11269-2 (2005) International Organization for Standardization ISO 11269-2. Soil quality: determination of the effects of pollutants on soil flora - Part 2: Effects of chemicals on the emergence and growth of higher plants. ISO 11269-2. Geneve, Switzerlan: International O. Geneva, Switz.
65. EC. Commission European (2003) Commission European. Technical Guidance Document on Risk Assessment. in support of Commission Directive 93/67/EEC on Risk Assessment for new notified substances Commission Regulation (EC) No 1488/94 on Risk Assessment for existing substances Directive 98/.
66. Sheppard S, Evenden W (1992) Bioavailability Indices for Uranium: Effect of Concentration in Eleven Soils. Arch Environ Contain Toxicol 23, 117–124.

67. Taylor J, Wilson B, Mills M, Burns R (2002) Comparison of microbial numbers and enzymatic activities in surface soils and subsoils using various techniques. 34.
68. Meyer MC, Paschke MW, McLendon T, Price D (1998) Decreases in Soil Microbial Function and Functional Diversity in Response to Depleted Uranium. J Environ Qual 27: 1306. Available: https://www.agronomy.org/publications/jeq/abstracts/27/6/JEQ0270061306. Accessed 31 March 2014.
69. Wang M, Markert B, Shen W, Chen W, Peng C, et al. (2011) Microbial biomass carbon and enzyme activities of urban soils in Beijing. Environ Sci Pollut Res Int 18: 958–967. Available: http://www.ncbi.nlm.nih.gov/pubmed/21287285. Accessed 25 September 2013.
70. Alvarenga P, Palma P, Gonçalves AP, Baião N, Fernandes RM, et al. (2008) Assessment of chemical, biochemical and ecotoxicological aspects in a mine soil amended with sludge of either urban or industrial origin. Chemosphere 72: 1774–1781. Available: http://www.ncbi.nlm.nih.gov/pubmed/18547605. Accessed 13 May 2014.
71. Winding A, Hund-Rinke K, Rutgers M (2005) The use of microorganisms in ecological soil classification and assessment concepts. Ecotoxicol Environ Saf 62: 230–248. Available: http://www.ncbi.nlm.nih.gov/pubmed/15925407.
72. Khan S, Cao Q, Hesham AE-L, Xia Y, He J-Z (2007) Soil enzymatic activities and microbial community structure with different application rates of Cd and Pb. J Environ Sci (China) 19: 834–840. Available: http://www.ncbi.nlm.nih.gov/pubmed/17966871. Accessed 15 October 2013.
73. Lee S, Kim E, Hyun S, Kim J (2009) Metal availability in heavy metal-contaminated open burning and open detonation soil: assessment using soil enzymes, earthworms, and chemical extractions. J Hazard Mater 170: 382–388. Available: http://www.ncbi.nlm.nih.gov/pubmed/19540045. Accessed 5 April 2013.
74. Papa S, Bartoli G, Pellegrino A, Fioretto A (2010) Microbial activities and trace element contents in an urban soil. Environ Monit Assess 165: 193–203. Available: http://www.ncbi.nlm.nih.gov/pubmed/19444636. Accessed 13 May 2013.
75. Coppolecchia D, Puglisi E, Vasileiadis S, Suciu N, Hamon R, et al. (2011) Soil Biology & Biochemistry Relative sensitivity of different soil biological properties to zinc. Soil Biol Biochem 43: 1798–1807. Available: http://dx.doi.org/10.1016/j.soilbio.2010.06.018.
76. Lee S-H, Park H, Koo N, Hyun S, Hwang A (2011) Evaluation of the effectiveness of various amendments on trace metals stabilization by chemical and biological methods. J Hazard Mater 188: 44–51. Available: http://www.ncbi.nlm.nih.gov/pubmed/21333442. Accessed 17 August 2013.
77. Hu B, Liang D, Liu J, Xie J (2013) Ecotoxicological effects of copper and selenium combined pollution on soil enzyme activities in planted and unplanted soils. Environ Toxicol Chem 32: 1109–1116. Available: http://www.ncbi.nlm.nih.gov/pubmed/23401089. Accessed 8 October 2013.
78. Gülser F, Erdoğan E (2008) The effects of heavy metal pollution on enzyme activities and basal soil respiration of roadside soils. Environ Monit Assess 145: 127–133. Available: http://www.ncbi.nlm.nih.gov/pubmed/18027096. Accessed 17 August 2013.
79. Thavamani P, Malik S, Beer M, Megharaj M, Naidu R (2012) Microbial activity and diversity in long-term mixed contaminated soils with respect to polyaromatic hydrocarbons and heavy metals. J Environ Manage 99: 10–17. Available: http://dx.doi.org/10.1016/j.jenvman.2011.12.030.
80. Alvarenga P, Palma P, de Varennes A, Cunha-Queda AC (2012) A contribution towards the risk assessment of soils from the São Domingos Mine (Portugal): chemical, microbial and ecotoxicological indicators. Environ Pollut 161: 50–56. Available: http://www.ncbi.nlm.nih.gov/pubmed/22230067. Accessed 5 October 2013.
81. Antunes S, Pereira R, Marques S, Castro B, Gonçalves F (2011) Impaired microbial activity caused by metal pollution A field study in a deactivated. Sci Total Environmen 410–411.
82. Sivakumar S, Nityanandi D, Barathi S, Prabha D, Rajeshwari S, et al. (2012) Selected enzyme activities of urban heavy metal-polluted soils in the presence and absence of an oligochaete, *Lampito mauritii* (Kinberg). J Hazard Mater 227–228: 179–184. Available: http://www.ncbi.nlm.nih.gov/pubmed/22658212. Accessed 17 August 2013.
83. Dai J, Becquer T, Rouiller JH, Reversat G, Bernhard-Reversat F, et al. (2004) Influence of heavy metals on C and N mineralisation and microbial biomass in Zn-, Pb-, Cu-, and Cd-contaminated soils. Appl Soil Ecol 25: 99–109. Available: http://www.sciencedirect.com/science/article/pii/S0929139303001355. Accessed 15 October 2013.
84. Kızılkaya R, Bayraklı B (2005) Effects of N-enriched sewage sludge on soil enzyme activities. Appl Soil Ecol 30: 192–202. Available: http://www.sciencedirect.com/science/article/pii/S0929139305000594. Accessed 15 October 2013.
85. Turner BL, Hopkins DW, Haygarth PM, Ostle N (2002) β-Glucosidase activity in pasture soils. Available: http://www.sciencedirect.com/science/article/pii/S0929139302000203. Accessed 15 October 2013.
86. Sheppard SC, Stephenson GL (2012) Ecotoxicity of aged uranium in soil using plant, earthworm and microarthropod toxicity tests. Bull Env Contam Toxicol 8843–47: 43–47. doi:10.1007/s00128-011-0442-5.
87. Chelinho S, Domene X, Campana P, Natal-da-Luz T, Scheffczyk A, et al. (2011) Improving ecological risk assessment in the Mediterranean area: selection of reference soils and evaluating the influence of soil properties on avoidance and reproduction of two oligochaete species. Environ Toxicol Chem 30: 1050–1058. Available: http://www.ncbi.nlm.nih.gov/pubmed/21305581. Accessed 17 October 2013.
88. Murphy S, Giménez D, Muldowney LS, Heckman JR (2012) Soil Organic Matter Level and Interpretation How is the Organic Matter Content of Soils Determined? How Is Organic Matter Level: 1–3.
89. Natal-da-Luz T, Ojeda G, Pratas J, Van Gestel CA, Sousa JP (2011) Toxicity to *Eisenia andrei* and *Folsomia candida* of a metal mixture applied to soil directly or via an organic matrix. Ecotoxicol Environ Saf 74: 1715–1720. Available: http://www.ncbi.nlm.nih.gov/pubmed/21683441. Accessed 13 March 2013.
90. Lourenço J, Silva A, Carvalho F, Oliveira J, Malta M, et al. (2011) b) Histopathological changes in the earthworm Eisenia andrei associated with the exposure to metals and radionuclides. Chemosphere 85: 1630–1634. Available: http://www.ncbi.nlm.nih.gov/pubmed/21911243. Accessed 4 October 2013.
91. Hobbelen PHF, Koolhaas JE, van Gestel CAM (2006) Bioaccumulation of heavy metals in the earthworms *Lumbricus rubellus* and *Aporrectodea caliginosa* in relation to total and available metal concentrations in field soils. Environ Pollut 144: 639–646. Available: http://www.ncbi.nlm.nih.gov/pubmed/16530310. Accessed 15 October 2013.
92. Li L, Wu J, Tian G, Xu Z (2009) Effect of the transit through the gut of earthworm (*Eisenia fetida*) on fractionation of Cu and Zn in pig manure. J Hazard Mater 167: 634–640. Available: http://www.ncbi.nlm.nih.gov/pubmed/19232822. Accessed 19 September 2013.
93. Peijnenburg W, Jager T (2003) Monitoring approaches to assess bioaccessibility and bioavailability of metals: matrix issues. Ecotoxicol Environ Saf 56: 63–77. Available: http://www.ncbi.nlm.nih.gov/pubmed/12915141. Accessed 15 October 2013.
94. Giovanetti A, Fesenko S., Cozzella ML, Asencio LD, Sansone U (2010) Bioaccumulation and biological effects in the earthworm *Eisenia fetida* exposed to natural and depleted uranium. J Environ Radioact 101: 509–516. Available: http://www.ncbi.nlm.nih.gov/pubmed/20362371. Accessed 15 October 2013.
95. Peijnenburg W, Baerselman R, de Groot C, Jager T, Posthuma L, et al. (1999) Relating environmental availability to bioavailability: soil-type-dependent metal accumulation in the oligochaete *Eisenia andrei*. Ecotoxicol Environ Saf 44: 294–310. Available: http://www.ncbi.nlm.nih.gov/pubmed/10581124.
96. Amorim M, Römbke J, Scheffczyk A, Soares AM (2005) Effect of different soil types on the enchytraeids *Enchytraeus albidus and Enchytraeus luxuriosus* using the herbicide Phenmedipham. Chemosphere 61: 1102–1114. Available: http://www.ncbi.nlm.nih.gov/pubmed/16263380.
97. Kuperman R (2004) Manganese toxicity in soil for *Eisenia fetida, Enchytraeus crypticus (Oligochaeta),* and *Folsomia candida* (Collembola). Ecotoxicol Environ Saf 57: 48–53. Available: http://linkinghub.elsevier.com/retrieve/pii/S0147651303001544. Accessed 22 June 2011.
98. Sheppard S, Sheppard M (2005) Derivation of ecotoxicity thresholds for uranium. J Environ 79: 55–83. Available: http://www.ncbi.nlm.nih.gov/pubmed/15571876. Accessed 19 November 2012.
99. Lock K, Janssen CR (2001) Cadmium Toxicity for Terrestrial Invertebrates: Taking Soil Parameters Affecting Bioavailability into Account. Environ Toxicol: 315–322.
100. Layinka T, Idowu B, Dedeke A, Akinloye A, Ademolu O, et al. (2011) Earthworm as bio-indicator of heavy metal pollution around Lafarge, Wapco Cement Factory, Ewekoro, Nigeria. Proceedings of the Environmental Management Conference, Federal University of Agriculture, Abeokuta, Nigeria: 489–496.
101. Amorim M, Rçmbke J, Scheffczyk A, Nogueira A, Soares A (2005) Effects of Different Soil Types on the Collembolans *Folsomia candida* and *Hypogastrura assimilis* using the Herbicide Phenmedipham. 352: 343–352. doi:10.1007/s00244-004-0220-z.
102. Jänsch S, Amorim MJ, Römbke J (2011) Identification of the ecological requirements of important terrestrial ecotoxicological test species. Environmental Reviews, 13(2): 51–83. Available: http://www.nrcresearchpress.com/doi/abs/10.1139/a05-007#.Ul0AxlDENPB. Accessed 15 October 2013.
103. Liu X, Zhang S, Shan X-Q, Christie P (2007) Combined toxicity of cadmium and arsenate to wheat seedlings and plant uptake and antioxidative enzyme responses to cadmium and arsenate co-contamination. Ecotoxicol Environ Saf 68: 305–313. Available: http://www.ncbi.nlm.nih.gov/pubmed/17239437. Accessed 15 October 2013.
104. Hubálek T, Vosáhlová S, Matejů V, Kovácová N, Novotný C (2007) Ecotoxicity monitoring of hydrocarbon-contaminated soil during bioremediation: a case study. Arch Environ Contam Toxicol 52: 1–7. Available: http://www.ncbi.nlm.nih.gov/pubmed/17106791. Accessed 13 August 2013.
105. Soudek P, Petrová S, Benešová D, Dvořáková M, Vaněk T (2011) Uranium uptake by hydroponically cultivated crop plants. J Environ Radioact 102: 598–604. Available: http://www.ncbi.nlm.nih.gov/pubmed/21486682. Accessed 13 September 2013.
106. Gopal R, Rizvi A (2008) Excess lead alters growth, metabolism and translocation of certain nutrients in radish. Chemosphere 70: 1539–1544. Available: http://www.ncbi.nlm.nih.gov/pubmed/17923149. Accessed 15 October 2013.
107. Sheppard S, Evenden W, Anderson A (1992) Multiple assays of uranium toxicity in soil. Environ. Toxicol. Water Qual., 7: 275–294.
108. Stojanović M, Stevanović D, Milojković J, Grubišić M, Ileš D (2009) Phytotoxic Effect of the Uranium on the Growing Up and Development the Plant of Corn.

Water, Air, Soil Pollut 209: 401–410. Available: http://www.springerlink.com/index/10.1007/s11270-009-0208-4. Accessed 25 September 2011.
109. Bednar A, Medina V, Ulmer-Scholle D, Frey B, Johnson B, et al. (2007) Effects of organic matter on the distribution of uranium in soil and plant matrices. Chemosphere 70: 237–247. Available: http://www.sciencedirect.com/science/article/pii/S0045653507008168. Accessed 15 October 2013.
110. Tunney H, Stojanovic M, Mrdakovic Popic J, McGrath D, Zhang C (2009) Relationship of soil phosphorus with uranium in grassland mineral soils in Ireland using soils from a long-term phosphorus experiment and a National Soil Database. J Plant Nutr Soil Sci 172: 346–352. Available: http://doi.wiley.com/10.1002/jpln.200800069. Accessed 13 September 2013.
111. Soudek P, Petrová Š, Benešová D, Vaněk T (2011) Uranium uptake and stress responses of in vitro cultivated hairy root culture of Armoracia rusticana. Agrochimical 1: 15–28.
112. Viehweger K, Geipel G (2010) Uranium accumulation and tolerance in *Arabidopsis halleri* under native versus hydroponic conditions. Environ Exp Bot 69: 39–46. Available: http://dx.doi.org/10.1016/j.envexpbot.2010.03.001.
113. Mitchell N, Pérez-Sánchez D, Thorne MC (2013) A review of the behaviour of U-238 series radionuclides in soils and plants. J Radiol Prot 33: R17–48. Available: http://www.ncbi.nlm.nih.gov/pubmed/23612607. Accessed 18 September 2013.
114. Laurette J, Larue C, Llorens I, Jaillard D, Jouneau P-H, et al. (2012) Speciation of uranium in plants upon root accumulation and root-to-shoot translocation: A XAS and TEM study. Environ Exp Bot 77: 87–95. Available: http://www.sciencedirect.com/science/article/pii/S0098847211002814. Accessed 15 October 2013.
115. Pereira A, Neves L (2012) Estimation of the radiological background and dose assessment in areas with naturally occurring uranium geochemical anomalies–a case study in the Iberian Massif (Central Portugal). J Environ Radioact 112: 96–107. Available: http://www.ncbi.nlm.nih.gov/pubmed/22694913. Accessed 14 September 2013.
116. Scott-Fordsmand JJ, Pedersen MB (1995) Soil quality criteria for selected inorganic compounds. Danish Environmental Protection Agency, Working Report No. 48, 200.

Permissions

All chapters in this book were first published in PLOS ONE, by The Public Library of Science; hereby published with permission under the Creative Commons Attribution License or equivalent. Every chapter published in this book has been scrutinized by our experts. Their significance has been extensively debated. The topics covered herein carry significant findings which will fuel the growth of the discipline. They may even be implemented as practical applications or may be referred to as a beginning point for another development.

The contributors of this book come from diverse backgrounds, making this book a truly international effort. This book will bring forth new frontiers with its revolutionizing research information and detailed analysis of the nascent developments around the world.

We would like to thank all the contributing authors for lending their expertise to make the book truly unique. They have played a crucial role in the development of this book. Without their invaluable contributions this book wouldn't have been possible. They have made vital efforts to compile up to date information on the varied aspects of this subject to make this book a valuable addition to the collection of many professionals and students.

This book was conceptualized with the vision of imparting up-to-date information and advanced data in this field. To ensure the same, a matchless editorial board was set up. Every individual on the board went through rigorous rounds of assessment to prove their worth. After which they invested a large part of their time researching and compiling the most relevant data for our readers.

The editorial board has been involved in producing this book since its inception. They have spent rigorous hours researching and exploring the diverse topics which have resulted in the successful publishing of this book. They have passed on their knowledge of decades through this book. To expedite this challenging task, the publisher supported the team at every step. A small team of assistant editors was also appointed to further simplify the editing procedure and attain best results for the readers.

Apart from the editorial board, the designing team has also invested a significant amount of their time in understanding the subject and creating the most relevant covers. They scrutinized every image to scout for the most suitable representation of the subject and create an appropriate cover for the book.

The publishing team has been an ardent support to the editorial, designing and production team. Their endless efforts to recruit the best for this project, has resulted in the accomplishment of this book. They are a veteran in the field of academics and their pool of knowledge is as vast as their experience in printing. Their expertise and guidance has proved useful at every step. Their uncompromising quality standards have made this book an exceptional effort. Their encouragement from time to time has been an inspiration for everyone.

The publisher and the editorial board hope that this book will prove to be a valuable piece of knowledge for researchers, students, practitioners and scholars across the globe.

List of Contributors

Antoine Stevens, Bas van Wesemael
Georges Lemaître Centre for Earth and Climate Research, Earth and Life Institute, UCLouvain, Louvain-la-Neuve, Belgium

Marco Nocita,Gergely Tóth and Luca Montanarella
SOIL Action, Land Resource Management Unit, Institute for Environment and Sustainability, Joint Research Centre of the European Commission, Ispra, Italy

Danying Wang, Song Chen, Chenglin Ji, Chunmei Xu and Xiufu Zhang
State Key Laboratory of Rice Biology, China National Rice Research Institute, Hangzhou, Zhejiang, China

Zaiman Wang
Key Laboratory of Key Technology on Agricultural Machine and Equipment, South China Agricultural University, Guangzhou, Guangdong, China

Bhagirath Singh Chauhan
Queensland Alliance for Agriculture and Food Innovation, University of Queensland, Queensland, Australia

Mohsen Ahmadi and Mohammad Kaboli
Department of Environmental Sciences, Faculty of Natural Resources, University of Tehran, Karaj, Iran

JoséVicente López-Bao
Research Unit of Biodiversity (UO/CSIC/PA), Oviedo University, Mieres, Spain
GrimsöWildlife Research Station, Dep. of Ecology, Swedish University of Agricultural Sciences (SLU), Riddarhyttan, Sweden

Haitao Liu, Liyue Guo, Jie Meng, Guanglei Wu and Gaoming Jiang
State Key Laboratory of Vegetation and Environmental Change, Institute of Botany, the Chinese Academy of Sciences, Beijing, China

Zhen Zhen
State Key Laboratory of Vegetation and Environmental Change, Institute of Botany, the Chinese Academy of Sciences, Beijing, China
Laboratory of Crop, Guangdong Ocean University, Zhanjiang, China, University of the Chinese Academy of Sciences, Beijing, China

Na Wang
Development and Promotion Center, Shandong Small and Mediumsized Enterprises, JI'nan, China

Na Ding
State Key Laboratory of Crop Biology, Shandong Agricultural University, Tai'an, China

Keke Hua, Daozhong Wang, Xisheng Guo and Zibin Guo
Soil and Fertilizer Research Institute, Anhui Academy of Agricultural Sciences, Hefei, China

Gary D. Bendin and Hendrik Schäfer
School of Life Sciences, Gibbet Hill Campus, University of Warwick, Coventry, United Kingdom

Lawrence O. Davies
School of Life Sciences, Gibbet Hill Campus, University of Warwick, Coventry, United Kingdom
Syngenta, Product Safety, Jealott's Hill International Research Centre, Bracknell, Berkshire, United Kingdom

Samantha Marshall, Irene Bramke and Robin G. Oliver
Syngenta, Product Safety, Jealott's Hill International Research Centre, Bracknell, Berkshire, United Kingdom

Bing Wang, Fenxiang Wen, Jiangtao Wu and Xiaojun Wang
College of Environmental Science and Resources, Shanxi University, Taiyuan, China

Yani Hu
Library, Hebei University of Science and Technology, Shijiazhuang, China

Huihua Zhang, Junjian Chen, Guoyi Yang and Dingqiang Li
Guangdong Institute of Eco-environmental and Soil Sciences, Guangzhou, China

Li Zhu2
Management School, Jinan University, Guangzhou, China

María Alonso-Ayuso, JoséLuis Gabriel, Miguel Quemada
School of Agriculture Engineering, Technical University of Madrid, Madrid, Spain

Jason Vleminckx and Olivier J. Hardy
Service d'Évolution Biologique et Écologie, Facultédes Sciences, UniversitéLibre de Bruxelles, Brussels, Belgium

Achille B. Biwolé, Jean-Franc¸ois Gillet and Jean-Louis Doucet
Laboratoire de Foresterie des Re´gions tropicales et subtropicales, Unitéde Gestion des Ressources forestiéres et des Milieux naturels, Gembloux Agro-Bio Tech, Universitéde Liége, Gembloux, Belgium

Julie Morin-Rivat
Laboratoire de Foresterie des Re´gions tropicales et subtropicales, Unitéde Gestion des Ressources forestiéres et des Milieux naturels, Gembloux Agro-Bio Tech, Universite´ de Liége, Gembloux, Belgium
Laboratory of Wood Biology & Xylarium, Royal Museum for Central Africa, Tervuren, Belgium

Kasso Daïnou
Laboratoire de Foresterie des Régions tropicales et subtropicales, Unitéde Gestion des Ressources forestieéres et des Milieux naturels, Gembloux Agro-Bio Tech, Universite´ de Lie`ge, Gembloux, Belgium
Laboratory of Applied Ecology, University of Abomey-Calavi, Cotonou, Benin

Thomas Drouet
Laboratoire d'Écologie Végétale et Biogéochimie, Facultédes Sciences, UniversitéLibre de Bruxelles, Brussels, Belgium

Marianne Stenrød, Sonja S. Klemsdal, Hans Ragnar Norli and Ole Martin Eklo
Norwegian Institute for Agricultural and Environmental Research (Bioforsk), Ås, Norway

Dongpu Wei, Shibao Chen, Jumei Li and Yibing Ma
National Soil Fertility and Fertilizer Effects Long-term Monitoring Network, Institute of Agricultural Resources and Regional Planning, Chinese Academy of Agricultural Sciences, Beijing, P. R. China

Bo Li
National Soil Fertility and Fertilizer Effects Long-term Monitoring Network, Institute of Agricultural Resources and Regional Planning, Chinese Academy of Agricultural Sciences, Beijing, P. R. China
Institute of Plant Nutrition and Environmental Resources, Liaoning Academy of Agricultural Sciences, Shenyang, P. R. China

Junxing Yang
Centre for Environmental Remediation, Institute of Geographic Sciences and Natural Resources Research, Chinese Academy of Sciences, Beijing, P. R. China

Junhui Li and Ying Lu
College of Natural Resources and Environment, South China Agricultural University, Guangzhou, China

Fei Dong
College of Natural Resources and Environment, South China Agricultural University, Guangzhou, China
Agricultural Bureau of Xiangfen County, Shanxi Province, Xiangfen, China

Qiuyan Yan
Institute of Wheat Research, Shanxi Academy of Agricultural Sciences, Linfen China

Hojae Shim
Department of Civil and Environmental Engineering, Faculty of
Science and Technology, University of Macau, Macau SAR, China

Myriam Dauzat and Denis Vile
Laboratoire d9Ecophysiologie des Plantes sous Stress Environnementaux (LEPSE), UMR759, Institut National de la Recherche Agronomique-SupAgro, Montpellier, France

Justine Bresson and Marc Labadie
Laboratoire d9Ecophysiologie des Plantes sous Stress Environnementaux (LEPSE), UMR759, Institut National de la Recherche Agronomique-SupAgro, Montpellier, France
Laboratoire des Symbioses Tropicales et Méditerranéennes (LSTM), UMR113, Université Montpellier 2-IRD-CIRAD-INRA-SupAgro, Montpellier, France

Fabrice Varoquaux and Bruno Touraine
Laboratoire des Symbioses Tropicales et Méditerranéennes (LSTM), UMR113, UniversitéMontpellier 2-IRD-CIRAD-INRA-SupAgro, Montpellier, France

Franc¸ois Vasseur
Laboratoire d9Ecophysiologie des Plantes sous Stress Environnementaux (LEPSE), UMR759, Institut National de la Recherche Agronomique-SupAgro, Montpellier, France
Max Planck Institute for Developmental Biology, Tübingen, Germany

Yilin Li and Weiming Shi
State Key Laboratory of Soil and Sustainable Agriculture, Institute of Soil Science, Chinese Academy of Sciences, Nanjing, China

Xingxiang Wang
State Key Laboratory of Soil and Sustainable Agriculture, Institute of Soil Science, Chinese Academy of Sciences, Nanjing, China
Jiangxi Key Laboratory of Ecological Research of Red Soil, Yingtan, Jiangxi, China

Helder Fraga, Aureliano C. Malheiro, José Moutinho-Pereira and João A. Santos
Centre for the Research and Technology of Agro-Environmental and Biological Sciences, Universidade de Trás-os-Montes e Alto Douro, UTAD, Vila Real, Portugal

Rita M. Cardoso
Instituto Dom Luiz, CGUL, Universidade de Lisboa, Lisbon, Portugal

Pedro M. M. Soares
Instituto Dom Luiz, CGUL, Universidade de Lisboa, Lisbon, Portugal
ADEC, Instituto Superior de Engenharia de Lisboa, Lisbon, Portugal

Javier J. Cancela
GI-1716, Proyectos y Planificación. Dpto. Ingeniería Agroforestal, Universidad de Santiago de Compostela, Escuela Politécnica Superior, Lugo, Spain

Joaquim G. Pinto
Department of Meteorology, University of Reading, Reading, United Kingdom
Institute for Geophysics and Meteorology, University of Cologne, Cologne, Germany

Xiang Li, Jianzhi Niu and Baoyuan Xie
Key Laboratory of Soil and Water Conservation and Desertification Combating of Education Ministry, Beijing Forestry University, Beijing, China

Shaohua Wu and Shenglu Zhou
School of Geographic and Oceanographic Science, Nanjing University, Nanjing, China

Zongqiang Wei
School of Geographic and Oceanographic Science, Nanjing University, Nanjing, China
School of Environmental and Land Resource Management, Jiangxi Agricultural University, Nanchang, China

Xiao Yan
School of Environmental and Land Resource Management, Jiangxi Agricultural University, Nanchang, China

Hai-Ming Tang, Xiao-Ping Xiao, Wen-Guang Tang, Ke Wang, Ji-Min Sun, Wei-Yan Li and Guang-Li Yang
Hunan Soil and Fertilizer Institute, Changsha, PR China

Rebecca T. Barnes, Morgan E. Gallagher, Caroline A. Masiello, Zuolin Liu and Brandon Dugan
Department of Earth Science, Rice University, Houston, Texas, United States of America

Xuelin Zhang, Qun Wang, Yilun Wang and Chaohai Li
The Incubation Base of the National Key Laboratory for Physiological Ecology and Genetic Improvement of Food Crops in Henan Province, Zhengzhou, China; Agronomy College of Henan Agricultural University, Zhengzhou, China

Frank S. Gilliam
Department of Biological Sciences, Marshall University, Huntington, West Virginia, United States of America

Feina Cha
Meteorological Bureau of Zhengzhou, Zhengzhou, China

Ana Luisa Caetano, Catarina R. Marques and Fernando Gonçalves
Department of Biology, University of Aveiro, Campus Universitário de Santiago, Aveiro, Portugal
CESAM, University of Aveiro, Campus Universitário de Santiago, Aveiro, Portugal

Fernando Carvalho
Nuclear and Technological Institute (ITN) Department of Radiological Protection and Nuclear Safety, Sacave´m, Portugal

Eduardo Ferreira da Silva
Department of Geosciences University of Aveiro, GeoBioTec Research Center, Campus Universitário de Santiago, Aveiro, Portugal

Ruth Pereira
Department of Biology, Faculty of Sciences of the University of Porto, Porto, Portugal
nterdisciplinary Centre of Marine and Environmental Research (CIIMAR/CIMAR), University of Porto, Porto, Portugal

Ana Gavina
Interdisciplinary Centre of Marine and Environmental Research (CIIMAR/CIMAR), University of Porto, Porto, Portugal CESAM, University of Aveiro, Campus Universitário de Santiago, Aveiro, Portugal

Index

www.ingramcontent.com/pod-product-compliance
Lightning Source LLC
Chambersburg PA
CBHW061303190326
41458CB00011B/3752

* 9 7 8 1 6 3 2 3 9 7 9 1 1 *